BARRON'S REVIEW COURSE SERIES

Let's Review: Sequential Mathematics, Course II

Lawrence S. Leff
Assistant Principal,
Mathematics Supervision
Franklin D. Roosevelt High School
Brooklyn, New York

BARRON'S EDUCATIONAL SERIES

New York • London • Toronto • Sydney

TO RHONA

For the patience and the understanding,
for the love and with love.

All inquiries should be addressed to:
Barron's Educational Series, Inc.
250 Wireless Boulevard
Hauppauge, New York 11788

Library of Congress Catalog Card No. 88-7753

International Standard Book No. 0-8120-4047-3

Library of Congress Cataloging-in-Publication Data

Leff, Lawrence S.
Let's review: sequential mathematics, course II / Lawrence S. Leff.
p. cm. – (Barron's review course series)
Includes index.
ISBN 0-8120-4047-3
1. Mathematics–1961– 2. Mathematics–Study and teaching
(Secondary)–New York (State) I. Title. II. Series.
QA39.2.L415 1989
510–dc19 88-7753
CIP

PRINTED IN THE UNITED STATES OF AMERICA

901 800 987654321

PREFACE

For which course can this book be used?

This book reflects the newly revised syllabus of study for Course II of the New York State Three-Year Sequence in High School Mathematics (1988 revision). It includes the topics that the current revision has introduced, including transformations in the coordinate plane, trigonometry of the right triangle, and expanded coverage of algebraic operations and fractions.

Teachers of traditional high school level courses in mathematics may find the material on logic, mathematical systems (groups and fields), probability and statistics, and transformation geometry a source of enrichment activities for their students.

What special features does this book have?

- *Ongoing Review for the Course II Regents Examination*
  Each chapter closes with a set of summary exercises that provides a comprehensive review of the chapter. In addition, to help the student get a head start in preparing for the New York State Course II Regents Examination, several sets of REGENTS TUNE-UP exercises are placed strategically in the book. These exercises include many actual test questions selected from past Regents examinations, thus providing the student with a cumulative review of the material covered in the preceding chapters, while previewing the types and the level of difficulty of questions found on Regents examinations. For a culminating activity, several full-length Course II Regents Examinations given in previous years have been reprinted at the end of the book.

- *Review of Selected Topics Introduced in Course I*
  Fundamental algebraic methods, elements of transformation geometry, and basic concepts of probability are reviewed before they are extended as required by the Course II syllabus. To facilitate understanding of the new material, the review of these Course I topics is integrated into the book rather than placed at the back in an appendix.

- *Clear Writing Style with Many Illustrative Diagrams and Examples*
  The book uses an easy-to-follow writing style that is enhanced by numerous demonstration examples and practice exercises designed to build skill and confidence. Each section of a chapter begins with a KEY IDEAS box that summarizes or highlights the new material that follows. The KEY IDEAS, together with the illustrative examples and their solutions, try to anticipate and then to answer the "why" types of questions that the student may have.

- *Inclusion of Mathematical Systems*
  In the most recent revision of the Course II syllabus, mathematical systems appears as an optional topic. To allow for maximum flexibility of course coverage, Chapter 2 of this book reviews this important topic. However, the book has been organized so that the student can skip Chapter 2 without compromising his or her understanding of the chapters that follow.

- *Answers to Summary Exercises*
  The answers to most of the CHAPTER REVIEW and REGENTS TUNE-UP exercises are provided at the back of the book so that the student can obtain additional feedback and guidance in progressing from one chapter to the next.

Who should use this book?

Students who wish to improve their classroom performance and test grades will benefit greatly from the concise explanations, helpful demonstration examples, and numerous practice exercises.

Teachers and *school systems* that desire an additional teaching and planning resource will find this book an ideal companion to any of the existing Course II textbooks. Indeed, the depth of coverage, the clarity of the explanations, and the wealth of illustrative material and practice exercises make the book a possible choice as the primary textbook for the course.

LAWRENCE S. LEFF

December 1988

TABLE OF CONTENTS

*Chapter includes material that is optional in the Revised New York State Course II syllabus.

UNIT I: LOGIC AND MATHEMATICAL SYSTEMS

CHAPTER 1

Logic

1.1 LOGICAL CONNECTIVES AND TRUTH VALUES

KEY IDEAS

Consider whether each of the following sentences is true or false:

1. Daffodils are prettier than roses.
2. It is a flower with five petals.
3. A daffodil is a flower.

The first sentence reflects an opinion and cannot be judged to be true or false. The second sentence is an **open sentence** since its truth or falsity cannot be determined until the word *it* is replaced by the name of an actual type of flower. The third sentence involves a fact that can be verified. This type of sentence is called a **statement**.

A statement may be *simple* or *compound*. A compound statement may be a *conjunction*, *disjunction*, *conditional*, or *biconditional*.

A statement has a **truth value** of either TRUE OR FALSE. Much of your work in logic will involve determining the truth values of statements.

SIMPLE AND COMPOUND STATEMENTS. "Today is Saturday" is an example of a **simple statement** since it expresses a single idea. A simple statement is usually represented by a single letter.

A **compound statement** is a statement formed by joining two (or more) simple statements with a logical connective. Examples of compound statements are given in Table 1.1, where p represents the statement "Today is Saturday" and q represents the statement "I sleep late."

TABLE 1.1 Some Examples of Compound Statements

Connective	Compound Statement	Symbolic Form
and	Today is Saturday *and* I sleep late.	$p \wedge q$
or	Today is Saturday *or* I sleep late.	$p \vee q$
if . . . then . . .	*If* today is Saturday, *then* I sleep late.	$p \rightarrow q$
if and only if	Today is Saturday *if and only if* I sleep late.	$p \leftrightarrow q$

CONJUNCTION AND DISJUNCTION. A compound statement formed by connecting two (or more) statements with the word *and* is called a **conjunction**. Each of the two statements that make up a conjunction is called a **conjunct**. The *conjunction* $p \wedge q$ is read as "p and q" and *is true when conjunct* p *and conjunct* q *are both true.*

A **disjunction** is a compound statement formed by connecting two (or more) statements with the word *or*. Each statement that forms a disjunction is called a **disjunct**. The *disjunction* $p \vee q$ is read as "p or q" and *is true when either disjunct* p *is true or disjunct* q *is true, or when both disjuncts are true.*

Example

1. If x is an integer, what value(s) of x will make the following statements true?

(a) $(x-1=4) \vee \left(\frac{x}{2}=1\right)$ (b) $(x>0) \wedge (x \leq 3)$

Solutions: (a) The disjunction is true when either disjunct is true or both disjuncts are true. The left disjunct is true if $x=5$, and the right disjunct is true if $x=2$. The disjunction is true when $x=\mathbf{5}$ or $x=\mathbf{2}$.

(b) The right conjunct is true for all integer values of x less than or equal to 3 (that is, $3, 2, 1, 0, -1, \ldots$). The left conjunct is true for all positive integers ($1, 2, 3, 4, \ldots$). The conjunction is true only for the values of x that make both conjuncts true. The conjunction is true when $x = 1, 2$, or 3.

CONDITIONAL STATEMENT. A compound statement of the form "If p, then q" is called a **conditional statement** (or *implication*) where statement p is the **antecedent** (*hypothesis*) and statement q is the **consequent** (*conclusion*). For example,

If today is Friday, then tomorrow is Saturday.

antecedent (hypothesis) — *consequent* (conclusion)

A conditional is true for all truth values of p *and* q *except in the single instance in which the antecedent* p *is true and the consequent* q *is false.*

The "If p, then q" form of a conditional statement may be written in any one of several equivalent forms.

Equivalent Form	Example
If p, *then* q	*If* the flower is a rose, *then* it is red.
p *implies* q	The flower is a rose *implies* it is red.
p *only if* q	The flower is a rose *only if* it is red.
q *if* p	The flower is red *if* it is a rose.

BICONDITIONAL STATEMENT. The biconditional of statements p and q is written as $p \leftrightarrow q$ and is read as "p if and only if q." *The biconditional of two statements is true when both statements have the same truth value.*

Table 1.2 summarizes the truth values for the logical connectives.

TABLE 1.2 Truth Values for the Logical Connectives

		Conjunction	Disjunction	Conditional	Biconditional
p	q	$p \wedge q$	$p \vee q$	$p \rightarrow q$	$p \leftrightarrow q$
T	T	T	T	T	T
T	F	F	T	F	F
F	T	F	T	T	F
F	F	F	F	T	T

Examples

2. Let p represent the statement "x is an even number," and let q represent the statement "x is a prime number." What is the truth value of each of the following statements when $x = 7$?

(*a*) $p \leftrightarrow q$ (*b*) $p \wedge q$ (*c*) $p \vee q$ (*d*) $q \rightarrow p$

Solutions: When $x = 7$, statement p is false and statement q is true.

(a) **False.** A biconditional is false when its left and right members have opposite truth values.

(b) **False.** A conjunction is false when one of the conjuncts is false.

(c) **True.** A disjunction is true when at least one of the disjuncts is true.

(d) **False.** A conditional is false when the antecedent is true and the consequent is false.

3. In Example 2, for what value of x is $p \leftrightarrow q$ true?

(1) 8 (2) 9 (3) 10 (4) 11

Solution: A biconditional is true when its left and right members have the same truth value. When $x = 9$, statement p is false and statement q is false, so $p \leftrightarrow q$ is true. The correct answer is **choice (2).**

4. If $h \rightarrow k$ is false, which statement must be true?

(1) $h \leftrightarrow k$ (2) $h \wedge k$ (3) $h \vee k$ (4) k

Solution: Since $h \rightarrow k$ is false, h is true and k is false. Therefore, $h \vee k$ is true. The correct answer is **choice (3).**

5. If $t \vee s$ is false and $r \rightarrow s$ is true, which statement is always true?

(1) $s \wedge r$ (2) $t \vee s$ (3) $t \vee r$ (4) $t \leftrightarrow r$

Solution: If the disjunction $t \vee s$ is false, then each disjunct must be false, so statement t is false and statement s is false. If the consequent (s) of a conditional is false, then, in order for the conditional to be true, its antecedent (r) must be false. Since statements t and r have the same truth value (both are false), their biconditional is true. The correct answer is **choice (4).**

TAUTOLOGY. A **tautology** is a compound statement that is always true regardless of the truth value of each of its component statements. A truth table can be constructed to determine whether a compound statement is a tautology.

Example

6. Determine whether the following statement is a tautology: $(p \wedge q) \rightarrow (p \vee q)$.

Solution: Construct a truth table.

p	q	$p \wedge q$	$p \vee q$	$(p \wedge q) \to (p \vee q)$
T	T	T	T	T
T	F	F	T	T
F	T	F	T	T
F	F	F	F	T

The last column of the truth table shows that the statement $(p \wedge q) \to (p \vee q)$ is always true, so **it is a tautology**.

EXERCISE SET 1.1

1. If p represents "Math is fun," q represents "Math is difficult," and r represents "Math is easy," write each of the following in symbol form, using p, q, and r.
(a) Math is fun or math is difficult.
(b) Math is fun and math is easy.
(c) If math is easy, then it is fun.
(d) Math is fun if and only if it is easy.

2. Let p represent "A square has three sides," and q represent "2 is a prime number." Determine the truth value of each statement.
(a) $p \wedge q$ (b) $p \vee q$ (c) $p \to q$ (d) $p \leftrightarrow q$ (e) $q \to p$

3. Let p represent "x is a prime number," and q represent "$x + 2$ is a prime number." Determine the truth value of each statement when $x = 17$.
(a) $p \wedge q$ (b) $p \vee q$ (c) $p \to q$ (d) $p \leftrightarrow q$ (e) $q \to p$

4. If p represents "x is divisible by 3," and q represents "x is the LCM (least common multiple) of 4 and 6," determine the truth value of each statement when $x = 24$.
(a) $p \wedge q$ (b) $p \vee q$ (c) $p \to q$ (d) $p \leftrightarrow q$ (e) $q \to p$

5. If p represents "$x > 5$," and q represents "x is divisible by 3," which statement is true if $x = 10$?
(1) $p \to q$ (2) $(p \vee q) \to q$ (3) $(p \wedge q) \to p$ (4) $p \leftrightarrow q$

6–10. If x is an integer, what is the smallest value of x, if any, that makes the given statement true?

6. $(x > 32) \wedge (x \text{ is prime})$

7. $(x > 5) \wedge (x < 8)$

8. $(x \text{ is even}) \wedge (x + 2 = 5)$

9. $(x < -4) \wedge (x > -9)$

10. $(x \text{ is divisible by } 3) \wedge (x \text{ is divisible by } 9)$

11. If $p \wedge q$ is true, which of the following statements is (are) *always* true?
(1) $p \leftrightarrow q$ (2) $p \rightarrow q$ (3) $p \vee q$ (4) All of these

12. If $p \rightarrow q$ is false, which statement is *never* true?
(1) $q \rightarrow p$ (2) p (3) $p \vee q$ (4) $p \leftrightarrow q$

13. If $p \rightarrow q$ is true and $p \leftrightarrow q$ is false, then:
(1) p is true, q is false
(2) p is false, q is true
(3) both p and q are true
(4) both p and q are false

14. If $p \leftrightarrow q$ is true, which statement is *always* true?
(1) p (2) $p \vee q$ (3) $p \wedge q$ (4) $p \rightarrow q$

15. If $r \vee s$ is true and $r \wedge s$ is false, which statement is *always* false?
(1) $r \rightarrow s$ (2) $r \leftrightarrow s$ (3) r (4) s

16–19. For each of the following statements, (a) construct a truth table, and (b) determine whether the statement is a tautology:

16. $[(p \rightarrow q) \wedge p] \rightarrow q$

17. $(p \rightarrow q) \vee (q \rightarrow p)$

18. $(p \rightarrow q) \leftrightarrow (p \vee q)$

19. $[(p \vee q) \wedge (p \wedge q)] \leftrightarrow (p \leftrightarrow q)$

1.2 STATEMENT NEGATIONS AND DE MORGAN'S LAWS

KEY IDEAS

The **negation** of a statement can be formed by inserting the word *not,* so that the original statement and its negation have opposite truth values. The negation of statement p is written as $\sim p$ and read as "not p." If statement p is true, then $\sim p$ is false; if p is false, then $\sim p$ is true.

FORMING THE NEGATION OF A STATEMENT. The negation of a statement may be written in more than one way. For example,

Statement: Two plus two is equal to four. (True)
Negation 1: Two plus two is *not* equal to four. (False)
or
Negation 2: It is *not* the case that two plus two is equal to four. (False)

NEGATIONS OF INEQUALITIES. The negation of the statement $\frac{6}{2}=3$ is the statement $\frac{6}{2}\neq 3$. Table 1.3 illustrates how inequality statements can be negated.

TABLE 1.3 Negations of Inequalities

p	$\sim p$
$x<4$	$x\geq 4$
$x\leq 4$	$x>4$
$x>7$	$x\leq 7$
$x\geq 7$	$x<7$

Examples

1. Write the negation of each statement.
(a) $-2<1$ (True) (b) Four is not a prime number. (True)

Solution: (a) $\mathbf{-2\geq 1}$ **(False)** (b) **Four is a prime number. (False)**

2. If $p \vee \sim q$ is true and p is false, which of the following statements is true?
(1) $\sim p \wedge q$ (2) $\sim q \rightarrow p$ (3) $p \leftrightarrow q$ (4) q

Solution: Since the disjunction is true and disjunct p is false, disjunct $\sim q$ must be true, and therefore statement q is false. Since statements p and q have the same truth value (both are false), their biconditional is true. The correct answer is **choice (3)**.

3. Let p represent: "Course II is studied after Course I." (True)
Let q represent: "Probability is not studied in Course I." (False)
Let r represent: "Elephants can fly." (False)

Using p, q, r and the proper connective, represent each of the following compound statements and determine its truth value.

(a) If elephants cannot fly, then Course II is *not* studied after Course I.

(b) Probability is not studied in Course I *or* elephants cannot fly.

(c) If Course II is studied after Course I, then probability is studied in Course I.

(d) It is not true that elephants can fly *and* Course II is studied after Course I.

(e) Course II is *not* studied after Course I, if elephants can fly.

Solutions: (a) $\sim r \rightarrow \sim p$. Since statement r is false, the antecedent $\sim r$ is true. Statement p is true, so the consequent $\sim p$ is false. Therefore, the conditional is **false**.

(b) $q \vee \sim r$. Since statement r is false, $\sim r$ is true, so the disjunction is **true**.

(c) $p \rightarrow \sim q$. The antecedent and the consequent are both true, so the conditional is **true**.

(d) $\sim(r \wedge p)$. Since r is false, the conjunction is false, so its negation is **true**.

(e) The original statement is a conditional having the form "consequent *if* antecedent," so it is represented as $r \rightarrow \sim p$. Since both parts of the conditional are false, the conditional is **true**.

LOGICALLY EQUIVALENT AND CONTRADICTORY STATEMENTS. Two statements are **logically equivalent** if they always have the *same* truth value, and are **logically contradictory** if they always have *opposite* truth values.

LAW OF DOUBLE NEGATION. The negation of $\sim p$ is written as $\sim(\sim p)$ and is referred to as the **double negation** of statement p. The first and second columns of the accompanying truth table illustrate that a statement and its negation are *logically contradictory*. The first and third columns of the truth table show that p and $\sim(\sim p)$ always have the same truth value and are, therefore, logically equivalent. This is sometimes referred to as the **Law of Double Negation**. The last column of the truth table illustrates that the *biconditional of two logically equivalent statements is always a tautology*.

p	$\sim p$	$\sim(\sim p)$	$p \leftrightarrow \sim(\sim p)$
T	F	T	T
F	T	F	T

NEGATION OF A COMPOUND STATEMENT. The negation of a compound statement involving a conjunction or a disjunction may be simplified by using one of De Morgan's laws.

De Morgan's Laws

- **$\sim(p \wedge q)$ is logically equivalent to $\sim p \vee \sim q$.**
- **$\sim(p \vee q)$ is logically equivalent to $\sim p \wedge \sim q$.**

Here is a simple way to apply De Morgan's laws to the negation of a conjunction (or disjunction):

1. Write the negation of each of the original conjuncts (or disjuncts).

2. Change the original logical connective so that a conjunction becomes a disjunction or a disjunction becomes a conjunction.

Examples

4. Using De Morgan's laws, write the negation of each compound statement.

(a) She is wealthy or she is wise.
(b) He is handsome and he is not famous.
(c) $\sim(\sim p \wedge q)$.

Solutions: (a) **She is not wealthy *and* she is not wise.**
(b) **He is not handsome *or* he is famous.**
(c) Apply De Morgan's law, so that $\sim(\sim p \wedge q)$ becomes $\sim(\sim p) \vee \sim q$. Then use the law of double negation to rewrite $\sim(\sim p)$ as p; then the negation of $\sim(\sim p \wedge q)$ is $\mathbf{p \vee \sim q}$.

5. Which of the following statements is logically equivalent to $\sim(k \vee \sim t)$?

(1) $\sim k \vee \sim t$ (2) $\sim k \wedge t$ (3) $\sim k \vee t$ (4) $\sim k \wedge \sim t$

Solution: Apply De Morgan's law, so that $\sim(k \vee \sim t)$ becomes $\sim k \wedge \sim(\sim t)$. Next, use the law of double negation and replace $\sim(\sim t)$ by t: $\sim k \wedge t$. The correct answer is **choice (2)**.

EXERCISE SET 1.2

1–5. Write the negation of each of the following inequalities:

1. $x > 3$ **2.** $x \leq 5$ **3.** $x \geq 1$ **4.** $x < 6$ **5.** $2 \leq x$

6. What is the truth value of $\sim p$ if p represents the statement "13 is a prime number"?

7. What is the truth value of $\sim q$ if q represents the statement "A triangle has three sides"?

8. What is the truth value of t if $\sim t$ represents the statement "Odd numbers are *not* evenly divisible by 2"?

9. Let p represent the statement "He is a snob," and q represent the statement "He likes caviar." Replacing the symbols by words, express each of the following as a sentence:

(a) $p \vee q$
(b) $\sim p \wedge q$
(c) $p \rightarrow q$
(d) $p \leftrightarrow q$
(e) $q \leftrightarrow p$
(f) $\sim q \rightarrow \sim p$
(g) $\sim p \rightarrow \sim q$
(h) $\sim p \leftrightarrow q$

10. In Exercise 9, assume that statement p is false and statement q is true. Find the truth value of each statement given in (a)–(h).

11. Let p represent the statement "The quotient of x and 4 has a remainder of 1," and q represent the statement "x is divisible only by itself and 1." Which statement is true when $x = 17$?
(1) $\sim(p \vee q)$ (2) $p \to \sim q$ (3) $\sim q \to p$ (4) $\sim(p \wedge q)$

12. Fill in the missing truth values.

(a)

p	q	$p \vee q$	$\sim(p \vee q)$
F	?	?	T

(b)

p	q	$\sim p$	$\sim p \vee q$
?	F	?	T

(c)

p	q	$\sim q$	$p \wedge \sim q$
?	?	?	T

(d)

p	q	$p \vee q$	$p \wedge q$
?	F	T	?

(e)

p	q	$\sim q$	$p \wedge \sim q$	$\sim(p \wedge \sim q)$
?	F	?	?	F

(f)

p	q	$\sim p$	$\sim p \wedge q$	$\sim q$	$p \vee \sim q$
?	?	?	?	?	F

13. Which statement is logically equivalent to $\sim(c \vee \sim d)$?
(1) $\sim c \vee d$ (2) $\sim c \vee \sim d$ (3) $\sim c \wedge d$ (4) $\sim c \wedge \sim d$

14. If p represents the statement "It is January," and q represents the statement "I have a cold," which statement is logically equivalent to $\sim(p \wedge q)$?
(1) It is January and I have a cold.
(2) It is not January and I do not have a cold.
(3) It is not January or I do not have a cold.
(4) It is January or I have a cold.

15. The statement $\sim r \wedge s$ is logically equivalent to:
(1) $\sim(r \wedge s)$ (2) $\sim(r \vee s)$ (3) $\sim(\sim r \vee s)$ (4) $\sim(r \vee \sim s)$

16. Let p represent the statement "A rhombus is a parallelogram," and q represent the statement "A circle has three sides." Which of the following statements is true?
(1) $p \rightarrow (p \vee q)$ (3) $(p \vee q) \rightarrow (p \wedge q)$
(2) $\sim q \rightarrow q$ (4) $\sim q \rightarrow \sim(p \vee q)$

17. Which of the following statements is a tautology?
(1) $\sim(p \rightarrow \sim p)$ (3) $q \rightarrow \sim q$
(2) $\sim(p \rightarrow q)$ (4) $(p \rightarrow q) \vee (q \rightarrow p)$

18. Let p represent the statement "x is divisible by a prime number that is less than x," and q represent the statement "x can be written as the product of two identical integers." Which of the following statements is true when $x = 15$?
(1) $p \rightarrow q$ (2) $\sim p \rightarrow q$ (3) $\sim p \vee q$ (4) $p \leftrightarrow q$

19. Which statement is true for all possible truth values of p and q?
(1) $p \rightarrow \sim q$ (3) $p \rightarrow (q \vee \sim q)$
(2) $(p \vee \sim p) \rightarrow q$ (4) $\sim(p \wedge \sim p) \rightarrow q$

20. Construct a truth table to verify that:
(a) $\sim(p \wedge q)$ is logically equivalent to $\sim p \vee \sim q$.
(b) $\sim(p \vee q)$ is logically equivalent to $\sim p \wedge \sim q$.

21. Construct a truth table to show that the following pairs of statements are logically equivalent:
(a) $(p \wedge q) \leftrightarrow p$ and $p \rightarrow q$ (b) $[(p \rightarrow q) \wedge \sim q]$ and $\sim p$

22–25. For each of the following compound statements, construct a truth table to determine whether the statement is a tautology:

22. $(p \vee \sim q) \rightarrow (\sim p \rightarrow \sim q)$
23. $(p \leftrightarrow q) \vee (q \rightarrow \sim p)$
24. $(p \rightarrow q) \leftrightarrow (\sim p \vee q)$
25. $[p \vee (\sim p \wedge q)] \leftrightarrow (p \vee q)$

1.3 FORMING THE CONVERSE, INVERSE, AND CONTRAPOSITIVE

KEY IDEA

By interchanging or negating both parts of a conditional, or by doing both, three conditionals of special interest—the *converse*, the *inverse*, and the *contrapositive*—can be formed.

RELATED CONDITIONALS. Let p represent the statement "$x = 2$," and q represent the statement "x is even." Then four related conditionals can be formed:

Statement	Symbolic Form	Example	
Original	$p \to q$	If $x = 2$, then x is even.	(True)
Converse	$q \to p$	If x is even, then $x = 2$.	(False)
Inverse	$\sim p \to \sim q$	If $x \neq 2$, then x is not even.	(False)
Contrapositive	$\sim q \to \sim p$	If x is not even, then $x \neq 2$.	(True)

LOGICALLY EQUIVALENT CONDITIONALS. As the preceding example suggests, the original conditional ($p \to q$) and its contrapositive ($\sim q \to \sim p$) will always have the same truth value and are, therefore, logically equivalent. The converse ($q \to p$) and inverse ($\sim p \to \sim q$) of a conditional are also logically equivalent to each other.

Examples

1. What is the inverse of $\sim p \to q$?

Solution: Forming the inverse requires negating both parts of the conditional: $\sim(\sim p) \to \sim(q)$. The hypothesis of this conditional may be simplified by keeping in mind that consecutive negations cancel out since one undoes the effect of the other. The correct answer is $\boldsymbol{p \to \sim q}$.

2. Let p represent "It is raining," and q represent "I have my umbrella." Express in words the contrapositive of $q \to p$.

Solution: The contrapositive of $q \to p$ is $\sim p \to \sim q$, so that:

$q \to p$: If I have my umbrella, then it is raining.

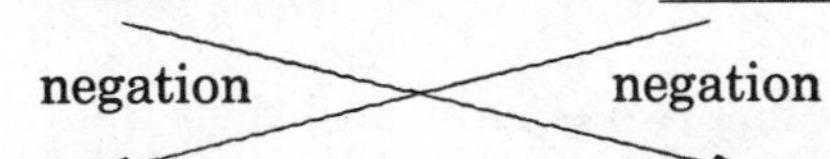

$\sim p \to \sim q$: **If it is not raining, then I do not have my umbrella.**

3. Which statement is logically equivalent to $r \to \sim s$?
(1) $\sim s \to r$ (2) $r \to s$ (3) $\sim s \to \sim r$ (4) $s \to \sim r$

Solution: A conditional and its contrapositive are logically equivalent. Choice (1) represents the converse. Negating both parts of the converse leads to the statement in choice (4), which is the contrapositive. The correct answer is **choice (4).**

4. Given the true statement "If a figure is a square, then the figure is a rectangle," which statement is also true?
(1) If a figure is not a rectangle, then it is a square.
(2) If a figure is a rectangle, then it is a square.
(3) If a figure is not a rectangle, then it is not a square.
(4) If a figure is not a square, then the figure is not a rectangle.

Solution: A conditional and its contrapositive always have the same truth value. Since the original statement is true, the contrapositive must also be true. The correct answer is **choice (3).**

EXERCISE SET 1.3

1–6. Fill in the blank in each of the following so that the resulting statement is true:

1. A conditional and its ____ are logically equivalent.
2. If the inverse of a statement is true, then the ____ must also be true.
3. If j represents the truth value of a conditional statement, and k represents the truth value of the contrapositive, then the truth value of $j \wedge k$ is ____.
4. The inverse of "If $x > 0$, then x is positive" is the statement ____.
5. If the converse of a statement is $p \to \sim q$, then the original statement is ____.
6. The truth value of the converse of the statement "If the sum of two numbers is negative, then the two numbers are negative" is ____.

7–11. Form the converse, inverse, and contrapositive of each of the following:

7. If a triangle has two equal sides, then the triangle is isosceles.

8. If n is an odd integer, then $n+1$ is an even integer.

9. If $a+b=b+a$, then $a-b=b-a$.

10. If $n \geq 9$, then $n > 4$.

11. $p \rightarrow {\sim}q$.

12. Which statement is logically equivalent to the statement "If it is sunny, then it is hot"?
(1) If it is hot, then it is sunny.
(2) If it is not hot, then it is not sunny.
(3) If it is not sunny, then it is not hot.
(4) If it is not hot, then it is sunny.

13. Which statement is logically equivalent to ${\sim}p \rightarrow q$?
(1) $p \rightarrow {\sim}q$ (2) ${\sim}q \rightarrow p$ (3) $q \rightarrow {\sim}p$ (4) ${\sim}q \rightarrow {\sim}p$

14. In which of the following pairs of statements are the statements logically equivalent?
(1) $k \rightarrow h$ and ${\sim}h \rightarrow {\sim}k$ (3) $k \rightarrow h$ and ${\sim}k \rightarrow {\sim}h$
(2) $h \rightarrow k$ and ${\sim}h \rightarrow {\sim}k$ (4) $h \rightarrow k$ and $k \rightarrow h$

15. Which statement is logically equivalent to the statement "If $n \leq 5$, then $n \leq 8$"?
(1) If $n > 8$, then $n > 5$. (3) If $n > 5$, then $n > 8$.
(2) If $n \leq 5$, then $n \leq 8$. (4) If $n \leq 8$, then $n \leq 5$.

16. Which statement is logically equivalent to the *inverse* of the statement "If I study hard, then I will pass"?
(1) If I do not study hard, then I will pass.
(2) If I do not pass, then I will not study hard.
(3) If I will pass, then I study hard.
(4) If I will pass, then I do not study hard.

17. Construct a truth table to show that the biconditional is logically equivalent to the conjunction of a conditional statement and its converse.

18. (a) Construct a truth table for the statement

$$[q \vee ({\sim}q \rightarrow {\sim}p)] \leftrightarrow [(p \leftrightarrow q) \rightarrow p].$$

(b) Determine whether the statement in part (a) is a tautology, and give a reason for your answer.

1.4 LAWS OF REASONING

KEY IDEAS

A **premise** is a sentence that, unless otherwise stated, is assumed to be true. **Logical inference** involves arriving at a *conclusion* as a result of applying valid methods of reasoning to a set of premises. This section presents some of the more commonly used laws of reasoning and illustrates how they can be applied in drawing conclusions.

FORMAT OF AN ARGUMENT. As an example, assume $p \leftrightarrow q$ is true (*premise 1*), and also assume statement q is true (*premise 2*). Then statement p is true (*conclusion*) because p and q must have the same truth value. A set of premises with a valid conclusion forms an *argument*, which can be concisely summarized using the following format:

Premise 1:	$p \leftrightarrow q$	Biconditional is true.
Premise 2:	q	Statement q is true.
Conclusion:	p	$\therefore$ p is true.

***Note*:** The symbol $\therefore$ is read as "therefore" and is sometimes used in place of the word *conclusion*.

LAW OF DISJUNCTIVE INFERENCE. If a disjunction is true, then at least one of the disjuncts must also be true. This means that, if a disjunction is true and one disjunct is false, you can conclude that the remaining disjunct must be true. This argument, referred to as the **Law of Disjunctive Inference**, takes the following form:

Premise 1:	$p \vee q$	Disjunction is true.
Premise 2:	$\sim p$	Negation of p is true.
Conclusion:	q	$\therefore$ other disjunct is true.

Note that the second premise states that $\sim p$ is true so p must be false. Since p is false, q must be true in order for $p \vee q$ to be true.

The law of disjunctive inference may be represented symbolically as follows:

$$[(p \vee q) \wedge \sim p] \rightarrow q \quad \text{or} \quad [(p \vee q) \wedge \sim q] \rightarrow p$$

LAW OF CONJUNCTIVE SIMPLIFICATION. If the conjunction of two statements is true, then each conjunct is true, that is,

Premise:	$p \wedge q$
Conclusion:	p (or q)

This is called the **Law of Conjunctive Simplification** and may be represented symbolically as follows:

$$(p \wedge q) \rightarrow p \qquad \text{or} \qquad (p \wedge q) \rightarrow q$$

Example

1. Draw a valid conclusion from the following set of true premises:

I will go skiing or I will stay at home.
I do not stay at home.

Solution: The second premise tells us that the right-hand disjunct of the first premise ("I will stay at home") is false. By the law of disjunctive inference, the left-hand disjunct must be true. Therefore, a valid conclusion is "**I will go skiing.**"

LAW OF CONTRAPOSITIVE INFERENCE. Starting with the true premise $p \rightarrow q$, you may conclude that $\sim q \rightarrow \sim p$ is also true since a conditional and its contrapositive always have the same truth value (are logically equivalent). This rule of logic is known as the **Law of Contrapositive Inference**. For example,

Premise:	If $3x = 12$, then $x = 4$. (True)
Conclusion:	If $x \neq 4$, then $3x \neq 12$. (True)

Example

2. Give the law of reasoning that can be used to justify each conclusion.

(a) $m \vee \sim h$
h
Conclusion: m

(b) $R \wedge T$
Conclusion: R

(c) $\sim t \vee \sim w$
t
Conclusion: $\sim w$

(d) $\sim r \rightarrow k$
Conclusion: $\sim k \rightarrow r$

Solutions: (a) **Law of Disjunctive Inference.** According to the second premise, statement h is true, so that the right-hand disjunct of the first premise ($\sim h$) is false. Therefore, the left-hand disjunct (m) must be true.

(b) **Law of Conjunctive Simplification.** Since the conjunction is true, each conjunct is true.

(c) **Law of Disjunctive Inference.** The second premise states that statement t is true. This means that the left-hand disjunct of the first premise ($\sim t$) is false, so the right disjunct ($\sim w$) is true.

(d) **Law of Contrapositive Inference.** A conditional ($\sim r \rightarrow k$) and its contrapositive ($\sim k \rightarrow r$) are logically equivalent. [***Note*:** r and $\sim(\sim r)$ are logically equivalent.]

THE CHAIN RULE (LAW OF THE SYLLOGISM). The **Chain Rule** states that if the consequent of one conditional and the antecedent of another conditional are related in such a way that the two conditionals take the form

$p \rightarrow q$

$q \rightarrow r$,

then the third conditional, $p \rightarrow r$, must also be true. For example,

Premise 1:	If it rains, then I will study.
Premise 2:	If I study, then I will do well in math.
Conclusion:	If it rains, then I will do well in math.

Example 3 shows that it is sometimes necessary to apply the Law of Contrapositive Inference before the Chain Rule can be used.

Example

3. For each set of premises draw a valid conclusion.

(a) $\sim h \rightarrow j$
$h \rightarrow f$

(b) If it rains, then I will go to the movies.
If I study, then I will not go to the movies.

Solutions: (a) Form the contrapositive of the first premise and then apply the chain rule:

	$\sim j \rightarrow h$	Contrapositive of $\sim h \rightarrow j$
	$h \rightarrow f$	Second premise
Conclusion:	$\sim j \rightarrow f$	Application of the chain rule

(b) Form the contrapositive of the second premise and then apply the chain rule:

	If it rains, then I will go to the movies.
	If I go to the movies, then I will not study.
Conclusion:	**If it rains, then I will not study.**

LAW OF DETACHMENT (*MODUS PONENS*). The **Law of Detachment** states that if a conditional ($p \rightarrow q$) and its antecedent (p) are true, then its consequent (q) must also be true. For example,

Premise 1:	If today is hot, then I will go swimming.
Premise 2:	Today is hot.
Conclusion:	I will go swimming.

LAW OF *MODUS TOLLENS*. This law of reasoning states that if a conditional ($p \rightarrow q$) is true and its consequent (q) is false, then its antecedent (p) must be false. For example,

Premise 1:	If today is hot, then I will go swimming.
Premise 2:	I will *not* go swimming.
Conclusion:	Today is *not* hot.

SUMMARY OF REASONING LAWS INVOLVING CONDITIONALS. Table 1.4 summarizes the laws of reasoning involving conditionals in symbolic and argument forms.

TABLE 1.4 Laws of Reasoning Involving Conditionals

Law	Symbolic Form	Argument Form
Contrapositive Inference	$(p \rightarrow q) \leftrightarrow (\sim q \rightarrow \sim p)$	$p \rightarrow q$ $\therefore\ \sim q \rightarrow \sim p$
Detachment (*Modus Ponens*)	$[(p \rightarrow q) \wedge p] \rightarrow q$	$p \rightarrow q$ p $\therefore\ q$
Modus Tollens	$[(p \rightarrow q) \wedge \sim q] \rightarrow \sim p$	$p \rightarrow q$ $\sim q$ $\therefore\ \sim p$
Chain Rule	$[(p \rightarrow q) \wedge (q \rightarrow r)] \rightarrow (p \rightarrow r)$	$p \rightarrow q$ $q \rightarrow r$ $\therefore\ p \rightarrow r$

Examples

4. For each valid argument, state the law(s) of reasoning being applied.

(a)
$$\begin{array}{ll} & \sim r \rightarrow s \\ & s \rightarrow p \\ \hline \therefore & \sim r \rightarrow p \end{array}$$

(b)
$$\begin{array}{ll} & \sim x \rightarrow y \\ & \sim x \\ \hline \therefore & \sim y \end{array}$$

(c)
$$\begin{array}{ll} & \sim x \rightarrow y \\ & y \\ \hline \therefore & \sim x \end{array}$$

(d)
$$\begin{array}{ll} & s \rightarrow t \\ & \sim t \\ \hline \therefore & \sim s \end{array}$$

(e)
$$\begin{array}{ll} & x \rightarrow y \\ & z \rightarrow \sim y \\ \hline \therefore & x \rightarrow \sim z \end{array}$$

(f)
$$\begin{array}{ll} & \sim m \rightarrow n \\ & \sim n \\ \hline \therefore & m \end{array}$$

Solutions: (a) **Chain Rule.** (b) **Law of Detachment.**

(c) **Argument is not valid** since the antecedent may be either true *or* false.

(d) **Law of *Modus Tollens*.**

(e) **Law of Contrapositive Inference and Chain Rule.**

(f) **Law of *Modus Tollens*** [*Note:* m and $\sim(\sim m)$ are logically equivalent.]

5. Let p represent the statement "Today is Saturday."
Let q represent the statement "I stay up late."
Let r represent the statement "I go to school."

Represent each of the following sets of premises symbolically, using the letters p, q, and r and the proper logical connectives. In each case, use both premises to draw a valid conclusion, if possible, and state the law(s) of reasoning being applied.

(a) I go to school or I stay up late.
I do not stay up late.

(b) If I go to school, then today is not Saturday.
Today is Saturday.

(c) If I stay up late, then I do not go to school.
I stay up late.

(d) If today is Saturday, then I do not go to school.
I do not go to school.

Solutions: (a)
$$\begin{array}{l} r \vee q \\ \sim q \\ \hline r \end{array}$$
Conclusion: r **(Law of Disjunctive Inference)**

(b)
$$\begin{array}{l} r \rightarrow \sim p \\ p \\ \hline \sim r \end{array}$$
Conclusion: $\sim r$ **(Law of *Modus Tollens*.** Since the consequent of the given conditional is false, the antecedent of the conditional must also be false.)

(c) $q \to \sim r$

$\underline{q}$

Conclusion: $\sim r$ (**Law of Detachment**. The antecedent of the given conditional is true, so its consequent must also be true.)

(d) $p \to \sim r$

$\underline{\sim r}$

No valid conclusion is possible. Since the consequent is true ($\sim r$), the antecedent may be either true or false.

6. The statement $[(a \to \sim b)] \wedge (c \to b)]$ is logically equivalent to:
(1) $a \to c$ (2) $a \to \sim c$ (3) $\sim a \to c$ (4) $c \to a$

Solution: Since a conditional and its contrapositive are logically equivalent, the original statement may be rewritten as $[(a \to \sim b)] \wedge (\sim b \to \sim c)]$. When the chain rule is applied, the conclusion $a \to \sim c$ follows. The correct answer is **choice (2)**.

EXERCISE SET 1.4

1. Which of the following is logically equivalent to the statement "If I watch TV, then I will do my homework"?
(1) If I do not watch TV, then I will do my homework.
(2) If I do my homework, then I will not watch TV.
(3) If I do not do my homework, then I will watch TV.
(4) If I do my homework, then I will watch TV.

2. Given these true statements: "If a boy plays high school football, he must be passing at least three subjects" and "Bob is not passing three subjects," it follows that:
(1) Bob plays on the football team.
(2) Bob does not play on the football team.
(3) Few boys try out for the team.
(4) No conclusion can be reached.

3. Given these true statements: "Mark goes shopping or he goes to the movies" and "Mark doesn't go to the movies," which statement *must* also be true?
(1) Mark goes shopping.
(2) Mark doesn't go shopping.
(3) Mark doesn't go shopping and he doesn't go the the movies.
(4) Mark stays home.

4. Given these true statements: "If you take a swim, then you don't catch a fish" and "If you row a boat, then you catch a fish," which statement *must* also be true?
(1) If you don't row a boat, then you don't take a swim.
(2) If you take a swim, then you don't row a boat.
(3) If you don't take a swim, then you catch a fish.
(4) If you don't catch a fish, then you don't take a swim.

5. Given the true statement $[(s \vee t) \wedge \sim s]$, which statement is true?
(1) t (2) $\sim t$ (3) s (4) $s \wedge \sim t$

6. *If* $\sim r \rightarrow s$ and $\sim s$ are given, which statement must be true?
(1) r (2) $\sim r$ (3) $r \rightarrow s$ (4) $s \wedge \sim s$

7. If $a \rightarrow b$ and $\sim c \rightarrow \sim b$ are true statements, which statement must also be true?
(1) $\sim c \rightarrow \sim a$ (2) $\sim a \rightarrow \sim c$ (3) $a \rightarrow \sim c$ (4) $\sim c \rightarrow a$

8–12. Given this true statement: "*If I do not pass this test, then I will eat my hat*," *assume each of the following statements is also true, and draw a valid conclusion, if possible, in each case.*

8. I do not eat my hat.
9. I fail this test.
10. I pass this test.
11. If I study hard, then I will not eat my hat.
12. I eat my hat.

13–42. For each set of premises, determine whether the conclusion is valid. If it is, then state the law(s) of reasoning being applied.

13.
$$\begin{array}{rl} & r \wedge \sim h \\ \hline \therefore & \sim h \end{array}$$

14.
$$\begin{array}{rl} & c \rightarrow g \\ & g \rightarrow h \\ \hline \therefore & c \rightarrow h \end{array}$$

15.
$$\begin{array}{rl} & s \vee \sim t \\ & t \\ \hline \therefore & s \end{array}$$

16.
$$\begin{array}{rl} & a \rightarrow \sim b \\ & a \\ \hline \therefore & \sim b \end{array}$$

17.
$$\begin{array}{rl} & x \rightarrow y \\ & \sim x \\ \hline \therefore & \sim y \end{array}$$

18.
$$\begin{array}{rl} & y \rightarrow \sim z \\ \hline \therefore & z \rightarrow \sim y \end{array}$$

19.
$$\begin{array}{rl} & E \rightarrow F \\ & F \\ \hline \therefore & E \end{array}$$

20.
$$\begin{array}{rl} & j \rightarrow \sim k \\ & k \\ \hline \therefore & \sim j \end{array}$$

21.
$$\begin{array}{rl} & w \wedge \sim f \\ & f \vee b \\ \hline \therefore & b \end{array}$$

22.
$$\begin{array}{rl} & \sim b \rightarrow a \\ & b \rightarrow c \\ \hline \therefore & \sim a \rightarrow c \end{array}$$

23. I am hungry or I am sleepy.
I am not sleepy.
∴ I am hungry.

24. If cows can fly, then turtles can dance.
Turtles cannot dance.
∴ Cows cannot fly.

25. If I do well in math, then I do well in science.
I do well in math.
∴ I do well in science.

26. If it snows, then I will not drive.
If I do not drive, then I will take a bus.
∴ If it snows, then I will take a bus.

27. I am tall and I am thin.
If I am tall, I play on the basketball team.
∴ I play on the basketball team.

28. If it is a weekday, then I go to school.
I do not go to school.
∴ It is the weekend.

29. If a rhombus contains a right angle, then it is a square.
The rhombus is not a square.
∴ The rhombus does not contain a right angle.

30. If $x < 5$, then it is cloudy.
$x = 2$
∴ It is cloudy.

31. I will wear a scarf, if it is windy.
I do not wear a scarf.
∴ It is not windy.

32. If I do not win the lottery, I do not quit my job.
I quit my job.
∴ I win the lottery.

*33–41. Without constructing a truth table, determine whether each of the following statements is a tautology. (*Hint*: If the statement illustrates a law of reasoning, or is the biconditional of two logically equivalent statements, then the statement is a tautology.)*

33. $[(r \vee {\sim}s) \wedge s] \rightarrow r$

34. $({\sim}r \rightarrow s) \leftrightarrow ({\sim}s \rightarrow r)$

35. $[(r \rightarrow s) \wedge (t \rightarrow {\sim}s)] \leftrightarrow (r \rightarrow t)$

36. $[(j \rightarrow c) \wedge j] \rightarrow c$

37. $[(j \rightarrow c) \wedge {\sim}j] \rightarrow {\sim}c$

38. $[(j \rightarrow c) \wedge c] \rightarrow c$

39. $[(j \rightarrow c) \wedge {\sim}c] \rightarrow {\sim}j$

40. $[({\sim}a \rightarrow g) \wedge {\sim}a] \rightarrow g$

41. $[({\sim}p \rightarrow q) \wedge {\sim}p] \rightarrow q$

42–47. For each set of premises, draw a valid conclusion and state the law(s) of reasoning being applied.

42. $\sim e \vee q$
e

43. $\sim w \to a$
$\sim a$

44. $h \to \sim b$
$c \to b$

45. $x \to \sim y$
y

46. $A \wedge \sim B$
$C \to B$

47. $\sim(r \vee s)$
$t \to s$

48. Given these premises:

$$(x - 3 = 1) \vee (x + 3 = 6)$$
$$x \neq 3$$

which statement is true?
(1) $x = 4$ (2) $x \neq 4$ (3) $x = 3$ (4) $x \geq 3$

49. For each of the following premises, follow the directions provided in Example 5 of this section:
(a) If I do not go to school, then I stay up late.
I do not stay up late.
(b) If I stay up late, then today is Saturday.
If I go to school, then today is not Saturday.
(c) Today is Saturday and I stay up late.
If I stay up late, then I do not go to school.
(d) Today is Saturday or I do not stay up late.
I do not stay up late.
(e) I stay up late and I do not go to school.
I stay up late, if today is Saturday.

50. Given:

$$\sim S \vee T$$
$$W \to S$$
$$\sim T$$

draw a valid conclusion.

1.5 LOGIC PROOFS

KEY IDEAS

An argument that includes a detailed explanation of how valid methods of reasoning can be used to reach a conclusion from a given set of premises is called a **logic proof**. Although a proof may be presented in different ways, only the statement-reasons, two-column proof will be illustrated in this section.

TWO-COLUMN FORMAT OF LOGIC PROOFS. More complicated arguments that involve several laws of reasoning are sometimes presented as a sequence of numbered statements with corresponding reasons that show, in step-by-step fashion, how principles of logic are used to reach a conclusion. For example,

$$\begin{array}{lll} \text{Given:} & \left.\begin{array}{c} \sim(p \wedge \sim q) \\ p \end{array}\right\} & \text{Premises} \\ \text{Prove:} & \quad q & \text{Conclusion} \end{array}$$

PROOF

Statement	Reason
1. $\sim(p \wedge \sim q)$	1. Given.
2. $\sim p \vee q$	2. De Morgan's laws.
3. p	3. Given.
4. $\sim p$ is false.	4. A statement and its negation have opposite truth values.
5. q	5. Law of Disjunctive Inference (2, 4).

Notice that in this form of logic proof:

- Statements and reasons are presented side by side in a two-column arrangement.
- Each numbered statement has a corresponding reason preceded by the same number.
- When a premise appears in the "Statement" column, the corresponding reason is "given."
- The last numbered statement of the proof is the desired conclusion.
- When a law of reasoning appears in the "Reason" column, it is followed by a pair of numbers, inside parentheses, which identify the numbers of the statements to which the law is being applied. For example, in the preceding proof the Law of Disjunctive Inference is followed by (2, 4), indicating that it is being applied to the second and the fourth statement.

Examples

1. Present a two-column proof.

Given: $r \to t$
$r \vee \sim s$
s
Prove: t

Solution: Since statement s is true, look in the "Given" for a compound statement involving statement s and then use a law of reasoning to draw a conclusion about the truth value of the other letter (simple statement). For example, consider the truth value of $r \vee \sim s$. Since statement s is true, $\sim s$ is false; therefore, by the Law of Disjunctive Inference, r is true. Next, look for a statement in the "Given" involving r. Continue this process until a conclusion is reached about the truth value of statement t.

PROOF

Statement	Reason
1. s	1. Given.
2. $\sim s$ is false.	2. A statement and its negation have opposite truth values.
3. $r \vee \sim s$	3. Given.
4. r	4. Law of Disjunctive Inference.
5. $r \to t$	5. Given.
6. t	6. Law of Detachment (4, 5).

2. Present a two-column proof.

Given: $l \to \sim j$
$\sim l \to k$
j
Prove: k

Solution: Begin with the premise j since it is a simple statement (single letter) and work "backward" (that is, look for a compound statement in the "Given" that contains letter j, and draw a conclusion about the truth value of the other simple statement). Continue this process until a conclusion can be made about the truth value of statement k.

PROOF

Statement	Reason
1. j	1. Given.
2. $l \rightarrow \sim j$	2. Given.
3. $\sim l$	3. Law of *Modus Tollens* (1, 2).
4. $\sim l \rightarrow k$	4. Given.
5. k	5. Law of Detachment (3, 4).

3. Present a two-column proof.

Given: $r \wedge s$
$s \rightarrow t$
$\sim(t \wedge k)$
Prove: $\sim k$

Solution: Since there is no simple statement in the set of premises, consider the conjunction $r \wedge s$. Statements r and s are both true. Since r does not appear in the remaining premises, look for a compound statement in the "Given" that involves s and work toward obtaining a conclusion about the truth value of $\sim k$.

PROOF

Statement	Reason
1. $r \wedge s$	1. Given.
2. s	2. Law of Conjunctive Simplification.
3. $s \rightarrow t$	3. Given.
4. t	4. Law of Detachment (2, 3).
5. $\sim(t \wedge k)$	5. Given.
6. $\sim t \vee \sim k$	6. De Morgan's Laws.
7. $\sim t$ is false.	7. A statement and its negation have opposite truth values.
8. $\sim k$	8. Law of Disjunctive Inference (6, 7).

4. Given:
If Jill studies, then she will know the work.
If Jill goes out to play, then she will not know the work.
Jill studies or her parents will not be happy.
Jill's parents are happy or she will not be allowed to go to the party.
Jill is allowed to go to the party.

Let S represent: "Jill studies."
Let W represent: "Jill will know the work."
Let P represent: "Jill goes out to play."
Let H represent: "Jill's parents are happy."
Let A represent: "Jill is allowed to go to the party."

(a) Using S, W, P, H, A, and proper connectives, express each statement in symbolic form.
(b) Prove: "Jill does not go out to play."

Solutions:

(a) Given: $S \to W$
$P \to \sim W$
$S \vee \sim H$
$H \vee \sim A$
A

Prove: $\sim P$

(b) Since letter A in the "Given" stands alone, work "backward" by finding a compound statement that involves statement A and draw a conclusion about the truth value of the other letter (simple statement). Continue this process until a conclusion can be reached about the truth value of $\sim P$.

PROOF

Statement	Reason
1. A	1. Given.
2. $H \vee \sim A$	2. Given.
3. H	3. Law of Disjunctive Inference (1, 2).
4. $S \vee \sim H$	4. Given.
5. S	5. Law of Disjunctive Inference (3, 4).
6. $S \to W$	6. Given.
7. W	7. Law of Detachment (5, 6).
8. $P \to \sim W$	8. Given.
9. $\sim P$	9. Law of *Modus Tollens* (7, 8).

EXERCISE SET 1.5

1–10. In each case, write a two-column proof.

1. Given: $B \wedge C$
$C \to D$
Prove: D

2. Given: $R \vee \sim T$
$W \to T$
W
Prove: R

3. Given: $x \vee \sim y$
$z \to y$
$\sim x$
Prove: $\sim z$

4. Given: $H \vee E$
$H \to K$
$\sim K$
Prove: E

5. Given: $r \to s$
$s \to t$
$\sim t$
Prove: $\sim r$

6. Given: $A \vee B$
$A \to \sim C$
$\sim D \to C$
$\sim D$
Prove: B

7. Given: $s \wedge t$
$\sim s \vee y$
$y \to z$
Prove: z

8. Given: $f \wedge g$
$f \to \sim h$
$i \to h$
Prove: $\sim i$

9. Given: $p \wedge r$
$\sim p \vee q$
$(r \wedge q) \to t$
Prove: t

10. Given: $i \to j$
$\sim j$
$i \vee k$
$k \to m$
Prove: m

11. Given:
Either I do the geometry question or I do the algebra question.
If I do the algebra question, I get it correct.
If I get the algebra question correct, I do not lose points.
However, it is known that I lost points.
Therefore, I did the geometry question.

Let G represent: "I do the geometry question."
Let A represent: "I do the algebra question."
Let C represent: "I get the algebra question correct."
Let P represent: "I lose points."

(a) Using G, A, C, P, and proper connectives, express each statement in symbolic form.
(b) Using laws of inference, show that a valid conclusion has been reached.

12. Given:
I study hard or I do not take Regents mathematics.
If I take Course II, then I take Regents mathematics.
If I do not take Course II, then I did not pass Course I.
I passed Course I.

Let P represent: "I study hard."
Let Q represent: "I take Regents mathematics."
Let R represent: "I take Course II."
Let S represent: "I passed Course I."

Using laws of inference, prove that I study hard.

13. Given:

If the programmer is skilled, the computer will be accurate.
Either the programmer is skilled, or the employees are lazy.
If the employees are lazy, John will be fired.
If the computer is accurate, Harry will get a raise.
John did not get fired.

Let P represent: "The programmer is skilled."
Let C represent: "The computer is accurate."
Let E represent: "The employees are lazy."
Let J represent: "John will be fired."
Let H represent: "Harry will get a raise."

(a) Using $P, C, E, J, H,$ and proper connectives, express each statement in symbolic form.
(b) Using laws of inference, prove that Harry will get a raise.

14. Given:

Either Al went to college or he joined the army.
If he joined the army, then his hair was cut short.
If his hair was cut short, then it does not cover his ears.
Al's hair covers his ears.

Let C represent: "Al went to college."
Let A represent: "He joined the army."
Let H represent: "His hair was cut short."
Let E represent: "His hair covers his ears."

(a) Using $C, A, H, E,$ and proper connectives, express each statement in symbolic form.
(b) Using laws of inference, prove that Al went to college.

15. Given:

Either a crime was committed or Arthur is not telling the truth.
If the dog howled, a crime was not committed.
If the dog did not howl, the butler did it.
Arthur is telling the truth.

Let A represent: "Arthur is telling the truth."
Let B represent: "The butler did it."
Let C represent: "A crime was committed."
Let D represent: "The dog howled."

(a) Using $A, B, C, D,$ and proper connectives, express each statement in symbolic form.
(b) Using the laws of inference, prove that the butler did it.

CHAPTER 1 REVIEW EXERCISES

1. The statement $\sim(p \wedge \sim q)$ is logically equivalent to:
   (1) $\sim p \wedge q$ (2) $\sim p \vee q$ (3) $\sim p \rightarrow q$ (4) $\sim p \vee \sim q$

2. If the temperature in my room is above 80 degrees, the air conditioner goes on. The air conditioner is not on. Which is a valid conclusion?
   (1) The temperature in my room is going up.
   (2) The temperature in my room is going down.
   (3) The temperature in my room is above 80 degrees.
   (4) The temperature in my room is not above 80 degrees.

3. Given the premises $a \rightarrow b$ and a, which of the following is a logical conclusion?
   (1) a (2) $\sim a$ (3) b (4) $\sim b$

4. If statement r is true and statement s is false, then which of the following is true?
   (1) $r \rightarrow s$ (2) $r \wedge s$ (3) $\sim s \rightarrow \sim r$ (4) $r \vee s$

5. Which conclusion logically follows from the true statements "If we do not save fuel, there will be an energy crisis" and "If there is an energy crisis, schools will close"?
   (1) If we save fuel, there will not be an energy crisis.
   (2) If we do not save fuel, schools will close.
   (3) If the schools close, there is an energy crisis.
   (4) If we save fuel, there is an energy crisis.

6. The negation of $p \vee \sim q$ is:
   (1) $\sim p \vee q$ (2) $\sim p \wedge q$ (3) $p \wedge \sim q$ (4) $\sim p \vee \sim q$

7. If $F \rightarrow \sim G$ and G, then which statement is true?
   (1) $\sim F$ (2) $\sim G$ (3) F (4) No conclusion is possible.

8. Which is logically equivalent to the statement "If I eat, then I live"?
   (1) If I do not eat, then I do not live.
   (2) If I do not live, then I do not eat.
   (3) If I live, then I eat.
   (4) If I do not eat, then I live.

9. If $a \rightarrow b$ and $\sim c \rightarrow \sim b$, then it follows logically that:
(1) $a \rightarrow c$ (2) $b \rightarrow a$ (3) $c \rightarrow a$ (4) $c \rightarrow b$

10. Which of the following is the negation of the statement "Larry is old and Mario is not here"?
(1) Larry is old and Mario is here.
(2) Larry is not old or Mario is here.
(3) Larry is not old and Mario is not here.
(4) Larry is old or Mario is here.

11. Given the true statements "If Paul catches fish today, then he will give me some" and "Paul will give me some fish," which statement *must* be true?
(1) Paul will not give me some fish.
(2) Paul will not catch some fish today.
(3) Paul will catch some fish today.
(4) No conclusion is possible.

12. Given the true statement $[(p \vee q) \wedge (\sim q)]$, which statement *must* also be true?
(1) p (2) $\sim p$ (3) q (4) $p \rightarrow q$

13. Given the true statements:
If the sun shines, I will play tennis.
If I play tennis, I will *not* study math.
I study math.

Which statement *must* also be true?
(1) I played tennis.
(2) The sun didn't shine.
(3) I didn't study math.
(4) The sun did shine.

14. Write the *number* of the valid conclusion, *chosen from list (1)–(6) below*, that can be deduced from each set of premises.

(1) $\sim r$ (3) $\sim s$ (5) $t \rightarrow r$
(2) r (4) s (6) $t \rightarrow \sim r$

(a) $r \rightarrow s$
$\sim s$

(b) $r \rightarrow s$
r

(c) $r \vee s$
$\sim s$

(d) $t \rightarrow \sim s$
$\sim r \rightarrow s$

(e) $\sim(s \wedge \sim r)$
$\sim r$

(f) $r \wedge \sim t$
$s \rightarrow t$

15. Write a valid conclusion for each set of premises. If no conclusion is possible, write "no conclusion."

(a) If it snows this weekend, we will go skiing.
We will not go skiing.

(b) Either it rains in April or flowers will not grow in May.
It did not rain in April.

(c) The person who borrowed this book owes the library a quarter.
Mary borrowed this book.

(d) If I play tennis, then I will not have time for dinner.
If I do not have time for dinner, then I will not go swimming.

(e) Either I have money or I will stay home.
I do not have money.

(f) $\sim r \rightarrow s$
$t \rightarrow \sim s$

(g) $\sim x \rightarrow y$
$\sim x$

(h) Jim will get an A if he passes this test.
Jim passes this test.

(i) $(x + 2 = 5) \vee (x - 7 = 1)$
$x \neq 3$

(j) All Nifty candies are good to eat.
Crunchies is a Nifty candy.

16. Given the following statements:

If José plays the radio too loud, then his father will be angry.
If his father is angry, then José won't go to the party.
José goes to the party or he stays home.
José did not stay home.

Let R represent: "José plays the radio too loud."
Let A represent: "His father will be angry."
Let P represent: "José goes to the party."
Let S represent: "José stays home."

(a) Using R, A, P, S, and proper connectives, express each statement in symbolic form.

(b) Prove that José did not play the radio too loud.

CHAPTER 2

Mathematical Systems

2.1 THE REAL NUMBERS AND MATHEMATICAL SYSTEMS*

KEY IDEAS

A **mathematical system** includes the following:

- A set.
- At least one operation.
- Rules for applying the operation(s) to members of the set.

The set of real numbers, together with any of the four arithmetic operations, is an example of a mathematical system. If $R = \{\text{real numbers}\}$, then the notation $(R, +)$ refers to the mathematical system in which the operation of addition is defined for the set of real numbers. Similarly, $(R, \cdot)$ refers to the mathematical system in which the operation of multiplication is defined for the set of real numbers.

*This section includes material that is optional in the *revised* New York State Syllabus. The formal definitions in this section and the material that involves interpreting tables that define new operations are not required.

THE SET OF REAL NUMBERS AND ITS SUBSETS. The set of rational numbers, together with the set of irrational numbers, forms the set of **real numbers**. As the Description column of Table 2.1 illustrates, natural numbers, whole numbers, and integers are subsets of the set of rational numbers; rational and irrational numbers are non-intersecting subsets of the set of real numbers.

BINARY OPERATIONS, UNARY OPERATIONS, AND CLOSURE. The ordinary operations of addition, subtraction, multiplication, and division are examples of *binary operations* since each of these operations must be applied to exactly *two* numbers at a time. A **binary operation** for a set of elements is a rule that takes exactly *two* elements of the set and produces an element that may or may not be a member of the same set. Finding the absolute value of a real number and evaluating the square root of a nonnegative real number are **unary** operations since they work on a *single* number at a time.

The set of real numbers is *closed* under the operations of addition and multiplication since the result of these binary operations is always another real number. On the other hand, the set of natural numbers is *not closed* under the operation of subtraction since the difference of two natural numbers may or may not be another natural number. For example,

$$5-2=3 \quad \textit{but} \quad 2-5 \text{ is } \textit{not} \text{ equal to a natural number.}$$

Example 1: The set of even integers is closed under addition since the sum of any pair of even integers is an even integer.

Example 2: The set of odd integers is *not* closed under addition since the sum of two odd integers is not an odd integer. For example, $3+5=8$.

Example 3: The set of integers is *not* closed under division since the quotient of two integers is not necessarily an integer. For example, $\frac{12}{6}$ is equal to an integer, but $\frac{4}{3}$ is *not* equal to an integer.

DEFINITION OF CLOSURE

A set S is **closed** under a binary operation $*$ if, for all a and b in S, $a * b$ is also a member of set S.

TABLE 2.1 Subsets of the Real Numbers

Set	Description
1. Natural numbers	1. $N = \{1, 2, 3, \ldots\}$
2. Whole numbers	2. $W = \{0, 1, 2, 3, \ldots\}$
3. Integers	3. $Z = \{\ldots, -3, -2, -1, 0, 1, 2, 3, \ldots\}$
4. Rational numbers	4. $Q = \{x \mid x = p/q$, where p and q are integers and $q \neq 0\}$ Rational numbers include: • Integers and fractions having integer numerators and denominators. • Decimals that terminate, such as 0.356 and 0.10. • Decimals that have one or more nonzero digits that repeat endlessly, such as 0.333 . . . (or $0.\overline{3}$) and 0.8636363 . . . (or $0.8\overline{63}$). • Square roots of perfect squares, such as $\sqrt{9}$; cube roots of perfect cubes, such as $\sqrt[3]{8}$; and so forth.
5. Irrational numbers	Irrational numbers are numbers that cannot be expressed as the ratio of two integers and include: • Decimal numbers, such as $\pi = 3.141592\ldots$, that never end and that do not have one or more nonzero digits that endlessly repeat. • Roots that do not evaluate to a rational number, such as $\sqrt{3}$ and $\sqrt[3]{5}$.

DEFINING OPERATIONS. A new binary operation may be defined in one of two ways:

- Writing a formula that expresses the new operation in terms of familiar operations. As an example, consider the operation $\square$, which, for all a, b in set S, is defined by the formula

$$a \square b = \sqrt{a^2 + b^2}.$$

To find the value of $3 \square 4$, replace a by 3 and b by 4 in the formula $\sqrt{a^2 + b^2}$:

$$\begin{aligned} a \square b &= \sqrt{a^2 + b^2} \\ 3 \square 4 &= \sqrt{3^2 + 4^2} \\ &= \sqrt{9 + 16} = \sqrt{25} \\ &= \mathbf{5} \end{aligned}$$

- Presenting a table that shows the result of performing the operation on every possible pair of elements in the set for which the operation is defined. To illustrate, the accompanying table defines the operation $\diamond$ for the set $\{A, C, T\}$. The table is read in the same way that an ordinary "times table" (multiplication table) is read.

$\diamond$	A	C	T
A	C	T	W
C	A	T	T
T	A	A	C

To find the value of $T \diamond C$, locate T in the column headed by the operation symbol $\diamond$ and follow the same row to the right until it meets the column headed by C. The table entry that represents the value of $T \diamond C$ is located by drawing imaginary lines from T and C, as shown in the accompanying table. Since these lines intersect at A, $T \diamond C = A$.

$\diamond$	A	C	T
A		$\vert$	
C		$\downarrow$	
T	$-$	$\rightarrow A$	

Here are some additional examples that use the same table.

Example 1: What is the value of A^2?

Rewrite A^2 as $A \diamond A$, which has a value of $\mathbf{C}$.

Example 2: Is the set $\{A, C, T\}$ closed under $\diamond$?
The set $\{A, C, T\}$ is *not closed* under operation $\diamond$ because $A \diamond T = W$ and W is *not* a member of the set.

Example 3: What is the value of $(C \diamond T) \diamond A$?

Evaluate $(C \diamond T) \diamond A$ by working from left to right:

$$\underbrace{(C \diamond T)} \diamond A = T \diamond A = \mathbf{A}$$

◇	A	C	T
A	- - →	(T) ↑	
C			
T			

Example 4: Solve for x: $A \diamond x = T$.

Locate A in the column headed by the operation symbol, ◇. Move along the A row until the value T is encountered. Then determine the corresponding column heading, which is C. Since $A \diamond C = T$, $x = C$. The solution set is $\{\mathbf{C}\}$.

Example 5: Solve for x: $C \diamond x = T$.

From the table, $C \diamond C = T$ and $C \diamond T = T$, so that $x = C$ or $x = T$. The solution set is $\{\mathbf{C}, \mathbf{T}\}$.

Example 6: Solve for x: $T \diamond x = T$.

Since the third row of table values does not contain a T, there is *no* value of x that makes the equation $T \diamond x = T$ true. Therefore, the solution set is the empty set, $\{\ \ \}$.

Note: Examples 4, 5, and 6 illustrate that an equation involving a newly defined operation may have more than one solution, exactly one solution, or no solutions.

Examples

1. If $a \otimes b$ is defined as $\left(\frac{a}{b}\right)^2$, what is the value of $6 \otimes 2$?

Solution: Let $a = 6$ and $b = 2$. Then

$$a \otimes b = \left(\frac{a}{b}\right)^2$$

$$6 \otimes 2 = \left(\frac{6}{2}\right)^2 = 3^2 = \mathbf{9}$$

2. Compute $(z * y) * x$ in the system defined by the accompanying table.

*	x	y	z
x	x	z	z
y	z	y	z
z	x	y	x

Solution: $\underbrace{(z * y)} * x$

Compute $z * y$: $\underbrace{y * x}$

Compute $y * x$: z

3. Using the accompanying table, find y if $a * y = c * d$.

$*$	a	b	c	d
a	b	c	d	a
b	c	d	a	b
c	d	a	b	c
d	a	b	c	d

Solution: First use the table to compute the right side of the equation.

$$a * y = \underbrace{c * d}_{= \ c}$$

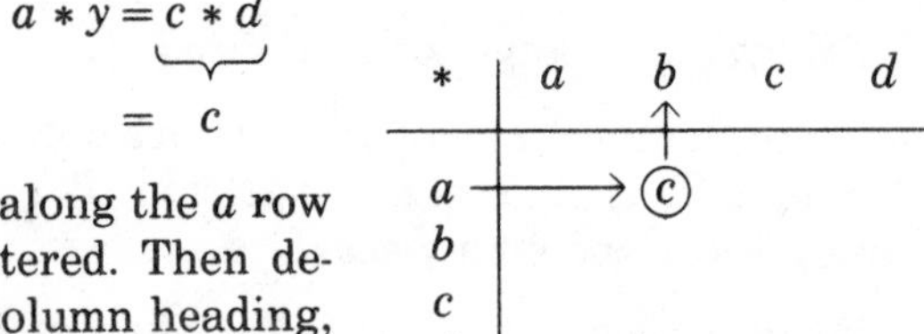

To find the value of y, move along the a row until the value c is encountered. Then determine the corresponding column heading, which is b. Therefore $y = \mathbf{b}$ since $a * b = c$.

IDENTITY AND INVERSE ELEMENTS. Adding 0 to a real number always results in the original number, so that 0 is the *identity element* for addition. Adding $-a$ to any real number a always produces 0 (the identity element for addition). For example, $3 + (-3) = 0$. Therefore, $-a$ is the *inverse* of element a under the operation of addition. Table 2.2 summarizes the identity and inverse elements for the operations of addition and multiplication for the real numbers.

TABLE 2.2 Identity and Inverse Elements for Real Numbers

Operation	Identity Element	Inverse Element
Addition	For all a, 0 is the additive identity since $a + \mathbf{0} = a$	For all a, $-a$ is the additive inverse since $a + (-\boldsymbol{a}) = 0$
Multiplication	For all a, 1 is the multiplicative identity since $a \cdot \mathbf{1} = a$	For all a ($a \neq 0$), $\frac{1}{a}$ is the multiplicative inverse since $a \cdot \left(\frac{\mathbf{1}}{\boldsymbol{a}}\right) = 1$

In general,

- A binary operation between a member of a set and an *identity* element of the set always produces the original member of the set. An identity element of a set for a particular binary operation is the *same* for *each* member of the set.

- A binary operation between a member of a set and its *inverse* always produces an identity element of the set for that operation. The inverse of an element may be *different* for *each* element of the set.

- In a mathematical system, an element may not have an inverse. For example, in $(R, \cdot)$, the multiplicative inverse (reciprocal) of 5 is $\frac{1}{5}$. However, 0 does not have a multiplicative inverse since $\frac{1}{0}$ is not defined.

DEFINITIONS OF IDENTITY AND INVERSE ELEMENTS

Let $*$ represent a binary operation on set S.

- For all x in S, e is an **identity element** for the system $(S, *)$ if

$$x * e = x \qquad \text{and} \qquad e * x = x.$$

- For all x in S with e the identity element for the system $(S, *)$, x^{-1} is the **inverse** of x if

$$x * x^{-1} = e \qquad \text{and} \qquad x^{-1} * x = e.$$

***Note*:** In the notation x^{-1}, -1 does *not* represent an exponent. When used in this context, x^{-1} is read as "the inverse of x."

Example

4. Using the accompanying table, find the following:
(a) the identity element for operation □
(b) the inverse of 2.

□	1	2	3	4
1	3	4	1	2
2	4	1	2	3
3	1	2	3	4
4	2	3	4	1

Solutions: (a) The identity element for a set always produces the same number as the one on which it operates.

Step 1. Find the row in the table that duplicates the column headings to the right of the operation symbol. Notice that:

	□	1	2	3	4
3 □ 1 = 1	1				
3 □ 2 = 2	2				
3 □ 3 = 3 ↦	3	1	2	3	4
3 □ 3 = 3	4				

Step 2. Find the column in the table that duplicates the column headed by the operation symbol, □. Notice that:

$1 \square 3 = 1$
$2 \square 3 = 2$
$3 \square 3 = 3$
$4 \square 3 = 4$

□	1	2	3	4
1			1	
2			2	
3			3	
4			4	

Step 3. Apply the information gained in Steps 1 and 2. Step 1 tells us that $3 \square x = x$ for all x in the set. Step 2 tells us that $x \square 3 = x$ for all x in the set. Therefore,

$$3 \square x = x \square 3 = x \quad \text{for all } x \text{ in the set.}$$

The identity element for the operation □ is **3**.

(b) Let $x =$ inverse of 2.
Then, $2 \square x = 3$ (the identity element).
From the table, $x = 4$ since $2 \square 4 = 3$.
Also, $4 \square 2 = 3$.

Therefore, the inverse of 2 is **4**.

Note: Keep in mind that before you can determine the inverse of an element of a set under a particular operation, you must know the identity element for that set under that operation.

COMMUTATIVE AND ASSOCIATIVE PROPERTIES. If a binary operation is *commutative* on a set, then, when performing the operation, the order in which two members of the set are written does not matter. If a binary operation is *associative* on a set, then, when performing the operation on three members of the set, the three members may be grouped in any order.

The set of real numbers R is commutative and associative under the operations of addition and multiplication. For example,

$3 + 4 = 4 + 3$ and $3 \cdot 4 = 4 \cdot 3$ (commutative)
$(3 + 4) + 5 = 3 + (4 + 5)$ and $(3 \cdot 4) \cdot 5 = 3 \cdot (4 \cdot 5)$ (associative)

The systems $(R, -)$ and $(R, \div)$ are not commutative and are not associative. For example,

$3 - 4 \neq 4 - 3$ and $\frac{3}{4} \neq \frac{4}{3}$ (not commutative)

$(5 - 4) - 3 \neq 5 - (4 - 3)$ and $\frac{12/3}{2} \neq \frac{12}{3/2}$ (not associative)

DEFINITION OF COMMUTATIVE AND ASSOCIATIVE PROPERTIES

Let $*$ represent a binary operation defined on set S.

- The system $(S, *)$ is **commutative** if, for all a and b in S,

$$a * b = b * a$$

- The system $(S, *)$ is **associative** if, for all a, b, and c in S,

$$(a * b) * c = a * (b * c)$$

Example

5. Let the binary operation $\triangle$ be defined on the set of real numbers such that $a \triangle b = a + b - ab$.

(a) Prove that the operation $\triangle$ is commutative on the set of real numbers.

(b) Prove that 0 is the identity element under this operation.

Solutions: (a) To prove that the operation is commutative on the set of real numbers, show that $a \triangle b = b \triangle a$ for all real numbers a and b:

(1) $a \triangle b = a + b - ab$
(2) $b \triangle a = b + a - ba = a + b - ab$

Since the right sides of equations (1) and (2) are equal for all a and b, the left sides of these equations must also be equal so that $a \triangle b = b \triangle a$. Hence, the operation is commutative.

(b) To prove that 0 is the identity element under the operation $\triangle$, show that $a \triangle 0 = 0 \triangle a = a$ for all real numbers a:

$$a \triangle b = a + b - ab$$
$$a \triangle 0 = a + 0 - a \cdot 0 = a - 0 = a$$
$$0 \triangle a = 0 + a - 0 \cdot a = a - 0 = a$$

Since $a \triangle 0 = 0 \triangle a = a$ for all real numbers a, 0 is the identity element under the operation $\triangle$.

DIAGONAL TEST FOR COMMUTATIVITY. A binary operation that is defined by a table is commutative if and only if the table values are symmetric with respect to a diagonal line drawn from the operation symbol to the opposite corner of the table. In the accompanying table, the operation Ω is commutative on $\{Q, U, A, D\}$ since it passes the diagonal test for commutativity. Notice that the same elements appear in corresponding positions (that is, positions that have their row and column numbers interchanged) on either side of the diagonal.

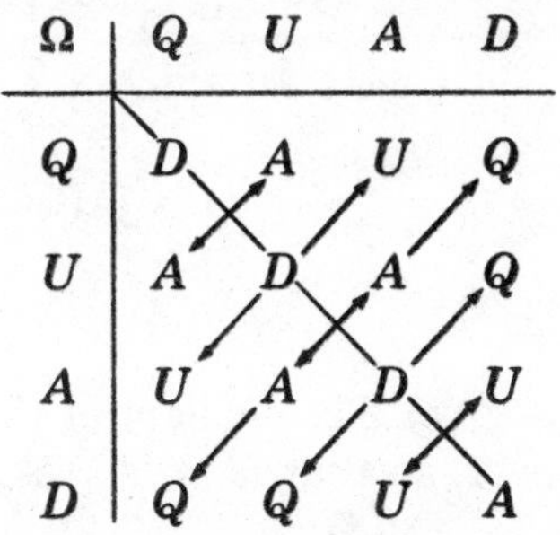

Ω	Q	U	A	D
Q	D	A	U	Q
U	A	D	A	Q
A	U	A	D	U
D	Q	Q	U	A

Example

6. Given the elements $\{M, A, T, H\}$ and the operation # as shown in the accompanying table:

#	M	A	T	H
M	A	T	H	M
A	T	H	M	A
T	H	M	A	T
H	M	A	T	H

(a) What is the identity element for the operation #?
(b) What is the inverse of M?
(c) Determine whether the set $\{M, A, T, H\}$ is commutative under #.
(d) Find the value of $M \# [A \# (T \# H)]$.
(e) Find x if $H \# x = A$.
(f) Which statement illustrates associativity?

(1) $H \# T = T$
(2) $H \# H = H$
(3) $M \# T = T \# M$
(4) $(M \# A) \# T = M \# (A \# T)$

Solutions: (a) ***H*** is the identity element for # since the row for H duplicates the column headings and the column for H duplicates the column headed by the operation symbol, #.

(b) The element at the intersection of the M row with the T column is H, so that $M \# T = H$, where H is the identity element. Also, the element at the intersection of the T row with the M column is H, so that $T \# M = H$. Since

$$M \# T = H \quad \text{and} \quad T \# M = H$$

the inverse of M is ***T***.

(c) As shown in the accompanying diagram, the table that defines # passes the diagonal test for commutativity, so the set $\{M, A, T, H\}$ **is commutative under #.**

#	M	A	T	H
M	A	T	H	M
A	T	H	M	A
T	H	M	A	T
H	M	A	T	H

(d) $M \# [A \# (T \# H)]$

$$M \# [A \# T]$$

$$M \# M$$

$$\mathbf{A}$$

(e) The element at the intersection of the H row with the A column is A, so that $H \# A = A$. Therefore, $x = \mathbf{A}$.

(f) **(4)**

THE DISTRIBUTIVE PROPERTY. The distributive property of real numbers links the operations of multiplication and addition so that an expression such as $3(2+5)$ can be evaluated in the following way:

$$\begin{aligned} 3(2+5) &= 3 \cdot 2 + 3 \cdot 5 \\ &= 6 + 15 \\ &= 21 \end{aligned}$$

Table 2.3 summarizes some of the properties of real numbers.

TABLE 2.3 Summary of Properties of Real Numbers

Property	Addition	Multiplication
Identity for a	$a + 0 = a$	$a \cdot 1 = a$
Inverse for a	$a + (-a) = 0$	$a \cdot \frac{1}{a} = 1 \quad (a \neq 0)$
Commutative	$a + b = b + a$	$ab = ba$
Associative	$(a + b) + c = a + (b + c)$	$(ab)c = a(bc)$
Distributive	$a \cdot (b + c) = a \cdot b + a \cdot c$	

Example

7. Using the set {0, 2, 4, 6, 8} and the operations # and @ as shown in the accompanying tables, verify that

$$4 \text{ @ } (6 \# 8) = (4 \text{ @ } 6) \# (4 \text{ @ } 8)$$

#	0	2	4	6	8
0	0	2	4	6	8
2	2	4	6	8	0
4	4	6	8	0	2
6	6	8	0	2	4
8	8	0	2	4	6

@	0	2	4	6	8
0	0	0	0	0	0
2	0	4	8	2	6
4	0	8	6	4	2
6	0	2	4	6	8
8	0	6	2	8	4

Solution: Evaluate and then compare each side of the given equation.

Evaluate left side of equation:

$$4 \text{ @ } \underbrace{(6 \# 8)}$$

$$\underbrace{4 \text{ @ } (\ 4\ )}_{6}$$

Evaluate right side of equation:

$$\underbrace{(4 @ 6)}_{4} \# \underbrace{(4 @ 8)}_{2}$$

$$\underbrace{(\ 4\ ) \# (\ 2\ )}_{6}$$

Since each side of the given equation evaluates to 6, the statement 4 @ (6 # 8) = (4 @ 6) # (4 @ 8) is **true**.

EXERCISE SET 2.1

1. Which is *not* a rational number?

(1) $\sqrt{36}$ (2) 1.25 (3) $\sqrt{200}$ (4) 0.121212. . .

2. For which operation is the set $\{-1, 0, 1\}$ closed?
(1) addition (2) subtraction (3) multiplication (4) division

3. Which set is *not* closed under the operation of multiplication?
(1) {odd integers} (3) {prime numbers}
(2) {even integers} (4) {rational numbers}

4. Excluding 0, which set does not have a multiplicative inverse for each of its elements?
(1) {integers} (3) {real numbers}
(2) {rational numbers} (4) $\left\{-1, \frac{1}{2}, 2, 1\right\}$

5. If $r = 2$ and $s = 6$, find the value of:
(a) $r \ \square\ s$ if $r \ \square\ s = 5r + 2s$
(b) $r \# s$ if $r \# s = \dfrac{rs}{s - r}$
(c) $r \ \Omega\ s$ if $r \ \Omega\ s = (r + s)^2 - (r^2 + s^2)$
(d) $r \diamond s$ if $r \diamond s = \dfrac{s^2 + r^2}{s - r}$

6. If $\diamond$ is a binary operation defined as $r \diamond s = \dfrac{r^2}{s}$, evaluate $6 \diamond 3$.

7. If $x * y$ is defined as $x^2 - 3y$, find the value of $4 * 2$.

8. If $a \circledast b$ is a binary operation defined as $\dfrac{a + b}{a}$, evaluate $2 \circledast 4$.

9. Find the value of $(B \# S) \# S$ within the system defined below.

#	B	E	S	T
B	T	S	E	B
E	S	T	B	E
S	E	B	T	S
T	B	E	S	T

10. What is the identity element in the system defined by the table below?

#	L	U	C	K
L	K	C	U	L
U	C	K	L	U
C	U	L	K	C
K	L	U	C	K

11. Solve the equation $3 * y = 1$ for y in the system defined below.

*	1	2	3	4
1	4	1	2	3
2	1	2	3	4
3	2	3	4	1
4	3	4	1	2

12. Using the accompanying table, find x if $x \oplus 4 = 3$.

$\oplus$	1	2	3	4
1	2	3	4	1
2	3	4	1	2
3	4	1	2	3
4	1	2	3	4

13. Using the accompanying table, find the inverse element of b.

$\square$	a	b	c	d
a	c	d	a	b
b	d	a	b	c
c	a	b	c	d
d	b	c	d	a

14. Using the accompanying table, find the inverse element of 2.

#	0	1	2
0	0	0	0
1	0	1	2
2	0	2	1

15 Using the accompanying tables, find the value of $(C \triangle A) * (A \triangle C)$.

$\triangle$	A	C	T
A	C	A	T
C	T	A	C
T	A	C	T

$*$	A	C	T
A	A	A	A
C	A	C	T
T	T	T	T

16. If the following table defines a commutative operation, find the value of $b \oplus c$.

$\oplus$	w	a	b	c
w	w		w	
a		c		
b			a	
c	w	a	b	c

17. A set contains the element a. If $a * x = x$ and $x * a = x$ for every element x in the set, it can be concluded that:

(1) a is the inverse of x.
(2) a is the identity of the set under $*$.
(3) The set is closed under $*$.
(4) x is the identity of the set under $*$.

18. Given the set $\{w, x, y, z\}$ and the operation # as shown in the table, which statement is *not* true?

(1) The identity for the system is y.
(2) The set is closed under #.
(3) The set is commutative under #.
(4) Every element of the set has an inverse under #.

#	w	x	y	z
w	x	z	w	x
x	z	y	x	w
y	w	x	y	z
z	x	w	z	y

19. Given the set $\{a, b, c, d\}$ and the operation $\triangle$ as shown in the accompanying table, except for the second row, which has been left out. If the operation is commutative, which could be the second row?

$\triangle$	a	b	c	d
a	b	c	d	a
b				
c	d	a	b	c
d	a	b	c	d

(1) $a\,b\,c\,d$ (2) $b\,c\,d\,a$ (3) $c\,d\,a\,b$ (4) $d\,a\,b\,c$

20. Given the set $\{0, 1, 2, 3\}$ and the accompanying table for the operation $\circledast$, which is true of the operation?

$\circledast$	0	1	2	3
0	0	1	2	3
1	1	2	3	4
2	4	5	6	7
3	9	10	11	12

(1) The operation is commutative.
(2) The identity element is 0.
(3) The operation $\circledast$ is not a binary operation.
(4) The set is not closed for the operation $\circledast$.

21. Using the accompanying table, which is the solution set for $y^2 = 9$?

$\bullet$	3	5	7	9
3	9	5	1	7
5	5	5	5	5
7	1	5	9	3
9	7	5	3	1

(1) $\{3\}$ (2) $\{7\}$ (3) $\{3, 7\}$ (4) $\{3, 7, 9\}$

22. The accompanying table represents the operation □ for the set $\{c, d, e, f\}$.

□	c	d	e	f
c	c	d	e	f
d	d	e	f	c
e	e	f	c	d
f	f	c	d	e

(a) What is the identity element of this system?
(b) What is the inverse of f?
(c) Find the value of $(e \ \square \ e) \ \square \ d$.
(d) Solve for x: $(f \ \square \ e) \ \square \ x = f$.

23. Given the elements d, e, f, and g and the operations ⊕ and ⊗ as defined by the accompanying tables.

⊕	d	e	f	g
d	g	d	e	f
e	d	e	f	g
f	e	f	g	d
g	f	g	d	e

⊗	d	e	f	g
d	f	g	d	e
e	g	d	e	f
f	d	e	f	g
g	e	f	g	d

(a) What is the identity element for ⊗?
(b) What is the inverse of element g under the operation ⊗?
(c) Find the value of $(d \oplus e) \otimes (f \oplus g)$.
(d) Solve for y: $(f \oplus g) \otimes y = e$.

24. Given: set $S = \{A, N, G, L, E\}$ and the commutative operation @ as shown in the accompanying table.

@	A	N	G	L	E
A	A	A		N	A
N		N			N
G	N	L	G	G	G
L		E		L	L
E	A	N	G	L	E

(a) Complete the table.
(b) What is the identity element for operation @?
(c) What is the inverse of L under operation @?
(d) Evaluate: $[G @ (L @ N)] @ A$.
(e) Find x if $(G @ N) @ x = N$.

2.2 CLOCK ARITHMETIC (OPTIONAL TOPIC)*

KEY IDEAS

In ordinary arithmetic, $8+5=13$. In clock arithmetic,

8 o'clock + 5 hours = **1 o'clock**.

In this section rules for performing arithmetic operations—addition, subtraction, and multiplication—based on using the numerals of an imaginary clock are presented. This type of arithmetic is referred to as **clock** or **modulo arithmetic**.

ADDITION AND SUBTRACTION IN CLOCK SYSTEMS. The accompanying figure shows a clock that "tells time" from 0 o'clock to 4 o'clock. The numbers that can be used in this clock system are, therefore, restricted to the set {0, 1, 2, 3, 4}. Since there are five elements in this set, this clock system is referred to as *clock 5* or *modulo 5*. The notation Z_n is sometimes used to identify a clock system, where n is the clock number and Z is the set of numbers that may be used in the clock arithmetic. For example, Z_5 refers to a clock 5 system where $Z = \{0, 1, 2, 3, 4\}$.

Clock Addition. As Figure 2.1 shows, in clock 5

$$\mathbf{2+3=0}$$

The result of starting at 2 and then moving 3 "hours" forward in the *clockwise* direction is 0.

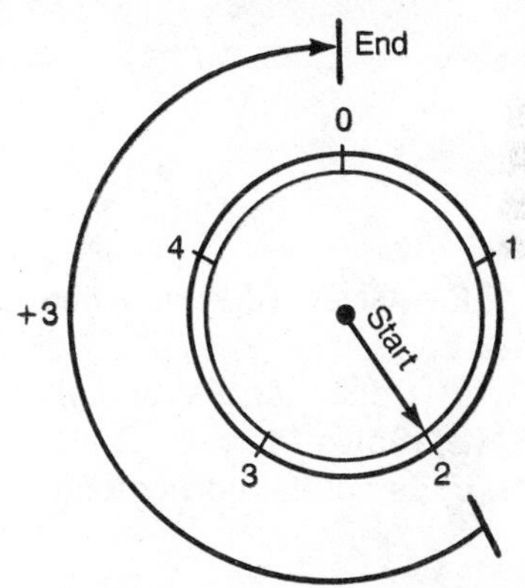

Figure 2.1 Clock Addition

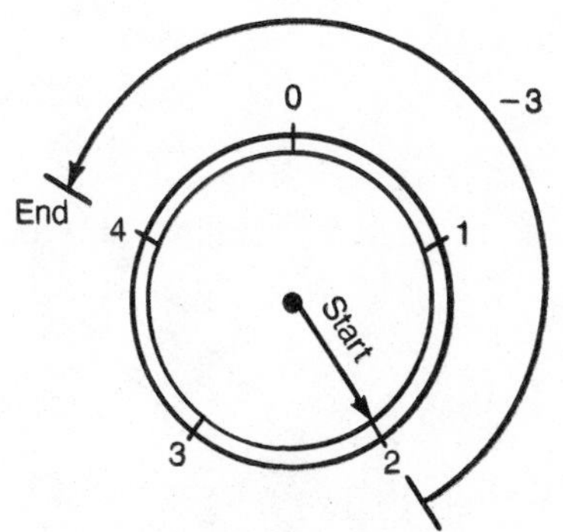

Figure 2.2 Clock Subtraction

Clock Subtraction. As Figure 2.2 shows, in clock 5

$$\mathbf{2-3=4}$$

The result of starting at 2 and then moving 3 "hours" backward in the *counterclockwise* direction is 4.

*This section covers a topic that is optional in the *revised* New York State Syllabus.

Examples

1. Given $(Z_5, +)$, where $Z_5 = \{0, 1, 2, 3, 4\}$ and $+$ is addition in clock 5.
   (a) Construct an addition table for Z_5.
   (b) What is the identity element for addition?
   (c) What is the inverse of 3?
   (d) Solve for x: $x + 4 = 3 + 2$.

Solutions: (a)

+	0	1	2	3	4
0	0	1	2	3	4
1	1	2	3	4	0
2	2	3	4	0	1
3	3	4	0	1	2
4	4	0	1	2	3

(b) The identity element is **0**.
(c) The inverse of 3 is **2** since $3 + 2 = 0$ and $2 + 3 = 0$.
(d) Use the table to evaluate the right side of the equation.

$$x + 4 = 3 + 2$$
$$= 0$$

Since $1 + 4 = 0$, $x = \mathbf{1}$.

2. Given $(Z_5, -)$, where $Z_5 = \{0, 1, 2, 3, 4\}$ and $-$ is subtraction in clock 5.
   (a) Construct a subtraction table for Z_5.
   (b) Solve for x: $x - 3 = 2$.
   (c) Is subtraction commutative for clock 5?

Solutions: (a)

−	0	1	2	3	4
0	0	4	3	2	1
1	1	0	4	3	2
2	2	1	0	4	3
3	3	2	1	0	4
4	4	3	2	1	0

(b) Since $0 - 3 = 2$, $x = \mathbf{0}$.
(c) **No**. For example, $0 - 1 = 4$ and $1 - 0 = 1$.

MULTIPLICATION IN CLOCK SYSTEMS. In clock systems multiplication is treated as repeated addition. In clock 5, for example, $2 \cdot 4$ may be interpreted as $4+4=3$ (see the table in Example 1). Here is another way of obtaining the same result:

Find the usual product: $2 \cdot 4 = 8$

Divide by the clock number, 5: $\frac{8}{5} = 1$ remainder 3

The remainder is the answer: $2 \cdot 4 = \mathbf{3}$

In general, to multiply numbers in clock n

- Find the usual arithmetic product of the two numbers.
- Divide this product by n.
- Write the remainder obtained in the division process as the product of the two numbers in clock n.

Example

3. Given $(Z_5, \cdot)$, where $Z_5 = \{0, 1, 2, 3, 4\}$ and $\cdot$ is multiplication in clock 5.

(a) Construct a multiplication table for Z_5.
(b) What is the identity element for multiplication?
(c) What is the inverse of 2?
(d) Solve for x: (1) $3x = 2$ (2) $x^2 = 4$

Solutions: (a)

·	0	1	2	3	4
0	0	0	0	0	0
1	0	1	2	3	4
2	0	2	4	1	3
3	0	3	1	4	2
4	0	4	3	2	1

(b) The identity element is **1**.
(c) From the table, $2 \cdot 3 = 1$ and $3 \cdot 2 = 1$, so the inverse of 2 is **3**.
(d) (1) Since $3 \cdot 4 = 2$, $x = \mathbf{4}$.
(2) $x^2 = x \cdot x = 4$. Since $2 \cdot 2 = 4$ *and* $3 \cdot 3 = 4$, $x = \mathbf{2}$ or $x = \mathbf{3}$.

EXERCISE SET 2.2

1. Given $(Z_4, +)$, where $Z_4 = \{0, 1, 2, 3\}$ and $+$ is addition in clock 4.
(a) Construct an addition table for Z_4.
(b) What is the identity element for addition?
(c) What is the inverse of 1?
(d) Solve for x: (1) $3 + x = 1$ (2) $x + 1 = 2 + 2$

2. Given $(Z_4, -)$ where $Z_4 = \{0, 1, 2, 3\}$ and $-$ is subtraction in clock 4.
   (a) Construct a subtraction table for Z_4.
   (b) Solve for x: $x - 1 = 3$.

3. Given $(Z_4, \cdot)$, where $Z_4 = \{0, 1, 2, 3\}$ and $\cdot$ is multiplication in clock 4.
   (a) Construct a multiplication table for Z_4.
   (b) What is the identity element for multiplication?
   (c) What is the inverse of 3?
   (d) Solve for x:
   (1) $3x = 2$ (2) $2x = 2$ (3) $2x = 3$ (4) $x^2 = 1$

4–9. Solve each of the following equations in a clock 5 system:

4. $4 + x = 2$
5. $x + x = 1$
6. $x + 2 = 3 + 4$
7. $3x = 4$
8. $2x = 3 + 4$
9. $x^2 = 4$

10–15. Solve each of the following equations in a clock 6 system:

10. $x + 4 = 1$
11. $x + x = 5$
12. $x^2 = 3$
13. $x^2 = 5$
14. $x + 5 = 4 + 3$
15. $x - 2 = 0$

2.3 GROUPS AND FIELDS (OPTIONAL TOPIC)*

KEY IDEAS

Groups and fields are special types of mathematical systems. A *group* involves one binary operation, while a *field* requires two binary operations.

GROUPS. A *group* is a special mathematical system that may contain a finite or an infinite number of elements.

DEFINITION OF GROUP

The mathematical system $(S, *)$ is a **group** if each of the following four conditions is met:

1. S is closed under the operation $*$.
2. The operation $*$ is associative.
3. An identity element exists under the operation.
4. Every element in S has an inverse under the operation $*$.

Example 1: The set of real numbers, rational numbers, and integers each form a group under the operation of addition.

*This section covers a topic that is optional in the Revised New York State Syllabus.

Example 2: The set of real numbers does *not* form a group under the operation of multiplication since 0 does not have an inverse under this operation.

Example 3: The set of whole numbers does *not* form a group under the operation of addition since each element does not have an inverse.

GROUPS AND CLOCK SYSTEMS. Here are some generalizations about groups and clock systems:

- In a group the *identity* element is unique so that there is exactly one identity element.
- In a group the *inverse* of each element is unique, that is, each element has exactly one inverse.
- To determine whether *every* element of a system has an inverse under an operation defined by a table:
    1. Find the identity element of the system.
    2. Scan the table to see whether each row *and* each column contains the identity element somewhere within it.
- All clock arithmetic systems are associative under the operations of addition and multiplication.
- All clock systems are groups under the operation of addition.
- If the clock number n is prime, then a clock n system is a group under the operation of multiplication provided that 0 is excluded.
- If the clock number n is not prime, then a clock n system is *not* a group under the operation of multiplication since at least one nonzero element would not have a multiplicative inverse.

FIELDS. A group involves one binary operation and may or may not be commutative. A mathematical system may involve two binary operations. If such a system forms a commutative group under each operation, and the distributive property holds, then the system is called a *field*.

DEFINITION OF FIELD

Let $+$ represent a binary operation (for example, a form of addition) and $*$ represent another binary operation (for example, a form of multiplication). The mathematical system $(S, +, *)$ is a **field** if each of the following three conditions is met:

1. $(S, +)$ is a commutative group.

2. $(S, *)$ is a commutative group when 0 is excluded.

3. For all a, b, and c in S,

$$a * (b + c) = a * b + a * c \quad \text{(distributive property)}$$

Example 1: The mathematical system $(R, +, \cdot)$, where $R = \{\text{real numbers}\}$, is a field since

1. $(R, +)$ is a commutative group.
2. When 0 is excluded, $(R, \cdot)$ is a commutative group.
3. The distributive property of multiplication over addition holds for the real numbers.

Example 2: The mathematical system $(Q, +, \cdot)$ is a field where $Q = \{\text{rational numbers}\}$ since it can be demonstrated that each of the three conditions of the definition is satisfied.

Example 3: The mathematical system $(Z, +, \cdot)$ is *not* a field where $Z = \{\text{integers}\}$. This is true because $(Z, \cdot)$ is *not* a group since every member of the set except ± 1 does *not* have a multiplicative inverse.

Example

1. Given $(Z_3, +, \cdot)$, where $Z_3 = \{0, 1, 2\}$, + is addition clock 3, and $\cdot$ is multiplication clock 3, prove that $(Z_3, +, \cdot)$ is a field.

Solution: Construct the addition and multiplication tables for clock 3.

+	0	1	2
0	0	1	2
1	1	2	0
2	2	0	1

·	0	1	2
0	0	0	0
1	0	1	2
2	0	2	1

Verify that the three conditions for a field are satisfied.

1. $(Z_3, +)$ is a commutative group since the system satisfies the definition for a group and is commutative:

- *Closure*: The table shows that S is closed under addition.
- *Associativity*: All clock arithmetic systems are associative under addition.
- *Existence of an identity element*: The identity element is 0.
- *Existence of inverses*: Every element in Z_3 has an inverse under + since 0 is in every row and in every column of the addition table.
- *Commutativity*: By the diagonal line test, the system is commutative.

2. $(Z_3, \cdot)$ is a commutative group since the system satisfies the definition for a group and is commutative:

- *Closure*: The table shows Z_3 is closed under multiplication.
- *Associativity*: All clock arithmetic systems are associative under multiplication.
- *Existence of an identity element*: The identity element is 1.
- *Existence of inverses*: Every element in Z_3 has an inverse under $\cdot$ since 1 is in every row and in every column (except those for 0) of the multiplication table.
- *Commutativity*: By the diagonal line test, the system is commutative.

3. The distributive property of multiplication over addition holds in all clock systems that include these operations.

Conclusion: Since each of the three parts of the definition of a field is satisfied, $(Z_3, +, \cdot)$ **is a field.**

IMPORTANCE OF GROUPS AND FIELDS. Algebraic systems that have the same structure must also have the same properties. In higher mathematics this is particularly important because any general property (theorem) that is proved to be true for one group (or field) must also be true for all groups (or fields).

Field properties provide the rationale for being able to perform many familiar arithmetic and algebraic operations within the set of real numbers. Here are three illustrations.

- Factoring is based on the distributive property.

Example: $xy + 5x = x(y + 5)$.

- Changing a fraction into an equivalent fraction having a desired denominator is based on the use of a multiplicative identity element.

Example:
$$\frac{2}{ax} + \frac{3}{a} = \frac{2}{ax} + \frac{3}{a} \cdot 1$$
$$= \frac{2}{ax} + \frac{3}{a} \cdot \frac{x}{x} = \frac{2 + 3x}{ax}$$

$\frac{x}{x} = 1 = \text{identity}$ (arrow pointing to $\frac{x}{x}$)

- Methods for solving various types of equations make use of the field properties of the real numbers.

Example: Solve the equation $3x = 7$.

Steps in Solution	Reasons
1. $3x = 7$	1. Given.
2. $\frac{1}{3} \cdot 3x = \frac{1}{3} \cdot 7$	2. If both sides of an equation are multiplied by the same nonzero number, an equivalent equation results.
3. $\left(\frac{1}{3} \cdot 3\right)x = \frac{1}{3} \cdot 7$	3. Associative property.
4. $1 \cdot x = \frac{1}{3} \cdot 7$	4. The product of a number and its multiplicative inverse is 1.
5. $x = \frac{1}{3} \cdot 7$	5. The product of a number and the multiplicative identity is the number.

EXERCISE SET 2.3

1. Given (C_3, +, ·), where $C_4 = \{0, 1, 2, 3\}$, + is addition clock 4, and · is multiplication clock 4.
(a) Construct an addition table and a multiplication table for C_4.
(b) Find all elements that do not have multiplicative inverses.
(c) Which statement is true?
  (1) (C_4, +) and (C_4, ·) are both groups.
  (2) Only (C_4, +) is a group.
  (3) Only (C_4, ·) is a group.
  (4) Neither (C_4, +) nor (C_4, ·) is a group.
(d) Give a reason why (C_4, +, ·) is not a field.

2. Given the set $S = \{G, R, O, U, P\}$ and the operation #, defined by the accompanying table.

#	G	R	O	U	P
G	R	O	P	G	U
R	O	P	U	R	G
O	P	U	G	O	R
U	G	R	O	U	P
P	U	G	R	P	O

(a) Verify that $(G \# O) \# U = G \# (O \# U)$.
(b) Assuming that the associative property holds, verify that the system $(G, \#)$ is a group.

3. For each of the following sets, determine whether the system $(S, +)$, where $+$ is the familiar operation of addition, is a group. If a system is not a group, then give the group properties that are lacking.
(a) $S = \{-1, 0, 1)\}$
(b) $S =$ set of all integral multiples of 4
(c) $S =$ set of nonnegative rational numbers
(d) $S =$ set of nonzero integers
(e) $S = \left\{\ldots, -\frac{1}{3}, -\frac{1}{2}, -1, 0, 1, \frac{1}{2}, \frac{1}{3}, \ldots\right\}$
(f) $S =$ set of numbers of the form $q\sqrt{2}$, where q is any integer

4. For each of the sets given in Exercise 3, determine whether the system (S, x), where x is the familiar operation of multiplication, is a group. If a system is not a group, then give the group properties that are lacking.

5. Let $Z = \{\text{integers}\}$ and the operation $\square$ be defined so that, for all a and b in set Z, $a \mathbin{\square} b = a^2 - b^2$.
(a) Determine whether $(Z, \square)$ is a group, and give a reason for your answer.
(b) Determine whether $(Z, \square)$ is a group if $\square$ is now defined as $a \mathbin{\square} b = a^2 + b^2$.

6. Let $S = \{0, 1, 2, 3\}$ and the operation $\diamond$ be defined so that, for all a and b in set S, $a \diamond b =$ the remainder obtained by dividing ab by 4.
(a) Construct a table of values that shows the set of all possible results of $a \diamond b$.
(b) Determine whether $(S, \diamond)$ is a group, and give a reason for your answer.

7. For each of the following sets, determine whether the system $(S, +, \cdot)$ is a field under the operations of addition and multiplication. If a system is not a field, give the field properties that are lacking.
(a) $S =$ set of irrational numbers
(b) $S =$ set of odd integers
(c) $S =$ set of all real numbers greater than -5 and less than 5
(d) $S =$ set of numbers of the form $q\sqrt{2}$, where q is any rational number
(e) $S =$ set of numbers of the form $p + q\sqrt{2}$, where p and q are rational numbers

8. Given $(U, \oplus, \otimes)$, where $U = \{0, 2, 4, 6, 8\}$. For any a and b in set U, the operations $\oplus$ and $\otimes$ are defined as follows:

$a \oplus b =$ the units digit of $a + b$. For example, $6 \oplus 8 = 4$.

$a \otimes b =$ the units digit of ab. For example, $6 \otimes 8 = 8$.

Prove or disprove each of the following:
(a) The elements of set U form a group under the operation $\oplus$.
(b) The nonzero elements of the set U form a group under the operation $\otimes$.
(c) $(U, \oplus, \otimes)$ is a field where 0 is excluded under $\otimes$.

9. Given $(Z_5, \oplus, \otimes)$, where $Z_5 = \{0, 1, 2, 3, 4\}$. For any a and b in set Z_5, the operations $\oplus$ and $\otimes$ are defined as follows:

$a \oplus b =$ the remainder obtained by dividing $a + b$ by 5.

$a \otimes b =$ the remainder obtained by dividing ab by 5.

Prove each of the following:
(a) $(Z_5, \oplus)$ is a commutative group.
(b) $(Z_5, \otimes)$ is a commutative group where 0 is excluded.
(c) $(Z_5, \oplus, \otimes)$ is a field where 0 is excluded under $\otimes$.

10. Fill in the missing reasons for the proof of the following theorem, which establishes the multiplication property of 0.

THEOREM *For all real numbers* a, $a \cdot 0 = 0$.

Statement	Reason
1. $a \cdot 0 = a \cdot 0$	1. Any quantity is equal to itself.
2. $0 = 0 + 0$	2. ?
3. $a \cdot (0 + 0) = a \cdot 0$	3. Substitution principle.
4. $a \cdot 0 + a \cdot 0 = a \cdot 0$	4. ?
5. $a \cdot 0 + a \cdot 0 = 0 + a \cdot 0$	5. ?
6. $a \cdot 0 = 0$	6. For all real numbers a, b, and c, if $a + c = b + c$, then $a = b$ (Cancellation Law for Addition).

11. The *Zero Product Theorem* is needed when solving quadratic equations by factoring.

THEOREM *For all real numbers* a *and* b, ab = 0 *if and only if* a = 0 *or* b = 0.

(a) Prove the *if* part of the theorem:

Given: $a = 0$ or $b = 0$.
Prove: $ab = 0$.

(b) Prove the *only if* part of the theorem:

Given: $ab = 0$.
Prove: $a = 0$ or $b = 0$.

CHAPTER 2 REVIEW EXERCISES

1. If @ is a binary operation defined as $p @ q = p^2 - 2pq + q^2$, evaluate 2 @ 3.

2. What is the inverse of a in the system defined below?

&	a	b	c	d
a	b	d	a	c
b	d	c	b	a
c	a	b	c	d
d	c	a	d	b

3. Which system below forms a group?

(1)

Φ	T	E	A	M
T	T	E	A	M
E	E	A	M	T
A	A	M	T	E
M	M	T	E	A

(2)

Φ	T	E	A	M
T	T	E	A	M
E	A	T	E	M
A	E	A	E	M
M	M	M	M	M

(3)

Φ	T	E	A	M
T	T	E	A	M
E	E	A	T	M
A	A	T	A	T
M	M	M	T	M

(4)

Φ	T	E	A	M
T	M	E	A	T
E	E	M	T	E
A	A	T	M	E
M	T	E	E	M

4. The accompanying table defines the operation @ on the set $\{N, I, T, A\}$.

@	N	I	T	A
N	I	T	A	N
I	T	A	N	I
T	A	N	I	T
A	N	I	T	A

(a) What is the identity element for @?
(b) What is the inverse of T?
(c) What is the value of N @ A @ T?
(d) Solve for x in the system: N @ $x = I$
(e) Solve for y in the system: y @ $y = A$

5. Given the elements 0, 1, 2, and 3 and the operations $\oplus$ and $\odot$ as defined below.

$\oplus$	0	1	2	3
0	0		2	3
1	1	2		0
2	2	3	0	1
3	3	0		2

$\odot$	0	1	2	3
0	0	0	0	0
1	0	1	2	3
2	0	2	0	2
3	0	3	2	1

(a) If $\oplus$ is a commutative operation, complete the table.
(b) What is the identity element for $\odot$?
(c) What is the inverse of element 1 under the operation $\oplus$?
(d) Find the value of $3 \oplus (1 \odot 2)$.
(e) Find x if:
  (1) $3 \oplus x = 1$ (2) $2 \oplus x = 3 \odot 2$ (3) $x \odot x = 1$
(f) For the set $S = \{0, 1, 2, 3\}$, why is the system $(S, \odot)$ not a group?
  (1) S is not closed under $\odot$.
  (2) There is no identity element.
  (3) Operation $\odot$ is not commutative.
  (4) Each element of S does not have an inverse.

UNIT II: REVIEW AND EXTENSION OF ALGEBRAIC METHODS

CHAPTER 3

Polynomials

3.1 VARIABLES AND EXPONENTS

KEY IDEAS

A symbol that represents a quantity whose value is not presently known or may change is called a **variable.**

The numbers 2 and 15 are factors of 30 since $2 \times 15 = 30$. Each of the numbers being multiplied to obtain a product is called a **factor** of the product. A number may be written as the product of more than two factors. For example, $2 \times 3 \times 5 = 30$. If the same number appears as a factor in a product more than once, it may be written using an *exponent*.

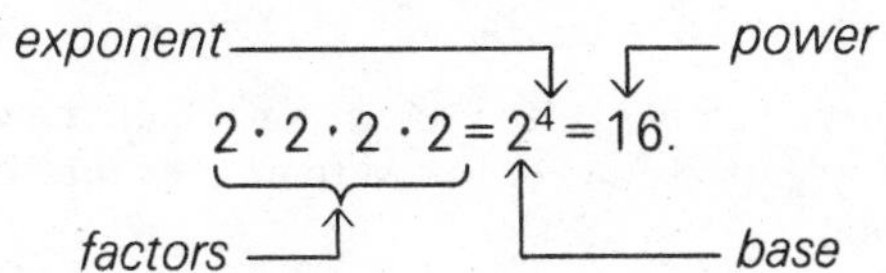

VARIABLES. Letters of the alphabet are generally used to represent variables, whose values either are not currently known or may change. Associated with each variable is a **replacement set** (sometimes called *domain*), which defines the set of all possible values that the variable may have. Unless otherwise stated, the replacement set for a variable that represents a numerical quantity will be assumed to be the set of real numbers. If a variable appears in the denominator of a fraction, then any value of the variable that makes the denominator have a value of 0 *must* be excluded from the replacement set.

EXPONENTS. Repeated multiplication of the same number may be expressed in a more compact form by writing the number only once and using an exponent. The exponent tells the number of times a number is repeated in the multiplication process. For example, $2 \cdot 2 \cdot 2 \cdot 2 = 2^4 = 16$. The number that appears as a factor (2) is called the **base**. The **exponent** (4) represents the number of times the base is used as a factor. The product of the repeated factors (16) is called the **power**. The expression 2^4 is read "2 raised to the fourth power."

SPECIAL INTEGER EXPONENTS. When an exponent is equal to 1, 0, or some negative integer, special rules apply. For example,

$5^1 = 5$ (a quantity raised to the first power is equal to itself);

$5^0 = 1$ (a nonzero quantity raised to the zero power is equal to 1);

$5^{-2} = \frac{1}{5^2} = \frac{1}{25}$ (a nonzero quantity having a negative-integer exponent is equal to the reciprocal of the same quantity raised to the exponent having the opposite sign as the original exponent).

These rules may be summarized as follows:

- $x^1 = x$
- $x^0 = 1 \quad (x \neq 0)$
- $x^{-a} = \frac{1}{x^a} \quad (x \neq 0)$

Example

Evaluate: (a) $(-2)^3$ (b) $2x^0$ (c) $(2x)^0$

Solutions: (a) $(-2)^3 = (-2)(-2)(-2) = (+4)(-2) = \mathbf{-8}$

***Note*:** A negative number raised to an *odd* integer power will always have a negative value. A negative number raised to an *even* integer power will always have a positive value.

(b) $2x^0 = 2 \cdot 1 = \mathbf{2}$ (c) $(2x)^0 = \mathbf{1}$

LAWS OF EXPONENTS. Table 3.1 reviews some laws of exponents. Keep in mind that, in order for the multiplication and division laws to be applied, the powers must have the *same* nonzero base.

TABLE 3.1 Some Laws of Exponents

Law	Rule	Example
Multiplication	$a^x \cdot a^y = a^{x+y}$	$n^5 \cdot n^2 = n^7$
Division	$a^x \div a^y = a^{x-y}$	$n^5 \div n^2 = n^3$
Power of a power	$(a^x)^y = a^{xy}$	$(n^5)^2 = n^{10}$
Power of a product	$(ab)^x = a^x b^x$	$(mn^5)^2 = m^2 n^{10}$

EXERCISE SET 3.1

1. Evaluate each of the following:

(a) 2^3 (c) $4^2 \cdot 3^2$ (e) $2^{-1} + 2^{-3}$ (g) $(2+3)^{-1}$ (i) $2^{-2} \cdot 3^{-1}$

(b) 2^{-3} (d) $3^0 + 4^0$ (f) $\left(\dfrac{2}{3}\right)^{-2}$ (h) $4(2+3)^0$ (j) $\left(1+\dfrac{1}{2}\right)^{-2}$

2–13. Find the product or quotient of each of the following:

2. $x \cdot x^5$
3. $y^9 \div y^3$
4. $a^3 \cdot a^2 \cdot a^5$
5. $(b^2b^3)^4$
6. $(-c^2d)^3$
7. $x^{10} \div x^{10}$
8. $(-a^3b^4)^2$
9. $y^5 \div y^4$
10. $t^6 \div t^7$
11. $a^2 \cdot b^5$
12. $x^5y^3 \div x^3$
13. $(ab^4)^2 \div ab^5$

***Note*:** Exercise 14 is based on an optional topic in the revised Course II syllabus.

14. Let $*$ represent a binary operation defined as $a * b = a^b$.
    (a) Give an example that illustrates the set of integers is not closed under the operation $*$.
    (b) Determine whether 1 is the identity for the set of integers under the operation $*$. Give a reason for your answer.
    (c) Show that the operation $*$ is not commutative on the set of integers.
    (d) Show that the set of whole numbers is closed under the operation $*$.

3.2 OPERATIONS WITH POLYNOMIALS

KEY IDEAS

A **monomial** is a single number or variable, or the product of a number and variable(s). A **polynomial** is a monomial or the sum (or difference) of monomials. **Like monomials** are monomials that have the same literal (letter) factors, as in $3x^2y$ and x^2y. The *numerical coefficient* of the monomial $3x^2y$ is 3, and the *literal factor* is x^2y. The numerical coefficient of the monomial xy^2 is 1 since $xy^2 = 1 \cdot xy^2$.

The addition of like monomials is based on the distributive property.

Example 1: $3x^2y + x^2y = (3+1)x^2y = \mathbf{4x^2y}$

Example 2: $2x + 3y \neq 5xy$ since $2x$ and $3y$ are *not* like monomials and cannot be combined.

Polynomials may be *added, subtracted, multiplied,* and *divided.*

CLASSIFYING POLYNOMIALS. A **binomial** is a polynomial having two monomial terms, as in $2x^2 + 3x$. A **trinomial** is a polynomial having three monomial terms, as in $2x^2 + 3x - 7$. The polynomial $3x^2 + x - y$ is a polynomial in *two* variables since it has two different variables, x and y. The polynomial $2x^2 + 3x - 7$ is a polynomial in *one* variable since x is the only variable that it contains.

A polynomial in one variable is in **standard form** when its terms are arranged so that the exponents decrease in value as the polynomial is read from left to right. To write a polynomial in standard form, it may be necessary to rearrange its terms. For example,

Given polynomial: $8x^3 + 6x^4 + 5 - 2x$
Polynomial in standard form: $6x^4 + 8x^3 - 2x + 5$

SIMPLIFYING POLYNOMIALS. A polynomial is in simplest form when no two of its terms can be combined. Sometimes like terms have to be rearranged, using the associative and commutative properties, so that they can be combined. For example,

$$\begin{aligned} 5c^4d - 3c + c^4d &= (5c^4d + c^4d) - 3c \\ &= \mathbf{6c^4d - 3c} \end{aligned}$$

Sometimes parentheses must be removed, using the distributive property, before like terms can be combined.

Example 1:
$$\begin{aligned} 3(xy + 5) + 4xy &= (3xy + 15) + 4xy \\ &= (3xy + 4xy) + 15 \\ &= \mathbf{7xy + 15} \end{aligned}$$

Example 2: $4x - (2x^2 - 5x + 3) = 4x - 1(2x^2 - 5x + 3)$
$$= 4x - 2x^2 + 5x - 3$$
$$= -2x^2 + (4x + 5x) - 3$$
$$= \mathbf{-2x^2 + 9x - 3}$$

ADDING POLYNOMIALS. To add *two* polynomials, collect like terms and then simplify. For example,

$$(4x^2 + 3y) + (5x^2 - 2y + 7) = (4x^2 + 5x^2) + (3y - 2y) + 7$$
$$= 9x^2 + y + 7$$

Sometimes it is easier to add polynomials by writing them on separate lines, one underneath the other, so that like terms are aligned in the same vertical columns. The numerical coefficients of like terms can then be added mentally. For example, the sum of $(4x^2 + 3y)$ and $(5x^2 - 2y + 7)$ may be found as follows:

Write the first term, using 0 as a placeholder:	$4x^2 + 3y + 0$
Write the second term, placing like terms in the same columns:	$+5x^2 - 2y + 7$
Combine like terms in each column:	$9x^2 + \ y + 7$

SUBTRACTING POLYNOMIALS. To *subtract* a polynomial from another polynomial, add the *opposite* of each term of the polynomial that is being subtracted. For example,

$$(5a + 7b - 4c) - (3a - 2b - 9c) = (5a + 7b - 4c) + (-3a + 2b + 9c)$$
Group like terms: $= (5a - 3a) + (7b + 2b) + (-4c + 9c)$
Combine like terms: $= 2a + 9b + 5c$

Example

1. How much greater is $29a^2 + 13b + 3c$ than $17a^2 - 6b + 8c$?

Solution: Subtract $(17a^2 - 6b + 8c)$ from $(29a^2 + 13b + 3c)$. It is sometimes convenient to perform the subtraction by writing the polynomials vertically, aligning like terms in the same columns. Write the polynomial being subtracted underneath the other polynomial. Change to an addition example by circling the sign of each term of the bottom polynomial and replacing it with its opposite sign. Then add like terms.

Original Subtraction Example	Equivalent Addition Example
$29a^2 + 13b + 3c$	$29a^2 + 13b + 3c$
$-$	$+$
$+17a^2 - \ 6b + 8c$	$\ominus - \quad + \quad -$ $\oplus 17a^2 \ominus 6b \oplus 8c$
	$\mathbf{12a^2 + 19b - 5c}$

MULTIPLYING MONOMIALS. To *multiply* monomials, group and then multiply like factors. For example, to multiply $3x^5$ by $2x^3$, proceed as follows:

Group like factors:	$(3x^5)(2x^3) = (3 \cdot 2)(x^5x^3)$
Multiply numeral factors:	$= 6(x^5x^3)$
Multiply literal factors with the same base:	$= 6x^{5+3} = \mathbf{6x^8}$

Examples

2. Multiply: $(3a^2b)(-4a^3)$.

Solution: $(3a^2b)(-4a^3) = (3)(-4)(a^2a^3)b$
$= -12(a^2a^3)b$
$= \mathbf{-12a^5b}$

3. Multiply: $(2x^5y)(4x^3)(-3xy)$.

Solution: $(2x^5y)(4x^3)(-3xy) = [(2)(4)(-3)](x^5x^3x)(yy)$
$= \mathbf{-24x^9y^2}$

DIVIDING MONOMIALS. As the following example illustrates, monomials are *divided* using a procedure similar to that for multiplying monomials. To divide $\dfrac{24a^5b^4c}{8a^2b}$, proceed as follows:

Group the quotients of like variables:	$\dfrac{24a^5b^4c}{8a^3b} = \left(\dfrac{24}{8}\right)\left(\dfrac{a^5}{a^3}\right)\left(\dfrac{b^4}{b^1}\right)c$
Divide numerical coefficients:	$= 3\left(\dfrac{a^5}{a^3}\right)\left(\dfrac{b^4}{b^1}\right)c$
Divide factors having the same base:	$= 3a^{5-3}b^{4-1}c$
	$= \mathbf{3a^2b^3c}$

Examples

4. Divide: $\dfrac{20y^3}{4y^7}$.

Solution: $\dfrac{20y^3}{4y^7} = \dfrac{5y^3}{y^7} = 5y^{-4} = \dfrac{5}{y^4}$

5. Divide $-15m^2np^4$ by $3m^5p^4$.

Solution: Write the quotient in fractional form and simplify:

$$\frac{-15m^2np^4}{+3m^5p^4} = \left(\frac{-15}{3}\right)\left(\frac{m^2}{m^5}\right)(n)\left(\frac{p^4}{p^4}\right)$$
$$= (-5)(m^{2-5})(n)(p^{4-4})$$
$$- (-5)(m^{-3})(n)(p^0)$$

Replace p^0 by 1: $\quad = (-5)(m^{-3})n$

Replace m^{-3} by $\frac{1}{m^3}$: $\quad = \boldsymbol{\frac{-5n}{m^3}}$

MULTIPLYING A POLYNOMIAL BY A MONOMIAL. To *multiply a polynomial by a monomial*, multiply *each* term of the polynomial by the monomial. For example,

$$4x^2(2x^3 + xy - 3y^2) = 4x^2(2x^3) + 4x^2(xy) + 4x^2(-3y^2)$$
$$= 8x^{2+3} + 4x^{2+1}y - 12x^2y^2$$
$$= \mathbf{8x^5 + 4x^3y - 12x^2y^2}$$

DIVIDING A POLYNOMIAL BY A MONOMIAL. To *divide a polynomial by a monomial*, divide *each* term of the polynomial by the monomial. For example,

$$\frac{72x^3 - 32x^2}{8x} = \frac{72x^3}{8x} - \frac{32x^2}{8x}$$
$$= \left(\frac{72}{8}\right)\left(\frac{x^3}{x}\right) - \left(\frac{32}{8}\right)\left(\frac{x^2}{x}\right)$$
$$= 9x^{3-1} - 4x^{2-1}$$
$$= \mathbf{9x^2 - 4x}$$

MULTIPLYING A POLYNOMIAL BY A POLYNOMIAL. When *multiplying polynomials*, it is sometimes convenient to write them vertically, one polynomial underneath the other. For example, to find the product of $(5a^4 - 6a^2 + 2a - 7)$ and $(3a - 4)$, first write the polynomial having the fewer number of terms underneath the other polynomial.

Use $0 \cdot a^3$ as a placeholder:	$5a^4 + 0 \cdot a^3 - 6a^2 + 2a - 7$
Polynomial with fewer terms:	$3a - 4$
$3a(5a^4 + 0 \cdot a^3 - 6a^2 + 2a - 7) =$	$15a^5 + 0a^4 - 18a^3 + 6a^2 - 21a$
$-4(5a^4 + 0a^3 - 6a^2 + 2a - 7) =$	$- 20a^4 + 0a^3 + 24a^2 - 8a + 28$
Combine like terms:	$= \mathbf{15a^5 - 20a^4 - 18a^3 + 30a^2 - 29a + 28}$

MULTIPLYING BINOMIALS USING FOIL. When the indicated product of a pair of binomials, such as $(3x + 7)(2x + 5)$, is written horizontally, special pairs of terms may be identified by their positions: $3x$ and $2x$ are the *First* terms of each binomial; $3x$ and 5 are the two *Outermost* terms; 7 and $2x$ are the two *Innermost* terms; and 7 and 5 are the *Last* terms of each binomial. The first letters of these four words form the word **FOIL**, which tells us the products to be taken in a shortcut method for multiplying two binomials. Here's how to find the product of $(3x + 7)$ and $(2x + 5)$ using FOIL:

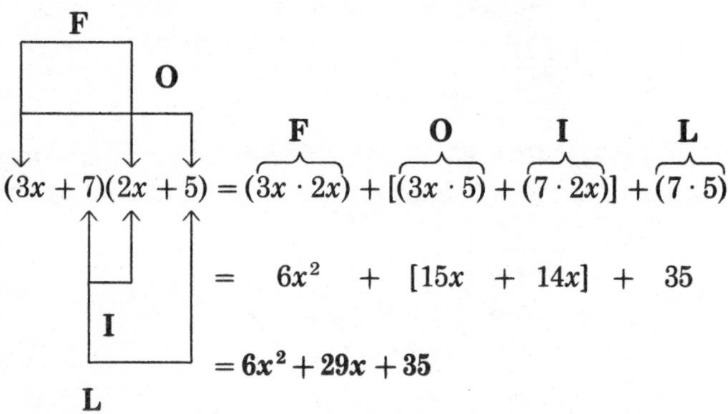

Examples

6. Use the FOIL method to find each of the following products:

(a) $(x-2)(x-6)$ (c) $(5x-9)(2x+3)$
(b) $(y+3)^2$ (d) $(w-6)(w+6)$

Solution:

F O I L

(a) $(x - 2)(x - 6) = x^2 + [(-6x) + (-2x)] + 12 = \mathbf{x^2 - 8x + 12}$

(b) Rewrite the square of the binomial as the product of two identical binomials.

F O I L

$$(y + 3)^2 = (y + 3)(y + 3) = y^2 + [3y + 3y] + 9 = \mathbf{y^2 + 6y + 9}$$

F O I L

(c) $(5x - 9)(2x + 3) = 10x^2 + [15x + (-18x)] - 27 = \mathbf{10x^2 - 3x - 27}$

F O I L

(d) $(w - 6)(w + 6) = w^2 + [6w + (-6w)] - 36$
$= w^2 + 0w - 36 = \mathbf{w^2 - 36}$

Notice that the product of two binomials is *not* always a trinomial.

EXERCISE SET 3.2

1. Remove the parentheses by applying the distributive property, and then simplify.

(a) $3(4x + 1)$ (c) $-2(3x - 5)$ (e) $(2x - 1)(-5)$ (g) $3m - 2(1 - m)$
(b) $2(3 + x)$ (d) $-(1 - x)$ (f) $3 - 4(2x + 1)$ (h) $-2y + 3(2 - 5y)$

2–11. Simplify.

2. $-2(x^2 - 7x - 9) + 10$
3. $2x - 3y + 5x$
4. $3(x + 2) - x$
5. $5 - 2(x - 1)$
6. $6x - 3(2 - 2x)$
7. $5ax - x^2 + 6ax$
8. $a^2b - 2ab + 3a^2b$
9. $-rs^3 + 5rs^3 - r^3s$
10. $3(2a^3 - 5a + 1) + 15a$
11. $3x^2y^3 - 7x^2y^3 + 5x^2y^3$

12. Write each of the following polynomials in standard form, and determine its degree.
(a) $x^2 - x^3 + x - 12$
(b) $5y^2 - 3y + y^3 - 2y^4$
(c) $3n - 5n^3 + 13$
(d) $4x - 3x^2 + 7 + 3x^2$
(e) $-(6x - 5 + x^2)$
(f) $(2x^2 - 5) - (x^2 + 3x + 1)$

13–18. After studying the following example, find, in each case, the value of y *for the indicated value of* x.

Example: Given $y = 3x^2 - 5x + 9$, find the value of y when $x = -2$.

$$\begin{aligned} y &= 3(-2)^2 - 5(-2) + 9 \\ &= 3 \cdot 4 \quad + 10 \quad + 9 \\ &= \mathbf{31} \end{aligned}$$

13. $y = x^2 - 3x + 7;\quad x = 3$
14. $y = -2x^2 + 5x - 8;\quad x = 2$
15. $y = 3x^2 - 4x - 1;\quad x = -2$
16. $y = (x - 2)^2 + 9;\quad x = 0$
17. $y = -3x^2 - 3x + 12;\quad x = -3$
18. $y = x^3 - 2x^2 + x - 8;\quad x = -2$

19. Given $y = -x^2 + 4$, for what value of x is the value of y the greatest?

20–29. Add.

20. $(5y - 8) + (3y - 2)$
21. $(4x^2 + 1) + (3x^2 - 7)$
22. $(2x - 5y + 4c) + (-3x + 2y - 3c)$
23. $(2n^2 - 7n + 8) + (8n^2 - 3n - 11)$
24. $(7p^3 - 4p^2 + 9) + (-5p^3 + 6p - 7)$
25. $(-8m^2 - 7m) + (8m^2 + 3m + 2)$
26. $(-x^3 + 7x^2 - 9) + (3x^3 + x^2 - 6x)$
27. $(7t^5 - 8t^3 + 2t) + (5t^3 - 2t^3 + 4)$
28. $(3x^2 - 5x) + (2x^2 + 7) + (9x - 10)$
29. $(5r^2s^2 - 7r^2s + 3rs) + (-8r^2s^2 + 2rs^2 - 3rs + 1)$

30–34. Subtract.

30. $(6x + 5) - (3x - 2)$
31. $(-2y - 9) - (-8y + 6)$
32. $(4n^2 + 11) - (10n^2 + 7)$
33. $(2x^3 - 4x^2 + 9x) - (5x^3 + x^2 - 2x)$
34. $(3y^4 - 2y^4 - 9) - (4y^3 + 3y^2 - 7y)$
35. Subtract $3x - 2$ from $4x + 3$.
36. From $5x^2 - 2x + 3$, subtract $3x^2 + 4x + 3$.

37–50. Multiply.

37. $3x(x^2 - 5x + 7)$
38. $5y^2(y^3 + 8y - 1)$
39. $-4a(a^3 + 6a - 9)$
40. $(3n^2 - 2n + 1)(7n)$
41. $b^2(-2b^3 + 8b + 4)$
42. $0.06c^3(0.5c^2 - 0.8)$
43. $xy^2(2x^2 - 3y^2 + 8xy)$
44. $a^2b(2a^3b - 5ab^2 - 3ab)$
45. $(9x^3 - 2x^2 + 7)(4x - 1)$
46. $(3a - 7a^3 + 5)(a^2 - 2)$
47. $(p^3 + 3p^2q + 6pq^2)(p - q)$
48. $(5w + 8)(5w - 8)$
49. $(2c - 2)^2$
50. $(3y^2 - 9)(3y^2 + 9)$

51. To the product of $(4x - 1)$ and $(3x - 5)$, add $2x^2 - 8x - 9$.

52. From the product of $(2x + 3)$ and $(x - 6)$, subtract $-3x^2 + 5x + 4$.

53. From the product of $(x^2 - 4)$ and $(2x + 1)$, subtract $4x^3 - 2x^2 + 7x - 2$.

54–63. Divide.

54. $\dfrac{-6w^8y^3z^5}{42w^9y^4z}$

55. $\dfrac{(-5x^2y^7)^{13}}{(-5x^2y^7)^{13}}$

56. $\dfrac{32a^5 - 8a^2}{4a}$

57. $\dfrac{15p^3 - 45p^2 + 9p}{3p}$

58. $\dfrac{t^4 + t^3 + 5t^2}{t^2}$

59. $\dfrac{18r^4 - 27r^3s^2}{9p^2}$

60. $\dfrac{30y^6 + 5y^3 - 10y^2}{-5y^2}$

61. $\dfrac{h^3k^2 + h^2k^3 - 7hk}{hk}$

62. $\dfrac{(a^2b)^2 - (ab)^3}{ab}$

63. $\dfrac{0.14a^3 - 1.05a^2b}{0.7a}$

64–93. Use FOIL to find each of the following products:

64. $(x - 4)(x - 7)$
65. $(y + 3)(y + 10)$
66. $(a + 2)(a + 2)$
67. $(t - 8)(t + 8)$
68. $(2x + 1)(x + 4)$
69. $(3x + 1)(2x + 5)$
70. $(2y - 1)(4y - 3)$
71. $(a + 7)(a - 7)$
72. $(n + 9)(n + 9)$
73. $\left(x - \dfrac{1}{4}\right)\left(x + \dfrac{1}{4}\right)$
74. $(8m - 7)(5m + 9)$
75. $(4x - 13)(x + 6)$
76. $(5w - 3)(4x + 7)$
77. $(6t + 7)(3t - 11)$
78. $(2x - 5)(2x + 5)$
79. $(5 - 6p)(8 - 3p)$
80. $(10 - x)(10 + x)$
81. $(0.6y - 5)(0.4y + 8)$
82. $(x + y)(x - y)$
83. $(1 - 4x)(9 - 5x)$
84. $(3 - 7z)(7 + 3z)$
85. $(p + 3q)(2p - 7q)$
86. $(x - 7)^2$
87. $(y + 6)^2$
88. $(2n - 3)^2$
89. $(4t + 7)^2$
90. $(x + 2y)^2$
91. $(0.3n - 5)^2$
92. $2(3x - 1)^2$
93. $-3(1 - 4y)^2$

3.3 FINDING THE GREATEST COMMON FACTOR (GCF)

KEY IDEAS

The **prime factorization** of a number factors the number as the product of two or more numbers that cannot be factored further. For example, the prime factorizations of 18 and 30 are

$$18 = 2 \cdot 3 \cdot 3 \quad \text{and} \quad 30 = 2 \cdot 3 \cdot 5$$

The Greatest Common Factor (GCF) of 18 and 30 is **6** (=2 × 3). The **GCF** of two or more terms is the product of their common prime factors.

FINDING THE GCF. To obtain the GCF (greatest common factor) of two or more terms, compare their prime factorizations and write the product of the prime factors that are common to each term. For example, to find the GCF of $15y^3$ and $10y$, proceed as follows:

Write the prime factorization of $15y^3$: $3 \cdot \mathbf{5} \cdot \mathbf{y} \cdot y \cdot y$

Write the prime factorization of $10y$: $2 \cdot \mathbf{5} \cdot \mathbf{y}$

Select common factors: The GCF of $15y^3$ and $10y$ is $\mathbf{5y}$.

Example

1. Determine the GCF of $21a^5$ and $-14a^3$.

Solution: Write the prime factorization of each monomial.

$$21a^5 = 3 \cdot \mathbf{7} \cdot \mathbf{a} \cdot \mathbf{a} \cdot \mathbf{a} \cdot a \cdot a$$

$$-14a^3 = -1 \cdot 2 \cdot \mathbf{7} \cdot \mathbf{a} \cdot \mathbf{a} \cdot \mathbf{a}$$

The GCF of $21a^5$ and $-14a^3$ is $\mathbf{7a^3}$.

FACTORING POLYNOMIALS. Compare the terms of the polynomial $2xy^2 - 6x^2y + 4xy$. Notice that each term includes similar factors. The GCF of the terms of the polynomial is $2xy$ since

$$2xy^2 = \mathbf{2xyy}$$
$$6x^2y = 3 \cdot \mathbf{2xxy}$$
$$4xy = 2 \cdot \mathbf{2xy}$$

What factor "remains" when $2xy$ is *factored out* of $2xy^2 - 6x^2y + 4xy$?

$$2xy^2 - 6x^2y + 4xy = 2xy(\underline{\ ?\ })$$

To find this factor, divide the original polynomial by $2xy$. (For example, if 3 is a factor of 21, the corresponding factor may be found by dividing 21 by 3, thereby obtaining 7 as the other factor.)

$$\frac{2xy^2 - 6x^2y + 4xy}{2xy} = \frac{2xy^2}{2xy} - \frac{6x^2y}{2xy} + \frac{4xy}{2xy} = y - 3x + 2$$

Therefore $2xy^2 - 6x^2y + 4xy = \mathbf{2xy(y - 3x + 2)}$.

Example

2. Factor: $24x^3y + 30x^2y^5$.

Solution:

Step 1. Find the GCF of the terms of the polynomial.

$$\begin{aligned} 24x^3y &= 2 \cdot 2 \cdot \mathbf{2} \cdot \mathbf{3xx}\,x\,\mathbf{y} \\ 30x^2y^5 &= \quad\;\; 5 \cdot \mathbf{2} \cdot \mathbf{3xx} \quad \mathbf{y}\,yyyy \end{aligned}$$

The GCF of $24x^3y$ and $30x^2y^5$ is $2 \cdot 3xxy = 6x^2y$.

Step 2. Find the remaining factor by dividing the polynomial by the GCF.

$$\begin{aligned} \frac{24x^3y + 30x^2y^5}{6x^2y} &= \frac{24x^3y}{6x^2y} + \frac{30x^2y^5}{6x^2y} \\ &= 4x + 5y^4 \end{aligned}$$

Therefore $24x^3y + 30x^2y^5 = \mathbf{6x^2y(4x + 5y^4)}$.

You can *check* that the factorization of a polynomial is correct by multiplying the factors on the right side of the equals sign, and then comparing the resulting product with the original polynomial on the left side of the equals sign.

As you gain experience in factoring, you will be able to use shortcut techniques. For example, in many cases the GCF of the terms of a polynomial can be found by making a visual comparison of the terms and thinking:

1. "What is the largest number, if any, that evenly divides the numerical coefficient of each term?"

2. "Which literal factors, if any, are contained in each term? What is the greatest *common* exponent of these literal factors?"

After the GCF is found, the remaining factor of each term can be obtained by mentally applying the reverse of the distributive property. Below are some additional examples that illustrate this idea. For each example, check the answer by multiplying the factors.

Examples

3. Factor: $21x^3 - 28x^5$.

Solution: The GCF of $21x^3$ and $28x^5$ is $7x^3$.

$$21x^3 - 28x^5 = 7x^3(\underline{\ ?\ }) - 7x^3(\underline{\ ?\ })$$

Use the reverse of the distributive property by thinking: "What term when multiplied by $7x^3$ gives $21x^3$? What term when multipled by $7x^3$ gives $28x^5$?"

$$\begin{aligned} 21x^3 - 28x^5 &= 7x^3(3) - 7x^3(4x^2) \\ &= \mathbf{7x^3(3 - 4x^2)} \end{aligned}$$

4. Factor: $20a^2c + 32a^3b^2$.

Solution: The GCF of $20a^2c$ and $32a^3b^2$ is $4a^2$.

$$\begin{aligned} 20a^2c + 32a^3b^2 &= 4a^2(\underline{\ ?\ }) + 4a^2(\underline{\ ?\ }) \\ &= 4a^2(5c) + 4a^2(8ab^2) \\ &= \mathbf{4a^2(5c + 8ab^2)} \end{aligned}$$

5. Factor: $9x^4 - 3x^3 + 12x$.

Solution: The GCF of $9x^4$, $-3x^3$, and $12x$ is $3x$.

$$\begin{aligned} 9x^4 - 3x^3 + 12x &= 3x(\underline{\ ?\ }) + 3x(\underline{\ ?\ }) + 3x(\underline{\ ?\ }) \\ &= 3x(3x^3) + 3x(-x^2) + 3x(4) \\ &= \mathbf{3x(3x^3 - x^2 + 4)} \end{aligned}$$

6. Factor: $-2rs - 2$.

Solution: $-2rs - 2 = -2(rs) - 2(1) = \mathbf{-2(rs + 1)}$.

7. Factor: $12x + 7y$.

Solution: This cannot be expressed in factored form since the terms have no common factors other than 1.

EXERCISE SET 3.3

1–5. Write the prime factorization of each integer.

1. 11 **2.** 24 **3.** 128 **4.** 196 **5.** 200

6–14. Find the GCF for each set of terms.

6. 48 and 190
7. 30 and 60
8. $11x$ and $22x^2$
9. a^2b and ab^2
10. $5n^3$ and $-15n^5$
11. $8x^2y^3$ and $20x^3y$
12. $-3x$ and $-3x^4$
13. a^3b, a^2b^3, and a^2b^4
14. $9h^6k^5$, $-27h^3k^2$, and $18h^3k^5$

15–32. Factor each of the following polynomials so that one of the factors is the GCF:

15. $5x^2 + 11x$
16. $6a^3 - 9a^2$
17. $4p^2q + 12p^2q$
18. $7n^2 + 7t^2$
19. $x^3 + x^2 + x$
20. $14x - 7x^2$
21. $n^4 - 2n^3 + 5n^2$
22. $4s^3 - 12s^2 + 8s - 20$

23. $3y^7 - 6y^5 + 12y^3 + 21$
24. $-3a - 3b$
25. $8u^5w^2 - 40u^2w^5$
26. $-14t^3 - 21t^5$
27. $p^2k^4 - p^3k^2 + (pk)^2$
28. $an^3 - 4n^2 + 8n$
29. $(p - q)^3 + (p - q)^2$
30. $(x + a)^2 - a(x + a)$
31. $a^3b^5c^2 - a^4b^2c^3 + a^5b^2c$
32. $18x^3y - 12x^2y^2 + 24x^4y$

33. If the area of a rectangle is $14x^3 - 21x^2$ and the width is $7x^2$, what is the length?

34. If the area of a rectangle is $45h^4 - 18h^2$ and the length is $5h^2 - 2$, what is the width?

3.4 FACTORING QUADRATIC TRINOMIALS

KEY IDEAS

Given two binomials, a polynomial that represents their product can be found using FOIL.

Starting with a polynomial that represents the product of two binomials, the binomial factors of the polynomial can be discovered using methods that are based on applying the *reverse* of FOIL.

FACTORING $ax^2 + bx + c$ ($a = 1$). FOIL can be used to verify that the product of $(x + 2)$ and $(x + 5)$ is $x^2 + 7x + 10$. The binomial factors of the quadratic trinomial $x^2 + 7x + 10$ may, therefore, be written as follows:

product of 2 and 5

$$x^2 + \underline{7x} + \overline{10} = (x + 2)(x + 5).$$

sum of 2 and 5

Observe that there is a relationship between the terms of the binomial factors and the values of the coefficients of the terms of the quadratic trinomial. For example, the product of 2 and 5 is 10 (c term), and the sum of 2 and 5 is 7 (bx term). This suggests a convenient method by which similar types of quadratic trinomials may be expressed in factored form. When attempting to factor a quadratic trinomial of the form $x^2 + bx + c$, think: "What *two* numbers when multiplied give c *and* when added give b?" If two such numbers exist, say p and q, then write

$$x^2 + bx + c = \mathbf{(x + p)(x + q)}, \quad \text{where } p + q = b \text{ and } pq = c.$$

Examples

1. Factor $x^2 + 11x + 18$ as the product of two binomials.

Solution: Step *1.* Write $x^2 + 11x + 18 = (x \quad \square)(x \quad \square)$.

Step 2. Think: "What two numbers when multiplied give 18, *and* when added give 11?" Since the product and the sum of these factors are both positive, each factor must be positive. List the set of all possible pairs of positive integers whose product is 18. Then choose the pair of factors that satisfy the additional condition that their sum is 11.

The desired numbers are 2 and 9, so there is no need to test other pairs of factors of 18.

Step 3. Write the binomial factors.

$$x^2 + 11x + 18 = \mathbf{(x + 2)(x + 9)}$$

Step 4. Check by multiplying the factors using FOIL.

$$\begin{aligned}(x + 2)(x + 9) &= x^2 + 9x + 2x + 18 \\ &= x^2 + 11x + 18\end{aligned}$$

2. Factor $n^2 - 5n - 14$ as the product of two binomials.

Solution: The numbers you are seeking have different signs since their product is negative. Write

$$n^2 - 5n - 14 = (n + \square)(n - \square).$$

Think: "What two numbers when multiplied give -14, *and* when added give -5?" The desired numbers are $+2$ and -7.

$x^2 - 5x - 14 = = \mathbf{(x + 2)(x - 7)}$. The check is left for you.

FACTORING $ax^2 + bx + c (a \neq 1)$. When factoring a quadratic trinomial in which the coefficient of the quadratic term is *not* 1, you must take into account the effect of the coefficient a in forming the products of the "first" and "outer" terms.

Example

3. Factor $4x^2 + 3x - 7$.

Solution:

Step *1.* List possible pairs of binomial factors of the quadratic polynomial whose first terms are factors of the quadratic coefficient a. Since the product of the first terms of each factor must be $4x^2$, possible binomial factors must take the form

$$(2x \quad ?)(2x \quad ?) \quad \textit{or} \quad (4x \quad ?)(x \quad ?).$$

Step 2. Find all possible pairs of factors. The possible pairs of factors of -7 are 1 and -7, and -1 and 7.

Step 3. Form all possible pairs of binomial factors. Find the pair of factors whose outer and inner products have a sum of $3x$.

Possible Binomial Factors	Outer Product + Inner Product = 3x?
$(2x-7)(2x+1)$ $(2x+7)(2x-1)$	$2x+(-14x) = -12x \neq 3x$ $-2x+14x = 12x \neq 3x$
$(4x-1)(x+7)$ $(4x+1)(x-7)$ $(4x-7)(x+1)$ $(4x+7)(x-1)$	$28x+(-x) = 27x \neq 3x$ $-28x+x = -27x \neq 3x$ $4x+(-7x) = -3x \neq 3x$ $-4x+7x = 3x$ *Success!*

Therefore $4x^2+3x-7 = \mathbf{(4x+7)(x-1)}$.

EXERCISE SET 3.4

1–24. Factor.

1. $x^2+8x+15$
2. $x^2-10x+21$
3. $x^2+3x-21$
4. y^2+6y+9
5. $n^2+3n-88$
6. $a^2-4a-45$
7. $w^2-13w+42$
8. $b^2+3b-40$
9. $t^2-7t-60$
10. y^2-9y+8
11. s^2-s-56
12. $x^2-19x+90$
13. y^2-2x+1
14. a^2+a-20
15. $2a^2+5a-3$
16. $2q^2-q-15$
17. $3x^2+2x-21$
18. $5s^2+14s-3$
19. $5t^2+18t-8$
20. $3n^2+29n-44$
21. $7x^2+52x-32$
22. $-x^2+x+12$
23. $-h^2-h+30$
24. x^4-3x^2-10

25. If $2x-3$ is a factor of $4x^2+4x-15$, what is the other binomial factor?

26. If $x+8$ is a factor of $4x^2+27x-40$, what is the other binomial factor?

27. If $2x+6$ is a factor of $6x^2+4x-42$, what is the other binomial factor?

28. If $6x+5$ is a factor of $12x^2-14x-20$, what is the other binomial factor?

29. If $5x-8$ is a factor of $30x^2-38x-16$, what is the other binomial factor?

30. If the binomial factors of $4x^2-36x+81$ are identical, what is each factor?

31–45. Factor as the product of two binomials.

31. $4x^2-10x-50$
32. $6m^2+m-12$
33. $10p^2-33p-7$
34. $8n^2+9n-14$
35. $3h^2+11h-42$
36. $12y^2-5y-28$
37. $8d^2-22d+15$
38. $6a^2-13a+5$
39. $16x^2-24x+9$
40. $8r^4-10r^2+3$
41. $x^2-(x+42)$
42. $t-2(t^2-5)$
43. $c(c-4)-45$
44. $y(y+10)+4(y+12)$
45. $6x(x+2)-(x-4)$

3.5 MULTIPLYING CONJUGATE BINOMIALS AND FACTORING THE DIFFERENCE OF TWO SQUARES

KEY IDEAS

Binomial pairs such as $(x+3)$ and $(x-3)$, $(m+7)$ and $(m-7)$, and $(2y+5)$ and $(2y-5)$ are examples of *conjugate binomials.* **Conjugate binomials** are binomials that take the sum and difference of the same two terms.

MULTIPLYING CONJUGATE BINOMIALS. Observe the pattern in the following examples, in which pairs of conjugate binomials are multiplied:

$$\begin{aligned} &\qquad\quad\ \ \text{F} \qquad \text{O} \qquad \text{I} \qquad \text{L} \\ (x+3)(x-3) &= x^2 - 3x + 3x - 9 = \mathbf{x^2 - 9}; \\ (m+7)(m-7) &= m^2 - 7m + 7m - 49 = \mathbf{m^2 - 49}; \\ (2y+5)(2y-5) &= 4y^2 - 10y + 10y - 25 = \mathbf{4y^2 - 25}. \end{aligned}$$

Notice that the sum of the outer and inner products will always be equal to 0, so that the product always lacks a "middle" term. *The product of a pair of conjugate binomials is a binomial formed by taking the difference of the squares of the first and last terms of each of the original binomials.* In general,

Multiplying Conjugate Binomials

$$(a+b)(a-b) = a^2 - b^2$$

Example

1. Find the product of $(x-3y)$ and $(x+3y)$.

Solution: $(x-3y)(x+3y) = x^2 - (3y)^2 = \mathbf{x^2 - 9y^2}$

FACTORING THE DIFFERENCE OF TWO SQUARES. By reversing the pattern observed in multiplying two conjugate binomials, we acquire a method for factoring a binomial that represents the difference of two squares. To illustrate, since $(x+5)(x-5) = x^2 - 25$, then $x^2 - 25$ can be factored by observing that

$$x^2 - 25 = (x)^2 - (5)^2 = \mathbf{(x+5)(x-5)}.$$

To factor a binomial that is the *difference* of two squares, write the product of the corresponding pair of conjugate binomials. In general,

Factoring the Difference of Two Squares

$$p^2 - q^2 = (p+q)(p-q)$$

Examples

2. Factor: $n^2 - 100$.

Solution: $n^2 - 100 = (n)^2 - (10)^2 = \mathbf{(n+10)(n-10)}$

3. Factor: $4a^2 - 25b^2$.

Solution: $4a^2 - 25b^2 = (2a)^2 - (5b)^2 = \mathbf{(2a+5b)(2a-5b)}$

4. Factor: $0.16y^2 - 0.09$.

Solution: $0.16y^2 - 0.09 = (0.4y)^2 - (0.3)^2 = \mathbf{(0.4y+0.3)(0.4y-0.3)}$

5. Factor: $p^2 - \frac{36}{49}$.

Solution: $p^2 - \frac{36}{49} = (p)^2 - \left(\frac{6}{7}\right)^2 = \mathbf{\left(p+\frac{6}{7}\right)\left(p-\frac{6}{7}\right)}$

EXERCISE SET 3.5

1–16. Multiply.

1. $(x+2)(x-2)$
2. $(y-10)(y+10)$
3. $(3a-1)(3a+1)$
4. $(-n+6)(-n-6)$
5. $(4-x)(4+x)$
6. $(1-2y)(1+2y)$
7. $(0.8n-7)(0.8n+7)$
8. $\left(x-\frac{2}{3}\right)\left(x+\frac{2}{3}\right)$
9. $\left(2y-\frac{1}{5}\right)\left(2y+\frac{2}{10}\right)$
10. $\left(0.2p-\frac{3}{8}\right)\left(0.2p+\frac{3}{8}\right)$
11. $(0.5n+0.3)(0.5n-0.3)$
12. $(a-7b)(a+7b)$
13. $(x^2-8)(x^2+8)$
14. $(y^3-z)(y^3+z)$
15. $(m^2-n^2)(m^2+n^2)$
16. $(2w-3)^2$

17–31. Factor

17. $x^2 - 144$
18. $y^2 - 0.49$
19. $25 - a^2$
20. $p^2 - \frac{1}{9}$
21. $16a^2 - 36$
22. $64x^2 - 1$
23. $h^2 - k^2$
24. $121w^2 - 25z^2$
25. $\frac{4}{9}x^2 - 49$
26. $0.36y^2 - 0.64x^2$
27. $0.09h^2 - 0.04$
28. $\frac{1}{4}y^2 - \frac{1}{9}$
29. $100a^2 - 81b^2$
30. $n^4 - 49$
31. $x^6 - y^4$

3.6 FACTORING COMPLETELY

KEY IDEAS

A polynomial is *factored completely* when *each* of its factors cannot be factored further. Sometimes it is necessary to apply more than one factoring technique in order to factor a polynomial completely.

A STRATEGY FOR FACTORING COMPLETELY. To factor a polynomial completely, proceed as follows:

1. Factor out the GCF, if any.

2. If there is a binomial, determine whether it can be factored as the difference of two squares.

3. If there is a quadratic trinomial, determine whether it can be factored as the product of two binomials by using the reverse of FOIL.

Examples

1. Factor completely: $3x^3 - 75x$.

Solution: First factor out the GCF of $3x$.

$$\begin{aligned} 3x^3 - 75x &= 3x(x^2 - 25) \\ &= \mathbf{3x(x-5)(x+5)} \end{aligned}$$

2. Factor completely: $t^3 + 6t^2 - 16t$.

Solution: First factor out the GCF of t.

$$\begin{aligned} t^3 + 6t^2 - 16t &= t(t^2 + 6t - 16) \\ &= \mathbf{t(t+8)(t-2)} \end{aligned}$$

3. Factor completely: $x^4 - y^4$

Solution: Factor as the difference of two squares.

$$\begin{aligned} x^4 - y^4 &= (x^2)^2 - (y^2)^2 \\ &= (x^2 - y^2)(x^2 + y^2) \\ \text{Factor } (x^2 - y^2)\text{:} &= \mathbf{(x-y)(x+y)(x^2+y^2)} \end{aligned}$$

EXERCISE SET 3.6

1–12. Factor completely.

1. $2y^2 - 50$

2. $b^3 - 49b$

3. $x^3 + x^2 - 56x$

4. $8w^3 - 32w$

5. $-x^2 - 7x - 10$

6. $3y^2 - 9y + 6$

7. $12s^3 - 2s^2 - 4s$

8. $10y^3 + 50y^2 - 500y$

9. $3t^4 + 12t^3 - 15t^2$

10. $p^4 - 1$

11. $9a^2w^2 - 12a^2w + 4a^2$

12. $-5t^2 + 5$

CHAPTER 3 REVIEW EXERCISES

1. Find the sum of $4a + 2b - c$ and $3a - 5b - 2c$.
2. From $5x^2 - 3x - 2$ subtract $3x^2 - 4x - 1$.
3. Express $3x^2(2x - 5)$ as a binomial.
4. If the length of a side of a square is represented by $2x + 1$, express the area of the square as a trinomial.
5. The length of a rectangle exceeds three times its width by 2. If x represents the width of the rectangle, express the perimeter of the rectangle as a binomial in terms of x.
6. Perform the indicated operations and express the result as a trinomial: $(3x + 2)^2 - (x - 1)^2$.
7. If the product of $(2x + 3)$ and $(x + k)$ is $2x^2 + 13x + 15$, find the value of k.
8. The length of a rectangle is $x + 3$, and the width is $x - 5$. Express the area as a trinomial in terms of x.
9. Find the product of $(2x - 1)$ and $(x^3 - 3x + 5)$.
10. Factor: (a) $y^2 - 3y - 28$ (b) $2x^2 - 2x - 84$ (c) $-t^2 + 3t + 10$
11. Factor: (a) $2p^2 - p - 6$ (b) $3q^2 + 10q - 8$ (c) $4w^2 - 4w + 1$
12. Factor completely:
    (a) $3b^2 - 12$ (b) $y^3 - 16y$ (c) $-r^3 - 3r^2 + 4r$
13. Divide $12x^2 - 20x$ by $4x$.
14. The product of two polynomials is $6a^3 - 15a^2$. If one of the polynomials is $3a^2$, find the other polynomial.
15. Simplify: $2x - (x^2 - 3x + 1)$.
16. Factor completely: $12m^3p - 4mp + 6m^2p^2$.

17–18. In each case, perform the indicated operations and express the result as a trinomial in simplest form.

17. $3x(x + 1) + 4(x - 1)$ 18. $(x - 3)^2 + 6(x - 1)$
19. The sum of two polynomials is 0. If one of the polynomials is $3x^2 + 5x - 7$, what is the other polynomial?
20. What is the sum of the quotients obtained by dividing $(14x^3 - 35x^2 + 7x)$ by $7x$ and $(15x^3 - 9x^2 + 3x)$ by $3x$?
21. The product of $3x^5$ and $4x^2$ is:
    (1) $7x^7$ (2) $12x^7$ (3) $7x^{10}$ (4) $12x^{10}$

22. The expression $18x^6 \div 3x^3$ is equivalent to:
(1) $15x^2$ (2) $15x^3$ (3) $6x^2$ (4) $6x^3$

23. The product of $9x^3$ and $2x^4$ is:
(1) $11x^7$ (2) $11x^{12}$ (3) $18x^7$ (4) $18x^1$

24. The expression $(3x^2y^3)^2$ is equivalent to:
(1) $9x^4y^6$ (2) $9x^4y^5$ (3) $3x^4y^6$ (4) $6x^4y^6$

25. The quotient of $\dfrac{-4a^6b^2}{2a^2b}$ is:
(1) $2a^3b^2$ (2) $-2a^4b^2$ (3) $-2a^4b$ (4) $-6a^3b$

26. The length of a rectangle is represented by $x - 5$ and the width by $x + 2$. The area of the rectangle is represented by:
(1) $x^2 + 3x - 10$ (2) $2x - 3$ (3) $x^2 - 3x - 10$ (4) $4x - 6$

27. The quotient of $(4x^3 - 3x^2) \div x^2$ is:
(1) 1 (2) $2x - 3$ (3) $4x - 3$ (4) $4 - 3x$

28. The expression $(2x)^3(4x^2)$ is equivalent to:
(1) $33x^5$ (2) $32x^5$ (3) $128x^5$ (4) $128x^6$

29. If $15a^2 - 3a$ is divided by $3a$, the quotient is:
(1) $5a$ (2) $2a$ (3) $5a - 3$ (4) $5a - 1$

30. The expression $(x + 2)^2 - 4x - 4$ is equivalent to:
(1) x^2 (2) $x^2 - 4x$ (3) $x^2 + 4x$ (4) $x^2 - 4x + 8$

31. The expression $(2a^2)^3$ is equivalent to:
(1) $2a^5$ (2) $2a^6$ (3) $8a^5$ (4) $8a^6$

32. The expression $x(x - y)(x + y)$ is equivalent to:
(1) $x^2 - y^2$ (2) $x^3 - y^3$ (3) $x^3 - xy^2$ (4) $x^3 - x^2y + y^2$

33. When $12x^4 - 3x^3 + 6x^2$ is divided by $3x^2$, the quotient is:
(1) $9x^2 - 3$ (2) $5x^2$ (3) $4x^2 - 3x + 2$ (4) $4x^2 - x + 2$

CHAPTER 4

Solving Linear and Quadratic Equations

4.1 SOLVING LINEAR EQUATIONS

KEY IDEAS

Equations that have the same solution set are **equivalent**. The equations

$$x+1=4, \qquad 2x=6, \qquad \text{and} \qquad x=3$$

are equivalent since $\{3\}$ is the solution set of each equation. When solving an equation, our goal is to try to obtain an equivalent equation in which the variable is "isolated" on one side of the equation, so that the solution set can be read from the opposite side, as in $x=3$.

An equation in which the greatest exponent of the variable is 1 is called a **first-degree** or **linear equation**. This section will illustrate how to solve various types of linear equations.

SOLVING EQUATIONS WITH TWO ARITHMETIC OPERATIONS. **In the equation $2n + 5 = -11$ the variable is involved in two operations: multiplication ($2n$) and addition ($2n + 5$). In solving this equation, undo the addition before the multiplication.**

Examples

1. Solve and check: $2n+5=-11$.

Solution:

$$\begin{aligned} 2n+5-5 &= -11-5 \\ 2n+0 &= -16 \\ \frac{2n}{2} &= \frac{-16}{2} \\ n &= \mathbf{-8} \end{aligned}$$

Check:

$$\begin{array}{r|l} 2n+5 & = -11 \\ 2(-8)+5 & \\ -16+5 & \\ -11 & = -11\checkmark \end{array}$$

2. Solve and check: $\frac{x}{3}-2=13$.

Solution:

$$\begin{aligned} \frac{x}{3}-2+2 &= 13+2 \\ \frac{x}{3}+0 &= 15 \\ \overset{1}{\cancel{3}}\left(\frac{x}{\cancel{3}}\right) &= 3(15) \\ x &= \mathbf{45} \end{aligned}$$

Check:

$$\begin{array}{r|l} \frac{x}{3}-2 & = 13 \\ \frac{45}{3}-2 & \\ 15-2 & \\ 13 & = 13\checkmark \end{array}$$

SOLVING EQUATIONS WITH PARENTHESES. If an equation contains parentheses, remove them by applying the distributive property of multiplication over addition (or subtraction).

Examples

3. Solve and check: $3(1-2x)=-15$.

Solution:

$$\begin{aligned} 3(1-2x) &= -15 \\ 3-6x &= -15 \\ 3-6x-3 &= -15-3 \\ \frac{-6x}{-6} &= \frac{-18}{-6} \\ x &= \mathbf{3} \text{ or } \{\mathbf{3}\} \end{aligned}$$

Check:

$$\begin{array}{r|l} 3(1-2x) & = -15 \\ 3(1-2\cdot 3) & \\ 3(1-6) & \\ 3(-5) & \\ -15 & = -15\checkmark \end{array}$$

4. Solve for x and check: $\frac{2}{3}=\frac{x+9}{21}$.

Solution: Cross-multiply and then simplify.

$$\begin{aligned} 3(x+9) &= 2\cdot 21 \\ 3x+27 &= 42 \\ 3x+27-27 &= 42-27 \\ 3x &= 15 \\ \frac{3x}{3} &= \frac{15}{3} \\ x &= \mathbf{5} \end{aligned}$$

Check:

$$\begin{array}{r|l} \frac{2}{3} & = \frac{x+9}{21} \\ & \frac{5+9}{21} \\ & \frac{14}{21} \\ & \frac{14\div 7}{21\div 7} \\ \frac{2}{3} & = \frac{2}{3}\checkmark \end{array}$$

SOLVING EQUATIONS WITH VARIABLE TERMS ON THE SAME SIDE. To solve an equation in which the variable appears in more than one term on the same side of the equation, first combine like terms. Then solve the resulting equation as usual.

Example

5. Solve and check: $5x + 3x = -24$.

Solution: Begin by combining like terms.

$$8x = -24$$
$$\frac{8x}{8} = \frac{-24}{8}$$
$$x = \mathbf{-3}$$

Check: $5x + 3x = -24$
$$5(-3) + 3(-3) \quad \Big| $$
$$-15 + (-9) \quad \Big|$$
$$-24 = -24\checkmark$$

SOLVING EQUATIONS WITH VARIABLE TERMS ON DIFFERENT SIDES. If variable terms appear on opposite sides of an equation, work toward collecting variables on the same side of the equal sign and constant terms (numbers) on the other side.

Examples

6. Solve and check: $5w + 14 = 3(w - 8)$.

Solution: $5w + 14 = 3(w - 8)$
Apply the distributive property: $5w + 14 = 3w - 24$
Subtract $3w$ from each side: $-3w + 5w + 14 = -3w + 3w - 24$
Simplify: $2w + 14 = -24$
Subtract 14 from each side: $2w + 14 - 14 = -24 - 14$
Divide each side by 2: $\frac{2w}{2} = \frac{-38}{2}$
$w = \mathbf{-19}$ *or* $\{\mathbf{-19}\}$

Check: $5w + 14 = 3(w - 8)$

$5(-19) + 14$	$3(-19 - 8)$
$-95 + 14$	$3(-27)$

$$-81 = -81\checkmark$$

7. Solve: $3x + 0.9 = 1.3 - 7x$.

Solution:

Method 1	Method 2
$10(3x)+10(0.9)=10(1.3)-10(7x)$	$3x+0.9=1.3-7x$
$30x+9=13-70x$	$7x+3x+0.9=1.3-7x+7x$
$70x+30x+9=13-70x+70x$	$10x+0.9=1.3$
$100x+9=13$	$10x+0.9-0.9=1.3-0.9$
$100x+9-9=13-9$	$10x=0.4$
$\frac{100x}{100}=\frac{4}{100}$	$\frac{10x}{10}=\frac{0.4}{10}=0.04$
$x=\frac{1}{25}$ *or* **0.04**	$x=$ **0.04**

The check is left for you.

EXERCISE SET 4.1

1–35. Solve for the variable in each equation, and check.

1. $2x-0.4=1.8$
2. $3x-1=-16$
3. $32=3w+5$
4. $2x+3=8$
5. $2(x+3)=8$
6. $3(2x-5)=-28$
7. $3x+4x=-21$
8. $5x-2x+1=-26$
9. $-2(1-x)=16$
10. $0.2x+0.3=8.1$
11. $-(8x-3)=19$
12. $8-5r=-7$
13. $-12+3k=-6$
14. $7t=t-42$
15. $0.54-0.07y=0.2y$
16. $3(x+4)=x$
17. $7-(3n-5)=n$
18. $4c-(c+7)=5$
19. $5(6-q)=-3(q+2)$
20. $\frac{3}{4}x-1=14$
21. $\frac{2t}{3}+5=13$
22. $\frac{x}{2}+5=-17$
23. $13-\frac{3x}{4}=-8$
24. $7-3(x-1)=-17$
25. $-5(2p+6)+3=-12$
26. $18-2(h+1)=0$
27. $\frac{a-2}{3}=a-6$
28. $\frac{n-5}{2}=3n-5$
29. $7(2p-1)=4(1-2p)$
30. $y-6=3(2y+9)+y$
31. $6\left(x+\frac{1}{2}\right)=33$
32. $6\left(\frac{x}{2}+1\right)=33$
33. $5(n+2)-2(n+2)=-9$
34. $-(11+5m)=4(7-2m)$
35. $0.7(x-0.2)+0.3(x-0.2)=0.4$

36–43. Solve each proportion for x and check.

36. $\frac{9}{x}=\frac{3}{4}$
37. $\frac{1}{x+1}=\frac{10}{5}$
38. $\frac{2}{3}=\frac{2-x}{12}$
39. $\frac{2x-5}{3}=\frac{9}{4}$

40. $\dfrac{x+5}{4}=\dfrac{x+2}{3}$

41. $\dfrac{4}{x+3}=\dfrac{1}{x-3}$

42. $\dfrac{10-x}{5}=\dfrac{7-x}{2}$

43. $\dfrac{4}{11}=\dfrac{x+6}{2x}$

44–50. Solve each problem algebraically.

44. The length of a rectangle exceeds twice its width by 5. If the perimeter of the rectangle is 52, find the dimensions of the rectangle.

45. The number 45 is 9 greater than twice the difference obtained by subtracting 7 from another number. What is the other number?

46. A 72-inch board is cut into two pieces so that the larger piece is seven times as long as the shorter piece. Find the length of each piece.

47. The product of 3 and 1 less than a number is the same as twice the number increased by 14. What is the number?

48. How old is David if his age 6 years from now will be twice his age 7 years ago?

49. Three years ago Rosita was one-half as old as she will be 2 years from now. What is Rosita's present age?

50. The denominator of a fraction is 2 less than the numerator. The fraction when expressed in simplest form is $\frac{6}{5}$. What is the numerator of the original fraction?

4.2 SOLVING QUADRATIC EQUATIONS BY FACTORING AND BY TAKING SQUARE ROOTS

KEY IDEAS

If a and b are real numbers and $a \cdot b = 0$, then

$$\boldsymbol{a \vee b = 0}$$

In words, if the product of two numbers equals 0, then the first number equals 0, or the second number equals 0, or both numbers equal 0. This is sometimes referred to as the **zero product rule**. The zero product rule provides a way of solving an equation that can be written in such a way that one side is 0, while the other side is written in factored form.

SOLVING QUADRATIC EQUATIONS BY FACTORING. A quadratic equation is an equation that can be written in the *standard form*:

$$ax^2 + bx + c = 0 \quad (a \neq 0)$$

The highest power of the variable in a linear equation is 1, whereas a quadratic equation must include the square of a variable. If the left side of a quadratic equation having the form $ax^2 + bx^2 + c = 0$ can be factored, then the zero product rule allows us to solve the equation by setting each factor equal to 0.

To solve the quadratic equation $x^2 + 4x = 5$, follow these steps:

Step	Example
1. Write the equation in standard form.	1. $x^2 + 4x - 5 = 0$
2. Factor the quadratic polynomial.	2. $(x + 5)(x - 1) = 0$
3. Set each factor equal to 0.	3. $(x + 5 = 0) \vee (x - 1 = 0)$
4. Solve each first-degree equation.	4. $x = -5 \vee x = 1$
5. Write the solution set.	5. $\{-5, 1\}$

Each member of the solution set must be checked by substituting for x in the *original* equation.

Let $x = -5$.		Let $x = 1$.	
$x^2 + 4x$	$= 5$	$x^2 + 4x$	$= 5$
$(-5)^2 + 4(-5)$		$(1)^2 + 4(1)$	
$25 - 20$		$1 + 4$	
	$5 = 5\checkmark$		$5 = 5\checkmark$

When solving quadratic equations, keep in mind that:

- The quadratic equation must be expressed in standard form *before* you attempt to factor the quadratic polynomial.

- Every quadratic equation has *two* solutions. Each solution is called a **root** of the equation. The two roots may be equal.

- The solution set of a quadratic equation can be checked by verifying that each different root makes the *original* equation a true statement.

- Not every quadratic equation can be solved by factoring.

Examples

1. Solve for a: $a^2 + 6a + 9 = 0$.

Solution: $(a + 3)(a + 3) = 0$

$$(a + 3 = 0) \vee (a + 3 = 0)$$

$$a = -3 \vee \quad a = -3$$

The two roots are equal, so the root does not have to be written twice in the solution set.

The solution set is $\{-3\}$. The check is left for you.

2. Solve for a: $6a^2+18a+12=0$.

Solution: Observe that 6 is a common factor of each term of the equation. Therefore the equation may be simplified *before* attempting to factor the quadratic polynomial by dividing each term of the equation by 6.

$$\frac{6a^2}{6}+\frac{18a}{6}+\frac{12}{6}=\frac{0}{6}$$
$$a^2+3a+2=0$$
$$(a+2)(a+1)=0$$
$$(a+2=0) \quad \vee (a+1=0)$$
$$a=-2 \vee \quad a=-1$$

The solution set is $\{-2, -1\}$. The check is left for you.

3. Solve for n: $\frac{n-4}{2}=\frac{3n}{n+4}$.

Solution: Cross-multiply. Then write the quadratic equation in standard form.

$$(n-4)(n+4)=2(3n)$$
$$n^2-16=6n$$
$$n^2-6n-16=0$$
$$(n-8)(n+2)=0$$
$$(n-8=0) \vee (n+2=0)$$
$$n=8 \vee \quad n=-2$$

The solution set is $\{-2, 8\}$. The check is left for you.

4. The length of a rectangular garden is twice its width. The garden is surrounded by a rectangular concrete walk having a uniform width of 4 feet. If the area of the garden and the walk is 330 square feet, what are the dimensions of the garden?

Solution: Let x = width of the garden.
Then $2x$ = length of the garden.

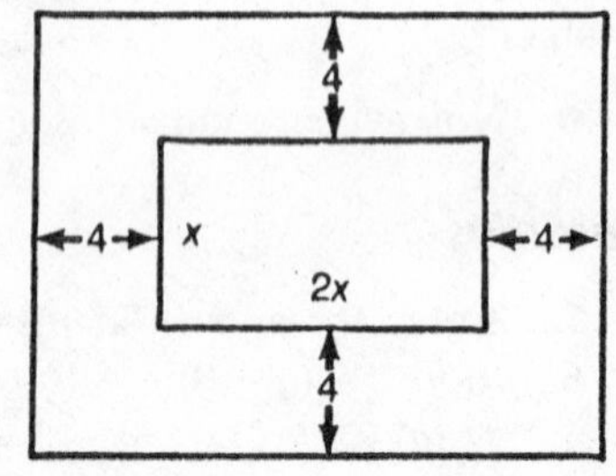

In the accompanying diagram the innermost rectangle represents the garden. Since the walk has a uniform width, the width of the larger (outer) rectangle is $4+x+4=x+8$. The length of the larger rectangle is $4+2x+4=2x+8$. The area of the larger rectangle is given as 330. Hence:

$$\begin{aligned}
\text{Length} \times \text{Width} &= \text{Area} \\
(2x+8)(x+8) &= 330 \\
2x^2+24x+64 &= 330 \\
2x^2+24x+64-330 &= 0 \\
2x^2+24x-266 &= 0 \\
\frac{2x^2}{2}+\frac{24x}{2}-\frac{266}{2} &= \frac{0}{2} \\
x^2+12x+133 &= 0 \\
(x-7)(x+19) &= 0 \\
(x-7=0) \vee (x+19 &= 0) \\
x=7 \vee \quad x &= -19
\end{aligned}$$

Reject -19 since the width must be a positive number.

The width of the garden is **7 feet** and the length is **14 feet**. The check is left for you.

SOLVING QUADRATIC EQUATIONS BY TAKING SQUARE ROOTS. If a quadratic equation takes the form $ax^2+c=0$, that is, does not have a first-degree term, then the roots of the equation may be obtained by solving for the square of the variable and then taking the square root of each side of the equation.

Example

5. Solve for x: $3x^2-15=0$.

Solution:

$$\begin{aligned}
3x^2-15 &= 0 \\
3x^2 &= 15 \\
\frac{3x^2}{3} &= \frac{15}{3} \\
x^2 &= 5
\end{aligned}$$

Take the square root of each side: $x=\pm\sqrt{5}$ or $\{-\sqrt{5}, \sqrt{5}\}$

A quadratic equation that involves the square of a binomial and takes the general form $(x+a)^2+c=0$ can also be solved by taking square roots.

Example

6. Solve for x: $(x-1)^2-3=0$

Solution:

$$\begin{aligned}
(x-1)^2-3 &= 0 \\
(x-1)^2 &= 3
\end{aligned}$$

Take the square root of each side.

$$\begin{aligned}
x-1 &= \pm\sqrt{3} \\
x &= 1\pm\sqrt{3} \quad \text{or} \quad \{1+\sqrt{3}, 1-\sqrt{3}\}
\end{aligned}$$

SOLVING HIGHER DEGREE EQUATIONS BY FACTORING COMPLETELY. Sometimes more than one factoring technique must be applied to express a polynomial in factored form, so that the zero product rule can be used.

Example

7. Solve for n: $n^3 + 5n^2 - 6n = 0$.

Solution:	$n^3 + 5n^2 - 6n = 0$
Factor out the GCF of n:	$n(n^2 + 5n - 6) = 0$
Factor the quadratic polynomial:	$n(n + 6)(n - 1) = 0$
Use the zero product rule:	$(n = 0) \vee (n + 6 = 0) \vee (n - 1 = 0)$
	$n = 0 \vee \quad n = -6 \quad \vee \quad n = 1$

The solution set is $\{\mathbf{0}, \mathbf{-6}, \mathbf{1}\}$. The check is left for you.

***Note*:** The solution set has three members since the highest power of the variable in the original equation is 3. The original equation is a cubic equation, also called a *third-degree equation*.

EXERCISE SET 4.2

1–30. Find the solution set and check.

1. $x(x - 2) = 0$
2. $y^2 + 3y + 2 = 0$
3. $x^2 + 14x + 49 = 0$
4. $x^2 - 5x + 4 = 0$
5. $x^2 - x = 12$
6. $x^2 - 7x = 0$
7. $x^2 + 2 = 6$
8. $q^2 - 6q = 27$
9. $11n - n^2 = 0$
10. $2r^2 = 5r + 3$
11. $6 = t^2 - t$
12. $y^3 - 9y = 0$
13. $2h^2 - 14 = 0$
14. $y(y + 9) = 36$
15. $0 = 2a^2 + 10a + 8$
16. $8p^2 = 6p - p^2$
17. $\dfrac{x^2}{2} - 15 = 1$
18. $t^3 - 90t = 10t$
19. $2b^2 - 18 = 5b$
20. $9x^2 - 12x + 4 = 0$
21. $6t^2 = 7t + 3$
22. $(x - 2)(x + 1) = 10$
23. $8x^2 + 18x = 5$
24. $(x - 5)^2 = 16$
25. $(x + 2)^2 = 7$
26. $n^2 = 10(n + 300)$
27. $b(b - 2) = b$
28. $(t + 2)^2 - 6(t + 2) - 27 = 0$
29. $5(x^2 - 2) + 23x = 0$
30. $(x + 3)^2 = 6x + 25$

31–34. Solve for x *and check.*

31. $\dfrac{x + 6}{x} = \dfrac{x + 4}{x - 6}$
32. $\dfrac{x + 5}{x + 1} = \dfrac{x - 1}{4}$
33. $\dfrac{x - 2}{x} = \dfrac{x + 4}{3x}$
34. $\dfrac{x - 7}{x + 1} = \dfrac{x - 10}{2x + 1}$

35–38. In the following table, a *and* b *represent the lengths of the legs of a right triangle, and* c *represents the length of the hypotenuse. In each case, find the value of* x.

	a	b	c
35.	x	x	18
36.	$x-1$	x	$x+1$
37.	$x+1$	$3x$	$5x-7$
38.	$x-1$	$\frac{x}{2}$	$x+1$

39–45. Solve algebraically.

39. The sum of the squares of two consecutive positive integers is 52. Find the integers.

40. Find three consecutive positive, even integers such that the difference in the squares of the first and the third is 48.

41. The perimeter of a right triangle is 132. If the length of the hypotenuse is 61, find the length of the shorter leg.

42. If a side of a square is doubled and an adjacent side is diminished by 3, a rectangle is formed whose area is numerically greater than the area of the square by twice the length of the original side of the square. Find the dimensions of the original square.

43. The perimeter of a certain rectangle is 24 inches. If the length is doubled and the width is tripled, then the area is increased by 160 square inches. Find the dimensions of the original rectangle.

44. A rectangular picture 30 cm wide and 50 cm long is surrounded by a frame having a uniform width. If the combined area of the picture and the frame is 2,016 square cm, what is the width of the frame?

45. A rectangular picture 24 inches by 32 inches is surrounded by a border of uniform width. If the area of the border is 528 square inches less than the area of the picture, find the width of the border.

4.3 SOLVING QUADRATIC EQUATIONS BY COMPLETING THE SQUARE

KEY IDEAS

A **perfect square trinomial** is a trinomial that can be factored as the square of a binomial. For example, $x^2 - 10x + 25$ is a perfect square trinomial since

$$x^2 - 10x + 25 = (x - 5)(x - 5) = \mathbf{(x - 5)^2}$$

Another method for solving quadratic equations is based on expressing the quadratic equation as an equivalent equation that has the square of a binomial on one side and a number on the other side. Taking the square root of each side of this equation leads to the two roots of the original quadratic equation.

FINDING PERFECT SQUARE TRINOMIALS. For what value of c is $x^2 + 14x + c$ a perfect square trinomial? To find the value of c that "completes the square," follow these steps:

Step	Example
1. Take $\frac{1}{2}$ of the coefficient of x.	1. $\frac{1}{2}(14) = 7$
2. Since the value of c is the square of the number obtained in step 1, substitute the square of this number for c.	2. $c = 7^2 = 49$
3. Write the perfect square trinomial.	3. $x^2 + 14x + 49 = (x + 7)^2$

This number is $\frac{1}{2}$ the coefficient of **x**.

Example

1. Find the value of k that will make $y^2 - 8y + k$ a perfect square trinomial, and then write the trinomial in factored form.

Solution: $y^2 - 8y + k$

Find $\frac{1}{2}$ the coefficient of y: $\frac{1}{2}(-8) = -4$

Square this number to obtain k: $k = (-4)^2 = 16$

Write the perfect square trinomial: $y^2 - 8y + 16 = (y - 4)^2$

SOLVING QUADRATIC EQUATIONS BY COMPLETING THE SQUARE. The quadratic equation $x^2 - 6x - 1 = 0$ cannot be factored. This quadratic equation can be solved, however, by completing the square. Follow these steps:

Step	Example
1. Write the original equation in the form $x^2 + bx = c$.	1. $x^2 - 6x = 1$
2. Find the number that will make a perfect square trinomial by taking half the coefficient of x and squaring it.	2. $x^2 - 6x + \underline{?} = 1 + \underline{?}$ $\frac{1}{2}(-6) = -3$ $(-3)^2 = 9$
3. Add this number to *both* sides of the original equation.	3. $x^2 - 6x + 9 = 1 + 9$
4. Write the perfect square trinomial as the square of a binomial.	4. $(x - 3)^2 = 10$
5. Take the square root of each side of the equation.	5. $x - 3 = \pm\sqrt{10}$
6. Solve the resulting equations and write the solution set.	6. $x = 3 \pm \sqrt{10}$ $\{3 + \sqrt{10}, 3 - \sqrt{10}\}$

Example

2. Solve by completing the square: $2x^2 - 8x + 5 = 0$.

Solution: Write the original equation in the form $x^2 + bx = c$ by writing the constant term on the right side and then dividing each member of the equation by the coefficient of x^2 (2):

$$2x^2 - 8x = -5$$

$$x^2 - 4x = \frac{-5}{2}$$

Find the number that makes the left side of the equation a perfect square trinomial.

$$x^2 - 4x + \underline{?} = \frac{-5}{2} + \underline{?}$$

$$\frac{1}{2}(-4) = -2 \qquad \text{and} \qquad (-2)^2 = 4$$

Add 4 to both sides: $x^2 - 4x + 4 = \frac{-5}{2} + 4$

Factor the trinomial and simplify: $(x-2)^2 = \frac{-5}{2} + \frac{8}{2} = \frac{3}{2}$

Take the square root of each side: $x - 2 = \pm\sqrt{\frac{3}{2}}$

$$x - 2 = \pm\frac{\sqrt{3}}{\sqrt{2}}$$

Express in simplest form by rationalizing the denominator: $x = 2 \pm \frac{\sqrt{3}}{\sqrt{2}} \cdot \frac{\sqrt{2}}{\sqrt{2}}$

$$= 2 \pm \frac{\sqrt{6}}{2}$$

$$= \frac{4}{2} \pm \frac{\sqrt{6}}{2}$$

Combine terms: $= \frac{4 \pm \sqrt{6}}{2}$

The solution set is $\left\{\frac{4+\sqrt{6}}{2}, \frac{4-\sqrt{6}}{2}\right\}$.

EXERCISE SET 4.3

1–9. Find the value of k *that completes the square, and then factor the trinomial as the square of a binomial.*

1. $x^2 + 12x + k$
2. $x^2 - 2x + k$
3. $y^2 - 18y + k$
4. $t^2 - 10t + k$
5. $y^2 + 6y + k$
6. $n^2 - 9n + k$
7. $w^2 - 5w + k$
8. $p^2 + 7p + k$
9. $x(x+3) + k$

10–21. Solve each quadratic equation by completing the square. Express irrational roots in simplest form.

10. $x^2 - 6x - 16 = 0$
11. $x^2 + 3x - 10 = 0$
12. $y^2 - 10y = 6$
13. $x(x+4) = 3$
14. $n^2 + 8n - 3 = 0$
15. $0 = x^2 + 6x + 4$
16. $p^2 - 7 = 2p$
17. $8t - 5 = t^2$
18. $2x^2 - 10x + 5 = 0$
19. $3x^2 + 4 = 12x$
20. $4 - 7s = -2s^2$
21. $2y^2 + 9y + 3 = 0$

4.4 SOLVING QUADRATIC EQUATIONS BY FORMULA

KEY IDEAS

Starting with a linear equation of the form $ax+b=c$, we can solve for variable x in terms of constants a, b, and c:

$$ax=c-b, \quad \text{which means} \quad x=\frac{c-b}{a} \quad (a \neq 0)$$

The resulting equation represents a general formula for solving linear equations having the form $ax+b=c$. For example, in the equation $2x+3=11$, $a=2$, $b=3$, and $c=11$, so that

$$x=\frac{c-b}{a}=\frac{11-3}{2}=\frac{8}{2}=4$$

Similarly, starting with a quadratic equation written in the standard form $ax^2+bx+c=0$ $(a \neq 0)$, variable x may be solved for by using the method of completing the square. The resulting equation is called the **quadratic formula**.

SOLVING QUADRATIC EQUATIONS BY APPLYING THE QUADRATIC FORMULA. *Any* quadratic equation (including those that cannot be factored) that is put into the standard form $ax^2+bx+c=0$ can be solved by applying the *quadratic formula*:

$$x=\frac{-b \pm \sqrt{b^2-4ac}}{2a} \quad (a \neq 0)$$

For example, to solve the equation $3x^2+2x=1$, follow these steps:

Step	Example
1. Put the equation into the standard form.	1. $3x^2 + 2x \underbrace{-1} = 0$
2. Identify the values for a, b, and c.	2. $ax^2 + bx + c = 0$ $a = 3$, $b = 2$, and $c = -1$
3. Write the quadratic formula, and replace the letters a, b, and c with their numerical values.	3. $x = \dfrac{-b \pm \sqrt{b^2 - 4ac}}{2a}$ $= \dfrac{-2 \pm \sqrt{(2)^2 - 4(3)(-1)}}{2(3)}$
4. Simplify.	4. $x = \dfrac{-2 \pm \sqrt{4+12}}{6}$ $= \dfrac{-2 \pm \sqrt{16}}{6}$ $= \dfrac{-2 \pm 4}{6}$ $x = \dfrac{-2+4}{6}$ *or* $x = \dfrac{-2-4}{6}$ $x = \dfrac{2}{6}$ $\qquad$ $x = \dfrac{-6}{6}$ $= \dfrac{1}{3}$ $\qquad$ $= -1$
5. Write the solution set.	5. $\left\{\mathbf{\dfrac{1}{3}, -1}\right\}$

In the preceding example the roots were rational, indicating that the original equation was factorable. In the next example the equation is *not* factorable, so its roots are *not* rational.

Example

Solve by formula: $x^2 + 8 = 7x$.

Solution: $\qquad x^2 + 8 = 7x$

Put the equation into standard form: $\quad x^2 - 7x + 8 = 0$

Identify a, b, and c: $\quad a = 1$, $b = -7$, and $c = 8$

Write the formula: $x = \dfrac{-b \pm \sqrt{b^2 - 4ac}}{2a}$

Evaluate the formula: $x = \dfrac{7 \pm \sqrt{(-7)^2 - 4(1)(8)}}{2(1)}$

$$= \frac{7 \pm \sqrt{49 - 32}}{2}$$

$$= \frac{7 \pm \sqrt{17}}{2}$$

Write the solution set: $\left\{\dfrac{7+\sqrt{17}}{2}, \dfrac{7-\sqrt{17}}{2}\right\}$

SELECTING A METHOD OF SOLUTION. The quadratic formula can be used to find the roots of *any* quadratic equation that is written in standard form, including quadratic equations that are not factorable and have irrational roots. However, some types of quadratic equations may be solved more easily by using a previously learned method.

- Type I ($c = 0$). If the constant term of a quadratic equation is missing, solve the equation by factoring.

Example: Solve: $3t^2 - 7t = 0$.

$$t(3t - 7) = 0$$
$$(t = 0) \vee (3t - 7 = 0)$$
$$t = 0 \vee \quad t = \frac{7}{3}$$

The solution set is $\left\{0, \dfrac{7}{3}\right\}$.

- Type II ($b = 0, c \neq 0$). If the linear term is missing, solve the equation by taking the square root of each side of the equation.

Example: Solve: $4w^2 - 24 = 0$.

$$4w^2 = 24$$
$$w^2 = 6$$
$$w = \pm\sqrt{6}$$

The solution set is $\{\sqrt{6}, -\sqrt{6}\}$.

- Type III (a, b, and $c \neq 0$). If, after writing a quadratic equation in standard form, a quadratic trinomial appears on one side of the equation, try first to solve the equation by factoring.

Example: Solve: $3y^2 + 5y = 2y^2 + 14$.

$$\begin{aligned} 3y^2 - 2y^2 + 5y - 14 &= 0 \\ y^2 + 5y - 14 &= 0 \\ (y+7)(y-2) &= 0 \\ (y+7=0) \vee (y-2&=0) \\ y = \quad -7 \vee y &= 2 \end{aligned}$$

The solution set is $\{\mathbf{-7, 2}\}$.

If the quadratic equation cannot be factored (or appears difficult to factor), solve the equation by using the quadratic formula.

Example: Solve: $2x + 4 = x^2$.

$$-x^2 + 2x + 4 = 0$$

$$x = \frac{-b \pm \sqrt{b^2 - 4ac}}{2a} \qquad a = -1, b = 2, c = 4$$

$$= \frac{-2 \pm \sqrt{(2)^2 - 4(-1)(4)}}{2(-1)}$$

$$= \frac{-2 \pm \sqrt{4 + 16}}{-2}$$

$$= \frac{-2 \pm \sqrt{20}}{-2}$$

Multiply numerator and denominator by -1:

$$= \frac{2 \mp \sqrt{20}}{2}$$

The solution set is $\left\{\frac{\mathbf{2 + \sqrt{20}}}{\mathbf{2}}, \frac{\mathbf{2 - \sqrt{20}}}{\mathbf{2}}\right\}$.

SIMPLIFYING AND CHECKING IRRATIONAL ROOTS. In the preceding example the roots $\frac{2 \pm \sqrt{20}}{2}$ may be expressed in *simplest form* by factoring the radicand 20 as the product of two positive integers, one of which is the highest perfect square factor of 20:

$$\frac{2 \pm \sqrt{20}}{2} = \frac{2 \pm \sqrt{4}\sqrt{5}}{2} = \frac{2 \pm 2\sqrt{5}}{2}$$

$$= \frac{2(1 \pm \sqrt{5})}{2} = \mathbf{1 \pm \sqrt{5}}$$

To *check* that $1 \pm \sqrt{5}$ are the roots of $2x + 4 = x^2$, first replace x by $1 + \sqrt{5}$ in the original equation:

$$\begin{array}{r|l} 2x + 4 = & x^2 \\ 2(1+\sqrt{5}) + 4 & (1+\sqrt{5})^2 \\ 2 + 2\sqrt{5} + 4 & (1+\sqrt{5})(1+\sqrt{5}) \\ 6 + 2\sqrt{5} & 1^2 + \sqrt{5} + \sqrt{5} + (\sqrt{5})^2 \\ & 1 + 2\sqrt{5} + 5 \\ & 6 + 2\sqrt{5} \checkmark \end{array}$$

Note **1.** The product $(1+\sqrt{5})(1+\sqrt{5})$ was obtained using FOIL.

2. To complete the check, you must also verify that the root $1 - \sqrt{5}$ makes the original equation a true statement.

EXERCISE SET 4.4

1–20. Solve each equation by using the quadratic formula. Express irrational roots in simplest radical form.

1. $x^2 - 6x - 16 = 0$
2. $y^2 - 10y = 6$
3. $x(x + 4) = 3$
4. $n^2 + 8n - 3 = 0$
5. $0 = x^2 + 6x + 4$
6. $p^2 = 2p + 7$
7. $0 = x^2 + 3x - 10$
8. $8t - 5 = t^2$
9. $2x(x - 5) + 5 = 0$
10. $3x^2 + 4 = 12x$
11. $4 - 7s = -2s^2$
12. $2y^2 + 3(3y + 1) = 0$
13. $n(n - 8) = -7$
14. $2x + 5 = x^2$
15. $4x^2 = x + 3$
16. $6x^2 + 2 = 9x$
17. $5x^2 - 10x + 3 = 0$
18. $4x + 1 = 8x^2$
19. $3x(x - 2) = 5$
20. $\dfrac{h^2}{2} = (3h + 1)$

21–32. Use any convenient method to find the solution set for each equation. Express irrational roots in simplest radical form.

21. $3y^2 - 7y + 2 = 0$
22. $w^2 - 3w + 5 = 5$
23. $3t^2 + 9 = 2t^2 + 6t$
24. $8n^2 = n + 3$
25. $4t^2 + 17t = 15$
26. $2r^2 - 5 = 9$
27. $10p - 7 = 3p^2$
28. $1 = 5x^2 + 9x$
29. $3y^2 = y^2 + 3y + 8$
30. $10s - 6s^2 = 3$
31. $6x^2 + 47x = 8$
32. $(x + 1)^2 = (2x + 3)^2$

33. In triangle ABC, $m\angle C = 90$, $\overline{BC}$ exceeds the length of $\overline{AC}$ by 3, and AB exceeds the length of AC by 5. Find AC. [*Answer may be left in radical form.*]

34. The perimeter of a rectangle is 40, and the length of a diagonal is 16. Find the width of the rectangle. [*Answer may be left in radical form.*]

35–38. Solve each proportion for x. Express irrational roots in simplest radical form.

35. $\dfrac{3}{x-1}=\dfrac{x+1}{x}$

36. $\dfrac{x+2}{2x-1}=\dfrac{2x+1}{x+3}$

37. $\dfrac{1-x}{x-2}=\dfrac{4x}{1+x}$

38. $\dfrac{x-5}{3x-1}=\dfrac{x+2}{x-2}$

4.5 THE RELATIONSHIP BETWEEN THE ROOTS AND THE COEFFICIENTS OF A QUADRATIC EQUATION

KEY IDEAS

Without actually solving a quadratic equation of the form $ax^2+bx+c=0$, the values of the coefficients *a*, *b*, and *c* can be used to provide information about the roots. The values of *a*, *b*, and *c* determine the *sum* and the *product* of the roots, as well as the *nature* of the roots. The term *nature of the roots* refers to whether the ιwo roots of a quadratic equation are real or not real, rational or irrational, equal or unequal.

SUM AND PRODUCT OF THE ROOTS. The quadratic formula tells us that the two roots of a quadratic equation are

$$x_1=\frac{-b+\sqrt{b^2-4ac}}{2a} \quad \text{and} \quad x_2=\frac{-b-\sqrt{b^2-4ac}}{2a}$$

Adding the expressions for x_1 and x_2 gives a relationship between the sum of the two roots and the coefficients *a*, *b*, and *c*, while multiplying x_1 by x_2 produces a relationship between the product of the two roots and the coefficients *a*, *b*, and *c*.

SUM AND PRODUCT OF ROOTS

If x_1 and x_2 represent the two roots of a quadratic equation having the form $ax^2+bx+c=0$ $(a\neq 0)$, then.

- **Sum of roots** $=x_1+x_2=\dfrac{-b}{a}$
- **Product of roots** $=x_1\cdot x_2=\dfrac{c}{a}$

Examples

1. Without solving the equation $2x^2-5x-3=0$, determine the sum and the product of its roots.

Solution: $a=2$, $b=-5$, and $c=-3$.

$$\text{Sum of roots} = \frac{-b}{a} = \frac{-(-5)}{2} = \mathbf{\frac{5}{2}}$$

$$\text{Product of roots} = \frac{c}{a} = \mathbf{\frac{-3}{2}}$$

2. One root of the equation $3x^2-10x=8$ is 4. Without solving the equation, find the other root.

Solution: Write the equation in standard form:

$3x^2-10x-8=0$; then $a=3$, $b=-10$, and $c=-8$

Let r = the unknown root.

Then $4r$ (product of the roots) $=\frac{c}{a}$.

$$4r=\frac{-8}{3}$$

Multiply each side by $\frac{1}{4}$: $\quad r=\frac{1}{4}\cdot\frac{-8}{3}=\frac{-2}{3}$

The other root is $\mathbf{\frac{-2}{3}}$.

3. If one root of the equation $x^2-7x+k=0$ is 2, what is the value of k?

Solution: Method 1: $a=1$, $b=-7$, and $c=k$

$$\text{Sum of roots} = \frac{-b}{a} = \frac{-(-7)}{1} = 7$$

$$\text{Product of roots} = \frac{c}{a} = \frac{k}{1} = k$$

If one root is 2 and the sum of the two roots is 7, then the other root must be 5. Since the two roots are 2 and 5, their product is 10, so $k=\mathbf{10}$.

Method 2 (Recommended): Since 2 is a root of the equation, a true statement results when x is replaced by 2 in the original equation:

$$\begin{aligned} x^2-7x+k&=0\\ 2^2-7(2)+k&=0\\ 4-14+k&=0\\ -10+k&=0\\ k&=\mathbf{10}\end{aligned}$$

NATURE OF THE ROOTS. In the quadratic formula

$$x = \frac{-b \pm \sqrt{b^2 - 4ac}}{2a} \quad a \neq 0$$

$b^2 - 4ac$, the expression underneath the radical sign, is called the **discriminant.** The discriminant determines the *type* of roots. If the discriminant is a perfect square greater than 0, then the radical in the quadratic formula evaluates to an integer greater than 0, so that the roots are *rational* and *unequal*; if the discriminant is a positive number that is *not* a perfect square, then the roots must include a radical, so that the roots are *irrational* and *unequal*. If the discriminant is equal to 0, then the $\pm$ radical term vanishes and the two roots are equal.

If the discriminant is negative, the roots are *not real* because the square root of a negative number is not real. Numbers that arise from taking square roots of negative numbers are called *imaginary numbers* and will be studied in future courses. Table 4.1 summarizes how the roots of a quadratic equation can be classified by looking at the discriminant.

TABLE 4.1 Classifying Roots by Looking at the Discriminant

Value of Discriminant	Type of Roots
1. $b^2 - 4ac < 0$ 2. $b^2 - 4ac = 0$ 3. $b^2 - 4ac > 0$ *and* (i) is *not* a perfect square (ii) is a perfect square	1. Not real (imaginary). 2. Real, rational, and equal. 3. Real, unequal, *and* (i) irrational (ii) rational

Examples

4. Find the discriminant and describe the nature of the roots of $3x^2 + 11x = 4$.

Solution: Write the equation in standard form:

$$3x^2 + 11x - 4 = 0; \text{ then } a = 3, b = 11, \text{ and } c = -4$$

Next, evaluate the discriminant:

$$\begin{aligned} b^2 - 4ac &= 11^2 - 4(3)(-4) \\ &= 121 + 48 \\ &= \mathbf{169} \end{aligned}$$

The discriminant 169 is a perfect square since $13 \times 13 = 169$. Therefore, the roots of the equation are **real, rational,** and **unequal.**

5. Find the positive value of k so that the roots of the equation $4x^2+kx+9=0$ are real and equal.

Solution: For the equation $4x^2+kx+9=0$, $a=4$, $b=k$, and $c=9$. If the roots are real and equal, then the discriminant must be equal to 0.

$$\begin{aligned} b^2-4ac&=0\\ k^2-4(4)(9)&=0\\ k^2-144&=0\\ k^2&=144\\ k&=\pm\sqrt{144}=\pm 12 \end{aligned}$$

The positive value of k that makes the roots real and equal is **12**.

6. Find the largest integral value of k so that the roots of the equation $x^2-5x+k=0$ are real.

Solution: For the equation $x^2-5x+k=0$, $a=1$, $b=-5$, and $c=k$. If the roots are real, then the discriminant must be greater than or equal to 0.

$$\begin{aligned} b^2-4ac&\geq 0\\ (-5)^2-4(1)(k)&\geq 0\\ 25-4k&\geq 0\\ -4k&\geq -25\\ k&\leq\frac{-25}{-4}\\ k&\leq 6\tfrac{1}{4} \end{aligned}$$

The largest integral value of k that makes the inequality true is **6**.

EXERCISE SET 4.5

1. If the discriminant of a quadratic equation with real coefficients is *not* negative, then the roots of the equation must be:
   (1) rational (2) irrational (3) real (4) not real

2. What is the product of the roots of the equation $x^2+x+2=0$?
   (1) 1 (2) -1 (3) 2 (4) -2

3. In which equation is the sum of the roots equal to $\frac{5}{3}$?
   (1) $3x^2+5x-9=0$ (3) $3x^2+7x-5=0$
   (2) $3x^2-5x+9=0$ (4) $3x^2+9x+5=0$

4. The roots of the equation $3x^2+5x+2$ are:
   (1) rational and unequal (3) irrational
   (2) rational and equal (4) not real

5. The roots of the equation $x - 1 = x^2 - 2x + 1$ are:
(1) rational and unequal (3) irrational
(2) rational and equal (4) not real

6. The solution set of the equation $2x^2 + x = 3$ contains two:
(1) integers
(2) positive rational numbers
(3) nonintegral rational roots
(4) rational numbers, one positive and one negative

7. What value of k will make the roots of the equation $x^2 - 2kx + 16 = 0$ real, rational, and equal?
(1) $-2\sqrt{2}$ (2) 2 (3) $4\sqrt{2}$ (4) -4

8. By what amount does the sum of the roots of the equation $3x^2 - 6x + 3 = 0$ exceed the product of its roots?
(1) -1 (2) 1 (3) 2 (4) 3

9. For what value of c will the roots of the equation $x^2 + 6x + c = 0$ be equal?

10. Find the sum of the roots of the equation $x^2 + 3x = 0$.

11. If one root of the equation $x^2 - 12x + k = 0$ is 4, what is the value of k?

12. One root of the equation $6x^2 + kx + 5 = 0$ is 1.
(a) What is the other root? (b) What is the value of k?

13. Given the equation $x^2 + bx + x + b = 0$; if one value of x that satisfies the equation is $x = -b$, find another value of x that also satisfies the equation.

14. If one of the roots of the equation $x^2 - x + q = 0$ is 3, what is the value of q?

15. The roots of an equation of the form $x^2 + px + q = 0$ are $2 + \sqrt{5}$ and $2 - \sqrt{5}$. Find the value of p.

16. For what value of b will the equation $bx^2 - bx - 2 = 0$ have 2 as a root?

17. In $ax^2 + bx + a = 0$, a and b are integers. If one root is $\frac{1}{2}$, find the other root.

18. If the roots of the equation $2x^2 - 3x + c = 0$ are real and irrational, the value of c may be:
(1) 1 (2) 2 (3) 0 (4) -1

19. For which of the equations given below is the product of the roots equal to 4?
(1) $2x^2 - 3x + 4 = 0$ (3) $x^2 - 4 = 0$
(2) $2x^2 - 8x + 5 = 0$ (4) $2x^2 - 3x + 8 = 0$

20–25. For each equation, find the value of k *that makes the roots real, rational, and equal.*

20. $x^2 + kx + 25 = 0$
21. $9x^2 + 6x + k = 0$
22. $kx^2 - 14x + 49 = 0$
23. $9x^2 + k = 12x$
24. $4x^2 - kx + 1 = 0$
25. $kx^2 - 11kx + 121 = 0$

26–29. For each equation, find the largest integral value of k *that makes the roots real.*

26. $x^2 - 4x + k = 0$
27. $kx^2 - 9x + 3 = 0$
28. $2x^2 - 7x + k = 0$
29. $kx^2 - 10x + k = 0$

30. What is the smallest integral value of k that makes the roots of the equation $3x^2 + kx + 3 = 0$ real?

4.6 FORMING A QUADRATIC EQUATION, GIVEN ITS ROOTS

KEY IDEAS

A quadratic equation having two given numbers as its roots can be formed by rewriting the equation $x^2 + bx + c = 0$ so that:

1. b is replaced by the negative of the sum of the roots.
2. c is replaced by the product of the roots.

FORMING QUADRATIC EQUATIONS. Dividing each term of the equation $ax^2 + bx + c = 0$ by a $(a \neq 0)$ results in the equation

$$x^2 + \left(\frac{b}{a}\right)x + \frac{c}{a} = 0$$

which may also be expressed as

$$x^2 - \left(\frac{-b}{a}\right)x + \frac{c}{a} = 0.$$

Since $\frac{-b}{a}$ represents the sum of the roots of a quadratic equation written in standard form and $\frac{c}{a}$ represents the product of the roots, another form of a quadratic equation is

$$\mathbf{x^2 - (\text{sum of roots})x + (\text{product of roots}) = 0}$$

Examples

1. Write the quadratic equation whose roots are 3 and −5.

Solution: First find the sum and the product of the roots.

Sum of roots $= 3 + (-5) = -2$

Product of roots $= (3)(-5) = -15$

$$x^2 - (\text{sum of roots})x + (\text{product of roots}) = 0$$
$$x^2 - (-2)x + (-15) = 0$$
$$\mathbf{x^2 + 2x - 15 = 0}$$

Another way of obtaining the quadratic equation is to "work backward." If 3 and -5 are roots, then, when the quadratic equation is written in standard form, $(x - 3)$ and $(x - (-5))$ are factors of the quadratic trinomial. Therefore,

$$(x - 3 = 0) \vee (x + 5 = 0)$$
$$(x - 3)(x + 5) = 0$$
$$\mathbf{x^2 + 2x - 15 = 0}$$

2. Write a quadratic equation having integral coefficients whose solution set is:

(a) $\left\{\frac{1}{3}, \frac{1}{2}\right\}$ (b) $\left\{2 \pm \sqrt{3}\right\}$

Solution: (a) First find the sum and the product of the roots.

$$\text{Sum of roots} = \frac{1}{3} + \frac{1}{2} = \frac{2}{6} + \frac{3}{6} = \frac{5}{6}$$

$$\text{Product of roots} = \left(\frac{1}{3}\right)\left(\frac{1}{2}\right) = \frac{1}{6}$$

$$x^2 - (\text{sum of roots})x + (\text{product of roots}) = 0$$
$$x^2 - \left(\frac{5}{6}\right)x + \frac{1}{6} = 0$$

Clear fractions by multiplying each member of the equation by 6: $\mathbf{6x^2 - 5x + 1 = 0}$

(b) Find the sum and the product of the roots.

Sum of roots $= (2 + \sqrt{3}) + (2 - \sqrt{3}) = 4$

Product of roots $= (2 + \sqrt{3})(2 - \sqrt{3})$

$= (2)^2 - (\sqrt{3})^2 = 4 - 3 = 1$

$$x^2 - (\text{sum of roots})x + (\text{product of roots}) = 0$$
$$\mathbf{x^2 - 4x + 1 = 0}$$

EXERCISE SET 4.6

1–12. In each case, write a quadratic equation with integral coefficients having the given solution set.

1. $\{-2, -1\}$
2. $\{5, -6\}$
3. $\left\{\frac{1}{4}, \frac{1}{2}\right\}$
4. $\left\{\frac{1}{6}, -\frac{1}{3}\right\}$
5. $\left\{-\frac{1}{2}, 3\right\}$
6. $\left\{\frac{1}{4}, -8\right\}$
7. $\left\{0, \frac{1}{5}\right\}$
8. $\{-4\}$
9. $\left\{-\frac{1}{3}, -\frac{1}{4}\right\}$
10. $\{1 \pm \sqrt{2}\}$
11. $\{3 \pm \sqrt{2}\}$
12. $\left\{\frac{1 \pm \sqrt{7}}{2}\right\}$

CHAPTER 4 REVIEW EXERCISES

1. Write an equation of the form $ax^2 + bx + c = 0$ for which the solution set is $\{-3, 4\}$.
2. Find the positive root of $2x^2 - x - 6 = 0$.
3. The measure of the length of a rectangle is three times the measure of the width, and the perimeter is 32. Find the area of the rectangle.
4. If one root of the equation $x^2 - 12x + k = 0$ is 4, what is the value of k?
5. What value(s) of x will make each of the following statements true?
   (a) $(x^2 = 9) \wedge (x + 2 = 5)$
   (b) $(x^2 + x = 2) \wedge (x + 3 \leq 5)$
   (c) $(x^2 - 3x = 0) \wedge (4x = x^2)$
   (d) $(10x^2 - 21 = 3x^2) \wedge (3x^2 + x = 14)$
6. What value of k makes the trinomial $x^2 - 10x + k$ a perfect square?
7. Find in radical form the roots of the equation $x^2 - 6x + 3 = 0$.
8. (a) The side of a square is represented by $x + 2$. Express the area of the square in terms of x.
   (b) The dimensions of a rectangle are represented by $2x + 1$ and $2x - 4$. Express the area of the rectangle in terms of x.
   (c) If the area of the square in part (a) equals the area of the rectangle in part (b), find x.
9. In rectangle $ABCD$, $AB = x$, $BC = x + 7$, and diagonal $BD = x + 8$. Find BD. [*Only an algebraic solution will be accepted.*]
10. The sum of the roots of the equation $x^2 - 3x + 2 = 0$ is:
    (1) -3 (2) 2 (3) 3 (4) -2
11. Which equation has irrational roots?
    (1) $x^2 - 4 = 0$
    (2) $x^2 - 2 = 0$
    (3) $x^2 - 2x + 1 = 0$
    (4) $x^2 - 2x = 0$

12. The solution to the quadratic equation $2x^2 - 5x - 1 = 0$ is:
(1) $\dfrac{5 \pm \sqrt{17}}{4}$
(2) $\dfrac{-5 \pm \sqrt{17}}{4}$
(3) $\dfrac{5 \pm \sqrt{33}}{4}$
(4) $\dfrac{-5 \pm \sqrt{33}}{4}$

13. Which statement about $3x^2 + 12x - 15 = 0$ is true?
(1) The sum of the roots is 4.
(2) The product of the roots is -4.
(3) The sum of the roots is -12.
(4) The product of the roots is -5.

14. Which quadratic equation has roots of -3 and 7?
(1) $x^2 - 4x - 21 = 0$
(2) $x^2 + 4x - 21 = 0$
(3) $x^2 + 4x + 21 = 0$
(4) $-x^2 - 4x + 21 = 0$

15. Which equation has $x = \dfrac{-6 \pm \sqrt{24}}{2}$ as its solution?
(1) $x^2 - 6x - 3 = 0$
(2) $x^2 - 6x + 3 = 0$
(3) $x^2 + 6x - 3 = 0$
(4) $x^2 + 6x + 3 = 0$

16. In right triangle ABC, $BC = x$, $AC = 8 - x$, and hypotenuse $AB = 6$.
(a) Write an equation that can be used to find x.
(b) Solve the equation for x. [*Answer may be left in radical form.*]

17. The perimeter of a rectangle is 28 centimeters. If the length of a diagonal of the rectangle is 10 centimeters, find the number of centimeters in the length and in the width of the rectangle. [*Only an algebraic solution will be accepted.*]

18. Solve for x. [*Answer may be left in radical form.*]

$$\frac{2x}{x+2} = \frac{x-3}{x-5}$$

CHAPTER 5

Operations with Algebraic Fractions

5.1 SIMPLIFYING ALGEBRAIC FRACTIONS

KEY IDEAS

A fraction may be simplified by dividing the numerator and the denominator by common factors (other than 1). For example,

$$\frac{10}{15}=\frac{\overset{1}{\cancel{5}}\cdot 2}{\underset{1}{\cancel{5}}\cdot 3}=\mathbf{\frac{2}{3}}.$$

The factor 5 appears in the numerator and the denominator, so that the quotient is 1. The fraction $\frac{2}{3}$ is in lowest terms since the numerator and the denominator do not have any common factors greater than 1.

WRITING FRACTIONS IN LOWEST TERMS. Algebraic fractions, like arithmetic fractions, are written in *lowest terms* by factoring out the GCF of the numerator and the denominator and then *canceling* common factors.

Examples

1. Write in lowest terms: $\frac{10x^6}{15x^2}$.

Solution: The GCF of $10x^6$ and $15x^2$ is $5x^2$.

$$\frac{10x^6}{15x^2}=\frac{\overset{1}{\cancel{5x^2}}\cdot 2x^4}{\underset{1}{\cancel{5x^2}}\cdot 3}=\mathbf{\frac{2x^4}{3}}$$

2. Write in lowest terms: $\frac{14a^5b}{35a^4b^3}$.

Solution: The GCF of $14a^5b$ and $35a^4b^3$ is $7a^4b$.

$$\frac{14a^5b}{35a^4b^3} = \frac{\overset{1}{\cancel{7a^4b}} \cdot 2a}{\underset{1}{\cancel{7a^4b}} \cdot 5b^2} = \mathbf{\frac{2a}{5b^2}}$$

Notice that the literal factor part of the answer may be obtained by finding the quotient of the literal factors of the numerator and denominator:

$$\frac{\overset{2}{\cancel{14}}a^5b}{\underset{5}{\cancel{35}}a^4b^3} = \frac{2a^{5-4}b^{1-3}}{5} = \frac{2ab^{-2}}{5} = \mathbf{\frac{2a}{5b^2}}$$

3. Write in lowest terms: $\frac{6a^5 - 20a^3}{8a^2}$.

Solution: *Step 1.* Factor out the GCF from the numerator.

$$\frac{6a^5 - 20a^3}{8a^2} = \frac{2a^3(3a^2 - 10)}{8a^2}$$

Step 2. Cancel common factors in the numerator and denominator. Observe that $8 \div 2 = 4$ and $a^3 \div a^2 = a$.

$$\frac{\overset{1a}{\cancel{2a^3}}(3a^2 - 10)}{\underset{4}{\cancel{8a^2}}}$$

Step 3. Write the remaining factors.

$$\mathbf{\frac{a(3a^2 - 10)}{4}}$$

4. Write in lowest terms: $\frac{-2x - 10}{x^2 - 25}$.

Solution: First factor the numerator and the denominator.

$$\frac{-2x - 10}{x^2 - 25} = \frac{-2(x + 5)}{(x + 5)(x - 5)}$$

Then cancel common factors: $\frac{-2\overset{1}{\cancel{(x + 5)}}}{\underset{1}{\cancel{(x + 5)}}(x - 5)} = \mathbf{\frac{-2}{(x - 5)}}$

EXERCISE SET 5.1

1-24. Write each fraction in lowest terms.

1. $\frac{28a^5}{4a^2}$
2. $\frac{-52x^3y}{-13xy}$
3. $\frac{100c^2}{25c^5}$
4. $\frac{32w^7z^6}{18w^2z^9}$
5. $\frac{2ab^2 - 2a^2b}{4ab}$
6. $\frac{-x-y}{x+y}$
7. $\frac{10y^3 - 5y^2}{15y}$
8. $\frac{12x^3 - 21x^4}{9x^2}$
9. $\frac{0.48xy - 0.16y}{0.8y}$
10. $\frac{21r^2s - 7r^3s^2}{14rs}$
11. $\frac{-3x-6}{x+2}$
12. $\frac{y^2+4y}{2y+8}$
13. $\frac{3x+15}{x^2+5x}$
14. $\frac{x^2-4}{x+2}$
15. $\frac{10y+30}{y^2-9}$
16. $\frac{-6a+18}{a^2-3a}$
17. $\frac{2x-16}{x^2-64}$
18. $\frac{x+1}{x^2-x-2}$
19. $\frac{x^2-9}{x^2+3x+4}$
20. $\frac{1-x^2}{3x+3}$
21. $\frac{2x^2-50}{2x^2+14x+20}$
22. $\frac{x^2-x-42}{x^2+7x+6}$
23. $\frac{a^2-4b^2}{(a+2b)^3}$
24. $\frac{x^2-y^2}{(x-y)^2}$

5.2 MULTIPLYING AND DIVIDING ALGEBRAIC FRACTIONS

KEY IDEAS

To **multiply** fractions, write the product of the numerators over the product of the denominators. Then write the resulting fraction in lowest terms.

$$\frac{4}{3} \cdot \frac{3}{10} = \frac{4 \cdot 3}{9 \cdot 10} = \frac{12}{90} = \mathbf{\frac{2}{15}}.$$

Sometimes the multiplication process can be made easier by canceling pairs of common factors in the numerator and the denominator *before* multiplying the fractions. For example,

$$\frac{4}{9} \cdot \frac{3}{10} = \frac{\overset{2}{\cancel{4}}}{\underset{3}{\cancel{9}}} \cdot \frac{\overset{1}{\cancel{3}}}{\underset{5}{10}} = \mathbf{\frac{2}{15}}.$$

To **divide** one fraction by another fraction, multiply the first fraction by the reciprocal of the second fraction. For example,

$$\frac{4}{9} \div \frac{3}{10} = \frac{4}{9} \cdot \frac{10}{3} = \frac{4 \cdot 10}{9 \cdot 3} = \frac{40}{27}$$

MULTIPLYING FRACTIONS. Algebraic fractions are *multiplied* in much the same way that fractions are multiplied in arithmetic.

Examples

1. Write the product: $\left(\frac{2x^3}{3y^5}\right)\left(\frac{4x^2}{7y}\right)$.

Solution: $\left(\frac{2x^3}{3y^5}\right)\left(\frac{4x^2}{7y}\right)=\frac{(2x^3)(4x^2)}{(3y^5)(7y)}=\frac{\mathbf{8x^5}}{\mathbf{21y^6}}$

2. Write the product in lowest terms: $\left(\frac{2a^7}{3b^2}\right)\left(\frac{12b^5}{5a^3}\right)$.

Solution: Cancel common factors in the numerator and denominator *before* multiplying.

$$\frac{2\overset{a^4}{\cancel{a^7}}}{\underset{1}{\cancel{3}}\cancel{b^2}}\cdot\frac{\overset{4\,b^3}{\cancel{12}\cancel{b^5}}}{\underset{5}{5\cancel{a^3}}}=\frac{(2a^4)(4b^3)}{5}=\frac{\mathbf{8a^4b^3}}{\mathbf{5}}$$

Fractions that contain polynomials should be factored, if possible, and then simplified *before* they are multiplied.

Examples

3. Write the product in lowest terms: $\frac{12y^2}{x^2+7x}\cdot\frac{x^2-49}{2y^5}$.

Solution: Factor the binomials, and then cancel pairs of common factors in the numerator and denominator before multiplying.

$$\frac{12y^2}{x^2+7x}\cdot\frac{x^2-49}{2y^5}=\frac{12y^2}{x(x+7)}\cdot\frac{(x+7)(x-7)}{2y^5}$$

$$=\frac{\overset{6}{\cancel{12}\cancel{y^2}}}{x\cancel{(x+7)}}\cdot\frac{\cancel{(x+7)}(x-7)}{\underset{1y^3}{\cancel{2}\cancel{y^5}}}$$

Write the products of the remaining factors in the numerator and denominator: $=\frac{\mathbf{6(x-7)}}{\mathbf{xy^3}}$

4. Write the product in lowest terms: $\frac{a^3-a^2b}{20b^3}\cdot\frac{5a+5b}{a^2}$.

Solution: Before multiplying, factor the numerators of each fraction. Cancel pairs of common factors in the numerators and denominators. Then multiply.

$$\frac{a^3 - a^2b}{20b^3} \cdot \frac{5a + 5b}{a^2} = \frac{\cancel{a^2}(a - b)}{\underset{4}{\cancel{20b^3}}} \cdot \frac{\overset{1}{\cancel{5}}(a + b)}{\cancel{a^2}}$$

$$= \frac{(a - b)(a + b)}{4b^3}$$

$$= \mathbf{\frac{a^2 - b^2}{4b^3}}$$

DIVIDING FRACTIONS. Algebraic fractions are *divided* in much the same way that fractions are divided in arithmetic.

Examples

5. Write the quotient in lowest terms: $\frac{8m^2}{3} \div \frac{6m^3}{3m - 12}$.

Solution: To begin, change from division to multiplication by taking the reciprocal of the second fraction. Then, where possible, factor

$$\frac{8m^3}{3} \div \frac{6m^3}{3m - 12} = \frac{8m^2}{3} \cdot \frac{3m - 12}{6m^3}$$

Cancel pairs of common factors in the numerator and denominator: $= \frac{\overset{4}{\cancel{8m^2}}}{\underset{1}{\cancel{3}}} \cdot \frac{\overset{1}{\cancel{3}}(m - 4)}{\underset{3m}{\cancel{6m^3}}}$

Multiply the remaining factors: $= \mathbf{\frac{4(m - 4)}{3m}}$

6. Divide and express in simplest form: $\frac{x^2 - 2x - 8}{x^2 - 25} \div \frac{x^2 - 4}{2x + 10}$.

Solution: Change to a multiplication example, and factor the numerators and denominators of each fraction.

$$\frac{x^2 - 2x - 8}{x^2 - 25} \div \frac{x^2 - 4}{2x + 10} = \frac{x^2 - 2x - 8}{x^2 - 25} \cdot \frac{2x + 10}{x^2 - 4}$$

$$= \frac{\cancel{(x + 2)}(x - 4)}{(x - 5)\cancel{(x + 5)}} \cdot \frac{2\cancel{(x + 5)}}{(x - 2)\cancel{(x + 2)}}$$

$$= \mathbf{\frac{2(x - 4)}{(x - 5)(x - 2)}}$$

EXERCISE SET 5.2

1–16. Write each product or quotient in simplest form.

1. $\frac{12a^2}{5c} \cdot \frac{15c^3}{8a}$

2. $\frac{3y}{4x} \cdot \frac{8x^2 - 4x}{9y}$

3. $\frac{5y + 10}{x^2} \cdot \frac{3x^2 - x^3}{15}$

4. $\frac{8}{rs} \cdot \frac{r^2s - rs^2}{12}$

5. $\frac{3x}{5y} \div \frac{12x^2 - 15x}{20y^2}$

6. $\frac{2b^2 - 2b}{3a} \cdot \left(\frac{2a}{3b}\right)^2$

7. $\frac{18}{x^2 - y^2} \div \frac{9}{x + y}$

8. $\frac{(a + b)^2}{4} \div \frac{a + b}{2}$

9. $\frac{2x + 6}{8} \div \frac{x + 3}{2}$

10. $\left(\frac{ay^2}{b^4}\right)^3 \cdot \left(\frac{b}{y^3}\right)^5$

11. $\left(\frac{ax^2}{b^3}\right)^3 \div \left(\frac{a^2x}{b}\right)^2$

12. $\left(\frac{3x}{4y}\right)^2 \div \frac{9x^2 - 6x}{8y}$

13. $\frac{(x - 7)^2}{x^2 - 6x - 7} \cdot \frac{5x + 5}{x^2 - 49}$

14. $\frac{a^2 - b^2}{2ab} \div \frac{a - b}{a^2}$

15. $\frac{x^2 - 9}{x^2 - 8x} \cdot \frac{x - 8}{x^2 - 6x + 9}$

16. $\frac{x^2 - 3x}{x^2 + 3x - 10} \div \frac{x^2 - x - 6}{x^2 - 4}$

17. The area of a rectangle is represented by $p^2 - 2p - 35$, and the length is represented by $p + 5$. In terms of p, what is the width of the rectangle?

18–20. Perform the indicated operations, and express the result in simplest form.

18. $\frac{7a^2}{12b^2}\left(\frac{2a}{5b^2}\right)^3 \div \frac{21a^3}{25b^7}$

19. $\frac{4x - 8ax^2}{3} \div \frac{8a^2x^2 - 2}{3 + 6ax}$

20. $\frac{t^2 - 4}{t^2 - 1} \div \frac{4t + 12}{9t + 9} \cdot \frac{t^2 + 2t - 3}{2 - t}$

5.3 ADDING AND SUBTRACTING ALGEBRAIC FRACTIONS

KEY IDEAS

To add (or subtract) arithmetic fractions that have the same denominator, write the sum (or difference) of the numerators over the common denominator:

$$\frac{2}{7} + \frac{3}{7} = \frac{2 + 3}{7} = \mathbf{\frac{5}{7}}$$

If the fractions have *different* denominators, then each fraction must first be changed to an equivalent fraction having the LCD (lowest common denominator) as its denominator. The addition and subtraction of algebraic fractions are handled in much the same way.

COMBINING FRACTIONS WITH THE SAME DENOMINATOR. To combine algebraic fractions having the same denominator, write the sum or difference of the numerators over the common denominator and then simplify.

Examples

1. Write the difference in lowest terms: $\frac{5a+b}{10ab} - \frac{3a-b}{10ab}$.

Solution:
$$\frac{5a+b}{10ab} - \frac{3a-b}{10ab} = \frac{5a+b-(3a-b)}{10ab}$$
$$= \frac{5a+b-3a+b}{10ab}$$
$$= \frac{(5a-3a)+(b+b)}{10ab}$$
$$= \frac{2a+2b}{10ab}$$
$$= \frac{\overset{1}{\cancel{2}}(a+b)}{\underset{5}{\cancel{10}}ab} = \mathbf{\frac{a+b}{5ab}}$$

2. Express as a single fraction in lowest terms: $\frac{4x}{x^2-4} + \frac{x+6}{4-x^2}$.

Solution: Rewrite $4 - x^2$ as $-x^2 + 4$. Then factor out -1 so that
$$-x^2 + 4 = -(x^2 - 4).$$
$$\frac{4x}{x^2-4} + \frac{x+6}{4-x^2} = \frac{4x}{x^2-4} + \frac{x+6}{-(x^2-4)}$$

Write the negative sign in front of the second fraction: $= \frac{4x}{x^2-4} - \frac{x+6}{x^2-4}$

Write the numerators over the common denominator: $= \frac{4x-(x+6)}{x^2-4}$

Simplify the numerator: $= \frac{4x-x-6}{x^2-4}$

Write the fraction in lowest terms: $= \frac{3x-6}{x^2-4}$

Factor, and cancel common factors: $= \frac{3\overset{1}{\cancel{(x-2)}}}{\underset{1}{\cancel{(x-2)}}(x+2)}$

$$= \mathbf{\frac{3}{x+2}}$$

COMBINING FRACTIONS WITH UNLIKE DENOMINATORS To combine fractions with unlike denominators, first determine the LCD of the fractions. Then compare the denominator of each fraction with the LCD. Multiply the denominator of each fraction by the number or algebraic expression that will give the LCD. To produce an equivalent fraction, you must also multiply the numerator of the fraction by the same number or expression as was used for the denominator. Simplify, and then follow the rules for adding fractions having like denominators.

Examples

3. Add: $\frac{2x+1}{6}+\frac{3x-5}{8}$.

Solution: The LCD of 6 and 8 is 24. Observe that $24 \div 6 = 4$. This means that multiplying the first fraction by $\frac{4}{4}$ will produce an equivalent fraction having the LCD of 24 as its denominator. Since $24 \div 8 = 3$, multiplying the second fraction by $\frac{3}{3}$ will also give an equivalent fraction having 24 as its denominator.

$$\frac{2x+1}{6}+\frac{3x-5}{8}=\left(\frac{2x+1}{6}\right)\frac{4}{4}+\left(\frac{3x-5}{8}\right)\frac{3}{3}$$

Write the numerators over the LCD: $=\frac{4(2x+1)+3(3x-5)}{24}$

Combine like terms in the numerator: $=\frac{8x+4+9x-15}{24}$

$$=\frac{\mathbf{17x-11}}{\mathbf{24}}$$

4. Write the difference in simplest form: $\frac{3}{10xy}-\frac{10x-y}{5xy^2}$.

Solution: Factor each denominator.

$$\begin{aligned} 10xy &= 2 \cdot 5xy \\ 5xy^2 &= \quad 5xy^2 \\ \text{LCD} &= 2 \cdot 5xy^2 \end{aligned}$$

Compare each of the factored denominators to the LCD. Multiply the numerator and denominator of each fraction by the factors contained in the LCD that are missing from the denominators of that fraction.

$$\left(\frac{3}{10xy}\right)\frac{y}{y}-\left(\frac{10x-y}{5xy^2}\right)\frac{2}{2}=\frac{3y}{10xy^2}-\frac{2(10x-y)}{10xy^2}$$
$$=\frac{3y-20x+2y}{10xy^2}$$
$$=\frac{5y-20x}{10xy^2}$$
$$=\frac{\overset{1}{\cancel{5}}(y-4x)}{\underset{2}{\cancel{10}}xy^2}$$
$$=\frac{y-4x}{2xy^2}$$

5. Subtract: $\frac{7x-2}{4}-x$.

Solution: Rewrite x as $\frac{x}{1}$. Since the LCD of 4 and 1 is 4, multiply the second fraction by $\frac{4}{4}$.

$$\frac{7x-2}{4}-x=\frac{7x-2}{4}-\left(\frac{x}{1}\right)\left(\frac{4}{4}\right)$$
$$=\frac{7x-2}{4}-\frac{4x}{4}$$
$$=\frac{7x-2-4x}{4}$$
$$=\frac{3x-2}{4}$$

6. Express the difference in simplest form: $\frac{a^2+1}{a^2-1}-\frac{a}{a+1}$.

Solution: Factor each denominator.

$$a^2-1=(a+1)(a-1)$$
$$a+1=a+1$$
$$\text{LCD}=(a+1)(a-1)$$

Compare each of the factored denominators to the LCD. The first fraction already has the LCD as its denominator, but the denominator of the second fraction is missing the factor $(a-1)$. Therefore, multiply the numerator and denominator of the second fraction by $(a-1)$.

$$\frac{a^2+1}{a^2-1}-\frac{a}{a+1}=\frac{a^2+1}{(a+1)(a-1)}-\frac{a}{a+1}\cdot\frac{a-1}{a-1}$$

$$=\frac{a^2+1-a(a-1)}{(a+1)(a-1)}$$

$$=\frac{a^2+1-a^2+a}{(a+1)(a-1)}$$

$$=\frac{\overset{1}{\cancel{(a+1)}}}{\underset{1}{\cancel{(a+1)}}(a-1)}$$

$$=\frac{\mathbf{1}}{\boldsymbol{a-1}}$$

7. Express the difference in simplest form: $\frac{5y+7}{y^2-4}-\frac{3}{4y+8}$.

Solution: Factor each denominator.

$$y^2-4=(y+2)(y-2)$$

$$4y+8=4(y+2)$$

$$\text{LCD}=4(y+2)(y-2)$$

Compare each of the factored denominators to the LCD. The denominator of the first fraction is missing a factor of 4, while the denominator of the second fraction lacks the factor $(y-2)$. Therefore, multiply the numerator and denominator of the first fraction by 4, and multiply the numerator and the denominator of the second fraction by $(y-2)$.

$$\frac{5y+7}{y^2-4}-\frac{3}{4y+8}=\frac{4}{4}\cdot\frac{5y+7}{(y+2)(y-2)}-\frac{3}{4(y+2)}\cdot\frac{y-2}{y-2}$$

Write the numerators over the LCD: $=\frac{4(5y+7)-3(y-2)}{4(y+2)(y-2)}$

Simplify: $=\frac{20y+28-3y+6}{4(y+2)(y-2)}$

$$=\frac{17y+34}{4(y+2)(y-2)}$$

Factor the numerator:
Cancel the common factor: $=\frac{17\overset{1}{\cancel{(y+2)}}}{4\underset{1}{\cancel{(y+2)}}(y-2)}$

Simplify: $=\frac{\mathbf{17}}{\mathbf{4(y-2)}}$

8. Combine and express the result in simplest form:

$$\frac{5}{x^2-9}-\frac{3}{x}+\frac{1}{2x+6}.$$

Solution: Factor each denominator.

$$\begin{aligned} x^2-9 &= (x+3)(x-3) \\ x &= x \\ 2x+6 &= 2(x+3) \\ \text{LCD} &= 2x(x+3)(x-3) \end{aligned}$$

Compare each of the factored denominators to the LCD. Multiply the numerator and denominator of each fraction by the factors contained in the LCD that are missing from the denominator of that fraction.

$$\left(\frac{5}{(x+3)(x-3)}\right)\frac{2x}{2x}-\left(\frac{3}{x}\right)\frac{2(x+3)(x-3)}{2(x+3)(x-3)}+\left(\frac{1}{2(x+3)}\right)\frac{x(x-3)}{x(x-3)}$$

$$=\frac{5(2x)-3(2)(x+3)(x-3)+1(x)(x-3)}{2x(x+3)(x-3)}$$

$$=\frac{10x-6(x^2-9)+x^2-3x}{2x(x+3)(x-3)}$$

$$=\frac{10x-6x^2+54+x^2-3x}{2x(x+3)(x-3)}=\mathbf{\frac{-5x^2+7x+54}{2x(x+3)(x-3)}}$$

EXERCISE SET 5.3

1–48. Write each of the following expressions as a single fraction in simplest form.

1. $\frac{5x}{6}+\frac{2x}{3}$

2. $\frac{a+1}{2}-\frac{a}{3}$

3. $\frac{x+7}{3}+\frac{x-2}{4}$

4. $\frac{3b}{8a^2}+\frac{5b}{12a^5}$

5. $\frac{5x+2}{3}-\frac{x+1}{3}$

6. $\frac{4}{a+b}+\frac{1}{-a-b}$

7. $\frac{2r+s}{r+2s}+\frac{r+5s}{r+2s}$

8. $\frac{3(n^2-2n)}{10n}+\frac{2n^2+n}{10n}$

9. $\frac{3(2b-1)}{4b}-\frac{5b+3}{4b}$

10. $\frac{2(3y+15)}{9y}-\frac{3(4y+8)}{9y}$

11. $\frac{2w-1}{8}-\frac{w+2}{6}$

12. $\frac{4a}{5x}-\frac{3a}{10x}$

13. $\frac{3y-2}{2}+\frac{2(y+5)}{9}$

14. $\frac{2c-9}{5x}-\frac{4c+3}{7x}$

15. $\frac{7(x+y)}{12xy}-\frac{3x-y}{12xy}$

16. $\frac{p}{p^2-9}+\frac{3}{9-p^2}$

17. $\dfrac{5a-b+c}{2a+b}+\dfrac{a+4b-c}{2a+b}$

18. $\dfrac{x+y}{11}-\dfrac{2x+4y}{22}$

19. $\dfrac{a^2-5}{a-b}-\dfrac{b^2-5}{a-b}$

20. $\dfrac{7t+3}{8rs}-\dfrac{3t+7}{8rs}+\dfrac{2(t+5)}{8rs}$

21. $\dfrac{x+2}{-3}+\dfrac{x-3}{2}$

22. $\dfrac{4x}{x-1}+2$

23. $\dfrac{2x}{3}+\dfrac{3x}{4}-\dfrac{x}{6}$

24. $\dfrac{3}{5a^2b}+\dfrac{1}{3a^2b}$

25. $\dfrac{3a+b}{10c}+\dfrac{5a-2b}{4c}$

26. $\dfrac{5t^2-9}{4t^2}+\dfrac{11}{6t}$

27. $\dfrac{7}{t}-\dfrac{3}{t^3}+\dfrac{5t+2}{t^2}$

28. $\dfrac{3}{8rs}+\dfrac{5}{6r^2}-\dfrac{1}{4s^2}$

29. $\dfrac{x-a^2}{a}+2a$

30. $\dfrac{x^2}{(x+1)^2}-\dfrac{x-1}{x+1}$

31. $\dfrac{3}{x+2}-\dfrac{2}{x-2}$

32. $\dfrac{5}{b-3}-\dfrac{4}{b}$

33. $\dfrac{y}{y^2-9}-\dfrac{1}{y+3}$

34. $\dfrac{3b}{2a}-\dfrac{2a}{3b}+1$

35. $\dfrac{x}{x+1}+\dfrac{3}{1-x}$

36. $\dfrac{6}{7p}-1+\dfrac{3}{4p}$

37. $\dfrac{2x-y}{3}+\dfrac{x-2z}{4}-\dfrac{y+3z}{2}$

38. $\dfrac{a}{s}-\dfrac{a}{r}$

39. $\dfrac{2}{x^2-1}+\dfrac{1}{2x+2}$

40. $\dfrac{4x}{a^2bc^3}-\dfrac{2x}{abc^2}+\dfrac{7x}{a^3b^2c}$

41. $\dfrac{2}{3}-\dfrac{x}{x-3}$

42. $\dfrac{a}{a-b}+\dfrac{b}{a+b}$

43. $\dfrac{6}{y^2-9}+\dfrac{4}{(y-3)^2}$

44. $\dfrac{3}{x^2-4}+\dfrac{2}{x^2+5x+6}$

45. $\dfrac{5}{r^2-s^2}-\dfrac{3}{(r+s)^2}$

46. $\dfrac{2x}{x^2+7x}-\dfrac{6x-5}{x^2-49}$

47. $w-y-\dfrac{1}{w+y}$

48. $\dfrac{5}{a^2-3a-10}+\dfrac{3a-2}{a^2-25}-\dfrac{2a}{3a+6}$

49–50. Find each of the following products by first simplifying the expressions within the parentheses:

49. $\left(2+\dfrac{2}{x}\right)\left(\dfrac{1}{x+1}-1\right)$

50. $\left(\dfrac{1}{x+1}-\dfrac{1}{x-1}\right)^2$

5.4. SOLVING EQUATIONS INVOLVING FRACTIONS

KEY IDEAS

The equation

$$\frac{x+1}{4}-\frac{2}{3}=\frac{1}{12}$$

has *fractional coefficients* since it is equivalent to

$$\frac{1}{4}(x+1)-\frac{2}{3}=\frac{1}{12},$$

but it is not a fractional equation.

The equation

$$\frac{2}{y}-\frac{9}{10}=\frac{1}{5y}$$

is an example of a **fractional equation** since some of the *denominators* contain a variable.

Each of these equations can be solved by first clearing the equation of fractions.

SOLVING EQUATIONS WITH FRACTIONS. To solve an equation that has fractional terms, multiply each member of *both* sides of the equation by the least common multiple (that is, the LCM) of the denominators. This will produce an equivalent equation that does not contain any fractions.

Examples

1. Solve for x and check: $\dfrac{x+1}{4}-\dfrac{2}{3}=\dfrac{1}{12}$.

Solution: The LCM of 4, 3, and 12 is 12.

Multiply each term by 12: $\overset{3}{\underset{1}{\cancel{12}}}\left(\dfrac{x+1}{\cancel{4}}\right)-\overset{4}{\underset{1}{\cancel{12}}}\left(\dfrac{2}{\cancel{3}}\right)=\overset{1}{\underset{1}{\cancel{12}}}\left(\dfrac{1}{\cancel{12}}\right)$

$$3(x+1)-4(2)=1$$
$$3x+3-8=1$$
$$3x-5=1$$
$$3x=5+1$$
$$\frac{3x}{3}=\frac{6}{3}$$
$$x=\mathbf{2}$$

Check: $\dfrac{x+1}{4} - \dfrac{2}{3} = \dfrac{1}{12}$

$$\dfrac{2+1}{4} - \dfrac{2}{3}$$
$$\dfrac{3}{4} - \dfrac{2}{3}$$
$$\dfrac{9}{12} - \dfrac{8}{12}$$
$$\dfrac{1}{12} = \dfrac{1}{12} \checkmark$$

2. Solve for y: $\dfrac{2}{y} - \dfrac{9}{10} = \dfrac{1}{5y}$.

Solution: The LCM of y, 10, and $5y$ is $10y$.

Multiply each term by $10y$: $\overset{1}{10\cancel{y}}\left(\dfrac{2}{\cancel{y}_{1}}\right) - \overset{1}{\cancel{10}y}\left(\dfrac{9}{\cancel{10}_{1}}\right) = \overset{2}{\cancel{10y}}\left(\dfrac{1}{\cancel{5y}_{1}}\right)$

$$20 - 9y = 2$$
$$-9y = 2 - 20$$
$$\frac{-9y}{-9} = \frac{-18}{-9}$$
$$y = 2$$

The check is left for you.

3. Solve for the positive value of n: $\dfrac{3}{5} + \dfrac{n-2}{3} = \dfrac{14}{5n}$.

Solution: The LCM of 5, 3, and $5n$ is $15n$.

Multiply each term by $15n$: $\overset{3}{\cancel{15}n}\left(\dfrac{3}{\cancel{5}_{1}}\right) + \overset{5}{\cancel{15}n}\left(\dfrac{n-2}{\cancel{3}_{1}}\right) = \overset{3}{\cancel{15n}}\left(\dfrac{14}{\cancel{5n}_{1}}\right)$

$$9n + 5n(n-2) = 42$$
$$9n + 5n^2 - 10n = 42$$
$$5n^2 - n = 42$$
$$5n^2 - n - 42 = 0$$
$$(5n + 14)(n - 3) = 0$$
$$(5n + 14 = 0) \vee (n - 3 = 0)$$
$$n = -\frac{14}{15} \textit{ or } n = 3$$

The check is left for you.

EXERCISE SET 5.4

1–26. Solve for the variable and check.

1. $\frac{x}{5} - 12 = 4$

2. $\frac{x}{3} - \frac{x}{4} = 1$

3. $\frac{n}{2} - 3 = \frac{n}{5}$

4. $\frac{1}{5} - \frac{1}{3x} = \frac{1}{15x}$

5. $\frac{y}{2} = 2 - \frac{y}{6}$

6. $\frac{b}{4} = \frac{2b}{5} + \frac{5}{2}$

7. $\frac{3}{5r} - \frac{2}{6r} = \frac{1}{15}$

8. $\frac{2r}{3} - \frac{5r}{12} = \frac{5}{4}$

9. $\frac{1}{x} + \frac{3}{2x} = \frac{x-4}{2}$

10. $\frac{5}{x} + 3x = \frac{17}{x}$

11. $\frac{x+2}{4x} - \frac{1}{x} = \frac{1}{12}$

12. $\frac{1}{x} - \frac{x+1}{8} = \frac{x-1}{4x}$

13. $\frac{3}{x} - \frac{1}{2x} = \frac{1}{2}$

14. $\frac{x+15}{5x} - \frac{1}{2} = \frac{x-2}{2x}$

15. $\frac{1}{2a} + \frac{5}{9} = -\frac{1}{18a}$

16. $\frac{x}{4} - \frac{x+2}{12} = \frac{x-2}{3}$

17. $n - 4 = \frac{n-1}{4}$

18. $\frac{x-2}{2} + \frac{2x-1}{20} = \frac{x}{4}$

19. $\frac{4}{b} + \frac{b+1}{2b} = 2$

20. $\frac{x-4}{3} - \frac{1}{x} = \frac{2}{3x}$

21. $\frac{y-3}{6} + \frac{y-25}{5} = 0$

22. $\frac{2x+1}{12} + \frac{x-3}{4} = \frac{x-1}{6}$

23. $2 + \frac{9}{x} = \frac{5}{x^2}$

24. $\frac{1}{x} - \frac{x-1}{14} = \frac{1}{7x}$

25. $\frac{x-1}{16} + \frac{5}{8x} = \frac{1}{x}$

26. $\frac{3}{m^2} - \frac{1}{3m} = \frac{m+5}{6m}$

27–30. Solve each of the following inequalities:

27. $\frac{3x-1}{4} > \frac{x+3}{2}$

28. $\frac{y+5}{8} < \frac{y-1}{4}$

29. $\frac{n+6}{3} - 2 \leq \frac{n-2}{2}$

30. $\frac{a+1}{4} - \frac{3a}{8} \geq \frac{1}{2}$

31. Solve for x: $\frac{1}{a} - \frac{1}{x} = \frac{1}{b}$.

32. Solve for r: $\frac{1}{r} - \frac{2}{s} = \frac{1}{2r}$.

33. Solve for y: $\frac{y}{a} - \frac{a+y}{y-a} = \frac{y}{2a}$.

34. Solve for x: $\frac{1}{b} + \frac{1}{a} = \frac{a}{bx} - \frac{b}{ax}$.

35. The first of three consecutive odd integers exceeds two thirds of the largest integer by 3. Find the three integers.

36. One half of the largest of three consecutive even integers exceeds one fourth of the second integer by 5. Find the integers.

37. Two numbers are in the ratio of 2 : 3. If the sum of their reciprocals is $\frac{5}{12}$, find the smaller of the two numbers.

38. If $\frac{1}{2}$ is added to the reciprocal of a number, the result is 1 less than twice the reciprocal of the original number. Find the number.

39. If the reciprocal of a number is multiplied by 3, the result exceeds the reciprocal of the original number by $\frac{1}{3}$. Find the number.

40. If the reciprocal of a number is multiplied by 1 less than the original number, the result exceeds one-half the reciprocal of the original number by $\frac{5}{8}$. Find the number.

CHAPTER 5 REVIEW EXERCISES

1–20. Write each of the following expressions as a single fraction in simplest form:

1. $\frac{5x+8}{3}-\frac{2(x+1)}{3}$
2. $\left(\frac{3b-3}{4ab}\right)\left(\frac{ab+a}{6}\right)$
3. $\frac{2a+6}{8}\cdot\frac{2}{a+3}$
4. $\frac{x-1}{3}\div\frac{4x-8}{9}$
5. $\frac{x+2}{4}-\frac{x-3}{3}$
6. $\frac{a+b}{3}+\frac{a-b}{2}$
7. $\frac{x}{x+4}-\frac{x^2+16}{x^2-16}$
8. $\frac{a}{a-5}+2$
9. $\frac{3}{x+2}-\frac{2}{x}$
10. $\frac{2x^2-8}{3x^2}\cdot\frac{9x}{4x^2+4x-8}$
11. $\frac{x^2-1}{x+2}\div\frac{x+1}{5x+10}$
12. $\frac{4x}{2x+6}+\frac{x}{x+3}$
13. $\frac{3a+1}{a^2-1}-\frac{1}{a+1}$
14. $\frac{x^2-y^2}{4}\div\frac{y-x}{12}$
15. $\frac{6a^2b}{x^2-x-72}\div\frac{2ab^3}{x^2-64}$
16. $\frac{x^2-9}{x}\cdot\frac{x^2+2x}{x^2+5x+6}$
17. $\frac{x-y}{x+y}-\frac{x+y}{x-y}$
18. $\frac{x^2-y^2}{2xy}\cdot\frac{6x^2}{3x-3y}$
19. $\frac{3}{1-a}-\frac{2}{a-1}$
20. $\frac{3p+6}{p^2-9}\div\frac{p^2+4p+4}{p^2+3p}$

21–28. Solve for the variable and check.

21. $\frac{t}{6} - \frac{t}{8} = 2$

22. $\frac{r}{3} + \frac{5r}{12} = \frac{9}{4}$

23. $\frac{1}{2x} = \frac{1}{x} - 2$

24. $\frac{3}{y} = 2 + \frac{5}{y}$

25. $\frac{x+2}{x} - \frac{3}{2x} = 5$

26. $\frac{a-2}{3} + \frac{2a-4}{4} = 5$

27. $\frac{3}{x} + \frac{1}{2x} = \frac{x-6}{2}$

28. $\frac{n-1}{6} - \frac{1}{n} = \frac{n-1}{12}$

REGENTS TUNE-UP: CHAPTERS 1–5

Here is an opportunity for you to review Chapters 1–5 and, at the same time, prepare for the Course II Regents Examination. Problems included in this section are similar in form and difficulty to those found on the New York State Regents Examination for Course II of the Three-Year Sequence for High School Mathematics. Problems preceded by an asterisk have actually appeared on a previous Course II Regents Examination.

***1.** What is the positive root of the equation $x^2 + 7x - 8 = 0$?

2. For what value of k are the roots of $2x^2 - 8x + k = 0$ equal?

3. Express in simplest form: $\frac{3a+2}{a^2+a} - \frac{1}{a+1}$.

4. Express the product $(y+1)\left(\frac{y}{1-y^2}\right)$ as a fraction in simplest form.

5. Perform the indicated operation and express the result in simplest form:

$$\frac{x^2-3x}{2x^2+x-6} \div \frac{x^2-5x+6}{x^2-4}.$$

6. Express $\frac{3}{x-1} - \frac{2}{x}$ as a single fraction in simplest form.

***7.** Given these true statements: $J \vee \sim N$ and N, which statement must also be true?
(1) J (2) $\sim J$ (3) $J \wedge \sim N$ (4) $\sim J \wedge N$

***8.** If $a \rightarrow b$ and $\sim c \rightarrow \sim b$ are both true statements, then which statement must also be true?
(1) $a \rightarrow c$ (2) $b \rightarrow c$ (3) $c \rightarrow a$ (4) $c \rightarrow b$

***9.** Which statement is logically equivalent to $\sim(p \vee \sim q)$?
(1) $p \wedge \sim q$ (2) $\sim p \wedge q$ (3) $\sim p \vee q$ (4) $\sim p \wedge \sim q$

***10.** Given these true statements: "Jay loves the math team" and "If the math team does not win, then Jay does not love the math team," which statement must also be true?
(1) The math team loses. (3) The math team loves Jay.
(2) The math team wins. (4) Jay does not love the math team.

***11.** Which conclusion logically follows from these true statements: "If the negotiations fail, the baseball strike will not end" and "If the World Series is played, the baseball strike has ended"?
(1) If negotiations fail, the World Series will not be played.
(2) If negotiations fail, the World Series will be played.
(3) If the baseball strike ends, the World Series will be played.
(4) If negotiations do not fail, the baseball strike will not end.

***12.** What are the roots of the equation $2x^2 + 4x - 5 = 0$?
(1) $\frac{4 \pm \sqrt{56}}{4}$ (2) $\frac{-4 \pm \sqrt{56}}{4}$ (3) $1 \pm \sqrt{14}$ (4) $\frac{-4 \pm \sqrt{14}}{4}$

13. If the sum of the roots of the equation $x^2 + kx + 2 = 0$ is 3, the value of k is:
(1) 1 (2) −1 (3) 3 (4) −3

14. The roots of the equation $3x^2 + x + 4 = 0$ are:
(1) real, rational, and equal
(2) real, rational, and unequal
(3) real and irrational
(4) imaginary (not real)

15. If $h \neq 0$, when the fraction $\frac{(x+h)^2 - x^2}{h}$ is simplified, the result is:
(1) h (2) 0 (3) $2x^2 + 2x + h$ (4) $2x + h$

***16.** In the clock 5 (mod 5) system of arithmetic, what is the solution set for the equation $x + 4 = 3$?
(1) {1} (2) {2} (3) {3} (4) {4}

***17.** If $*$ is a binary operation defined by $c * d = 3c + d$, what is the value of x when $2 * x = 8$?
(1) 1 (2) 2 (3) 3 (4) 4

***18.** The table below defines the operation $\otimes$ for the set $S = \{1, 3, 5, 7, 9\}$. According to this table, which statement is *false*?

$\otimes$	1	3	5	7	9
1	1	3	5	7	9
3	3	9	5	1	7
5	5	5	5	5	5
7	7	1	5	9	3
9	9	7	5	3	1

(1) The identity element is 1.
(2) The set S is closed under $\otimes$.
(3) The operation $\otimes$ is commutative.
(4) Every element has an inverse.

***19.** Given: If laws are good and strictly enforced, then crime will diminish.
If laws are not strictly enforced, then the problem is critical.
Crime has not diminished.
Laws are good.
Let G represent: "Laws are good."
Let S represent: "Laws are strictly enforced."
Let D represent: "Crime has diminished."
Let P represent: "The problem is critical."
Using G, S, D, and P, prove: "The problem is critical."

***20.** Given: set $A = \{1, 2, 3, 4\}$ and operation # and @ as defined by the following tables:

#	1	2	3	4
1	1	2	3	4
2	2	3	4	1
3	3	4	1	2
4	4	1	2	3

@	1	2	3	4
1	1	1	1	1
2	1	2	3	4
3	1	3	4	2
4	1	4	2	3

(a) Evaluate: $(3 \# 4)$ @ $(3$ @ $2)$.
(b) Find the inverse of 4 under the operation #.
(c) Find x: $(2 \# x)$ @ $3 = 2$.
(d) State one reason why (A, @) is *not* a group.

UNIT III: EUCLIDEAN GEOMETRY

CHAPTER 6

Basic Geometric Terms and Proving Triangles Congruent

6.1 BASIC GEOMETRIC TERMS AND CONCEPTS

KEY IDEAS

Geometry is an example of a *postulational system*, in which postulates (axioms), undefined terms, and defined terms are used to prove new relationships that are expressed as statements called *theorems*. These theorems, in turn, and the postulates and defined and undefined terms are used to prove other theorems. The result is an ever-expanding set of geometric properties and relationships. This section reviews some basic geometric terms and concepts.

UNDEFINED AND DEFINED TERMS. The "first" terms of geometry are *point*, *line*, and *plane*. They are undefined since they can be described but cannot be defined using previously defined words. A **point** indicates position but has no physical dimensions. A **line** is a continuous set of points that extends indefinitely in two opposite directions. A **plane** is a flat surface that extends indefinitely in all directions.

Other terms are defined in order to make communication easier and more precise. *Line segment*, *ray*, *opposite rays*, and *angle* are basic terms that were defined in Course I. Here are some additional terms that you should already know:

Definition of Collinear Points

Collinear points are points that lie on the same straight line.

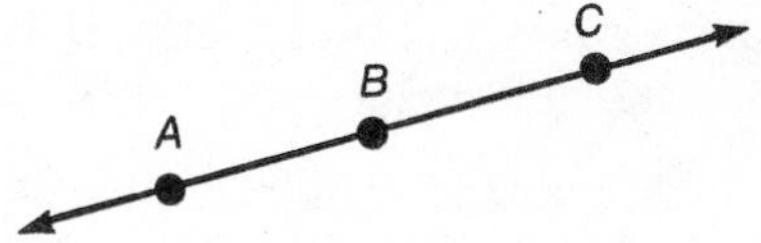

Points A, B, and C are collinear.

Definition of Betweenness of Points

A point P is **between** points A and B if (1) points A, P, and B are collinear and (2) $AP + PB = AB$.

8

3 5

A P B

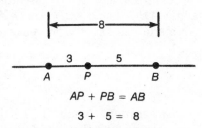

$AP + PB = AB$

$3 + 5 = 8$

Definitions of Congruent Angles and Congruent Line Segments

Congruent angles are angles that have the same degree measure. For example, if $m\angle A = 60$ (read as "measure of angle A is 60″) and $m\angle B = 60$, then $\angle A \cong \angle B$ (read as "angle A is congruent to angle B").

Congruent line segments are segments that have the same length. For example, if $AB = 10$ inches and $CD = 10$ inches, then $\overline{AB} \cong \overline{CD}$ (read as "line segment AB is congruent to line segment CD").

Definition of Midpoint

The **midpoint** of a line segment is the point on the line segment that divides the segment into two segments that have the same length. If point M is the midpoint of $\overline{AB}$, then each of the following statements is true:

1. $AM = MB$ or $\overline{AM} \cong \overline{MB}$.
2. $AM = \frac{1}{2}AB$.
3. $BM = \frac{1}{2}AB$.

A M B

Note: Matching vertical bars drawn through segments indicate that the segments are congruent.

Definition of Adjacent Angles

Adjacent angles are angles that have the same vertex, share a common side, and have no interior points in common (they do *not* overlap). In the accompanying diagram, angles 1 and 2 are adjacent.

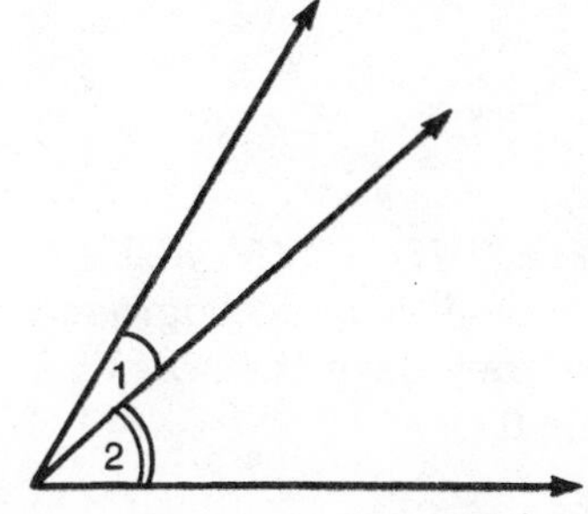

DEFINITIONS OF ANGLE BISECTOR AND SEGMENT BISECTOR

An **angle bisector** is a line (segment or ray) that divides an angle into two angles that have equal degree measures. In the accompanying diagram, $\overrightarrow{BD}$ is the bisector of angle ABC if m$\angle ABD$ = m$\angle CBD$. An angle has exactly one bisector.

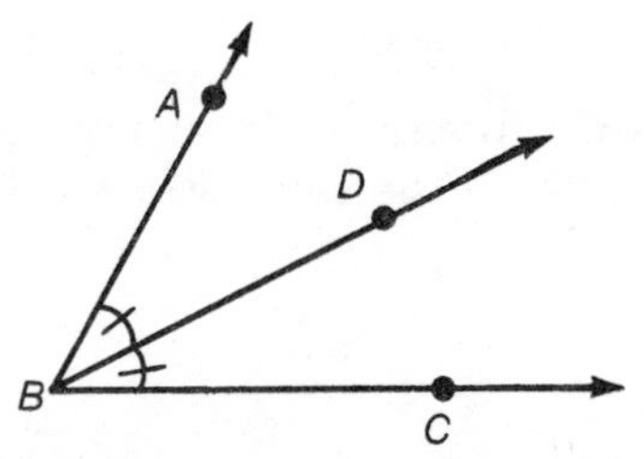

Note: Matching arcs, each with a dash through it, indicate that the angles have the same degree measure.

A **segment bisector** is a line (segment or ray) that divides a line segment into two segments that have the same length. In the accompanying diagram, $\overleftrightarrow{XM}$ and $\overleftrightarrow{YM}$ are bisectors if $AM = MB$. A line segment has an infinite number of bisectors.

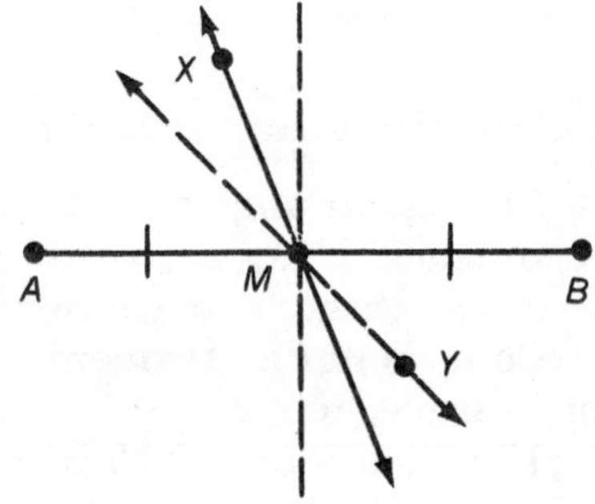

ANGLE DEFINITIONS

1. An **acute angle** is an angle whose measure is between 0 and 90.

2. A **right angle** is an angle whose measure is 90.

3. An **obtuse angle** is an angle whose measure is between 90 and 180.

4. A **straight angle** is an angle whose measure is 180. The accompanying diagram illustrates that the sides of a straight angle are opposite rays and therefore lie on a straight line.

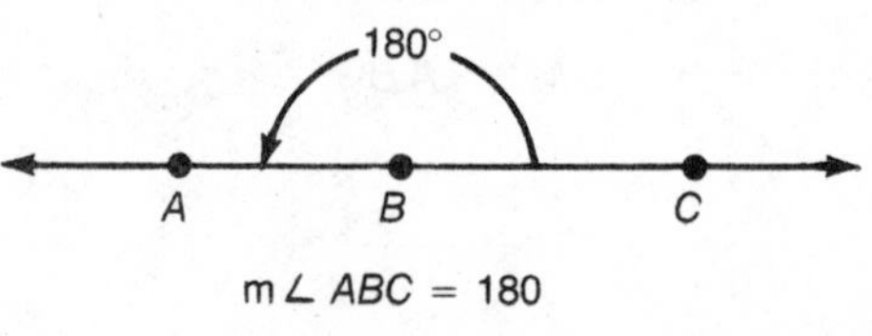

m$\angle ABC$ = 180

REFLEXIVE, SYMMETRIC, AND TRANSITIVE PROPERTIES. The equality and congruence symbols represent *equivalence relations* since they have the reflexive, symmetric, and transitive properties (see Table 6.1).

TABLE 6.1 Equality and Congruence as Equivalence Relations

Property	Relation Expressed in Words	Example
Reflexive	Any thing is equal (congruent) to itself.	$m\angle A = m\angle A$, $\overline{RS} \cong \overline{RS}$.
Symmetric	The members on either side of an equal sign (congruence symbol) may be interchanged.	If $m\angle A = m\angle B$, then $m\angle B = m\angle A$.
Transitive	If two quantities are equal (congruent) to the same quantity, then they are equal (congruent) to each other.	If $\overline{AB} \cong \overline{BC}$ and $\overline{XY} \cong \overline{BC}$, then $\overline{AB} \cong \overline{XY}$.

POSTULATES AND THEOREMS. A **theorem** is a generalization that can be proved to be true. Not everything can be proved, however, since there must be some basic assumptions, called *postulates*, that are needed as a beginning. A **postulate** (or axiom) is a statement that is accepted without proof.

A familiar theorem that was proved informally in Course I is as follows: *The sum of the measures of the angles of any triangle is 180.*

Here is our first postulate:

***Postulate 1*:** Exactly one line can be drawn between two given points. (Two points *determine* a line.)

PROOFS IN GEOMETRY. Geometric proofs are often presented using the statement-reason two-column format introduced in writing logic proofs. The "Reason" column of a geometric proof may contain only statements that fall into one of the following categories:

- The "Given" (the initial set of true premises).
- Postulates.
- Properties of equality and congruence.
- Definitions.
- Theorems that were previously proved.

Example

Given: $\angle A$ is a right angle,
$\angle B$ is a right angle.
Prove: $\angle A \cong B$.

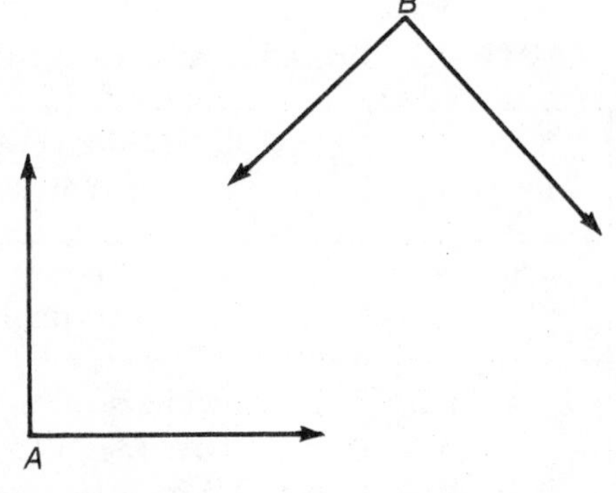

Solution: Use the two-column format.

PROOF

Statement	Reason
1. $\angle A$ is a right angle, $\angle B$ is a right angle.	1. Given
2. $m\angle A = 90$, $m\angle B = 90$	2. A right angle is an angle whose measure is 90.
3. $m\angle A = m\angle B$	3. Transitive property of equality.
4. $\angle A \cong \angle B$	4. If two angles are equal in measure. then they are congruent.

ANGLE ADDITION POSTULATE. Here is another postulate:

Postulate 2: A ray that lies in the interior of an angle divides the angle into two adjacent angles such that the measure of the angle is equal to the sum of the measures of the two adjacent angles.

For example, in the accompanying diagram if $\overrightarrow{OP}$ lies in the interior of $\angle AOB$, then

$m\angle 1 + m\angle 2 = m\angle AOB$.

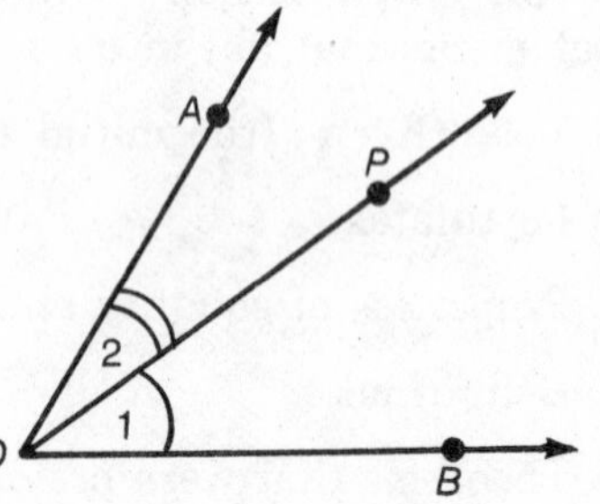

PERPENDICULAR LINES. Lines that meet at 90° angles are of special interest.

DEFINITION OF PERPENDICULAR LINES

Two lines are **perpendicular** if they intersect to form a right angle. If, in the accompanying diagram, line l is perpendicular to line m, we may write $l \perp m$, where the symbol "$\perp$" is read as "is perpendicular to." A square box drawn at the point of intersection indicates that the lines are perpendicular.

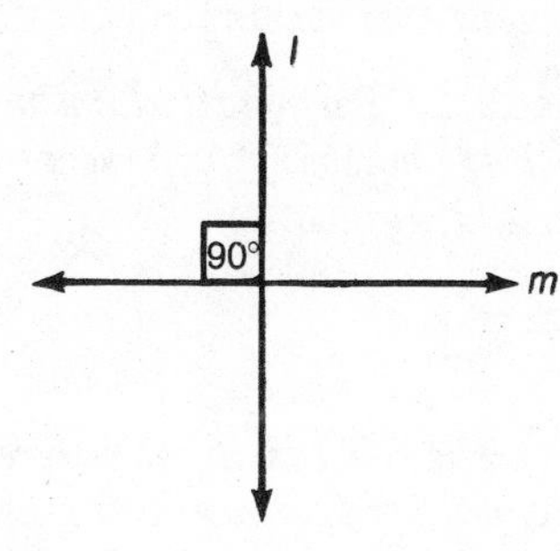

DEFINITION OF PERPENDICULAR BISECTOR

The **perpendicular bisector** of a segment is a line (or segment or ray) that is perpendicular to the segment at its midpoint. In the accompanying diagram, line l is the perpendicular bisector of $\overline{AB}$ since $l \perp \overline{AB}$ and $\overline{AM} \cong \overline{BM}$.

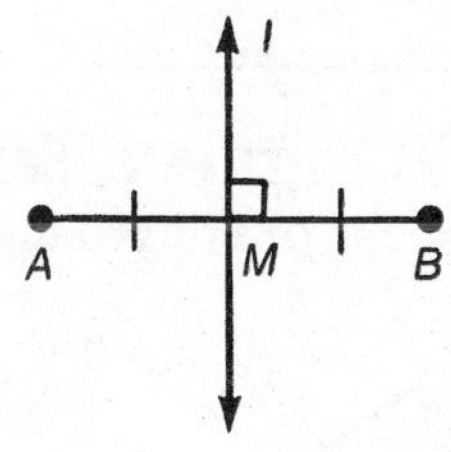

The following postulate guarantees the existence of perpendicular lines.

***Postulate 3*:** Given a point and a line in a plane, there is exactly one line that contains the point and is perpendicular to the original line (see the accompanying diagram).

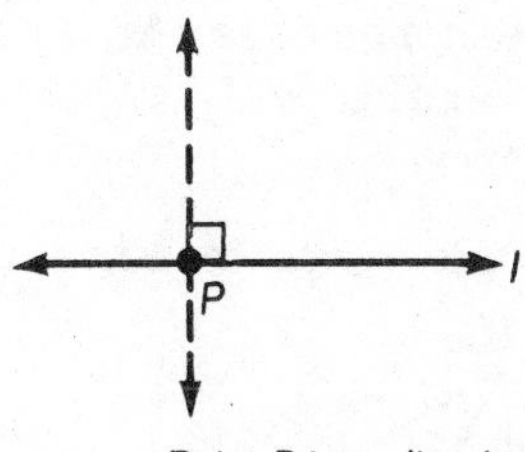

Point P is on line l.

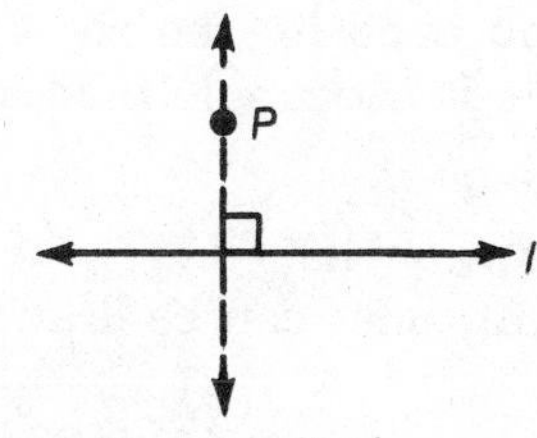

Point P is *not* on line l.

Here are some additional theorems about right angles and perpendicular lines that you should know:

THEOREMS INVOLVING RIGHT ANGLES AND PERPENDICULAR LINES

THEOREM 1: *All right angles are congruent* (see Example 1).

THEOREM 2: *Perpendicular lines (extended, if necessary) intersect to form four right angles.*

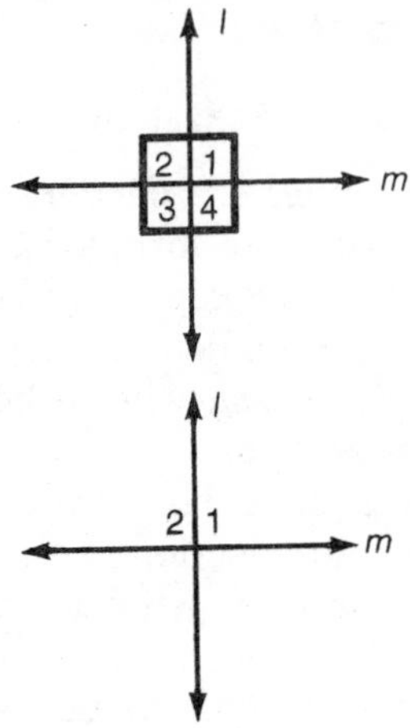

THEOREM 3: *If two lines intersect to form congruent adjacent angles, then the lines are perpendicular.* In the accompanying diagram, if $\angle 1 \cong \angle 2$, then $l \perp m$.

CLASSIFYING TRIANGLES. A triangle may be classified according to the number of congruent sides that it contains (see Figure 6.1).

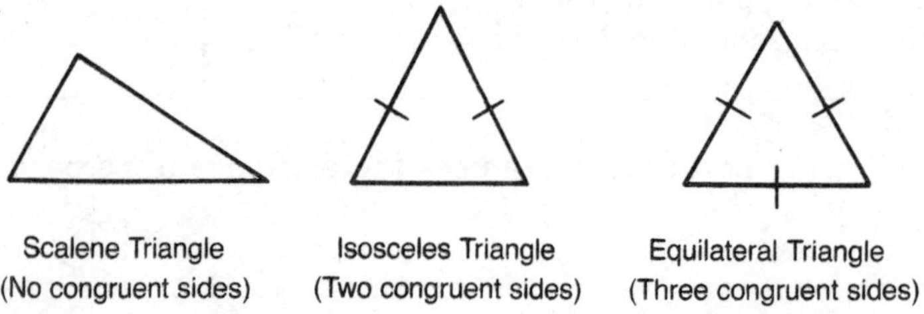

Figure 6.1 Triangles Classified by Number of Congruent Sides

A triangle may also be classified by the measure of its greatest angle. An **obtuse** triangle is a triangle that contains an obtuse angle. A **right** triangle is a triangle that contains a right angle, meaning that two sides of a right triangle, called *legs*, are perpendicular to each other; the side opposite the right angle is called the *hypotenuse*. An **acute** triangle is a triangle that contains three acute angles.

MEDIAN AND ALTITUDE. In every triangle three medians and three altitudes can be drawn.

DEFINITIONS OF MEDIAN AND ALTITUDE

A **median** of a triangle is a segment drawn from any vertex of a triangle to the midpoint of the opposite side.

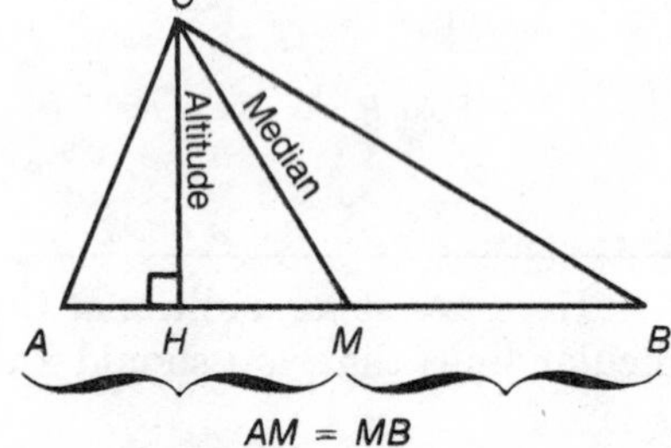

An **altitude** of a triangle is a segment drawn from any vertex of a triangle perpendicular to the opposite side or, as in the lower diagram, to the opposite side extended.

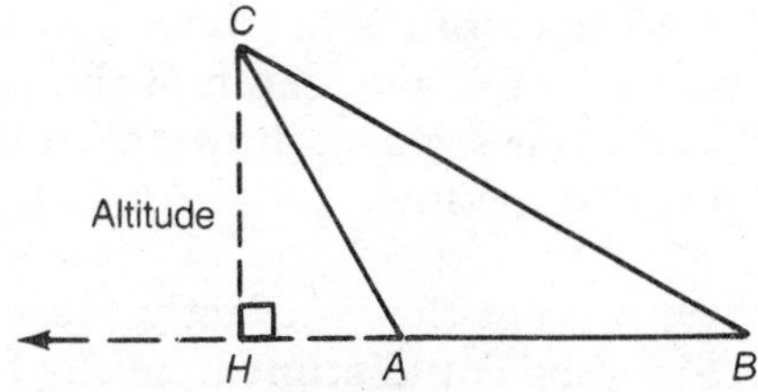

CONCURRENCY OF MEDIANS AND ALTITUDES. If two or more lines meet at the same point, these lines are said to be *concurrent* at this point. In every triangle, the three medians are concurrent at a point that lies inside the triangle (see Figure 6.2). Figure 6.3 illustrates that the three altitudes of a triangle are concurrent at a point in such a way that:

- If the triangle is *acute*, the altitudes meet at a point inside the triangle.

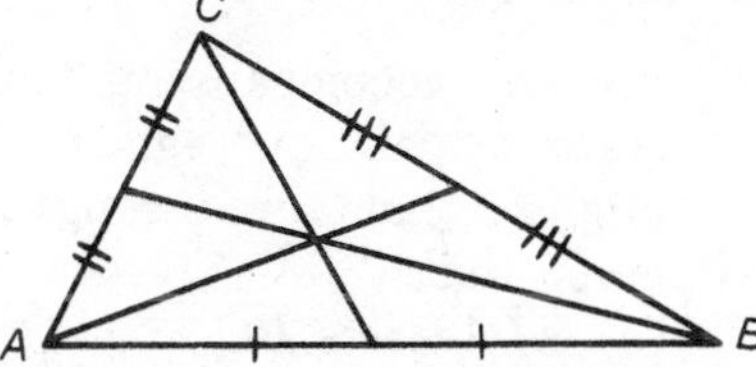

Figure 6.2 Concurrency of Medians

- If the triangle is a *right* triangle, the altitudes meet at the vertex of the right angle. (***Note*:** Two of the altitudes coincide with the legs of the right triangle.)

- If the triangle is *obtuse*, the altitudes (when extended) meet at a point outside the triangle.

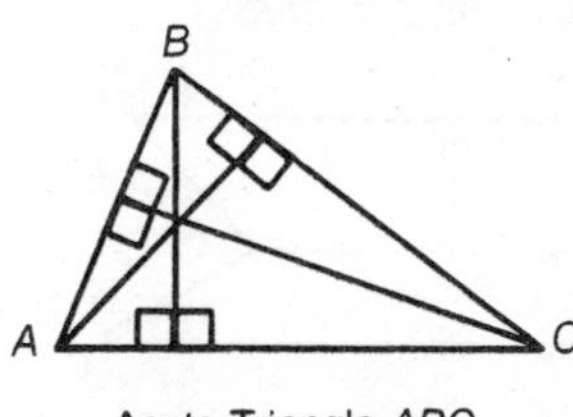

Acute Triangle *ABC*

A

C B

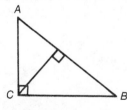

Right Triangle *ABC*

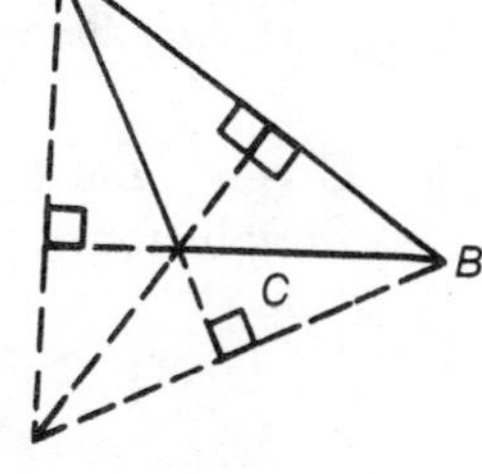

Obtuse Triangle *ABC*

Figure 6.3 Concurrency of Altitudes

DISTANCE. The term *distance* in geometry is always interpreted as the *shortest* path. In the figures accompanying the definition below, the distance between points P and Q is the length of $\overline{PQ}$, while the distance between point P and line l is the length of $\overline{PA}$, where $\overline{PA} \perp l$ at point A.

DEFINITIONS OF DISTANCE

The **distance between two points** is the length of the segment determined by the two points.

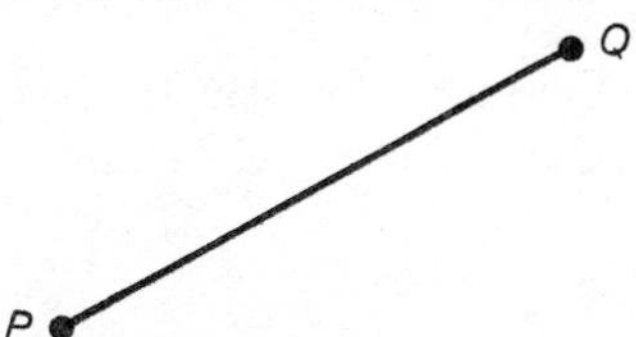

The **distance between a point and a line** is the length of the perpendicular segment drawn from the point to the line.

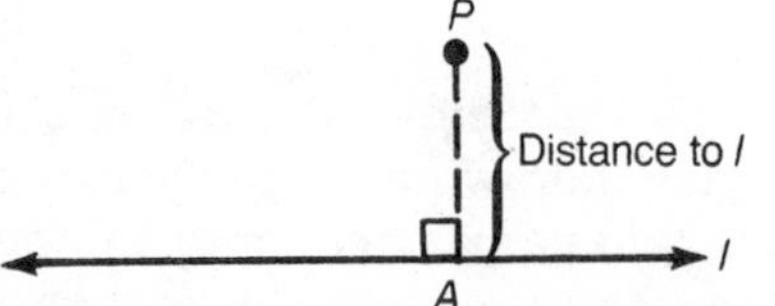

A point that is exactly the same distance from two other points is said to be **equidistant** from the two points. The midpoint of a segment, for example, is equidistant from the endpoints of the segment.

EXERCISE SET 6.1

1. In the accompanying diagram, point M is the midpoint of $\overline{RS}$. If $RM = 18$ and the length of $\overline{SM}$ is represented by $3x - 6$, find the value of x.

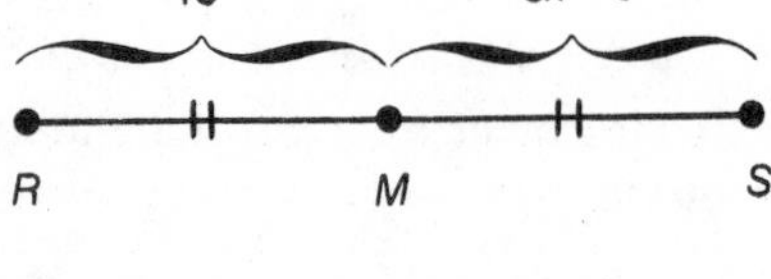

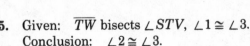

2. For the accompanying figure, state a pair of segments or angles that are congruent given that:
   (a) $\overline{AL}$ bisects $\overline{BC}$.
   (b) $\overline{BK}$ bisects $\angle ABC$.
   (c) $\overline{BK}$ bisects $\overline{AL}$.
   (d) AL bisects $\angle CAB$.

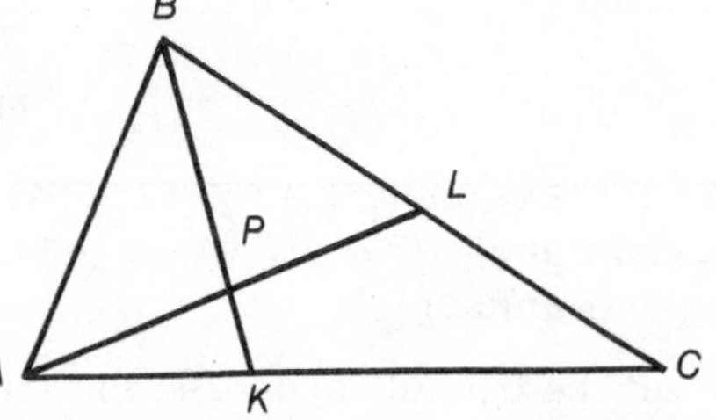

3–5. Justify the conclusion drawn in each case by identifying the property used to draw the conclusion as reflexive, transitive, or symmetric.

3. Given: $\overline{LM} \cong \overline{GH}$, $\overline{GH} \cong \overline{FV}$.
   Conclusion: $\overline{LM} \cong \overline{FV}$.

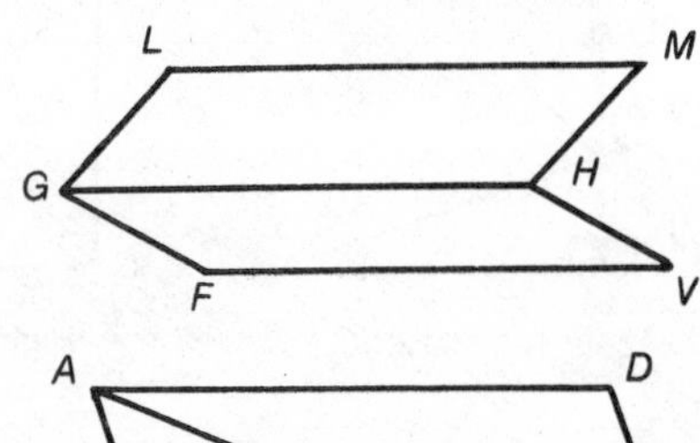

4. Given: Quadrilateral $ABCD$.
   Conclusion: $\overline{AC} \cong \overline{AC}$.

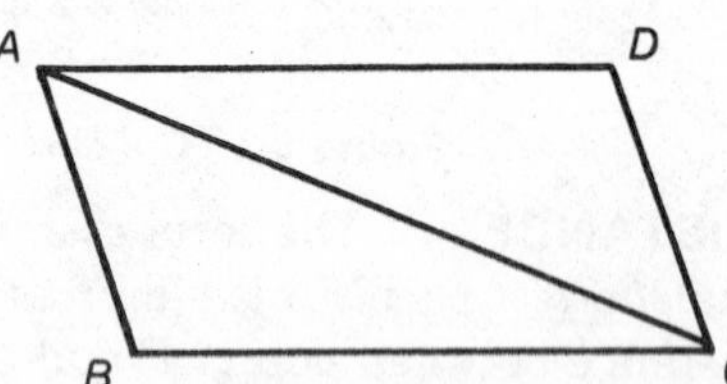

5. Given: $\overline{TW}$ bisects $\angle STV$, $\angle 1 \cong \angle 3$.
   Conclusion: $\angle 2 \cong \angle 3$.

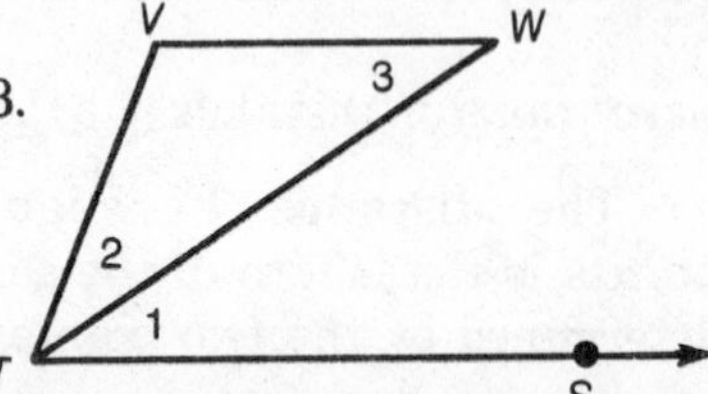

6.2 USING PROPERTIES OF EQUALITY IN PROOFS

KEY IDEAS

In solving equations the addition, subtraction, multiplication, and division *properties of equality* are used. The substitution principle (a quantity may be substituted for its equal in any expression) is used in checking the roots of equations. These properties may also be applied to the measures of angles and segments.

DRAWING CONCLUSIONS FROM DIAGRAMS. Many proofs provide the "Given" (what you may assume to be true), the "Prove" (the conclusion that you have to demonstrate is true), and a diagram. In general, when looking at a diagram you may assume only "obvious" properties such as betweenness of points and collinearity of points, and that angles drawn as adjacent are, in fact, adjacent. Generally speaking, you may *not* assume just from looking at a diagram that segments or angles are congruent, or that lines are parallel or perpendicular.

APPLYING THE ADDITION AND SUBTRACTION PROPERTIES OF EQUALITY. The examples that follow use the property that, if equals are added to (or subtracted from) equals, then the results are equal.

Examples

1. Given: $AD = FC$.
Prove: $AC = FD$.

Solution:

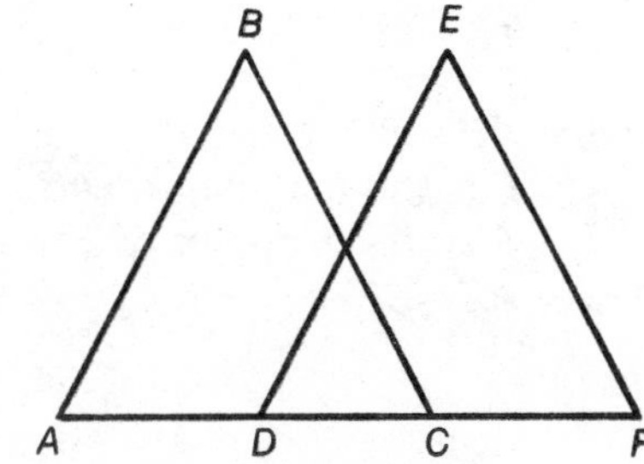

PROOF

Statement	Reason
1. $AD = FC$	1. Given.
2. $DC = DC$	2. Reflexive property of equality.
3. $AD + DC = FC + DC$	3. Addition property of equality.
4. $AC = FD$	4. Substitution principle.

2. Given: $\overline{WX} \perp \overline{XY}$, $\overline{ZY} \perp \overline{XY}$,
m$\angle WXH$ = m$\angle ZYH$.
Prove: m$\angle 1$ = m$\angle 2$.

Solution:

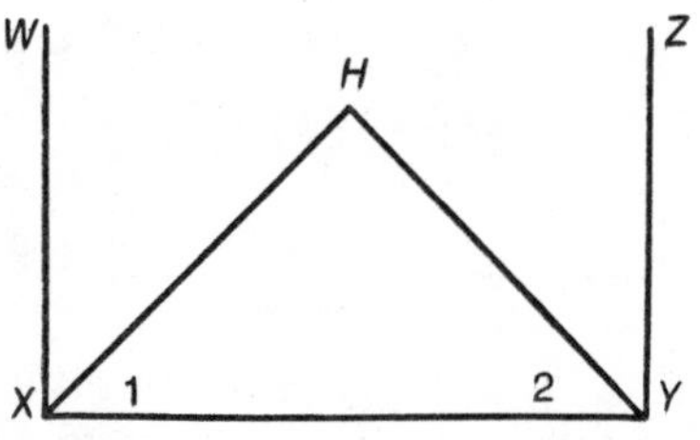

PROOF

Statement	Reason
1. $\overline{WX} \perp \overline{XY}$, $\overline{ZY} \perp \overline{XY}$	1. Given.
2. $\angle WXY$ is a right angle, $\angle ZYX$ is a right angle.	2. Perpendicular lines intersect to form right angles.
3. m$\angle WXY$ = m$\angle ZYX$	3. All right angles are equal in measure.
4. m$\angle WXH$ = m$\angle ZYH$	4. Given.
5. m$\angle WXY$ − m$\angle WXH$ = m$\angle ZYX$ − m$\angle ZYH$	5. Subtraction property.
6. m$\angle 1$ = m$\angle 2$	6. Substitution principle.

APPLYING THE MULTIPLICATION PROPERTY OF EQUALITY. If each side of an equation is multiplied (or divided) by the same nonzero quantity, then the results are equal. In the special case when the multiplying factor is $\frac{1}{2}$, we say that "halves of equals are equal."

Example

3. Given: $AB = BC$,
E is the midpoint of $\overline{AB}$,
F is the midpoint of $\overline{BC}$.
Prove: $AE = CF$.

Solution:

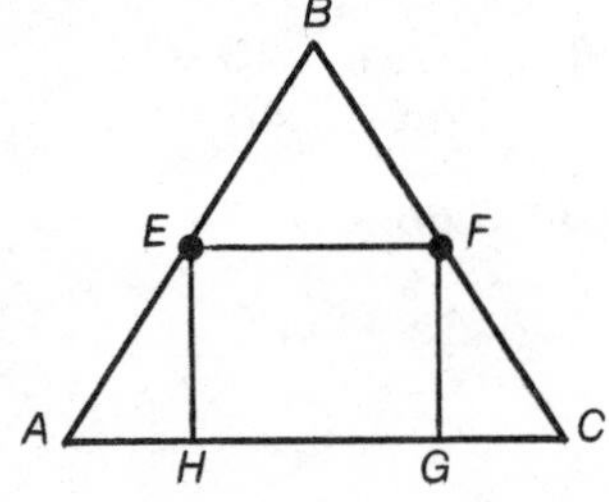

PROOF

Statement	Reason
1. $AB = BC$	1. Given.
2. E is the midpoint of $\overline{AB}$, F is the midpoint of $\overline{BC}$.	2. Given
3. $AE = \frac{1}{2}AB$ and $CF = \frac{1}{2}BC$	3. Definition of midpoint.
4. $AE = CF$	4. Halves of equals are equal.

EXERCISE SET 6.2

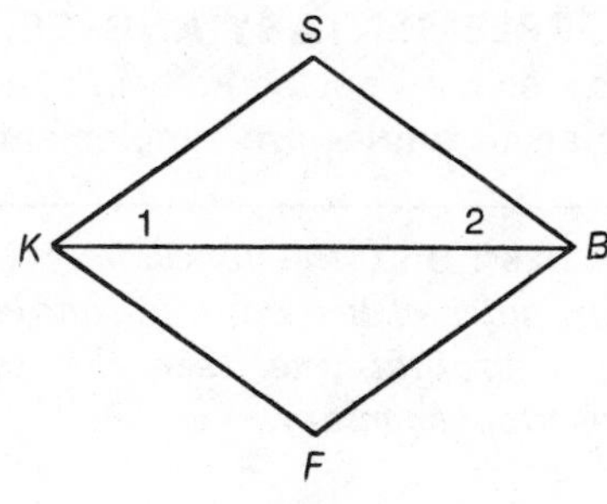

1. Given: $\overline{KB}$ bisects $\angle SBF$,
   $\overline{KB}$ bisects $\angle SKF$,
   $m\angle SKF = m\angle SBF$.
   Prove: $\angle 1 \cong \angle 2$.

2. Given: $m\angle TOB = m\angle WOM$.
   Prove: $m\angle TOM = m\angle WOB$.

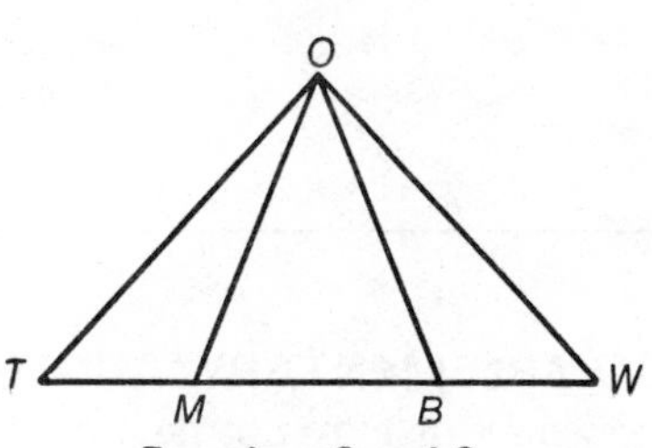

Exercises 2 and 3

3. Given: $\overline{TB} \cong \overline{WM}$.
   Prove: $\overline{TM} \cong \overline{WB}$.

4. Given: $BD = BE$, $DA = EC$.
   Prove: $AB = CB$.

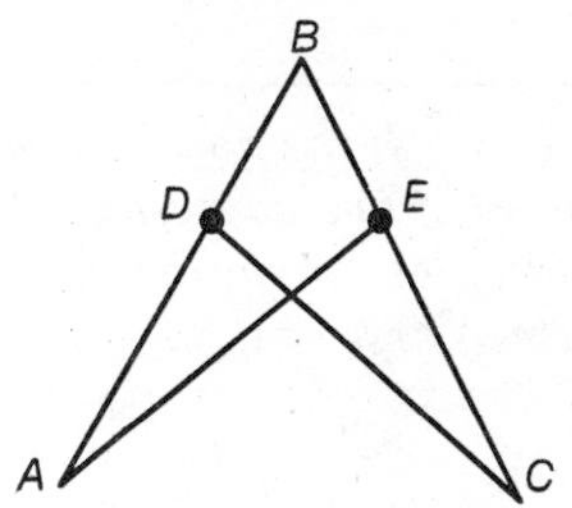

5. Given: $\angle WXY$ is a right angle,
   $\angle ZYX$ is a right angle,
   $m\angle 1 = m\angle 2$.
   Prove: $m\angle 3 = m\angle 4$.

6. Given: $m\angle 1 = m\angle 2$,
   $m\angle 3 = m\angle 4$
   $m\angle H = m\angle WXY$.
   Prove: $m\angle H = m\angle ZYH$.

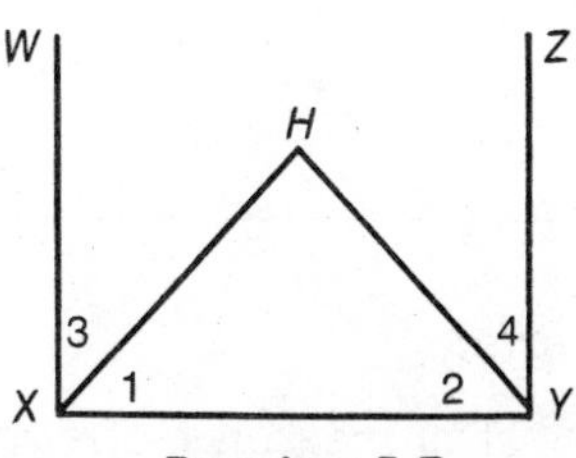

Exercises 5–7

7. Given: $\angle WXY$ is a right angle,
   $\angle ZYX$ is a right angle,
   $\overline{XH}$ bisects $\angle WXY$,
   $\overline{YH}$ bisects $\angle ZYX$.
   Prove: $m\angle 1 = m\angle 2$.

6.3 SPECIAL PAIRS OF ANGLES

KEY IDEAS

Two angles are **supplementary** if the sum of their measures is 180, and are **complementary** if the sum of their measures is 90.

The pairs of nonadjacent angles formed when two lines intersect are called *vertical angles* and are congruent.

SUPPLEMENTARY ANGLES. Two angles that are supplementary may or may not be adjacent angles. Theorem 1 tells us when a pair of adjacent angles are supplementary.

THEOREM 1: *If the noncommon (exterior) sides of two adjacent angles lie on a straight line, then the angles are supplementary.*

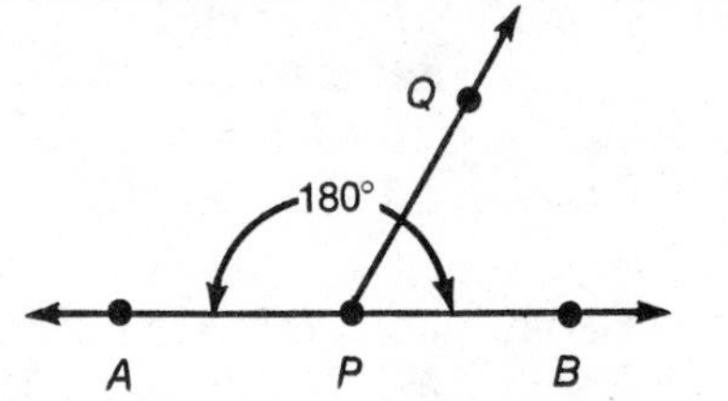

Angles *APQ* and *BPQ* are supplementary.

COMPLEMENTARY ANGLES. Two angles that are complementary may or may not be adjacent angles. Theorem 2 tells us when a pair of adjacent angles are complementary.

THEOREM 2: *If the noncommon (exterior) sides of two adjacent angles are perpendicular, then the angles are complementary.*

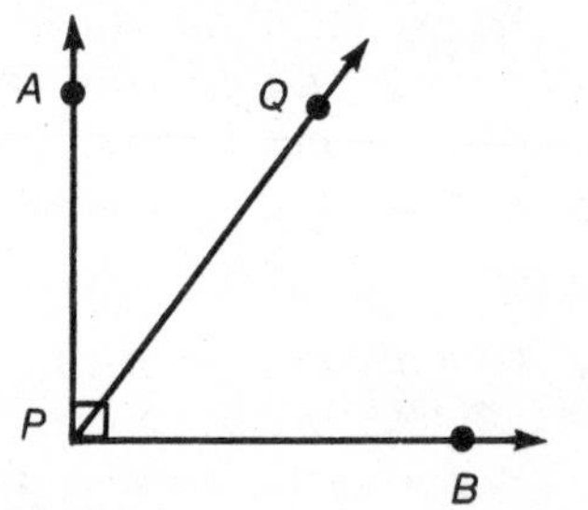

Angles *APQ* and *BPQ* are complementary.

Examples

1. The measures of two complementary angles are in the ratio of 2:13. Find the measure of the smaller angle.

Solution: Let $2x$ = measure of the smaller angle.
Then $13x$ = measure of the larger angle.

$$2x + 13x = 90$$
$$15x = 90$$
$$\frac{15x}{15} = \frac{90}{15}$$
$$x = 6$$
$$2x = 2(6) = 12.$$

The measure of the smaller angle is **12**.

2. If the measure of an angle exceeds twice its supplement by 30, find the measure of the angle.

Solution: Let x = measure of the angle.
Then $180 - x$ = measure of the angle's supplement.

$$x = 2(180 - x) + 30$$
$$x = 360 - 2x + 30$$
$$x + 2x = 390$$
$$\frac{3x}{3} = \frac{390}{3}$$
$$x = 130$$

MORE THEOREMS ON SUPPLEMENTARY AND COMPLEMENTARY ANGLES. Suppose $m\angle C = 50$. If angles A and B are both complementary to $\angle C$, then each must have a measure of 40 and, therefore, must be congruent to the other. This observation suggests the following two theorems.

THEOREM 3: *If two angles are* complementary *to the* same *angle (or to congruent angles), then they are congruent.*

THEOREM 4: *If two angles are* supplementary *to the* same *angle (or to congruent angles), then they are congruent.*

VERTICAL ANGLES. **Vertical angles** are pairs of nonadjacent (opposite) angles formed when two lines intersect. In Figure 6.4, angles 1 and 3 are vertical angles and, by Theorem 4, are congruent since both are supplementary to $\angle 2$ (also $\angle 4$). Similarly, angles 2 and 4 are vertical angles and are congruent since both are supplementary to the same angle ($\angle 1$ and also $\angle 3$). This analysis provides an informal proof of Theorem 5.

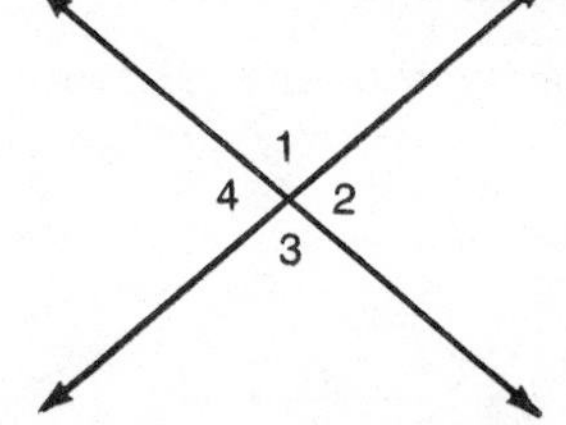

Figure 6.4 Vertical Angles

THEOREM 5: *Vertical angles are congruent and, therefore, are equal in measure.*

Examples

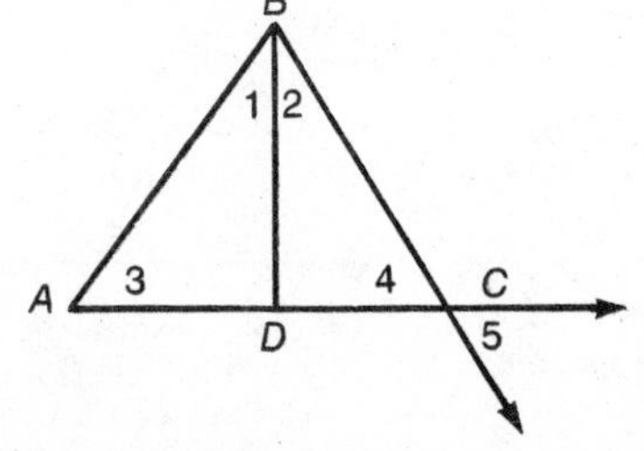

3. Given: $\overline{BD}$ bisects $\angle ABC$,
$\angle 3$ is complementary to $\angle 1$,
$\angle 4$ is complementary to $\angle 2$.
Prove: $\angle 3 \cong \angle 5$.

Solution:

PROOF

Statement	Reason
1. $\overline{BD}$ bisects $\angle ABC$.	1. Given.
2. $\angle 1 \cong \angle 2$	2. Definition of angle bisector.
3. $\angle 3$ is complementary to $\angle 1$, $\angle 4$ is complementary to $\angle 2$.	3. Given.
4. $\angle 3 \cong \angle 4$	4. If two angles are complementary to congruent angles, then they are congruent.
5. $\angle 4 \cong \angle 5$	5. Vertical angles are congruent.
6. $\angle 3 \cong \angle 5$	6. Transitive property of congruence.

4. In the accompanying diagram, find the value of y.

Solution: Since vertical angles are equal in measure,

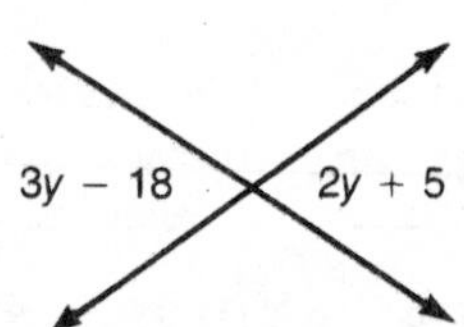

$$3y - 18 = 2y + 5$$
$$3y = 2y + 5 + 18$$
$$3y - 2y = 23$$
$$y = \mathbf{23}$$

EXERCISE SET 6.3

1. For each of the following, find the value of x:

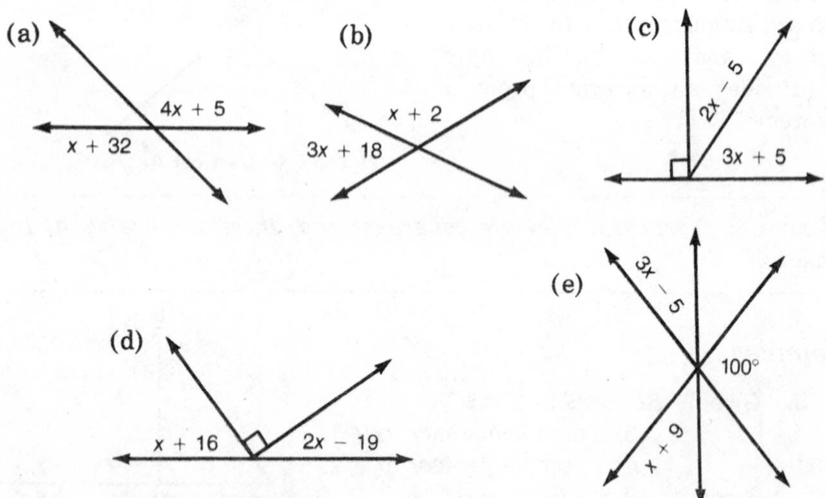

2. If two angles are supplementary, find the measure of the smaller angle if the measures of the two angles are in the ratio of:
(a) 1:8 (b) 3:5 (c) 3:1 (d) 5:7 (e) 7:11

3. If two angles are complementary, find the measure of the smaller angle if the measures of the two angles are in the ratio of:
(a) 2:3 (b) 1:2 (c) 4:5 (d) 3:1 (e) 1:5

4. The measure of an angle exceeds three times its supplement by 4. Find the measure of the angle.

5. The measure of an angle exceeds four times the measure of its complement by 6. Find the measure of the angle.

6. The measure of an angle is 22 less than three times the measure of the complement of the angle. Find the measure of the angle.

7. The measure of the supplement of an angle is three times as great as the measure of the angle's complement. What is the measure of the angle?

8. Find the measure of an angle if it is 12 less than twice the measure of its complement.

9. The difference between the measures of an angle and its complement is 14. Find the measure of the smaller of the two angles.

10. The difference between the measures of an angle and its supplement is 22. Find the measure of the smaller of the two angles.

11. The measure of the supplement of an angle is three times as great as the measure of the complement of the same angle. What is the measure of the angle?

12–14. Lines AB *and* CD *intersect at point* H. *Fill in the missing values in the accompanying table, solving for* x *where necessary*:

	$m\angle AHC$	$m\angle AHD$	$m\angle DHB$	$m\angle CHB$
12.	42	?	?	?
13.	?	$3x - 7$	?	$2x + 17$
14.	$5x - 2$	$2x + 21$	?	?

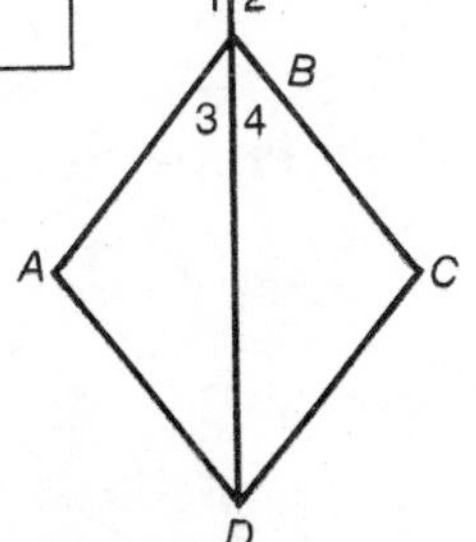

15. Given: $\overline{BD}$ bisects $\angle ABC$.
Prove: $\angle 1 \cong \angle 2$.

16. Given: $\angle 3$ is complementary to $\angle 1$, $\angle 4$ is complementary to $\angle 2$.
Prove: $\angle 3 \cong \angle 4$.

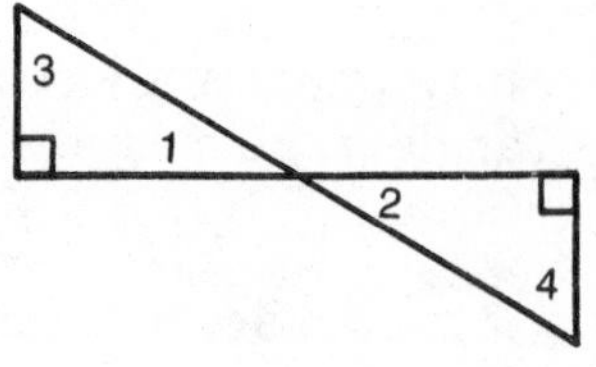

17. Given: $\overline{AB} \perp \overline{BD}$, $\overline{CD} \perp \overline{BD}$, $\angle 1$ is complementary to $\angle 3$, $\overline{BE}$ bisects $\angle ABD$.
Prove: $\angle 2 \cong \angle 4$.

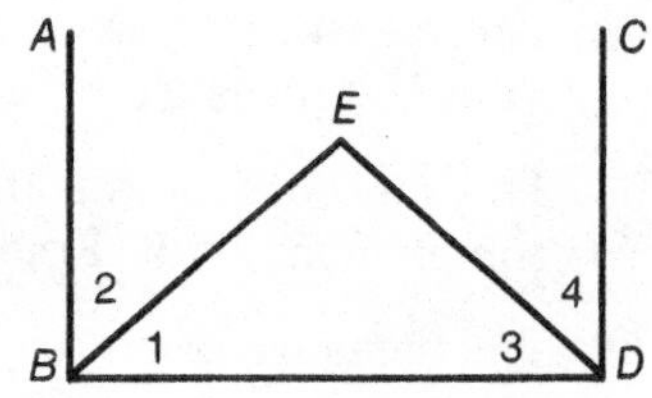

18. Given: $\overline{KL} \perp \overline{JM}$, $\overline{KL}$ bisects $\angle PLQ$.
Prove: $\angle 1 \cong \angle 4$.

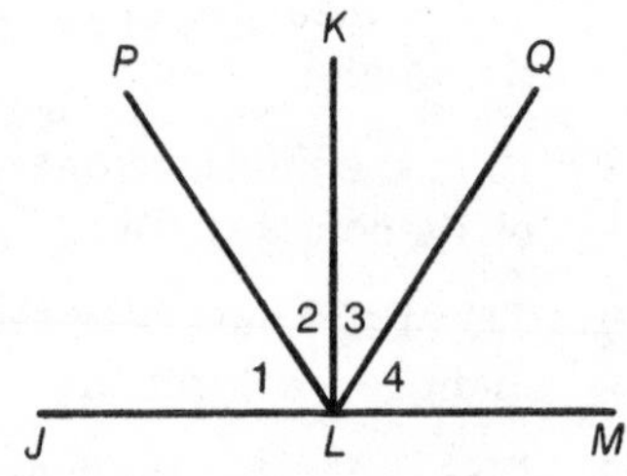

19. Given: $\overline{NW} \perp \overline{WT}$, $\overline{WB} \perp \overline{NT}$, $\angle 4 \cong \angle 6$.
Prove: $\angle 2 \cong \angle 5$.

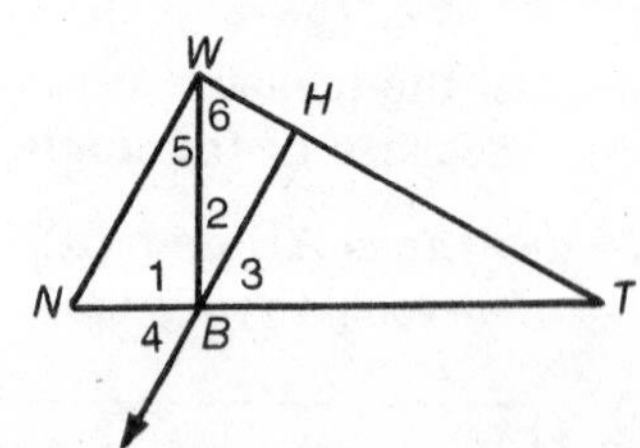

20. Given: $\overline{MT}$ bisects $\angle ETI$, $\overline{KI} \perp \overline{TI}$, $\overline{KE} \perp \overline{TE}$, $\angle 3 \cong \angle 1$, $\angle 5 \cong \angle 2$.
Prove: $\angle 4 \cong 6$.

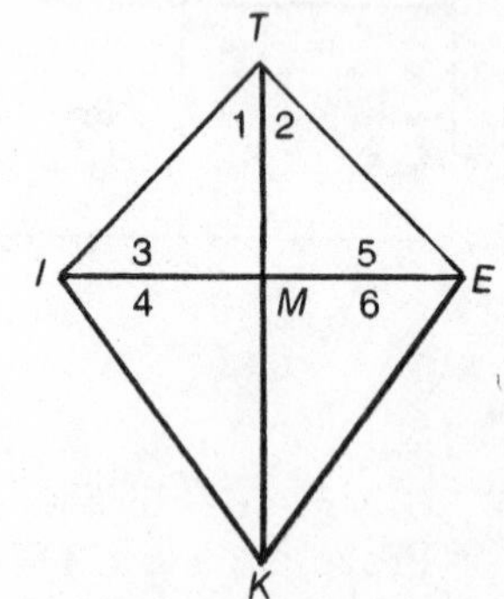

6.4 PROVING TRIANGLES CONGRUENT

KEY IDEAS

The size and shape of a triangle are determined by the measures of its six parts (three angles and three sides). In general, two polygons having the same number of sides are congruent if their vertices can be paired off so that corresponding angles are congruent and corresponding sides are congruent. If the polygons are triangles, then it is sufficient to show that only *three* pairs of parts are congruent, provided that they are a particular set of congruent parts.

CONGRUENT TRIANGLES. If, in Figure 6.5, $\triangle ABC \cong \triangle RST$, then the vertices of the two triangles can be paired off so that three pairs of corresponding angles are congruent and three pairs of corresponding sides are congruent.

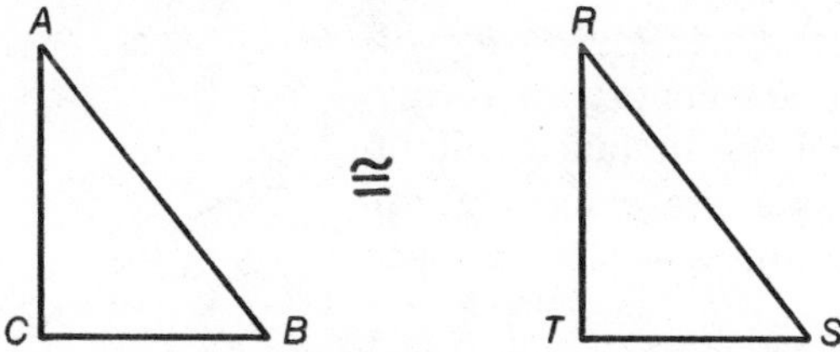

Figure 6.5 Congruent Triangles

Corresponding Angles	Corresponding Sides
$\angle A \cong \angle R$ $\angle B \cong \angle S$ $\angle C \cong \angle T$	$\overline{AB} \cong \overline{RS}$ $\overline{BC} \cong \overline{ST}$ $\overline{AC} \cong \overline{RT}$

Note that:

- Corresponding sides lie opposite corresponding angles.
- In naming pairs of congruent triangles, corresponding vertices are written in the same order.

$$\triangle ABC \cong \triangle RST$$

Here are some other ways in which this correspondence may be written:

$$\triangle CAB \cong \triangle TRS \qquad \triangle BAC \cong \triangle SRT \qquad \triangle CBA \cong \triangle TSR.$$

- Congruence of triangles satisfies the *reflexive*, *symmetric*, and *transitive* properties (see page 131).

POSTULATES FOR PROVING TRIANGLES CONGRUENT. Here are some ways of proving that triangles are congruent:

- ***SSS ≅ SSS Postulate*:** Two triangles are congruent if the three sides of the first triangle are congruent to the corresponding parts of the second triangle.

- ***SAS ≅ SAS Postulate*:** Two triangles are congruent if any two sides and the included angle of the first triangle are congruent to the corresponding parts of the second triangle.

- ***ASA ≅ ASA Postulate*:** Two triangles are congruent if any two angles and the included side of the first triangle are congruent to the corresponding parts of the second triangle.

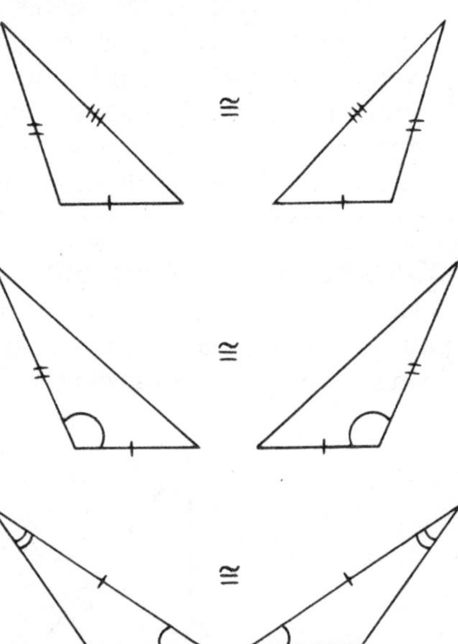

A STRATEGY FOR PROVING TRIANGLES CONGRUENT. To prove that a pair of triangles are congruent, proceed as follows:

1. Mark off the diagram with:
   - the "Given";
   - common sides or angles, if any, that are parts of both triangles;
   - congruent pairs of angles, if any, resulting from vertical angles or perpendicular lines.
2. Select the congruence method (for example, SSS, SAS, or ASA).
3. Write the formal proof.

Examples

1. Given: $\overline{AB} \cong \overline{BC}$,
   $\overline{BD}$ bisects $\angle ABC$.
   Prove: $\triangle ADB \cong \triangle CDB$.

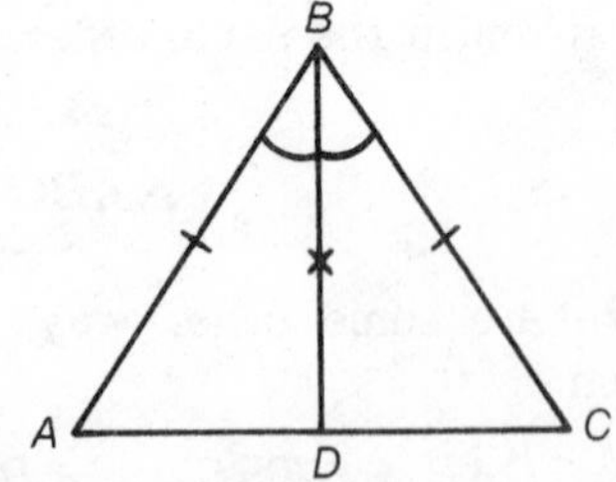

Solution:

PLAN. Since $\overline{AB} \cong \overline{BC}$, $\angle ABD \cong \angle CBD$, and $\overline{BD}$ is a side of both triangles (note that on the diagram " × " is used to indicate that $\overline{BD}$ is congruent to itself), prove the triangles congruent by using the SAS postulate.

PROOF

Statement		Reason
1. $\overline{AB} \cong \overline{BC}$	*Side*	1. Given.
2. $\overline{BD}$ bisects $\angle ABC$.		2. Given.
3. $\angle ABD \cong \angle CBD$	*Angle*	3. A bisector divides an angle into two congruent angles.
4. $\overline{BD} \cong \overline{BD}$	*Side*	4. Reflexive property of congruence.
5. $\triangle ADB \cong \triangle CDB$		5. SAS postulate.

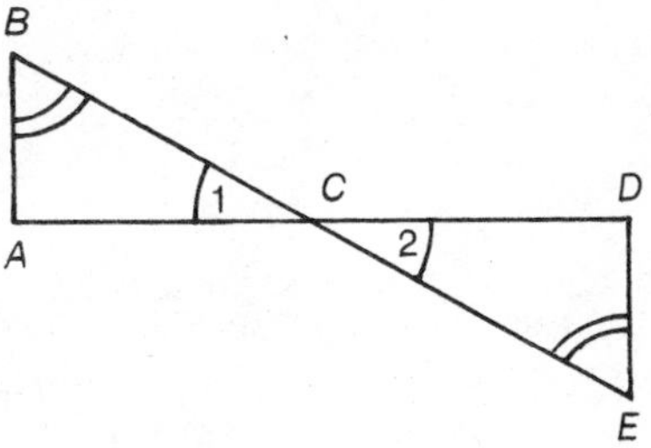

2. Given: $\overline{AD}$ bisects $\overline{BE}$, $\angle B \cong \angle E$.
Prove: $\triangle ABC \cong \triangle DEC$.

Solution:

PLAN. Since $\angle B \cong \angle E$, $\overline{BC} \cong \overline{EC}$, and $\angle 1 \cong \angle 2$ (vertical angles), prove the triangles congruent by using the ASA postulate.

PROOF

Statement		Reason
1. $\angle B \cong \angle E$	*Angle*	1. Given.
2. $\overline{AD}$ bisects $\overline{BE}$.		2. Given.
3. $\overline{BC} \cong \overline{EC}$	*Side*	3. A bisector divides a segment into two congruent segments.
4. $\angle 1 \cong \angle 2$	*Angle*	4. Vertical angles are congruent.
5. $\triangle ABC \cong \triangle DEC$		5. ASA postulate.

PROVING OVERLAPPING TRIANGLES CONGRUENT. Sometimes triangles are drawn so that one triangle overlaps the other. When this happens, examine the diagram closely in order to determine whether the same angle or the same side is contained in both triangles.

Example

3. Given: $\overline{MK} \perp \overline{JL}$, $\overline{LP} \perp \overline{JM}$, $\overline{JK} \cong \overline{JP}$.
   Prove: $\triangle JMK \cong \triangle JLP$.

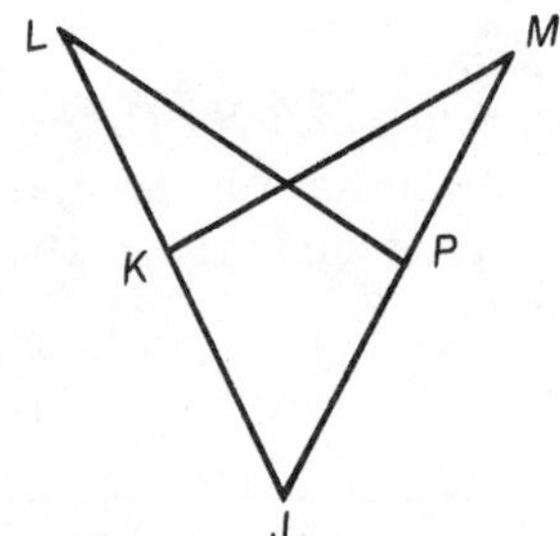

Solution: To help distinguish between the parts of the two triangles, outline the sides of one triangle with a thick, bold line (or, if handy, use a colored pencil). Notice that each triangle in the diagram has angle J as one of its angles.

PLAN. Since $\angle J \cong \angle J$ (*Angle*), $\overline{JK} \cong \overline{JP}$ (*Side*), and $\angle 1 \cong \angle 2$ (Angle), prove the triangles congruent by using the ASA postulate.

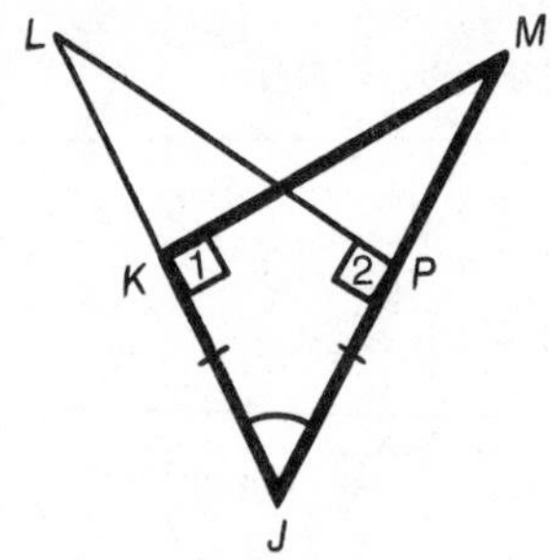

PROOF

Statement		Reason
1. $\overline{MK} \perp \overline{JL}$, $\overline{LP} \perp \overline{JM}$		1. Given.
2. Angles 1 and 2 are right angles.		2. Perpendicular lines intersect to form right angles.
3. $\angle 1 \cong \angle 2$	*Angle*	3. All right angles are congruent.
4. $\overline{JK} \cong \overline{JP}$	*Side*	4. Given.
5. $\angle J \cong \angle J$	*Angle*	5. Reflexive property of congruence.
6. $\triangle JMK \cong \triangle JLP$		6. ASA postulate.

EXERCISE SET 6.4

1. Given: C is the midpoint of $\overline{AD}$, C is the midpoint of $\overline{BE}$.
Prove: $\triangle ABC \cong \triangle DEC$.

2. Given: $\overline{AB} \perp \overline{BE}$, $\overline{ED} \perp \overline{BE}$, $\overline{AD}$ bisects $\overline{BE}$.
Prove: $\triangle ABC \cong \triangle DEC$.

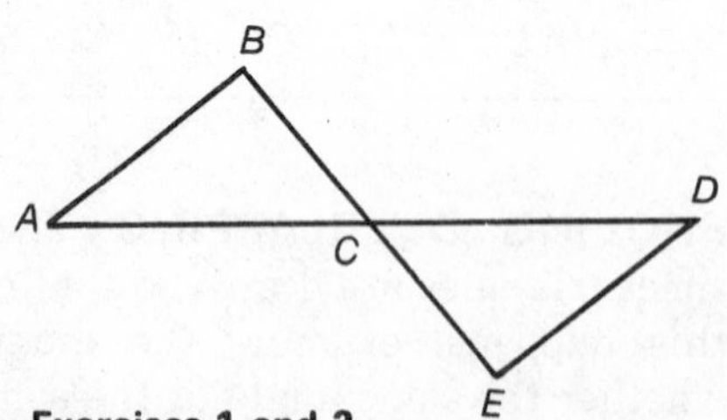

Exercises 1 and 2

3. Given: $\overline{BD}$ is an altitude to side $\overline{AC}$, $\overline{AE}$ is an altitude to side $\overline{BC}$, $\overline{CE} \cong \overline{CD}$.
   Prove: $\triangle AEC \cong \triangle BDC$.

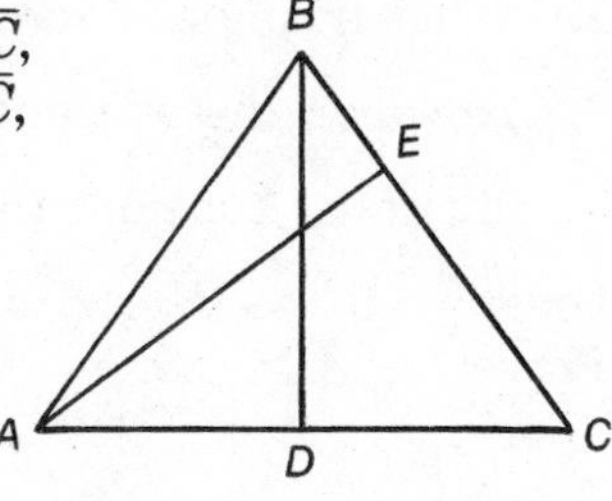

4. Given: $\overline{AB} \cong \overline{BC}$, $\angle A \cong \angle C$.
   Prove: $\triangle AEB \cong \triangle CDB$.

5. Given: $\overline{AE} \perp \overline{BC}$, $\overline{CD} \perp \overline{AB}$, $\overline{BD} \cong \overline{BE}$.
   Prove: $\triangle ABE \cong \triangle CBD$.

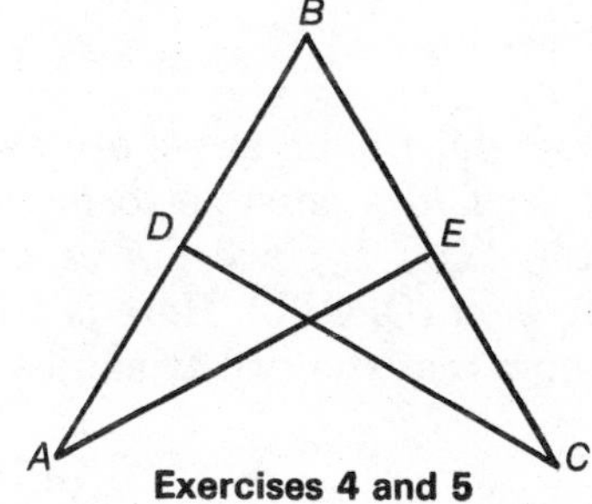

Exercises 4 and 5

6. Given: $\angle ABC \cong \angle DCB$, $\angle 1 \cong \angle 2$, $\overline{BE} \cong \overline{CE}$.
   Prove: $\triangle ABE \cong \triangle DCE$.

7. Given: $\overline{AB} \perp \overline{BE}$, $\overline{DC} \perp \overline{CE}$, $\angle 1 \cong \angle 2$.
   Prove: $\triangle ABC \cong \triangle DCB$.

8. Given: $\overline{DE} \cong \overline{AE}$, $\overline{BE} \cong \overline{CE}$, $\angle 1 \cong \angle 2$.
   Prove: $\triangle DBC \cong \triangle ACB$.

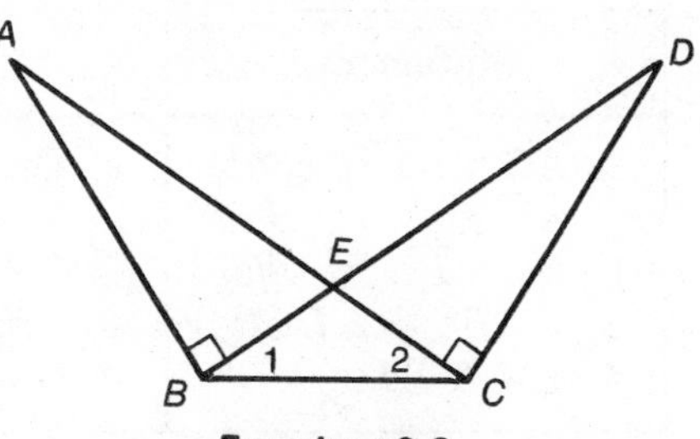

Exercises 6–8

6.5 USING CONGRUENT TRIANGLES TO PROVE ANGLES AND SEGMENTS CONGRUENT AND LINES PERPENDICULAR; AAS AND HY-LEG THEOREMS

KEY IDEA

> If it can be demonstrated that two triangles are congruent, then the definition of congruent triangles permits us to conclude that any pair of corresponding angles or pair of corresponding sides are congruent. Congruent triangles can also be used to prove that two lines are perpendicular.

PROVING SEGMENTS AND ANGLES CONGRUENT. To prove that a pair of segments or angles are congruent using congruent triangles, proceed by answering these questions:

1. What triangles contain these segments or angles as parts?
2. How can these triangles be proved congruent?

To illustrate, consider the following problem:

Given: $\overline{AC} \cong \overline{DB}$, $\overline{AB} \cong \overline{DC}$.
Prove: $\angle A \cong \angle D$.

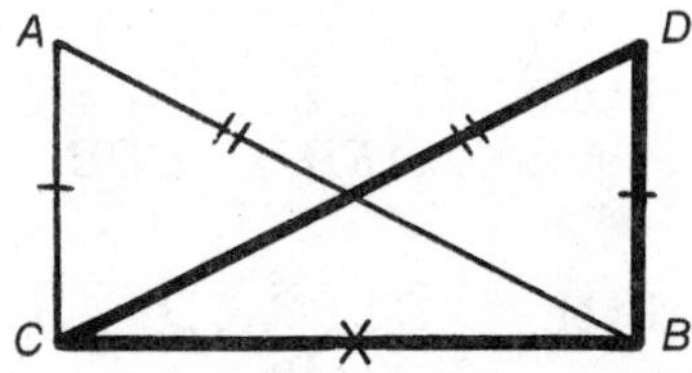

Angle A is an angle of $\triangle ABC$, while $\triangle DCB$ has $\angle D$ as one of its angles. If it is possible to prove that $\triangle ABC \cong \triangle DCB$, then it follows that $\angle A \cong \angle D$ since *Corresponding Parts of Congruent Triangles Are Congruent* (CPCTC). Here is a formal proof that uses the SSS postulate to show that the two triangles are congruent.

PROOF

Statement	Reason
1. $\overline{AC} \cong \overline{DB}$, *Side* $\overline{AB} \cong \overline{DC}$ *Side*	1. Given.
2. $\overline{BC} \cong \overline{BC}$ *Side*	2. Reflexive property of congruence.
3. $\triangle ABC \cong \triangle DCB$	3. SSS postulate.
4. $\angle A \cong \angle D$	4. CPCTC.

Example

1. Prove: If a point lies on the perpendicular bisector of a line segment, then it is equidistant from the endpoints of the line segment.

Solution: The "Given" is taken from the *if* clause of the statement to be proved, while the "Prove" is taken from the *then* clause.

Given: Line $l \perp \overline{AB}$ at M,
$\overline{AM} \cong \overline{BM}$,
point P is any point on l,
$\overline{AP}$ and $\overline{BP}$ are drawn.
Prove: $AP = BP$.

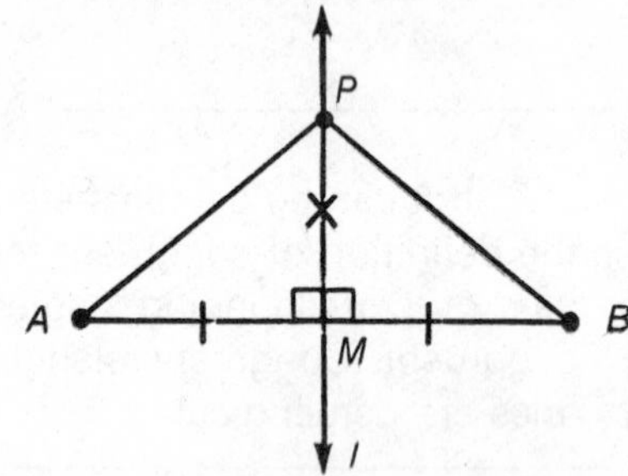

PLAN. Show that $\triangle AMP \cong \triangle BMP$ by SAS. By CPCTC, $\overline{AP} \cong \overline{BP}$. Therefore, $AP = BP$.

The formal proof is left for you to complete.

ADDITIONAL METHODS FOR PROVING TRIANGLES CONGRUENT. Here are two theorems that can be used to prove that a pair of triangles are congruent.

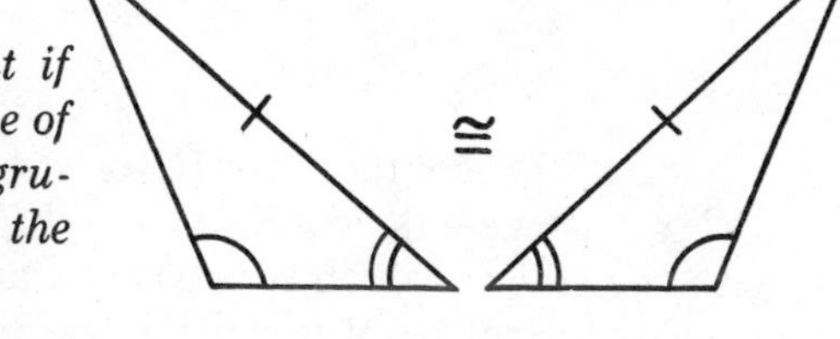

- AAS ≅ AAS THEOREM:
  Two triangles are congruent if two angles and a side opposite one of them in the first triangle are congruent to the corresponding parts of the second triangle.

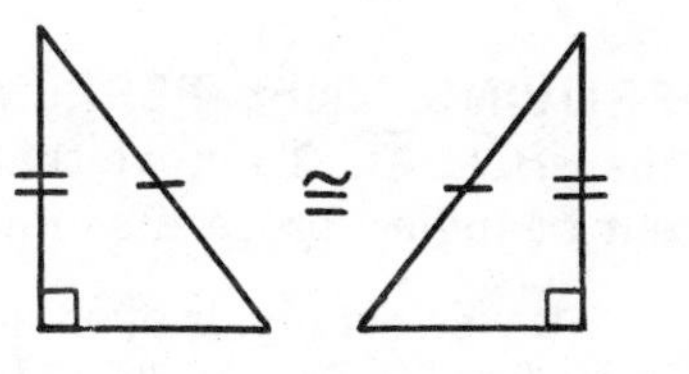

- HY-LEG ≅ HY-LEG THEOREM:
  Two triangles are congruent if the hypotenuse and a leg of one right triangle are congruent to the corresponding parts of the second right triangle. Note that this method applies only to *right* triangles.

Examples

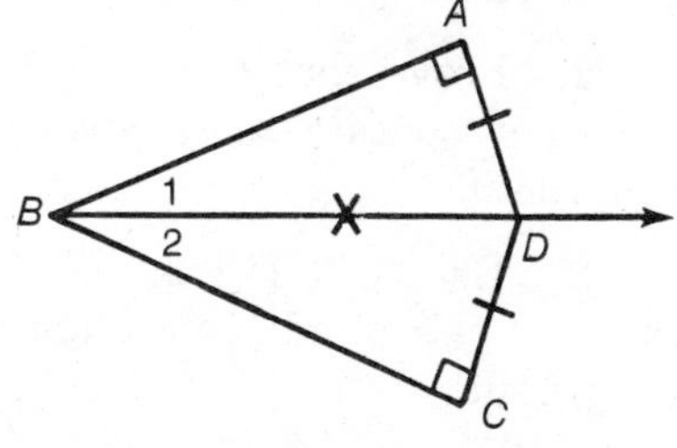

2. Given: $\overline{BA} \perp \overline{AD}$, $\overline{BC} \perp \overline{CD}$, $\overline{AD} \cong \overline{CD}$.
Prove: $\overrightarrow{BD}$ bisects $\angle ABC$.

Solution:

PLAN. Show that $\angle 1 \cong \angle 2$ by proving that $\triangle BAD \cong \triangle BCD$, using the Hy-Leg theorem.

PROOF

Statement	Reason
1. $\overline{BA} \perp \overline{AD}$, $\overline{BC} \perp \overline{CD}$	1. Given.
2. Triangles BAD and BCD are right triangles.	2. A triangle two of whose sides are perpendicular is a right triangle.
3. $\overline{BD} \cong \overline{BD}$ *Hyp*	3. Reflexive property of congruence.
4. $\overline{AD} \cong \overline{CD}$ *Leg*	4. Given.
5. $\triangle BAD \cong \triangle BCD$	5. Hy-Leg theorem.
6. $\angle 1 \cong \angle 2$	6. CPCTC.
7. $\overrightarrow{BD}$ bisects $\angle ABC$.	7. A ray that divides an angle into two congruent angles bisects the angle.

3. Given: $\angle B \cong \angle C$,
$\overline{AH}$ is the altitude to side $\overline{BC}$.
Prove: $\overline{AB} \cong \overline{AC}$.

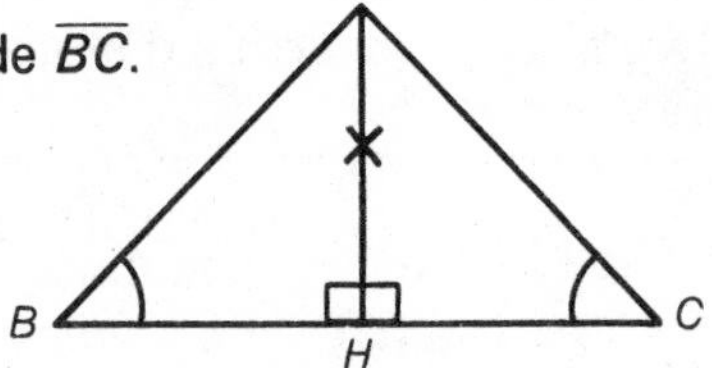

Solution:

PLAN. Prove that $\triangle AHB \cong \triangle AHC$ by the AAS theorem since $\angle B \cong \angle C$ (given), $\angle AHB \cong \angle AHC$ (an altitude forms right angles with the side to which it is drawn), and $\overline{AH} \cong \overline{AH}$ (reflexive property).

The formal proof is left for you to complete.

PROVING LINES PERPENDICULAR. If in Figure 6.6, $\angle 1 \cong \angle 2$, then $\overline{BD} \perp \overline{AC}$. To prove that two lines are perpendicular using congruent triangles, proceed as follows:

1. Select a pair of adjacent angles formed by the two lines required to be proved perpendicular.

2. Prove that a pair of triangles that contain these angles are congruent.

3. Use CPCTC to conclude that the pair of adjacent angles are congruent.

4. State that the desired lines are perpendicular, giving, as a reason, "If two lines intersect to form congruent adjacent angles, then the lines are perpendicular." See Exercise 10 at the end of this section.

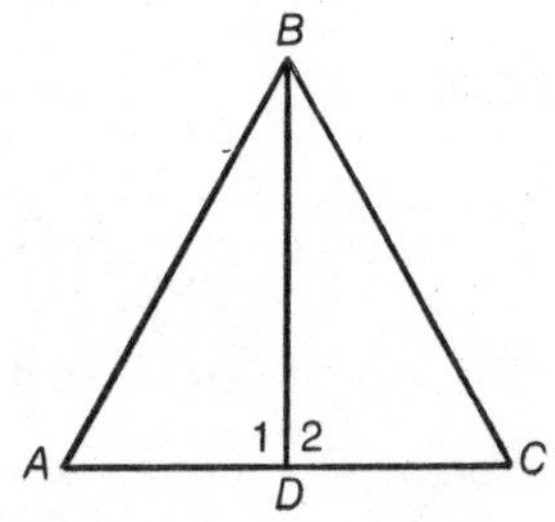

Figure 6.6 Proving Lines Are Perpendicular

EXERCISE SET 6.5

1. Given: $\angle J \cong \angle M$,
$\overline{JK} \perp \overline{KL}$, $\overline{ML} \perp \overline{KL}$.
Prove: $\overline{JL} \cong \overline{MK}$.

2. Given: $\overline{JK} \perp \overline{KL}$, $\overline{ML} \perp \overline{KL}$,
$\overline{JL} \cong \overline{MK}$.
Prove: $\angle J \cong \angle M$.

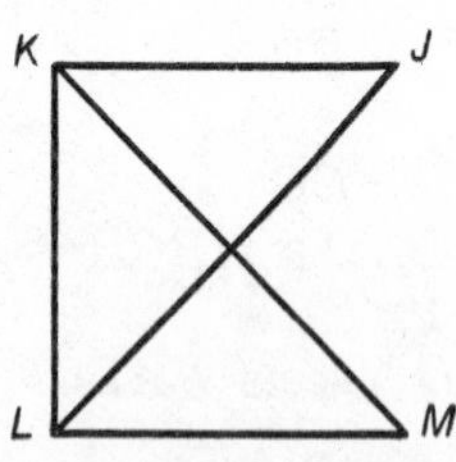

Exercises 1 and 2

3. Given: $\overline{JK} \cong \overline{PM}$,
$\overline{JK} \perp \overline{KM}$, $\overline{PM} \perp \overline{KM}$.
Prove: $\overline{KM}$ bisects $\overline{JP}$.

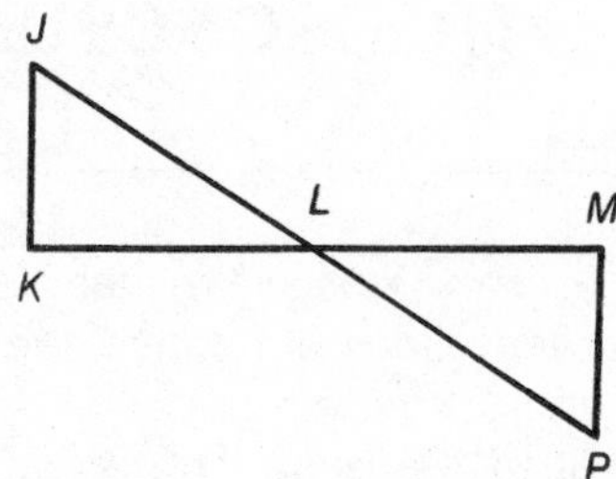

4. Given: $\overline{AB} \cong \overline{BC}$,
$\overline{BD}$ bisects $\angle ABC$.
Prove: $\triangle ADC$ is isosceles.

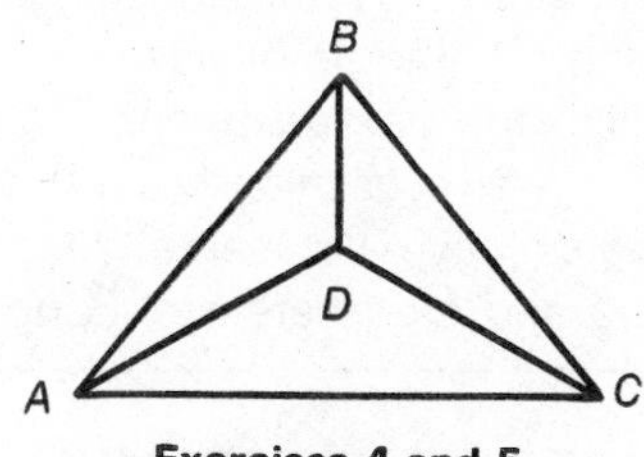

Exercises 4 and 5

5. Given: $\triangle ADC$ is equilateral,
$\overline{AD} \cong \overline{BC}$.
Prove: $\overline{BD}$ bisects $\angle ABC$.

6. Given: $\overline{AB} \perp \overline{BC}$, $\overline{AD} \perp \overline{DC}$,
$\angle 1 \cong \angle 2$.
Prove: $\overline{AC}$ bisects $\angle DAB$.

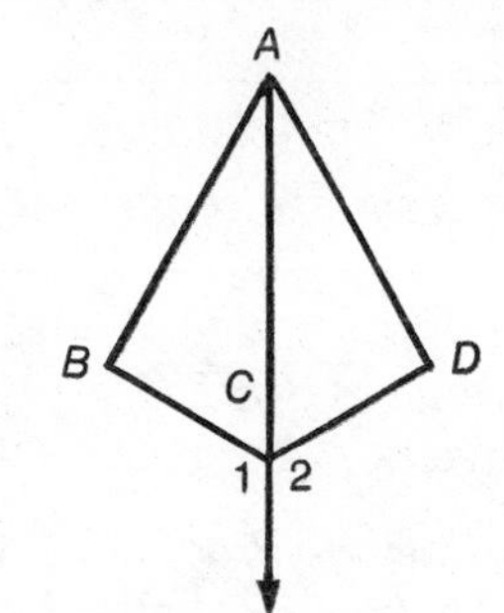

7. Given: $\overline{BE} \perp \overline{AC}$, $\overline{DF} \perp \overline{AC}$,
$\overline{AF} \cong \overline{CE}$, $\overline{BE} \cong \overline{DF}$,
Prove: $\overline{AB} \cong \overline{CD}$.

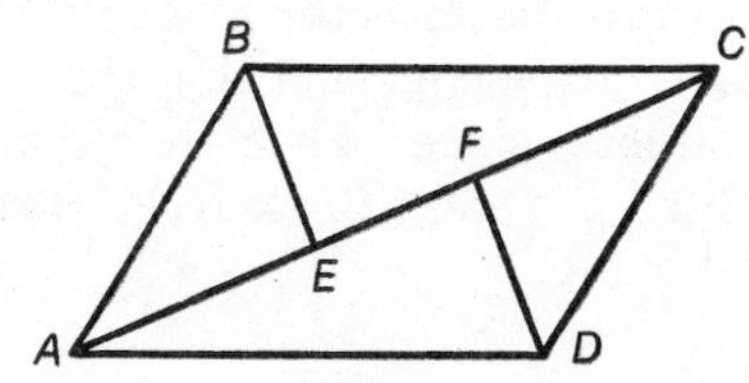

8. Given: $\overline{TL} \perp \overline{RS}$, $\overline{SW} \perp \overline{RT}$,
$\overline{TL} \cong \overline{SW}$.
Prove: $\overline{SL} \cong \overline{TW}$.

9. Given: $\overline{TL}$ is the altitude to $\overline{RS}$,
$\overline{SW}$ is the altitude to $\overline{RT}$,
$\overline{RS} \cong \overline{RT}$.
Prove: $\overline{RW} \cong \overline{RL}$.

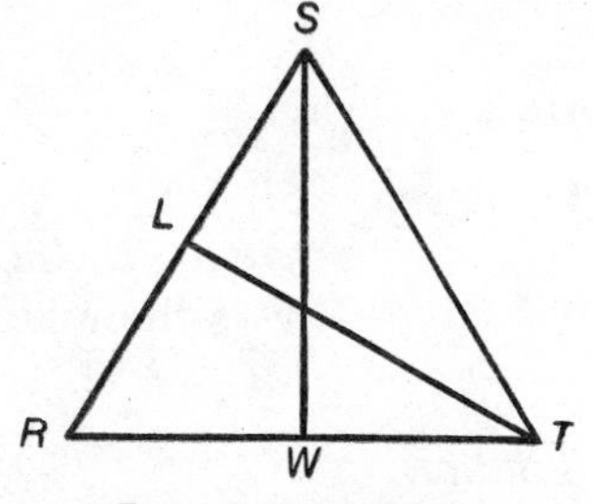

Exercises 8 and 9

10. Prove that the median drawn to the base of an isosceles triangle is perpendicular to the base.

6.6 THE ISOSCELES TRIANGLE

KEY IDEAS

In an isosceles triangle, the congruent sides are called the **legs**. The angles that are opposite the legs are called the **base angles**, and the side that they include is called the **base**. The angle opposite the base is called the **vertex angle**.

In an isosceles triangle the base angles have the same degree measure and are, therefore, congruent.

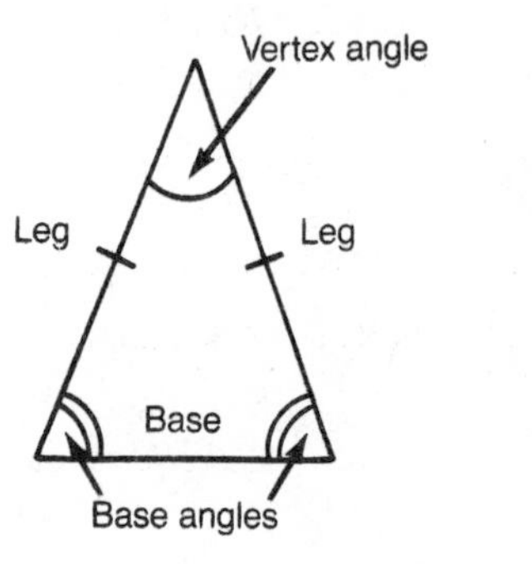

CONGRUENT SIDES IMPLY CONGRUENT ANGLES. Congruent triangles can be used to prove the base angles theorem.

THEOREM 1: *If two sides of a triangle are congruent, then the angles opposite them are congruent.*

Given: $\overline{AB} \cong \overline{BC}$.
Prove: $\angle A \cong \angle C$.

Outline of Proof: In the diagram, the two triangles formed by drawing the bisector of the vertex angle are congruent by the SAS postulate since $\overline{AB} \cong \overline{BC}$ (given), $\angle 1 \cong \angle 2$, and $\overline{BD} \cong \overline{BD}$. Hence, $\angle A \cong \angle C$.

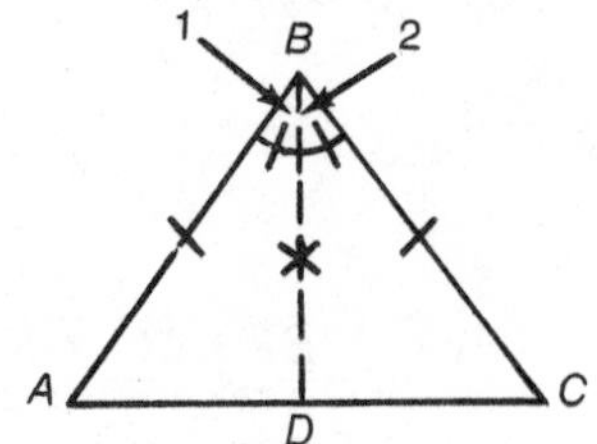

Example

1. Given: $\overline{SR} \cong \overline{ST}$,
$\overline{MP} \perp \overline{RS}$, $\overline{MQ} \perp \overline{ST}$,
M is the midpoint of $\overline{RT}$.
Prove: $\overline{MP} \cong \overline{MQ}$.

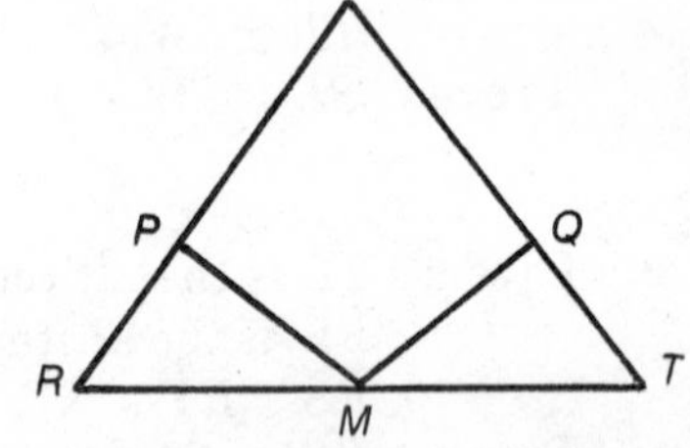

Solution:

PLAN. By the application of the base angles theorem, $\angle R \cong \angle T$. Marking off the diagram suggest that triangles *MPR* and *MQT* may be proved congruent by using the AAS theorem.

Solution:

Given: $\overline{CD}$ is the altitude to $\overline{AB}$, $\overline{AE}$ is the altitude to $\overline{BC}$, $\overline{CD} \cong \overline{AE}$.

Prove: $\triangle ABC$ is isosceles.

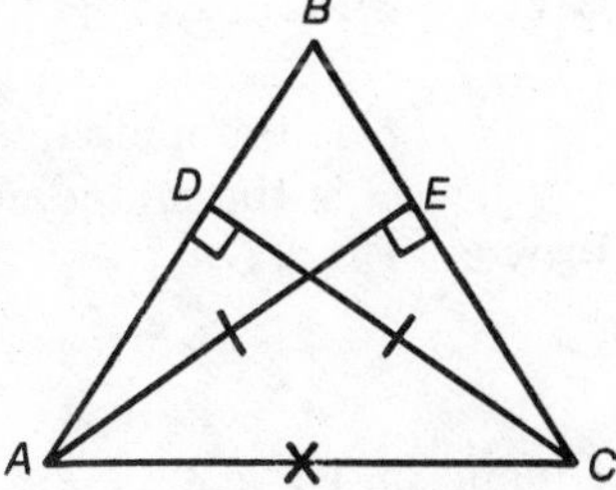

PLAN. Show that $\angle BAC \cong \angle BCA$ by proving $\triangle ADC \cong \triangle CEA$. Marking off the diagram suggests that the Hy-Leg method be used:

$$\overline{AC} \cong \overline{AC} \text{ (Hy)} \quad \text{and} \quad \overline{CD} \cong \overline{AE} \text{ (Leg)}$$

PROOF

Statement	Reason
1. $\overline{CD}$ is the altitude to $\overline{AB}$, $\overline{AE}$ is the altitude to $\overline{BC}$.	1. Given.
2. Triangles ADC and CEA are right triangles.	2. A triangle that contains a right angle is a right triangle. (***Note***: This step consolidates several obvious steps.)
3. $\overline{CD} \cong \overline{AE}$ (Leg)	3. Given.
4. $\overline{AC} \cong \overline{AC}$ (Hy)	4. Reflexive property of congruence.
5. $\triangle ADC \cong \triangle CEA$	5. Hy-Leg theorem.
6. $\angle BAC \cong \angle BCA$	6. CPCTC.
7. $\triangle ABC$ is isosceles.	7. A triangle that has a pair of congruent angles is isosceles.

EXERCISE SET 6.6

1. Given: $\angle 2 \cong \angle 4$, $\angle BDA \cong \angle BDC$.

Prove: $\triangle ABC$ is isosceles.

2. Given: $\angle 1 \cong \angle 3$, $\overline{AB} \cong \overline{BC}$.

Prove: $\triangle ADC$ is isosceles.

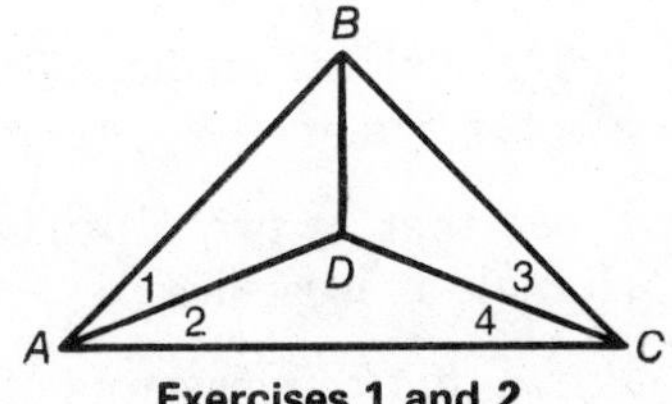

Exercises 1 and 2

3. Given: $\angle 1 \cong \angle 2$,
$\overline{AB} \cong \overline{BC}$,
F is the midpoint of $\overline{AB}$,
G is the midpoint of $\overline{BC}$.
Prove: $\overline{FD} \cong \overline{GE}$.

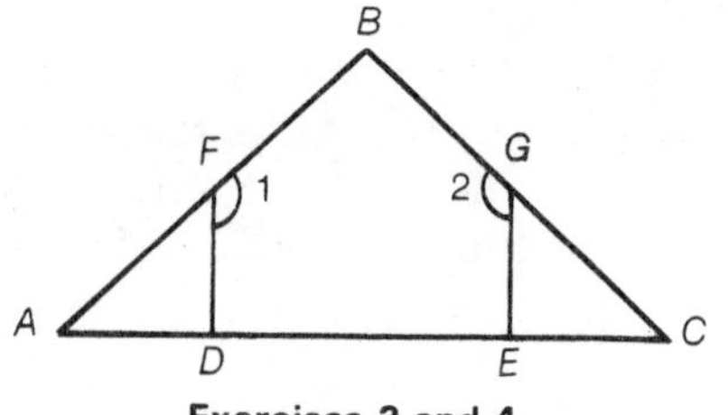

Exercises 3 and 4

4. Given: $\angle 1 \cong \angle 2$,
$\overline{FD} \perp \overline{AC}$, $\overline{GE} \perp \overline{AC}$,
$\overline{AE} \cong \overline{CD}$.
Prove: $\triangle ABC$ is isosceles.

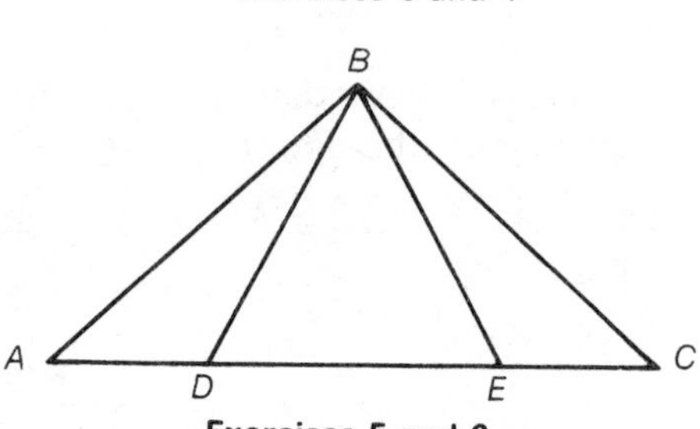

Exercises 5 and 6

5. Given: $\overline{AB} \cong \overline{BC}$, $\overline{AE} \cong \overline{CD}$.
Prove: $\triangle DBE$ is isosceles.

6. Given: $\angle BDE \cong \angle BED$,
$\angle ABE \cong \angle CBD$.
Prove: $\triangle ABC$ is isosceles.

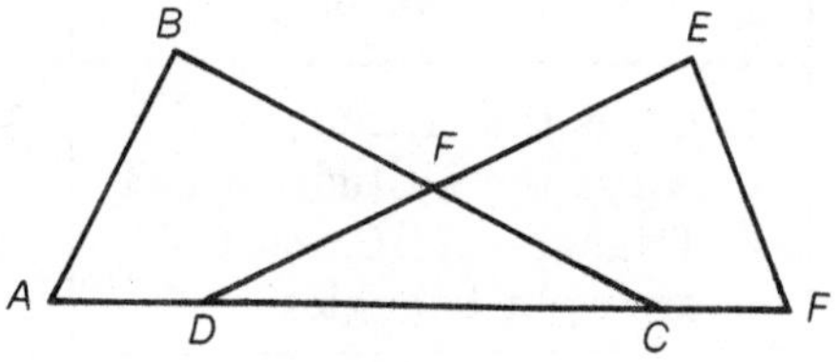

7. Given: $\overline{DF} \cong \overline{CF}$, $\overline{AD} \cong \overline{FC}$,
$\overline{BC} \cong \overline{ED}$.
Prove: $\angle B \cong \angle E$.

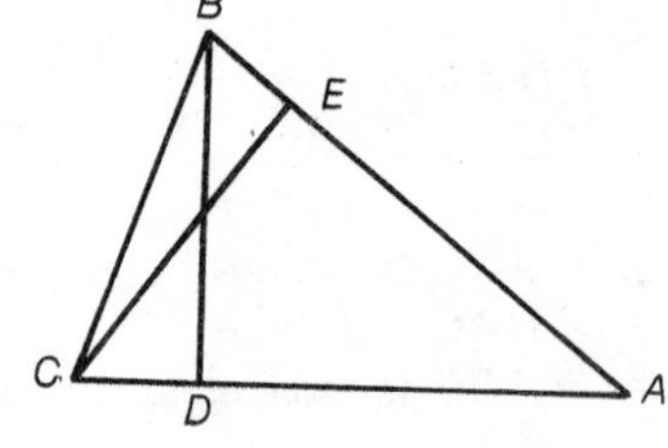

8. Given: $\overline{AB} \cong \overline{AC}$,
$\overline{CE} \perp \overline{AB}$, $\overline{BD} \perp \overline{AC}$.
Prove: $\overline{CE} \cong \overline{BD}$.

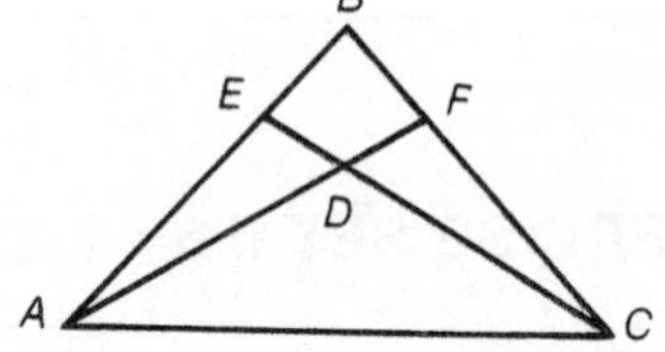

9. Given: $\overline{AB} \cong \overline{BC}$, $\overline{BE} \cong \overline{BF}$.
Prove: $\triangle ADC$ is isosceles.

10. Prove that the medians drawn to the legs of an isosceles triangle are congruent.

11. Prove that, if two altitudes of a triangle are congruent, then the triangle is isosceles.

12. Prove that the bisectors of the base angles of an isosceles triangle are congruent.

13. Prove that, if the perpendicular bisector of a side of a triangle passes through the opposite vertex, then the triangle is isosceles.

6.7 PROOFS INVOLVING TWO PAIRS OF CONGRUENT TRIANGLES

KEY IDEAS

In some problems it may appear that the information provided in the "Given" is insufficient to prove a pair of triangles congruent. Upon closer examination, however, it may be possible to prove a *different* pair of triangles congruent, and then use corresponding parts of these triangles to prove the desired pair of triangles congruent.

A mathematical proof may be presented in paragraph, rather than two-column, form.

DOUBLE-CONGRUENCE PROOFS. Consider the following proof:

Given: $\overline{AB} \cong \overline{CB}$,
E is the midpoint of $\overline{AC}$.
Prove: $\triangle AED \cong \triangle CED$.

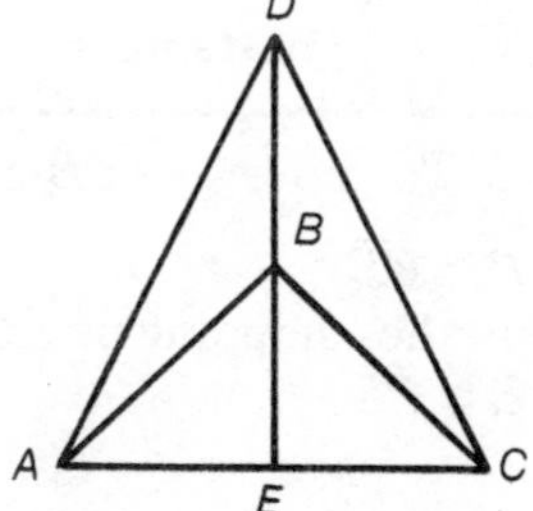

In analyzing this example, there does not appear to be enough information provided in the "Given" to prove that triangles *AED* and *CED* are congruent. Therefore it is necessary to prove a second pair of triangles (triangles *AEB* and *CEB*) congruent in order to obtain an additional pair of congruent parts (angles *AED* and *CED*) that can be used to prove the original pair of triangles (triangles *AED* and *CED*) congruent. When confronted by problems of this type, follow these steps:

Step	Example
1. Identify which additional pairs of angles or segments would have to be known to be congruent so that the desired pair of triangles could be proved congruent.	1. Based on the "Given," $\triangle AED$ could be proved congruent to $\triangle CED$ if it was known that angles AED and CED were congruent.
2. Look for a *different* pair of triangles that contain these parts and that you can prove congruent.	2. Triangles AEB and CEB contain these angles and can be proved congruent.
3. Prove the second pair of triangles congruent so that, by CPCTC, the needed pair of parts are congruent.	3. $\triangle AEB \cong \triangle CEB$ by the SSS postulate so that, by CPCTC, $\angle AED \cong \angle CED$.
4. Use this additional pair of congruent parts to help prove the original pair of triangles congruent.	4. $\triangle AED \cong \triangle CED$ by the SAS postulate.

The actual formal proof follows.

PROOF

Statement		Reason
Part I. To prove $\triangle AEB \cong \triangle CEB$:		
1. $\overline{AB} \cong \overline{CB}$	*Side*	1. Given.
2. E is the midpoint of $\overline{AC}$.		2. Given.
3. $\overline{AE} \cong \overline{CE}$	*Side*	3. A midpoint divides a segment into two congruent segments.
4. $\overline{BE} \cong \overline{BE}$		4. Reflexive property of congruence.
5. $\triangle AEB \cong \triangle CEB$		5. SSS postulate.
Part II. To prove $\triangle AED \cong \triangle CED$:		
6. $\angle AED \cong \angle CED$	*Angle*	6. CPCTC.
7. $\overline{DE} \cong \overline{DE}$	*Side*	7. Reflexive property of congruence.
8. $\triangle AED \cong \triangle CED$		8. SAS postulate.

PROOFS IN PARAGRAPH FORM. A valid proof in mathematics may take several different forms. Although there is no standard format for a mathematical proof, the logic and correctness of the proof should be readily apparent. The following example presents a proof in paragraph rather than two-column form.

Example

Given: $\overline{AB} \cong \overline{AC}$, $\overline{BD} \cong \overline{CE}$, $\overline{BF}$ and $\overline{CG}$ are drawn perpendicular to $\overline{AD}$ and $\overline{AE}$, respectively.
Prove: $\overline{DF} \cong \overline{EG}$.

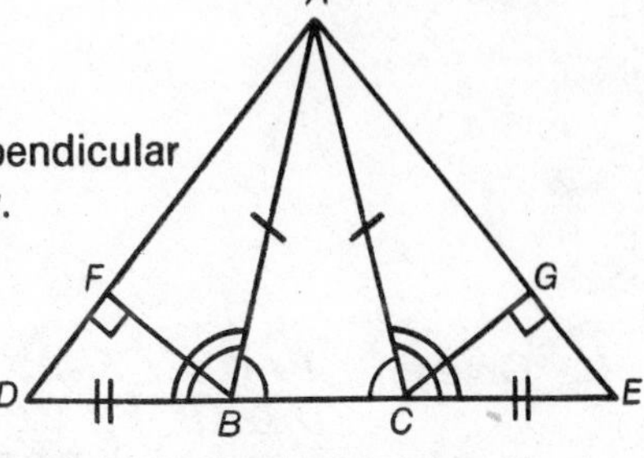

Solution:
PLAN. First prove $\triangle ABD \cong \triangle ACE$ in order to obtain an additional pair of congruent parts needed to prove that $\triangle DFB \cong \triangle EGC$.

PROOF

Part I. $\triangle ABD \cong \triangle ACE$ by SAS since $\overline{AB} \cong \overline{AC}$, $\angle ABD \cong \angle ACE$ (supplements of the congruent base angles are congruent), and $\overline{BD} \cong \overline{CE}$. Therefore, $\angle D \cong \angle B$.

Part II. $\triangle DFB = \triangle EGC$ by AAS since $\angle D \cong \angle B$, $\angle DFB \cong \angle EGC$ (right angles are congruent), and $\overline{BD} \cong \overline{CE}$. Therefore, $\overline{DF} \cong \overline{EG}$ by CPCTC.

EXERCISE SET 6.7

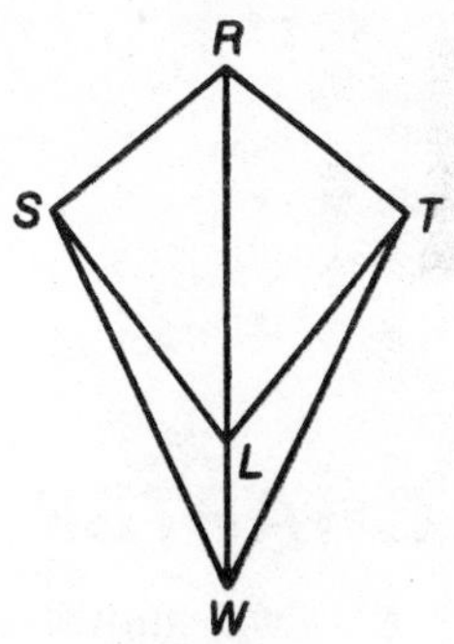

1. Given: $\overline{RS} \perp \overline{SL}$, $\overline{RT} \perp \overline{LT}$, $\overline{RS} \cong \overline{RT}$.
   Prove: (a) $\triangle RLS \cong \triangle RLT$.
   (b) $\overline{WL}$ bisects $\angle SWT$.

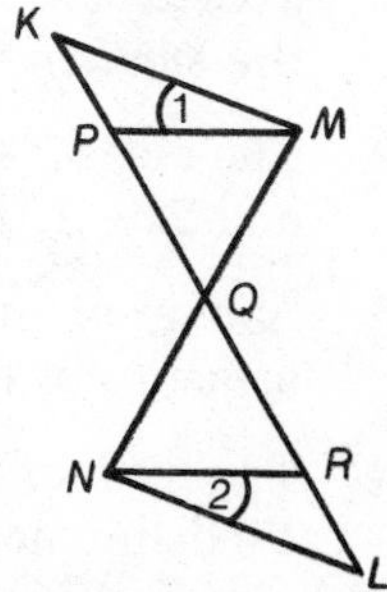

2. Given: $\overline{KPQRL}$, $\overline{MQN}$, $\overline{KM}$, $\overline{NL}$, $\overline{MP}$, $\overline{NR}$, $\overline{KL}$ and $\overline{MN}$ bisect each other at Q, $\angle 1 \cong \angle 2$.
   Prove: $\overline{PM} \cong \overline{NR}$.

3. Given: $\overline{AB} \cong \overline{AD}$,
$\overline{EA}$ bisects $\angle DAB$.
Prove: $\triangle BCE \cong \triangle DCE$.

4. Given: $\overline{BE} \cong \overline{DE}$, $\overline{BC} \cong \overline{DC}$.
Prove: $\overline{CA}$ bisects $\angle DAB$.

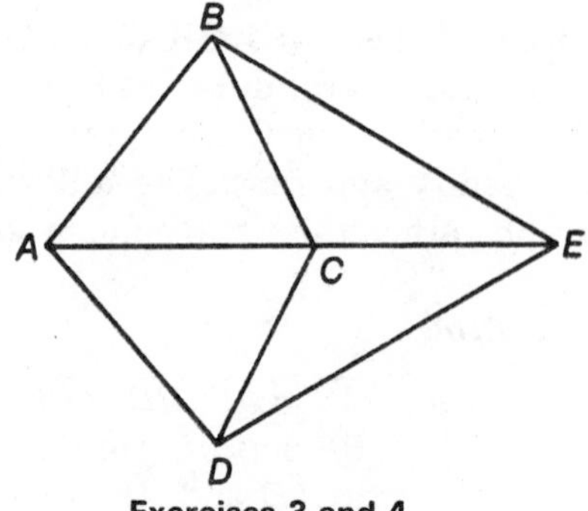

Exercises 3 and 4

5. Given: $\angle FAC \cong \angle FCA$,
$\overline{FD} \perp \overline{AB}$, $\overline{FE} \perp \overline{BC}$.
Prove: $\overline{BF}$ bisects $\angle DBE$.

6. Given: $\overline{BD} \cong \overline{BE}$,
$\overline{FD} \cong \overline{FE}$.
Prove: $\triangle AFC$ is isosceles.

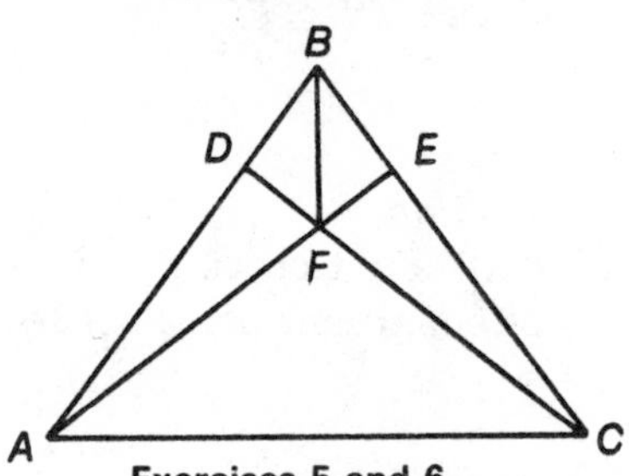

Exercises 5 and 6

7. Given: $\triangle ABC$,
$\overline{AC} \cong \overline{BC}$, $\overline{CE} \cong \overline{CD}$,
in $\triangle BCD$, $\overline{DF}$ is a median to $\overline{BC}$,
in $\triangle ACE$, $\overline{EG}$ is a median to $\overline{AC}$.
Prove: $\overline{EG} \cong \overline{DF}$.

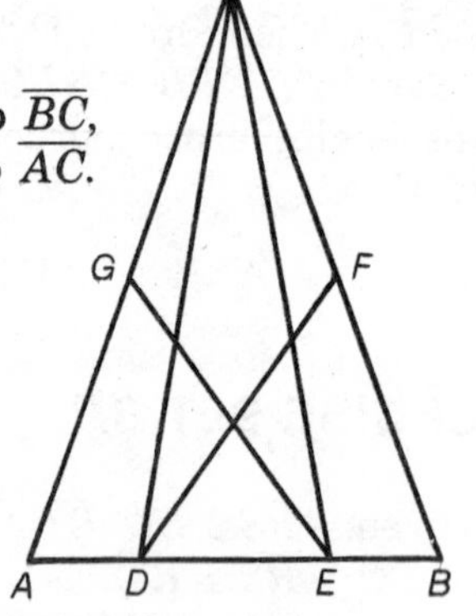

CHAPTER 6 REVIEW EXERCISES

1. Two angles are complementary, and the measure of one angle exceeds twice the measure of the other by 15. Find the measure of the *smaller* angle.

2. The measures of a pair of vertical angles are represented by $x + 12$ and $2x - 1$. Find the value of x.

3. A pair of angles are supplementary and congruent. What is the measure of each angle?

4. The measures of two supplementary angles are in the ratio of 2:3. Find the measure of the *complement* of the smaller angle.

5. The supplement of every acute angle *must* be:
(1) an acute angle
(2) a right angle
(3) an obtuse angle
(4) a straight angle

6. The median drawn to the base of an isosceles triangle divides the isosceles triangle into two triangles that are:
(1) always congruent
(2) sometimes congruent
(3) never congruent
(4) sometimes equal in perimeter

7. Two right triangles that have the same perimeter are:
(1) always congruent
(2) sometimes congruent
(3) never congruent
(4) always equal in area

8. In the accompanying diagram, lines a and b intersect. If $\angle 1 \cong \angle 3$, $m\angle 2 = 8x$, and $m\angle 3 = 10x - 30$, find the value of x.

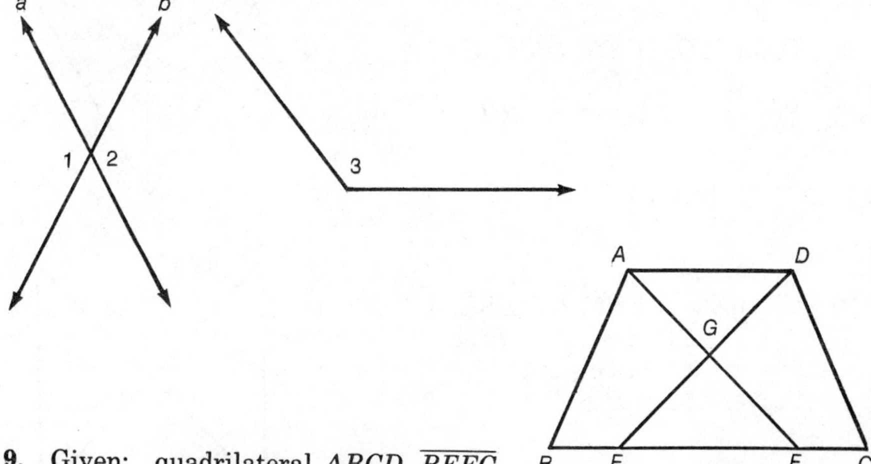

9. Given: quadrilateral $ABCD$, $\overline{BEFC}$, $\overline{BE} \cong \overline{FC}$, $\overline{AF} \cong \overline{DE}$, $\overline{AF}$ and $\overline{DE}$ bisect each other at G.
Prove: $\overline{AB} \cong \overline{DC}$.

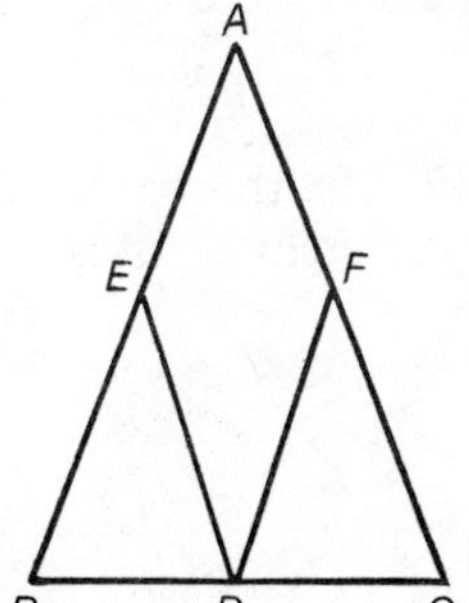

10. Given: $\triangle ABC$, $\overline{AEB}$, $\overline{AFC}$, D is the midpoint of $\overline{BC}$, $\overline{ED} \cong \overline{FD}$, $\angle EDC \cong \angle FDB$.

Prove: $\triangle ABC$ is isosceles.

11. Given: $\triangle ABC$, $\overline{ADB}$, $\overline{CEB}$,
$\overline{AE}$ and $\overline{DC}$ intersect at F,
$\overline{DF} \cong \overline{EF}$,
$\angle EFB \cong \angle DFB$.
Prove: $\triangle ABC$ is isosceles.

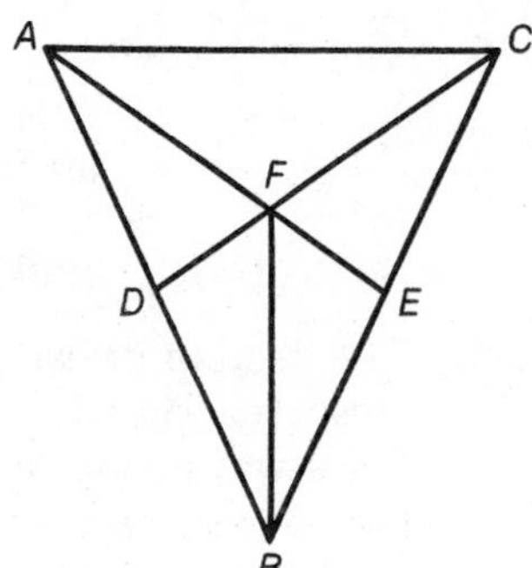

12. Given: $\overline{EHF}$,
$\overline{HK}$ intersects $\overline{FG}$ at R,
$\overline{FG} \perp \overline{EGK}$,
$\angle E \cong \angle K$.
Prove: $\triangle HFR$ is isosceles.

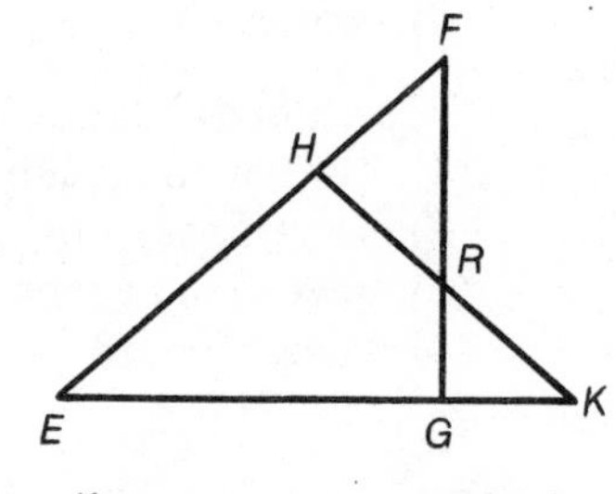

13. Given: $\overline{OR} \perp \overline{KG}$, $\overline{OL} \perp \overline{JG}$,
$\overline{OK} \cong \overline{OJ}$.
Prove: $\overline{OG}$ bisects $\angle RGL$.

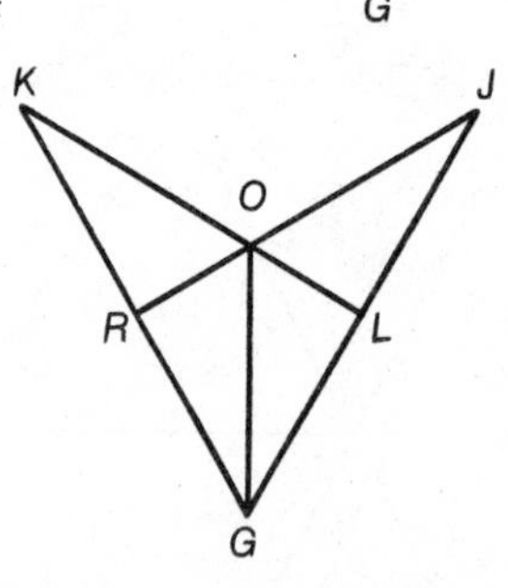

14. Given: $\overline{AB} \perp \overline{BC}$, $\overline{DC} \perp \overline{BC}$,
$\overline{DB}$ bisects $\angle ABC$,
$\overline{AC}$ bisects $\angle DCB$,
Prove: $\overline{EA} \cong \overline{ED}$.

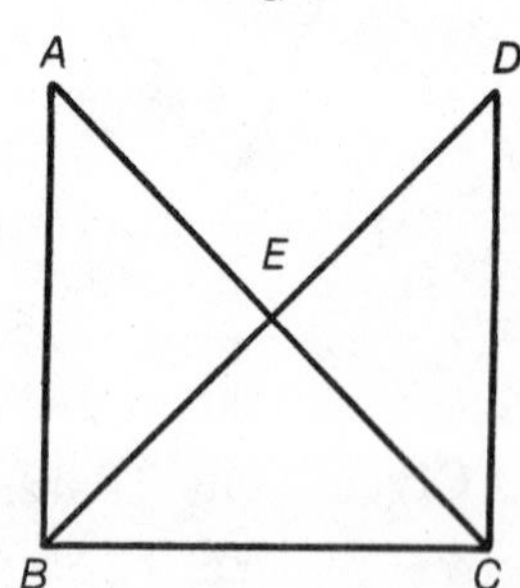

15. In the figure, $\overline{AE} \cong \overline{AF}$, $\angle 1 \cong \angle 4$, $\angle 2 \cong \angle 3$.
Prove: $\overline{AD} \cong \overline{AB}$.

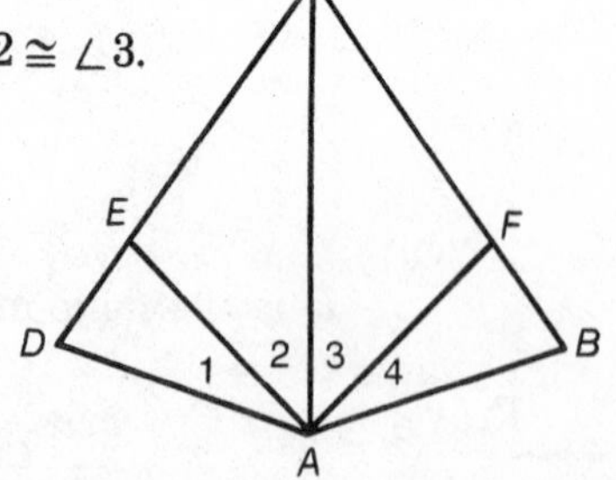

CHAPTER 7

Parallel Lines and Special Quadrilaterals

7.1 PROPERTIES OF PARALLEL LINES

KEY IDEAS

Lines that lie in the same plane and never meet are called **parallel lines**. The symbol ∥ means "is parallel to," so that $l \parallel m$ is read as "line l is parallel to line m." To identify parallel lines, mark them with arrowheads that point in the same direction. In the accompanying diagram, the corresponding pairs of arrowheads indicate that $\overline{AD} \parallel \overline{BC}$ and $\overline{AB} \parallel \overline{CD}$.

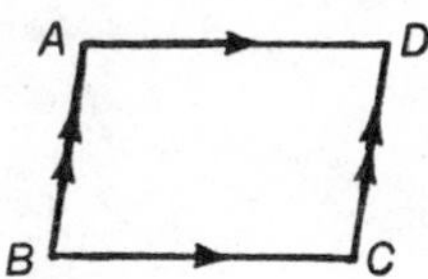

TRANSVERSALS AND SPECIAL ANGLE PAIRS. A line that intersects two or more lines in different points is called a **transversal.** In Figure 7.1, line t represents a transversal since it intersects lines l and m at two different points. Angles 1, 2, 5, and 6 lie *between* lines l and m and are called **interior angles**. Angles 3, 4, 7, and 8 lie *outside* lines l and m and are called **exterior angles.**

Table 7.1 further classifies special pairs of these angles.

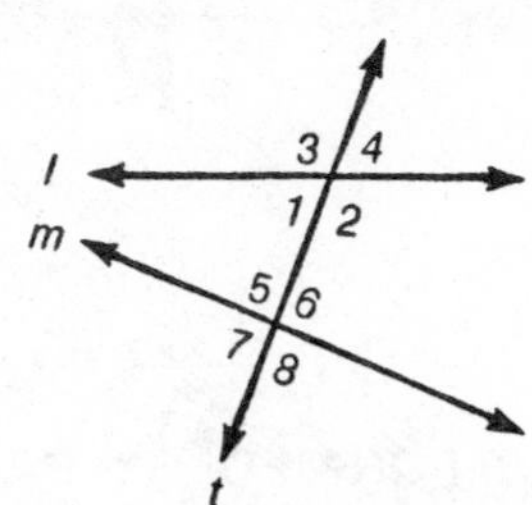

Figure 7.1 Transversal and Angle Pairs

TABLE 7.1 Special Angle Pairs

Type of Angle Pair	Distinguishing Features	Examples (Figure 7.1)
Alternate interior angles	• Angles are interior angles. • Angles are on opposite sides of the transversal. • Angles do not have the same vertex.	Angles 1 and 6; angles 2 and 5.
Corresponding angles	• One angle is an interior angle; the other angle is an exterior angle. • Angles are on the same side of the transversal. • Angles do not have the same vertex.	Angles 3 and 5; angles 4 and 6; angles 1 and 7; angles 2 and 8.

In analyzing diagrams, alternate interior angle pairs may be identified by their Z shape, while corresponding angles form an F shape (Figure 7.2). The Z and F shapes, however, may be positioned in various ways so that the letter may appear reversed or upside down.

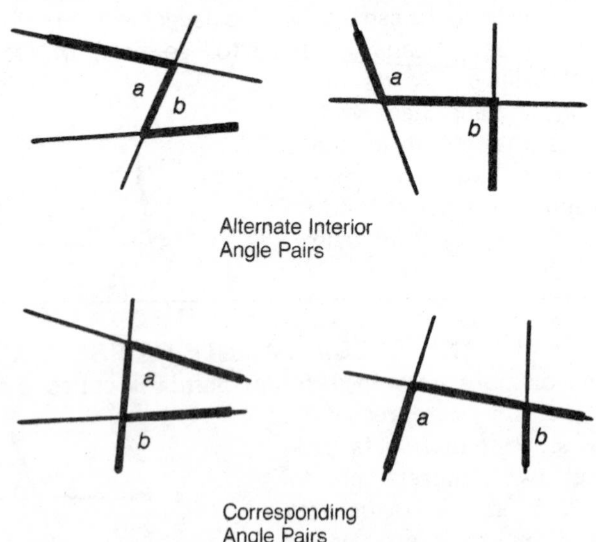

Figure 7.2 Z and F Shapes Formed by Angle Pairs

ANGLES FORMED BY PARALLEL LINES. When a transversal intersects a pair of parallel lines, the angles in every pair that is formed are either equal in measure or are supplementary.

Postulate 1: If two lines are parallel, then alternate interior angles are congruent.

If $l \parallel m$, then
$\angle 1 \cong \angle 2$ and $\angle 3 \cong \angle 4$.

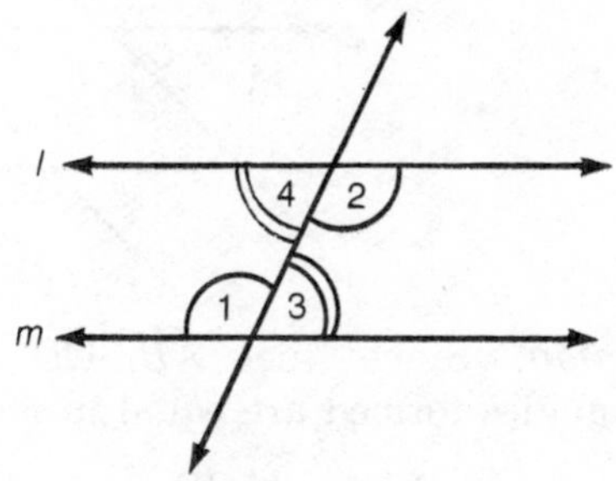

THEOREM 1: *If two lines are parallel, then corresponding angles are congruent.*

If $l \parallel m$, then
$\angle 1 \cong \angle 5$ and $\angle 3 \cong \angle 6$,
$\angle 7 \cong \angle 4$ and $\angle 8 \cong \angle 2$.

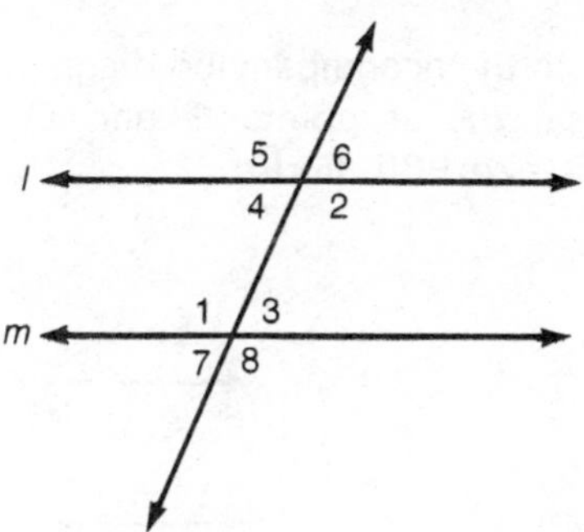

THEOREM 2: *If two lines are parallel, then interior angles on the same side of the transversal are supplementary.*

If $l \parallel m$, then
$m\angle 1 + m\angle 4 = 180$ and
$m\angle 3 + m\angle 2 = 180$.

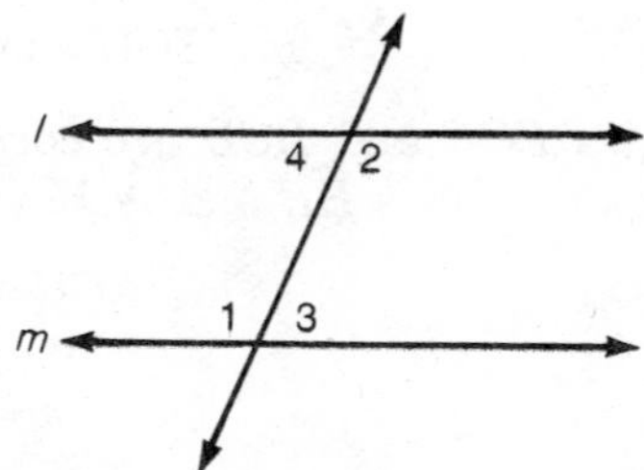

Examples

1. In the accompanying diagram, $\overleftrightarrow{AC}$ and $\overleftrightarrow{BDE}$ are parallel. Parallel lines $\overleftrightarrow{AB}$, $\overleftrightarrow{CD}$, and $\overleftrightarrow{EF}$ are drawn. If $m\angle 1 = 45$, find $m\angle 4$.

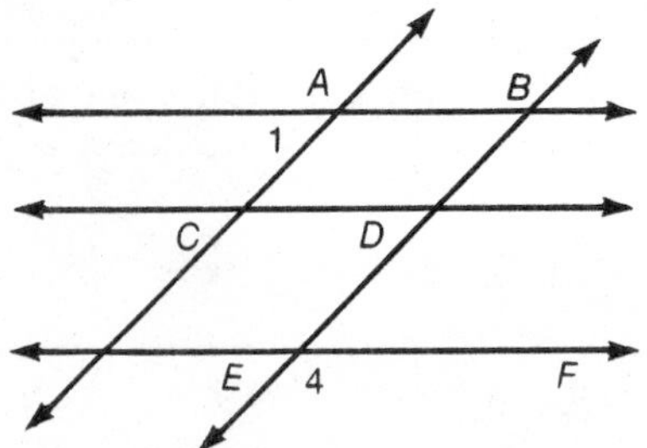

Solution: Since lines $\overleftrightarrow{AB}$, $\overleftrightarrow{CD}$, and $\overleftrightarrow{EF}$ are parallel, alternate interior angles formed are equal in measure. Therefore,

$$m\angle 1 = m\angle ACD = m\angle CDE = m\angle DEF = 45.$$

Since $\angle DEF$ and $\angle 4$ are supplementary,

$$\begin{aligned} m\angle 4 &= 180 - m\angle DEF \\ &= 180 - 145 = \mathbf{135} \end{aligned}$$

2. In the accompanying diagram, parallel lines $\overleftrightarrow{HE}$ and $\overleftrightarrow{AD}$ are cut by transversal $\overleftrightarrow{BF}$ at points G and C, respectively. If $m\angle HGF = 5n$ and $m\angle BCD = 2n + 66$, find n.

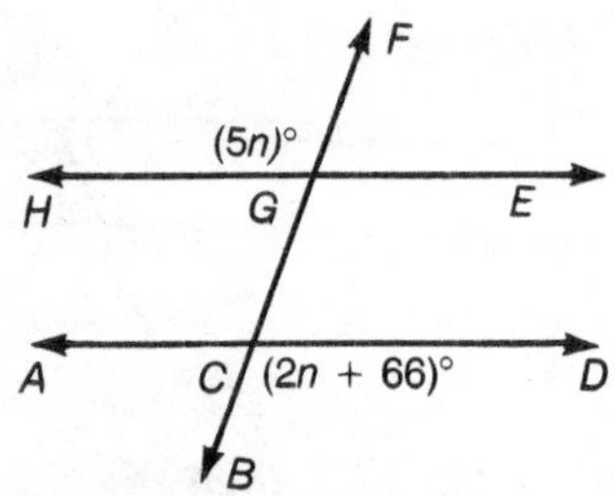

Solution: Since vertical angles are equal in measure,

$$m\angle EGC = m\angle FGH = 5n.$$

Angles EGC and BCD are corresponding angles and are equal in measure since lines $\overleftrightarrow{HE}$ and $\overleftrightarrow{AD}$ are parallel. Therefore,

$$\begin{aligned} m\angle EGC &= m\angle BCD \\ 5n &= 2n + 66 \\ 5n - 2n &= 66 \\ \frac{\cancel{3}n}{\cancel{3}} &= \frac{66}{3} \\ n &= \mathbf{22} \end{aligned}$$

PROVING LINES PARALLEL. Although the converse of a true statement is not necessarily true, the converse of each statement that gives a property of parallel lines is true. These converses provide methods that can be used in proving lines parallel.

In a plane, two lines are parallel if:

- A pair of alternate interior angles are congruent (*converse of Postulate 1*).
- A pair of corresponding angles are congruent (*converse of Theorem 1*).
- A pair of interior angles on the same side of the transversal are supplementary (*converse of Theorem 2*).

In addition, you may conclude that, in a plane, two lines are parallel if the lines are:

- perpendicular to the same line (see Figure 7.3);

If $l \perp p$ and $m \perp p$, then $l \parallel m$.

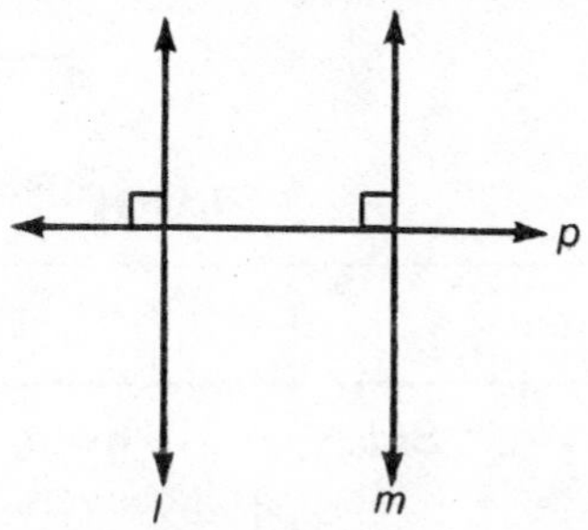

Figure 7.3 Lines Perpendicular to the Same Line

- each parallel to the same line (see Figure 7.4).

If $l \parallel p$ and $m \parallel p$, then $l \parallel m$.

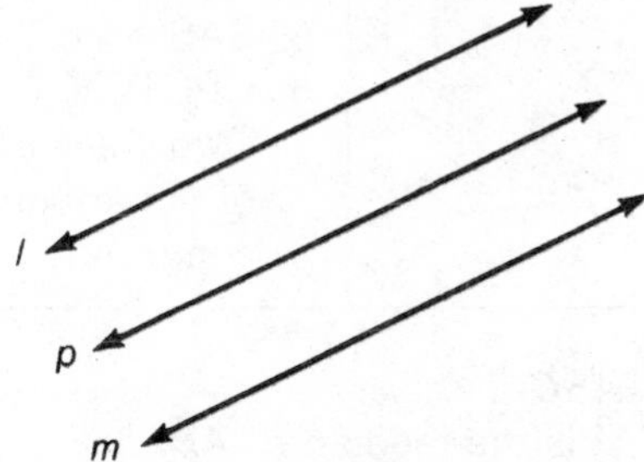

Figure 7.4 Lines Parallel to the Same Line

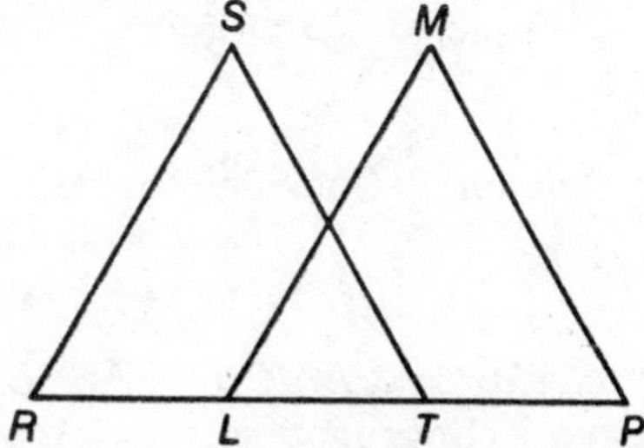

Examples

3. Given: $\overline{MP} \cong \overline{ST}$, $\overline{PL} \cong \overline{RT}$, $\overline{MP} \parallel \overline{ST}$.
   Prove: $\overline{RS} \parallel \overline{LM}$.

Solution:

PLAN. To prove a pair of line segments parallel, it is usually necessary to prove than an appropriate pair of angles are congruent.

Step 1. Look at the diagram. If $\overline{RS}$ is to be proved parallel to $\overline{LM}$, then it must be proved that $\angle SRT \cong \angle MLP$. This implies that the triangles that contain these angles must be proved congruent. Therefore, it is necessary to prove that $\triangle RST \cong \triangle LMP$.

Step 2. Mark off the diagram with the "Given."

Step 3. Use the SAS postulate.

Step 4. Write the proof.

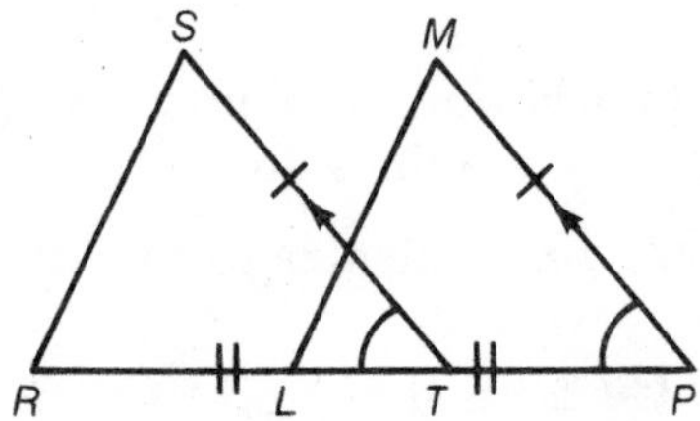

PROOF

Statement		Reason
1. $\overline{MP} \cong \overline{ST}$	*Side*	1. Given.
2. $\overline{MP} \parallel \overline{ST}$		2. Given.
3. $\angle MPL \cong \angle STR$	*Angle*	3. If two lines are parallel, then their corresponding angles are congruent.
4. $\overline{PL} \cong \overline{RT}$	*Side*	4. Given.
5. $\triangle RST \cong \triangle LMP$		5. SAS postulate.
6. $\angle SRT \cong \angle MLP$		6. CPCTC.
7. $\overline{RS} \parallel \overline{LM}$		7. Two lines are parallel if a pair of corresponding angles are congruent.

4. Given: $\triangle ABC$,
$\overline{CM}$ is the median to $\overline{AB}$,
$\overline{CM}$ is extended to point P
so that $\overline{CM} \cong \overline{MP}$,
$\overline{AP}$ is drawn.

Prove: $\overline{AP} \parallel \overline{CB}$.

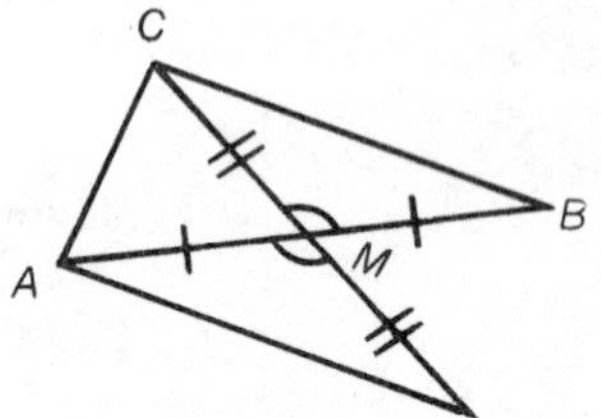

Solution: Consider the following proof in paragraph form:

PROOF

$\triangle AMP \cong \triangle BMC$ by SAS. $\overline{BM} \cong \overline{AM}$ (a median divides a segment into two congruent segments), $\angle BMC \cong \angle AMP$ (vertical angles are congruent), and $\overline{CM} \cong \overline{MP}$ (given). Therefore, $\angle P \cong \angle BCM$ by CPCTC. It follows that $\overline{AP} \parallel \overline{CB}$ since two lines are parallel if a pair of alternate interior angles are congruent.

EXERCISE SET 7.1

1. In the accompanying diagram, $l \parallel m$. Find the value of x if:
(a) $m\angle 1 = x$ and $m\angle 7 = 68$
(b) $m\angle 2 = 53$ and $m\angle 7 = x$
(c) $m\angle 3 = 64$ and $m\angle 8 = x$
(d) $m\angle 3 = 76$ and $m\angle 5 = 3x - 5$
(e) $m\angle 6 = 128$ and $m\angle 2 = 4x$
(f) $m\angle 2 = 3x - 15$ and $m\angle 5 = x + 29$
(g) $m\angle 8 = 2x - 11$ and $m\angle 5 = 3x - 47$
(h) $m\angle 4 = 3x + 7$ and $m\angle 3 = x + 5$
(i) $m\angle 1 = 3x$ and $m\angle 5 = 7x$
(j) $m\angle 4 = x$ and $m\angle 8 = 5x$
(k) $m\angle 3 = 4x - 35$ and $m\angle 8 = x + 7$

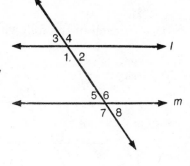

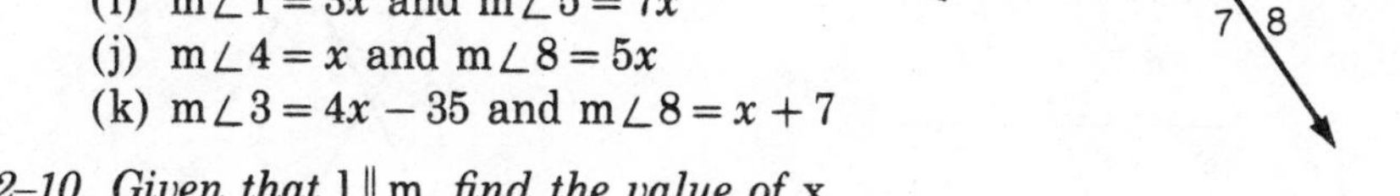

2–10. Given that $l \parallel m$, *find the value of* x.

2.

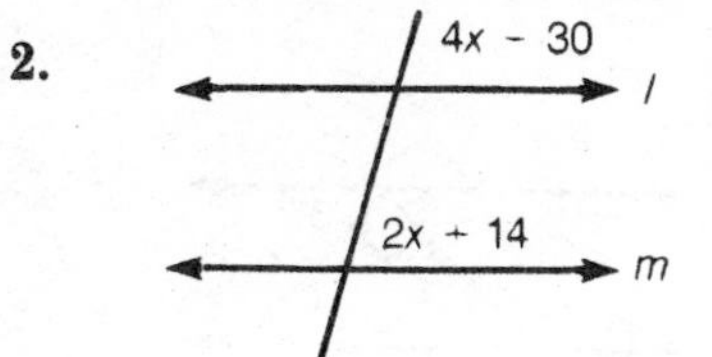

3.

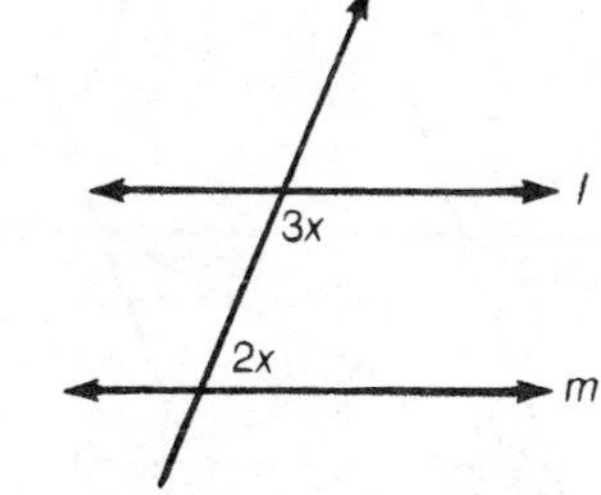

4.

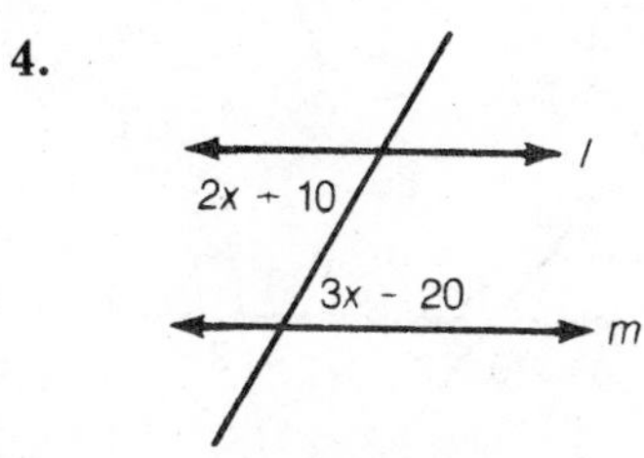

5.

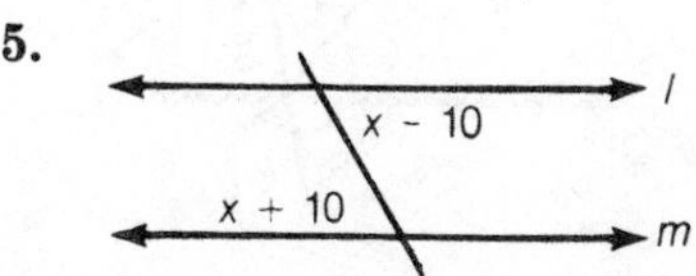

6.

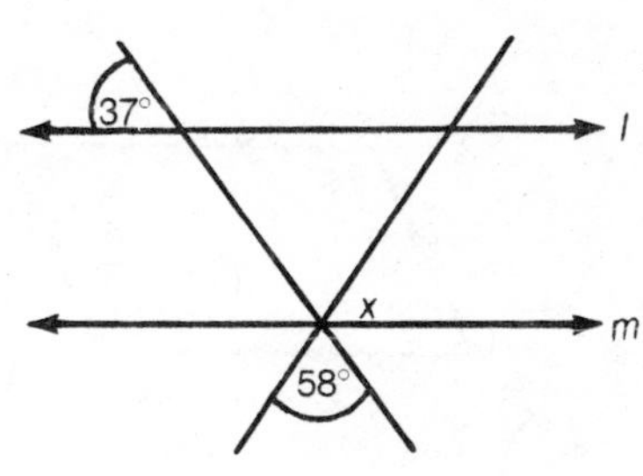

7.

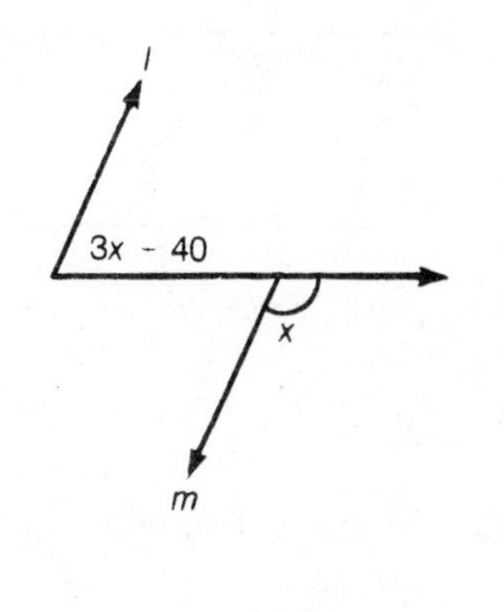

8.

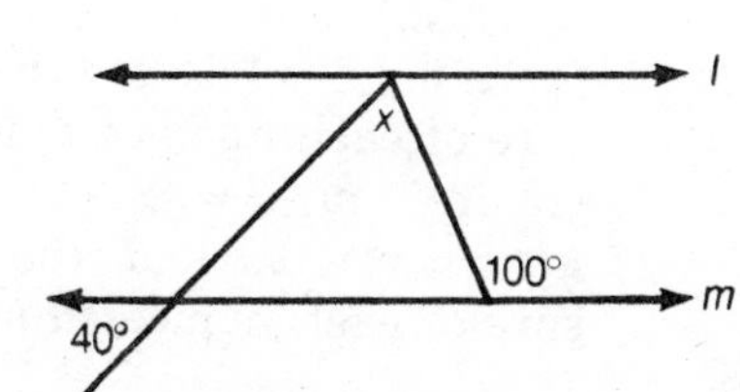

9.

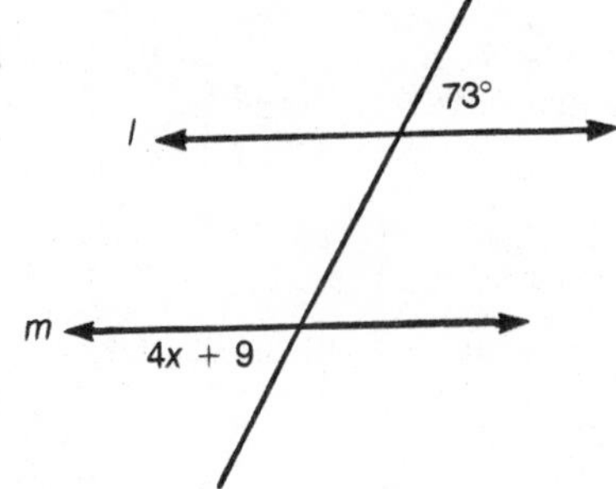

10.

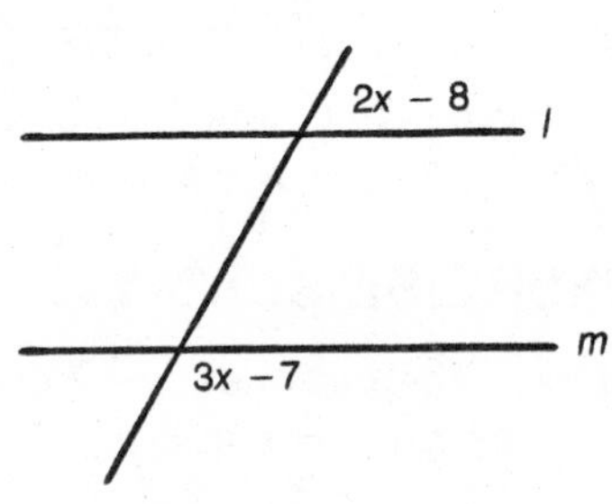

11. In each of the following, find the values of x and y:

(a)

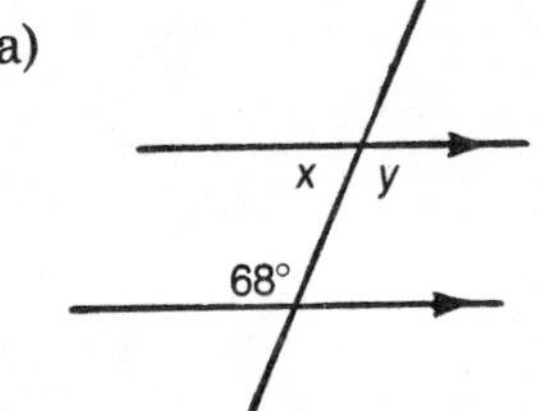

(b)

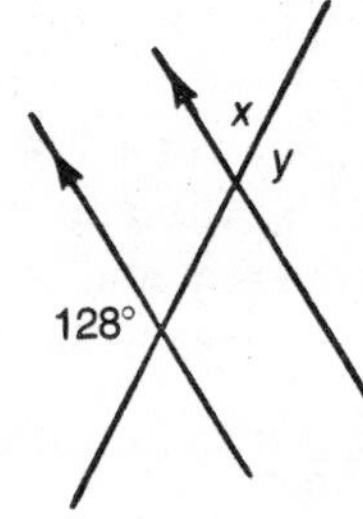

(c)

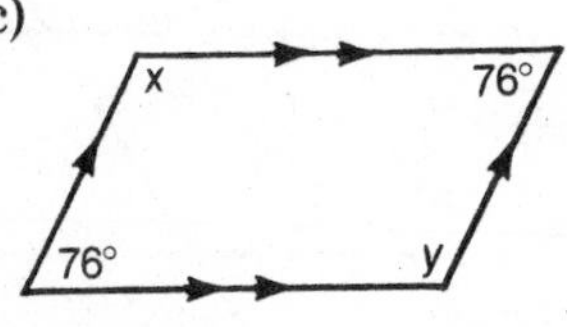

(d)

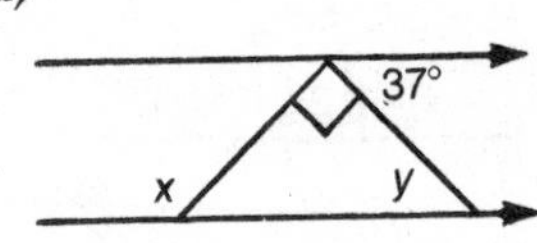

(e)

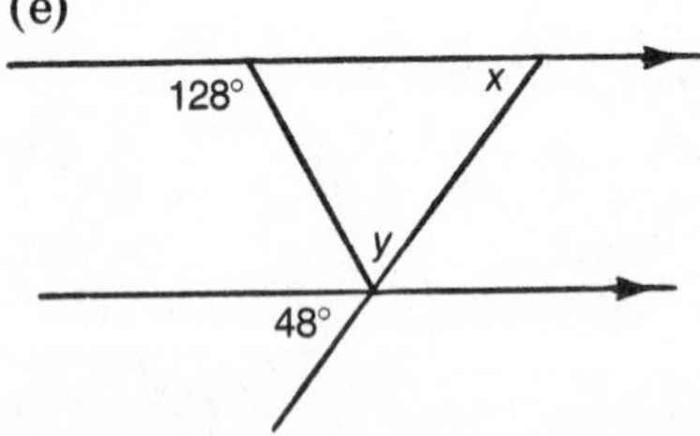

(f)

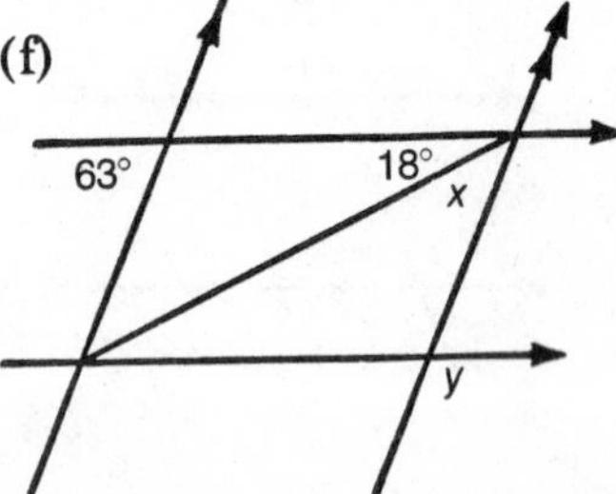

12. In the accompanying diagram, $\overleftrightarrow{AB} \parallel \overleftrightarrow{CD}$ and $\overline{EF}$ bisects $\angle AFG$.

(a) If $m\angle 1 = 100$, find the measure of each numbered angle.

(b) If $m\angle 3 = 4x - 9$ and $m\angle 5 = x + 19$, find the measure of each numbered angle.

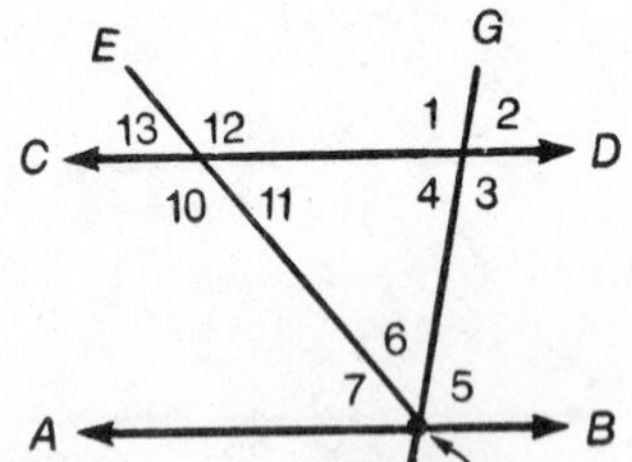

13. Two parallel lines are cut by a transversal. Find the measures of a pair of interior angles on the same side of the transversal if the angles:

(a) are represented by $5x - 32$ and $x + 8$

(b) have measures such that the measure of one angle is four times the measure of the other.

14. In the accompanying diagram, $\overleftrightarrow{AB} \parallel \overleftrightarrow{CD}$ and $\overrightarrow{FG}$ bisects $\angle EFD$. If $m\angle EFG = x$ and $m\angle FEG = 4x$, find x.

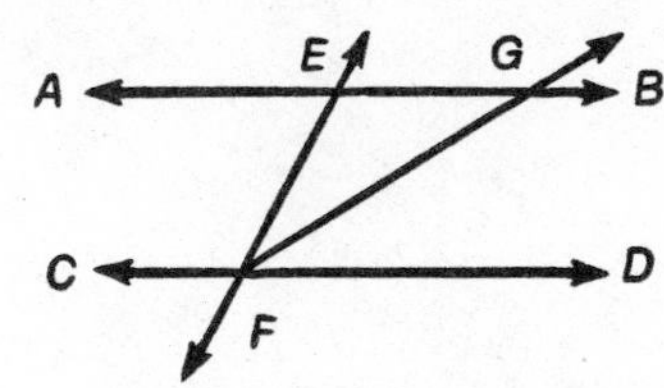

15. In the accompanying diagram, lines l_1 and l_2 are parallel and $m\angle 1 = 70$. What must $m\angle 2$ be so that lines l_3 and l_4 will be parallel?

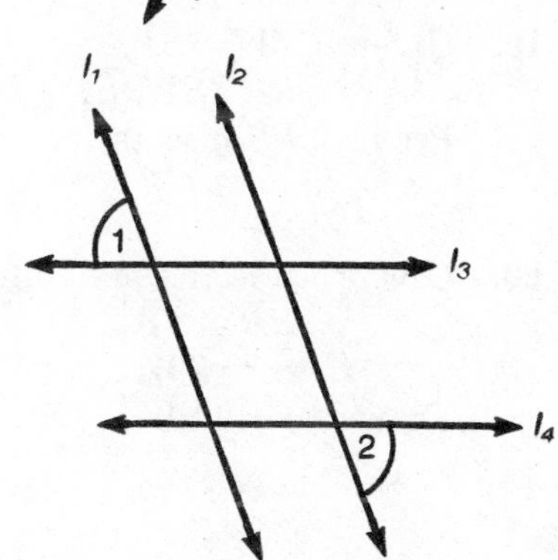

16. If, in the accompanying diagram, $L_1 \parallel L_2$ and $L_3 \parallel L_4$, then angle x is *not* always congruent to which angle?

(1) a (2) b (3) c (4) d

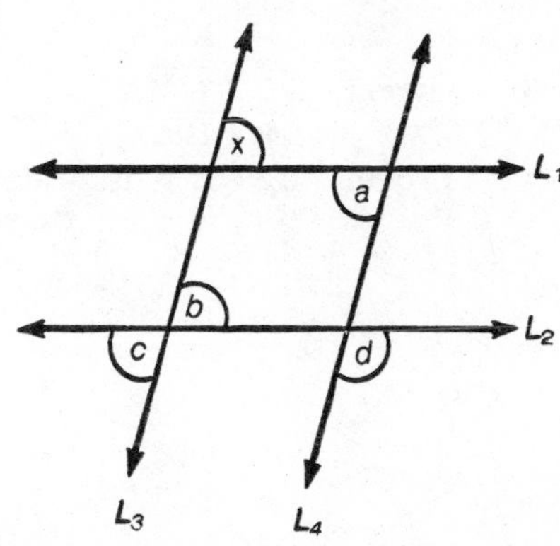

17. Given that lines p and q are parallel, determine whether line l is parallel to line m.

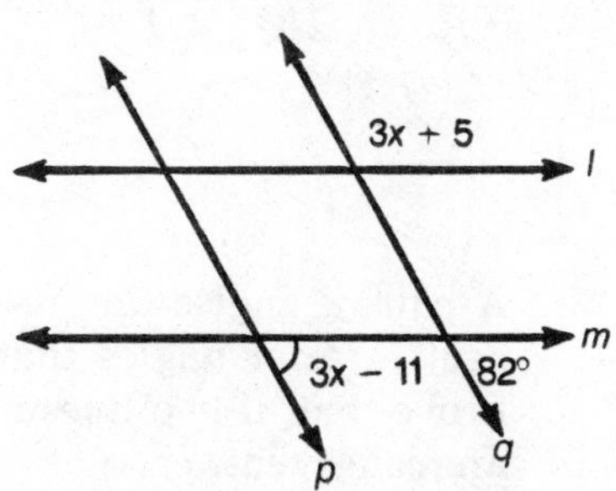

18. Given: $\overline{EF} \parallel \overline{AB}$, $\overline{ED} \parallel \overline{BC}$, $\overline{AD} \cong \overline{FC}$.
Prove: $\overline{AB} \cong \overline{EF}$.

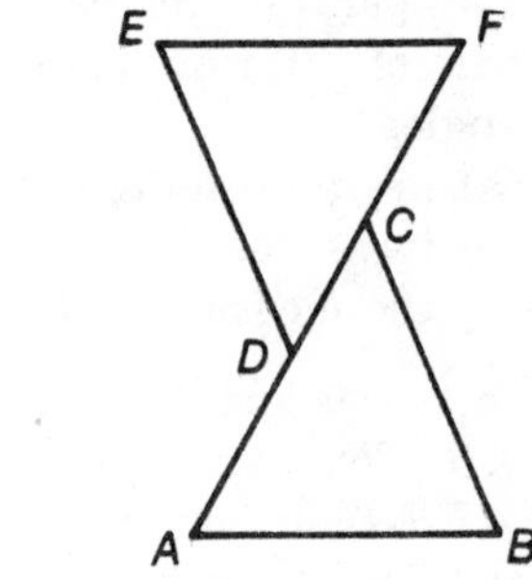

19. Given: $\overline{TW} \cong \overline{SP}$, $\overline{RP} \parallel \overline{SW}$, $\overline{SP} \parallel \overline{TW}$.
Prove: $\overline{PS}$ bisects $\overline{RT}$.

20. Given: S is the midpoint of $\overline{RT}$, $\overline{RP} \cong \overline{SW}$, $\overline{SP} \parallel \overline{TW}$.
Prove: $\overline{RP} \parallel \overline{SW}$.

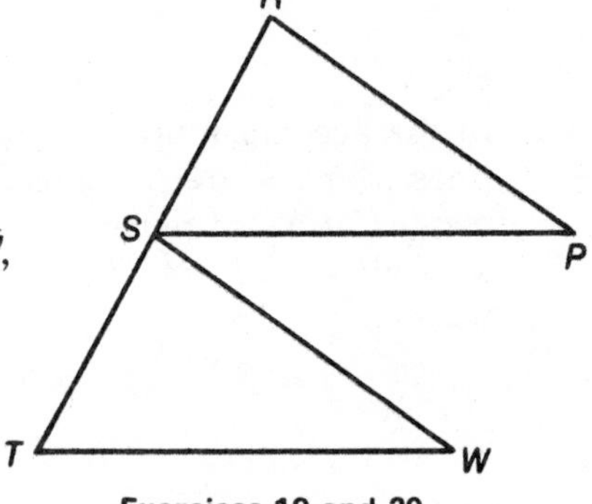

Exercises 19 and 20

21. Given: $\overline{QL} \cong \overline{QM}$, $\overline{LM} \parallel \overline{PR}$.
Prove: $\triangle PQR$ is isosceles.

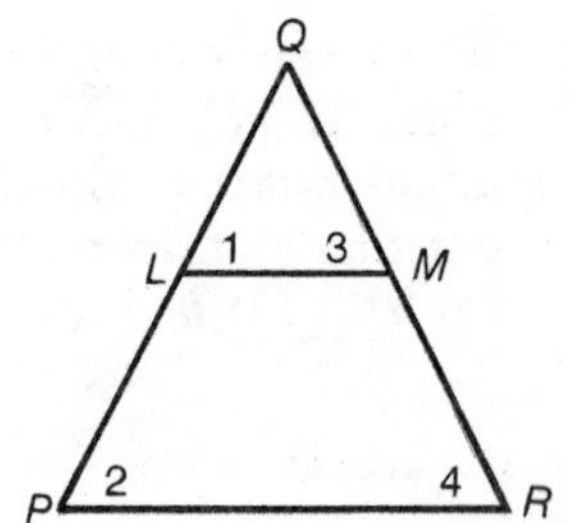

22. Given: $\overline{UT} \parallel \overline{DW}$, $\overline{UT} \cong \overline{DW}$, $\overline{QW} \cong \overline{AT}$.
Prove: $\overline{UQ} \parallel \overline{AD}$.

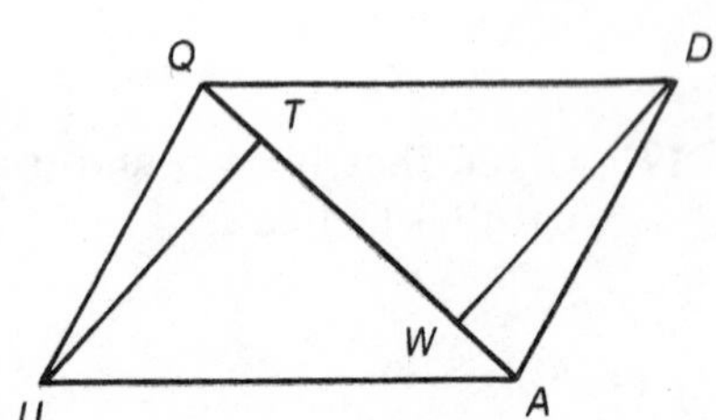

23. A pair of angles are *alternate exterior angles* if they are nonadjacent exterior angles that lie on opposite sides of the transversal. Prove that, if two lines are parallel, then alternate exterior angles are congruent.

24. Prove that, if one of two parallel lines is perpendicular to a third line, then the other parallel line is also perpendicular to the third line.

25. Given: $\overline{RS}$ intersects $\overleftrightarrow{ARB}$ and $\overleftrightarrow{CST}$,
$\overleftrightarrow{ARB} \parallel \overleftrightarrow{CST}$,
$\overline{RT}$ bisects $\angle BRS$,
M is the midpoint of $\overline{RT}$,
$\overline{SM}$ is drawn.
Prove: (a) $\overline{RS} \cong \overline{ST}$.
(b) $\overline{SM}$ bisects $\angle RST$.

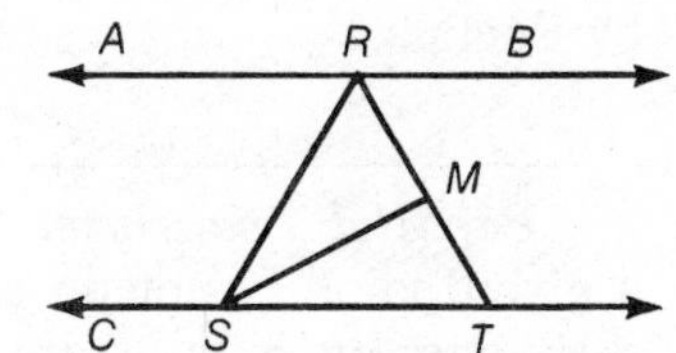

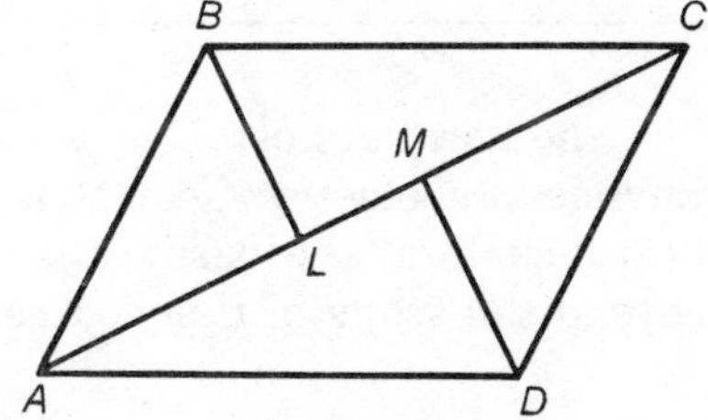

26. Given: $\overline{AB} \parallel \overline{CD}$,
$\overline{AB} \cong \overline{CD}$, $\overline{AL} \cong \overline{CM}$.
Prove: $\angle CBL \cong \angle ADM$.

27. Given: $\overline{AB} \parallel \overline{GCD} \parallel \overline{FE}$ with transversals $\overline{ACE}$ and $\overline{BCF}$,
$\overline{AC} \cong \overline{BC}$,
C is the midpoint of $\overline{GD}$.
Prove: $\overline{GF} \cong \overline{DE}$.

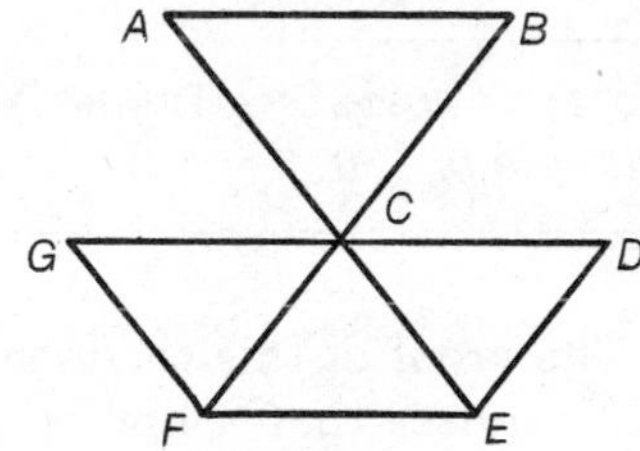

7.2 ANGLES OF A TRIANGLE

KEY IDEAS

In the accompanying diagram, can a line be drawn through B *and* parallel to $\overline{AC}$? Can a line be drawn from B to the midpoint of $\overline{AC}$? Can a line be drawn from B so that it is perpendicular to $\overline{AC}$?

The answer to each of these questions is yes since in each case the line is *determined*. A line is **determined** if there exists exactly one line that can be drawn to satisfy a given condition or set of conditions.

THE PARALLEL POSTULATE. Euclid, a Greek mathematician who lived in approximately 300 B.C., is credited with collecting and organizing the postulates and theorems that we study in plane geometry. The Parallel Postulate represents one of the fundamental assumptions made by Euclid.

***Parallel Postulate*:** Through a given point P, not on a line, exactly one line may be drawn parallel to the original line.

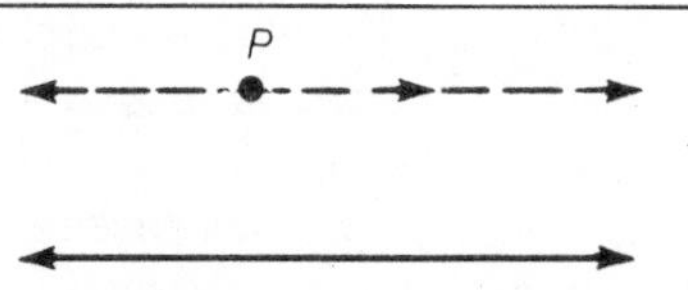

The Parallel Postulate is one of the most controversial postulates in Euclidean geometry. Geometric systems that do not accept the Parallel Postulate form the basis of *non-Euclidean* geometries, which are well beyond the scope of this course.

ANGLES OF A TRIANGLE THEOREM. You are already familiar with the following theorem:

THEOREM: ANGLES OF A TRIANGLE. *The sum of the measures of the angles of a triangle is 180.*

The proof of this theorem rests on the Parallel Postulate since the proof requires that a line be drawn through one of the vertices of the triangle and parallel to the opposite side. In the accompanying diagram, line l is drawn through B and parallel to $\overline{AC}$. Since angles 1, B, and 2 form a straight angle,

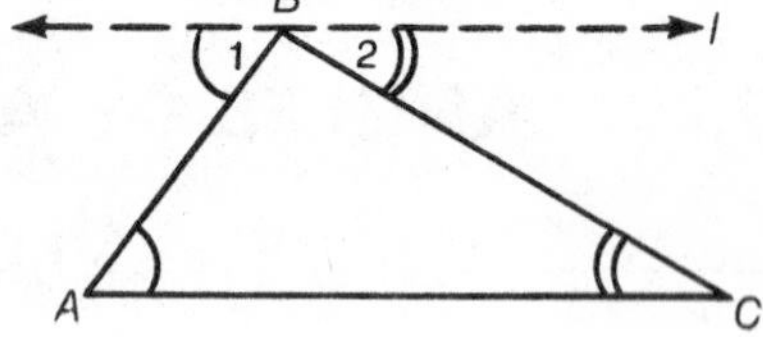

$$m\angle 1 + m\angle B + m\angle 2 = 180.$$

Angles 1 and A and angles 2 and C are equal in measure since parallel lines form congruent alternate interior angles. By substitution,

$$m\angle A + m\angle B + m\angle C = 180.$$

COROLLARIES. A **corollary** is a theorem that follows directly from a previously proved theorem.

Corollaries to Angles of a Triangle Theorem

Corollary 1: *A triangle may not have more than one right angle or more than one obtuse angle.*

Corollary 2: *If a triangle is equilateral, then it is also equiangular, so that each angle has a measure of 60.*

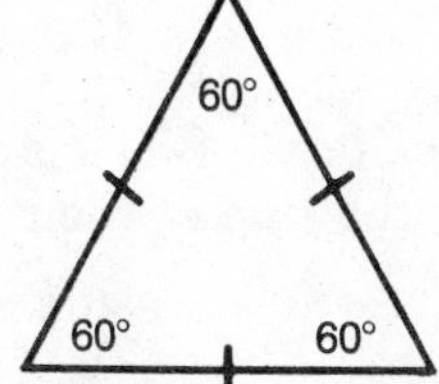

Corollary 3: *If two angles of one triangle are congruent to two angles of another triangle, then the third pair of angles must be congruent.*

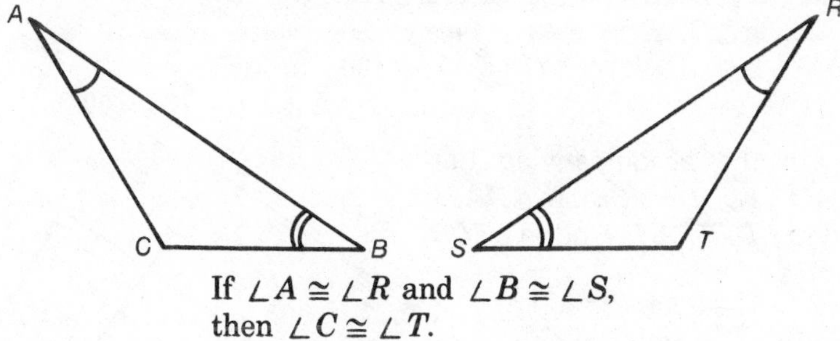

If $\angle A \cong \angle R$ and $\angle B \cong \angle S$,
then $\angle C \cong \angle T$.

Corollary 4: *The actue angles of a right triangle are complementary.*

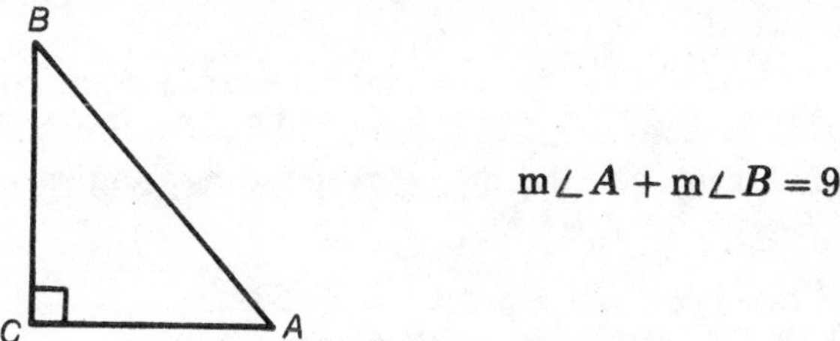

$$m\angle A + m\angle B = 90$$

Examples

1. The measures of the angles of a triangle are in the ratio of 2:3:5. What is the measure of the smallest angle of the triangle?

Solution: Let $2x$ = measure of the smallest angle of the triangle. Then $3x$ and $5x$ = measures of the remaining angles.

$$2x + 3x + 5x = 180$$
$$10x = 180$$
$$\frac{10x}{10} = \frac{180}{10}$$
$$x = 18$$
$$2x = 2(18) = 36$$

The measure of the smallest angle of the triangle is **36**.

2. In the accompanying diagram, $\overline{DE} \perp \overline{AEC}$. If $m\angle ADB = 80$ and $m\angle CDE = 60$, what is $m\angle DAE$?

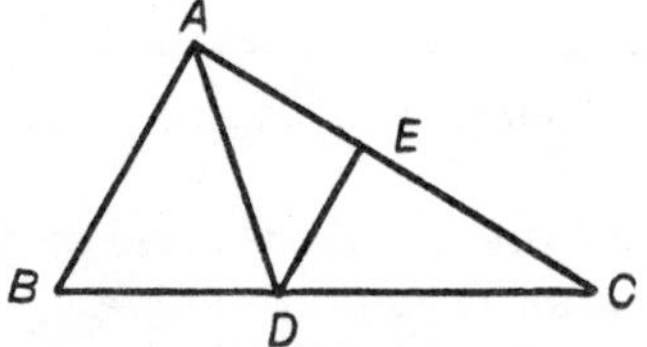

Solution: Since angles ADB, ADE, and CDE form a straight angle, the sum of their measures is 180. Hence

$$\begin{aligned} 80 + m\angle ADE + 60 &= 180 \\ m\angle ADE &= 180 - 140 = 40 \end{aligned}$$

In $\triangle ADE$,

$$\begin{aligned} m\angle DAE + m\angle ADE + m\angle AED &= 180 \\ m\angle DAE \quad 40 \quad + \quad 90 &= 180 \\ m\angle DAE + \quad 130 &= 180 \\ m\angle DAE &= 180 - 130 = \mathbf{50} \end{aligned}$$

3. In the accompanying diagram, $\overline{AD} \parallel \overline{EC}$, $\overline{DF} \parallel \overline{CB}$, $m\angle DAE = 34$, and $m\angle DFE = 57$. Find $m\angle ECB$.

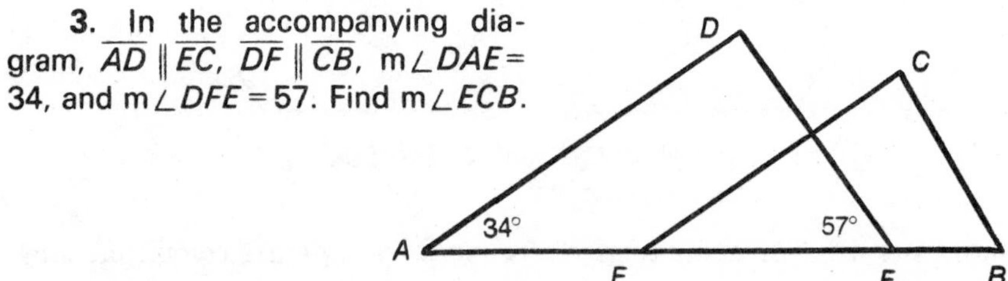

Solution: Since $\overline{AD} \parallel \overline{EC}$, transversal $\overline{AEFB}$ forms congruent corresponding angles, so that $m\angle CEB = m\angle DAE = 34$. Since $\overline{DF} \parallel \overline{CB}$, transversal $\overline{AEFB}$ forms congruent corresponding angles, so that $m\angle CBE = m\angle DFE = 57$. In $\triangle CEB$,

$$\begin{aligned} m\angle ECB + m\angle CEB + m\angle CBE &= 180 \\ m\angle ECB + \quad 34 \quad + \quad 57 &= 180 \\ m\angle ECB + \quad 91 &= 180 \\ m\angle ECB &= 180 - 91 = \mathbf{89} \end{aligned}$$

4. The degree measure of a base angle of an isosceles triangle exceeds twice the degree measure of the vertex angle by 15. Find the measure of the vertex angle.

Solution: Let x = measure of the vertex angle.
Then $2x + 15$ = measure of a base angle.

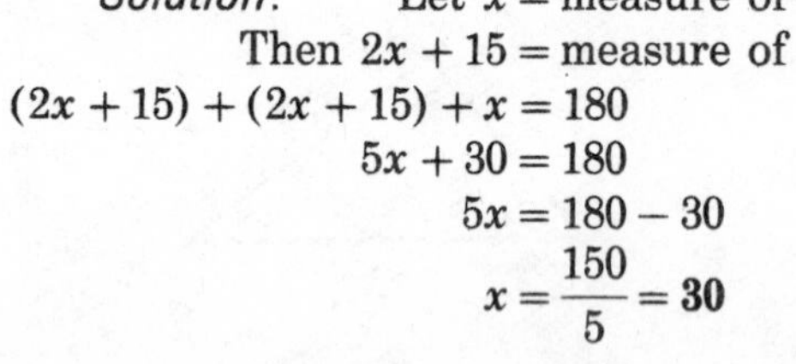

$$\begin{aligned} (2x + 15) + (2x + 15) + x &= 180 \\ 5x + 30 &= 180 \\ 5x &= 180 - 30 \\ x &= \frac{150}{5} = \mathbf{30} \end{aligned}$$

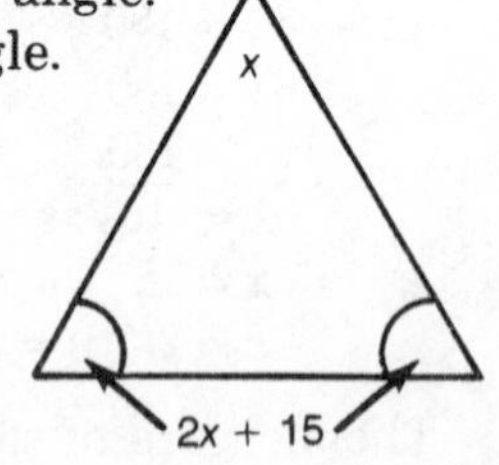

EXTERIOR ANGLES OF A TRIANGLE. At each vertex of a triangle an *exterior* angle of the triangle may be formed by extending one of the sides of the triangle. Notice in Figure 7.5 that each pair of adjacent angles consisting of an exterior and interior angle of the triangle are supplementary. For example,

(1) $m\angle 1 + c = 180$.

Also note that:

(2) $(a + b) + c = 180$.

Comparing the left sides of equations (1) and (2) leads to the conclusion that

$$m\angle 1 = a + b.$$

This result is generalized in the next theorem.

Figure 7.5 Exterior Angles of a Triangle

THEOREM: EXTERIOR ANGLE OF A TRIANGLE. *The measure of an exterior angle of a triangle is equal to the sum of the measures of the two nonadjacent (most remote) interior angles of the triangle.*

Examples

5. For each of the accompanying diagrams, find the value of *x*.

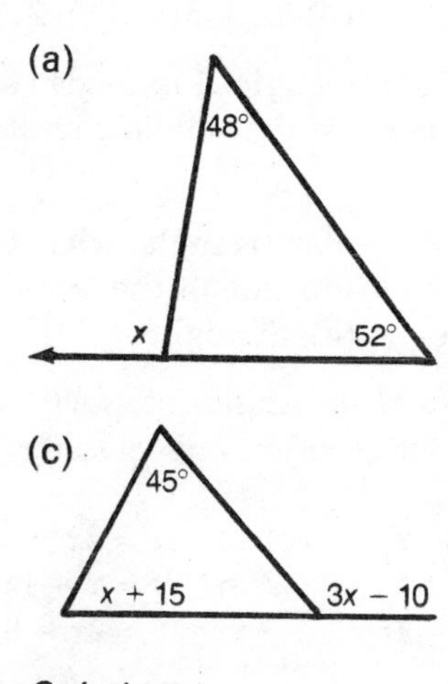

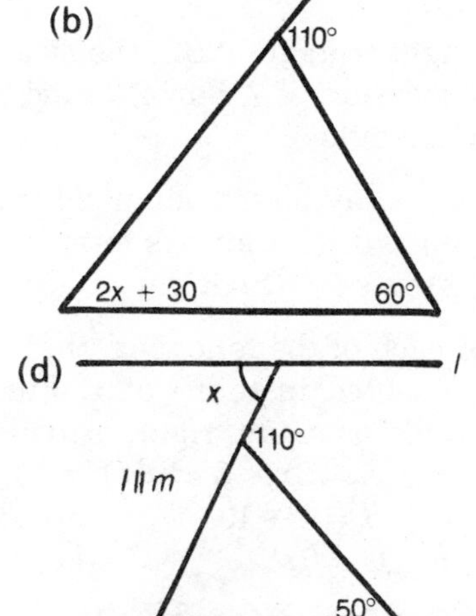

Solutions:

(a) $x = 48 + 52 = \mathbf{100}$

(b) $110 = 2x + 30 + 60$
$110 = 2x + 90$
$20 = 2x$
$x = \mathbf{10}$

(c) $3x - 10 = (x + 15) + 45$
$3x - 10 = x + 60$
$3x = x + 70$
$2x = 70$
$x = \mathbf{35}$

(d) $x = \mathbf{60}$

6. If the measure of an exterior angle formed by extending the base of an isosceles triangle is 112, what is the degree measure of the vertex angle?

Solution: $m\angle A = 180 - 112 = 68$
Therefore $m\angle C$ must also equal 68.

$$68 + 68 + m\angle B = 180$$
$$136 + m\angle B = 180$$
$$m\angle B = 180 - 136 = \mathbf{44}$$

EXERCISE SET 7.2

1. Find the measure of the vertex angle of an isosceles triangle if:
(a) the measure of a base angle is 43
(b) the measure of an exterior angle formed by extending the base is 117
(c) the measure of the vertex angle is three times the measure of a base angle
(d) the measure of the vertex angle exceeds the measure of a base angle by 15
(e) the measure of a base angle exceeds the measure of the vertex angle by 15

2. Find the measure of the smallest angle of a triangle if the measures of the three angles of the triangle are in the ratio:
(a) 1:2:6 (2) 2:3:10 (c) 1:1:2 (d) 3:4:5 (e) 2:7:7

3. In right triangle ABC, the measure of acute angle A exceeds twice the measure of $\angle B$ by 27. Find the measure of the smallest angle of the triangle.

4. When a ray bisects an angle of an equiangular triangle, what type of angle does it always form with the opposite side of the triangle?
(1) Acute (2) Right (3) Obtuse (4) Straight

5. For each of the following, the measures of the angles of $\triangle ABC$ are represented in terms of x. Find the value of x, and classify the triangle as acute, right, or obtuse.

(a)	(b)	(c)
$m\angle A = 3x + 8$	$m\angle A = x + 24$	$m\angle A = 3x - 5$
$m\angle B = x + 10$	$m\angle B = 4x + 17$	$m\angle B = x + 14$
$m\angle C = 5x$	$m\angle C = 2x - 15$	$m\angle C = 2x - 9$

6. In $\triangle ABC$, $\overline{BD} \perp \overline{AC}$. If $m\angle A = 72$ and $m\angle ABC = 54$, find $m\angle CBD$.

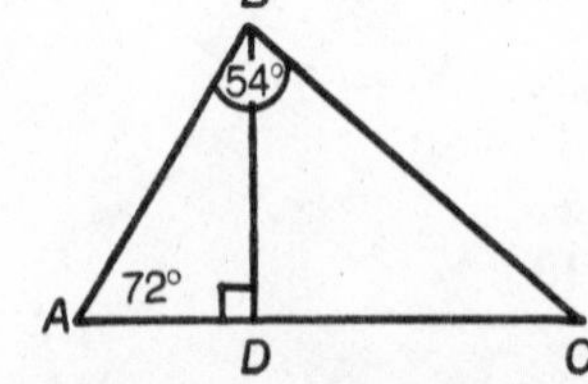

7. Given that $l \parallel m$, $m\angle 2 = 110$, and $m\angle 6 = 70$, find each remaining numbered angle.

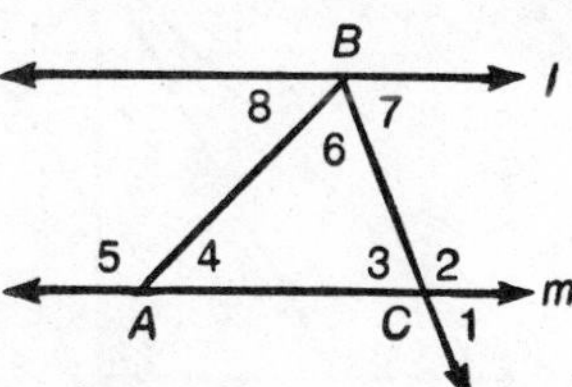

8–32. In each of the following, find the value of x:

8.

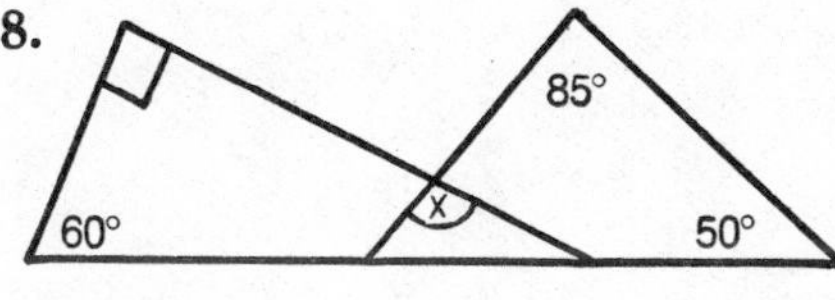

13.

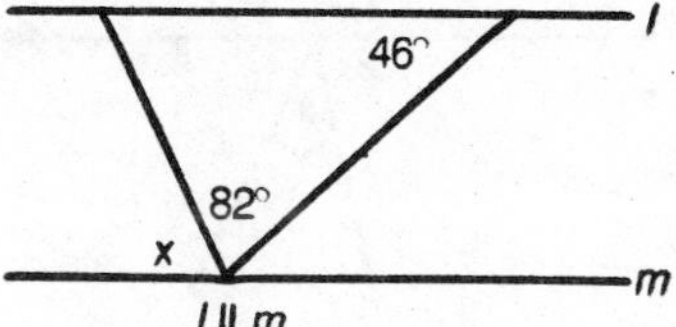

9.

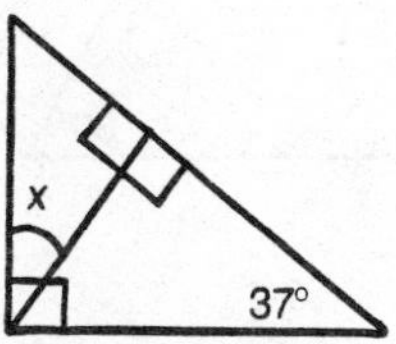

14.

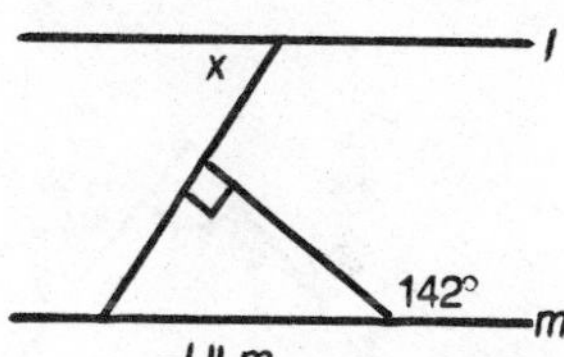

10. $\overline{AB} \parallel \overline{DE}$

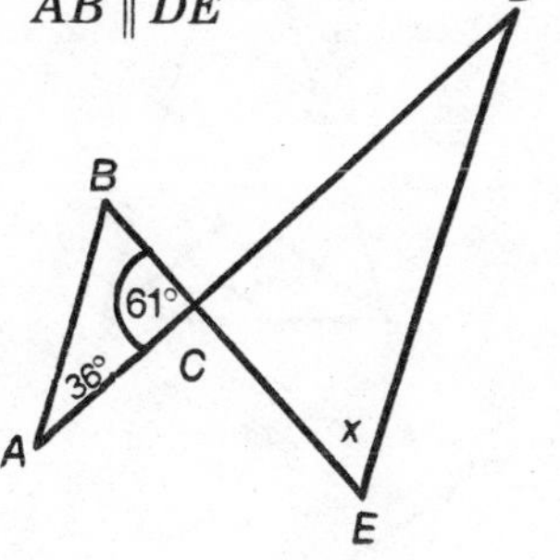

15.

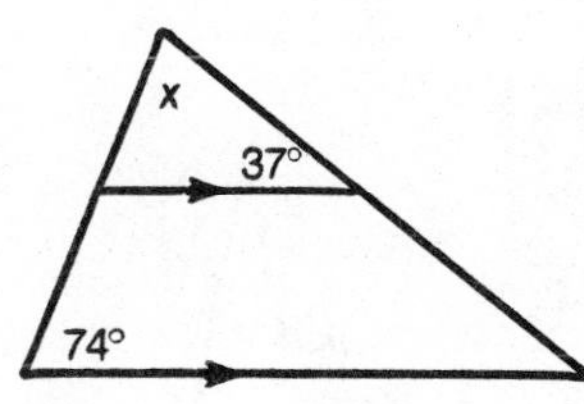

11.

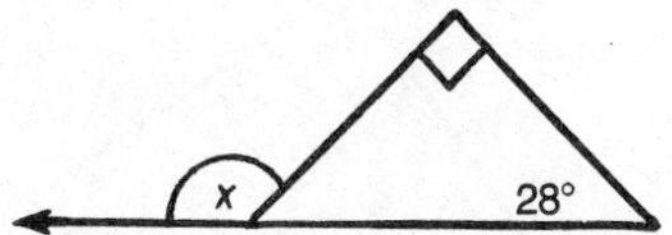

16.

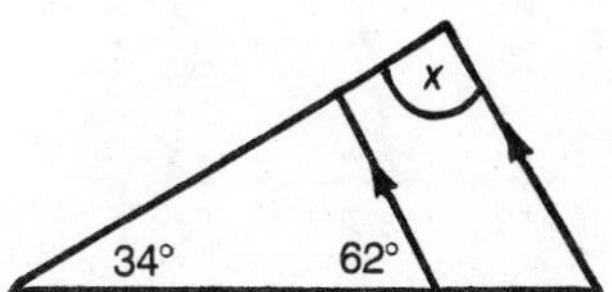

12.

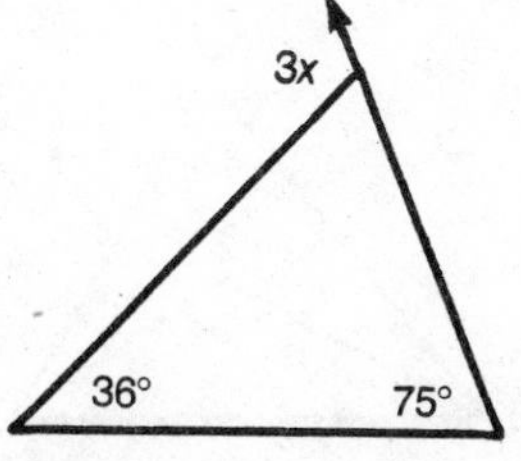

17.

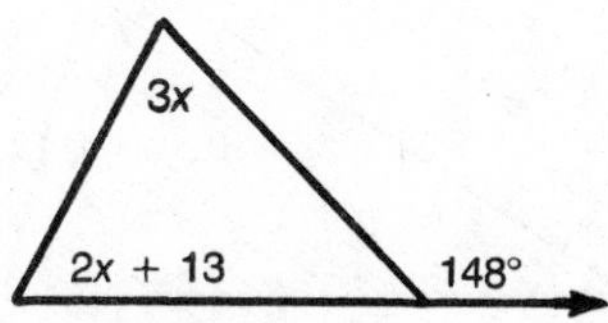

18.

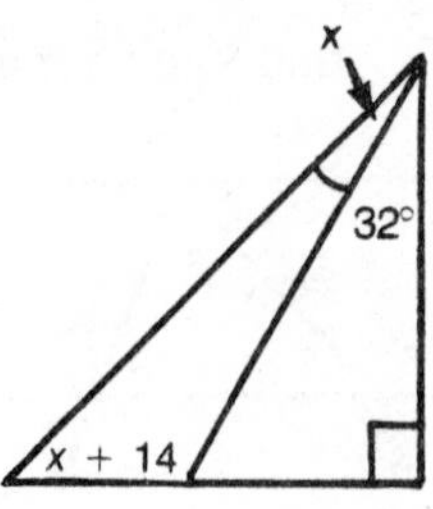

23.

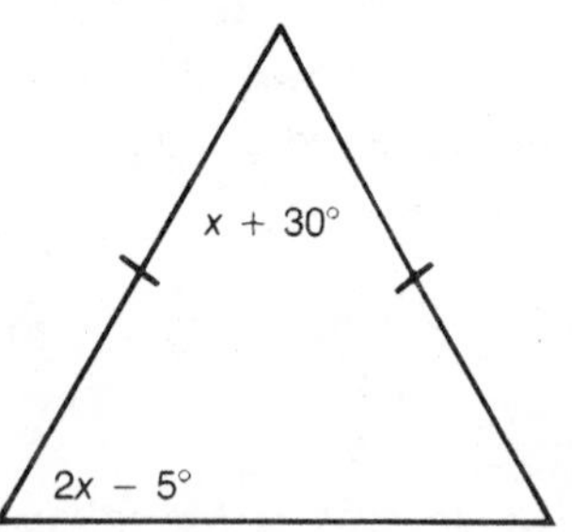

19.

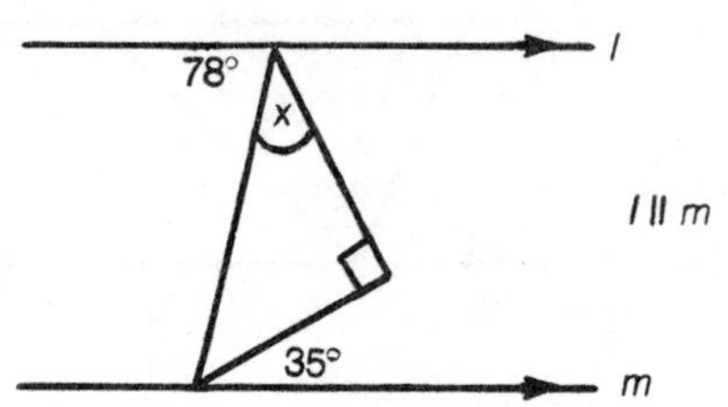

24.

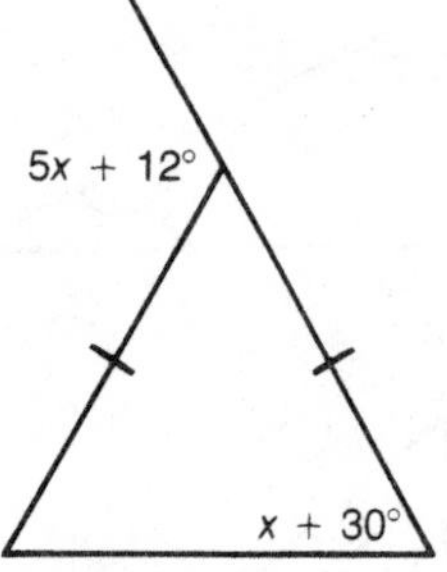

20.

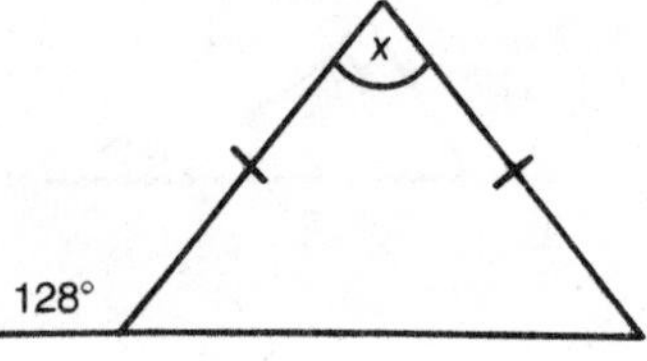

25.

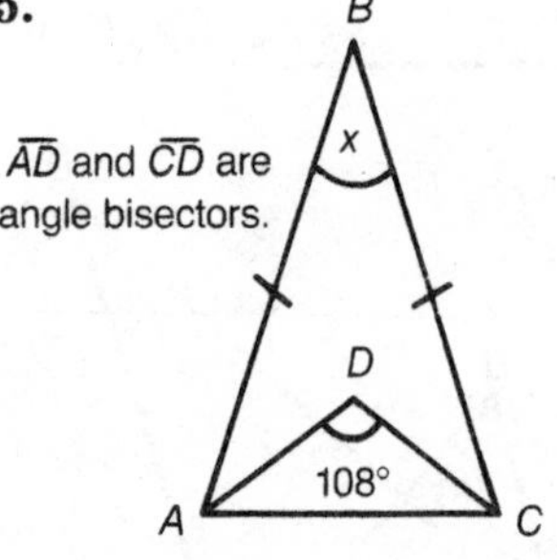

21.

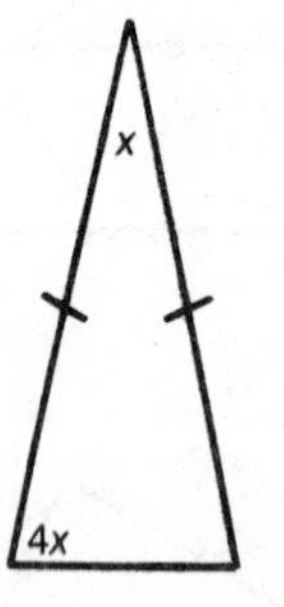

26.

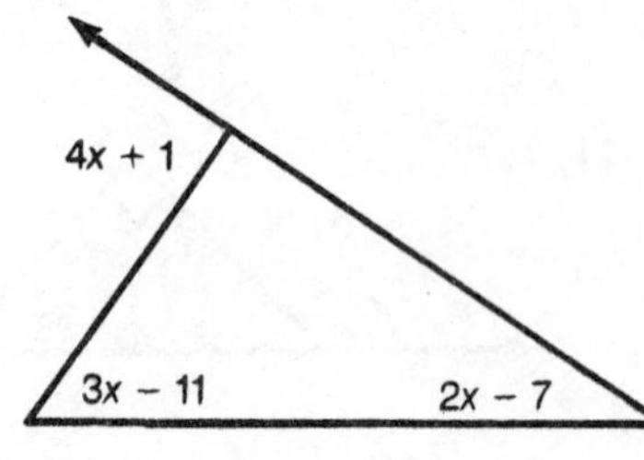

22.

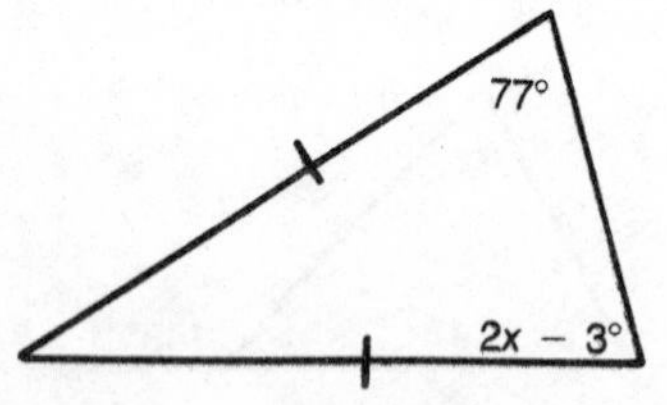

27.

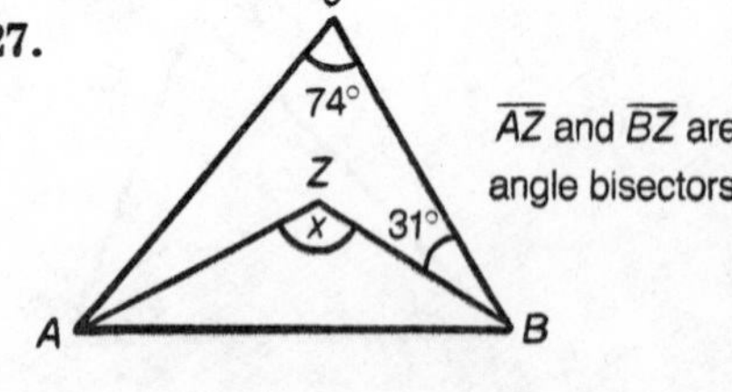

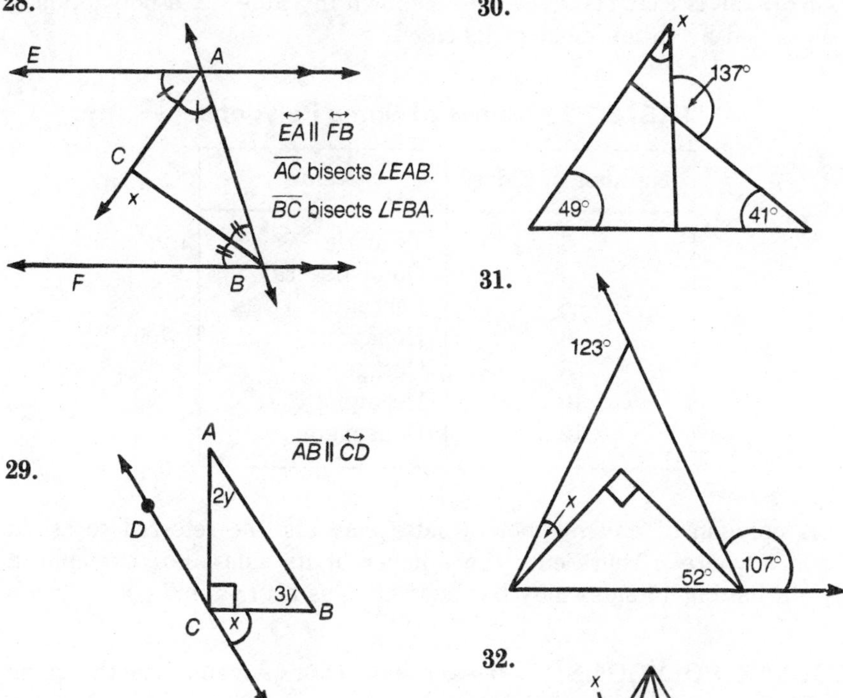

7.3 ANGLES OF A POLYGON

KEY IDEAS

A simple "closed" figure can be thought of as any figure that can be formed by stretching and/or twisting a rubber band so that it does not cross over itself.

A **polygon** is a simple closed curve whose sides are line segments that lie in the same plane. Each point of the polygon at which two sides intersect is called a **vertex** of the polygon. At each vertex there is an (interior) angle of the polygon. If the polygon has n sides, then the sum S of the measures of these n interior angles is given by the formula

$$\mathbf{S = (n - 2) \cdot 180}$$

CLASSIFYING POLYGONS. As shown in Table 7.2, a polygon may be classified by the number of its sides.

TABLE 7.2 Names of Some Polygons

Number of Sides	Name
3	Triangle
4	Quadrilateral
5	Pentagon
6	Hexagon
8	Octagon
10	Decagon
12	Dodecagon

A polygon of any number of sides may also be referred to as an "n-gon," where n represents the number of its sides. For example, a polygon having 13 sides may be referred to as a "13-gon."

REGULAR POLYGONS. If each side of a polygon has the same length, the polygon is **equilateral**. If each angle of a polygon has the same measure, the polygon is **equiangular**. If a polygon is both equilateral *and* equiangular, it is called a **regular polygon**.

THEOREMS FOR THE ANGLES OF A QUADRILATERAL AND A POLYGON. In Figure 7.6, $\overline{BD}$ is a *diagonal*. A **diagonal** of a polygon is a line segment joining two nonconsecutive vertices of the polygon. Notice that $\overline{BD}$ divides quadrilateral $ABCD$ into two triangles the sum of whose angles is the same as the sum of all the angles of the polygon. Thus, the sum of the measures of the four angles of the quadrilateral is equal to 2×180 or 360.

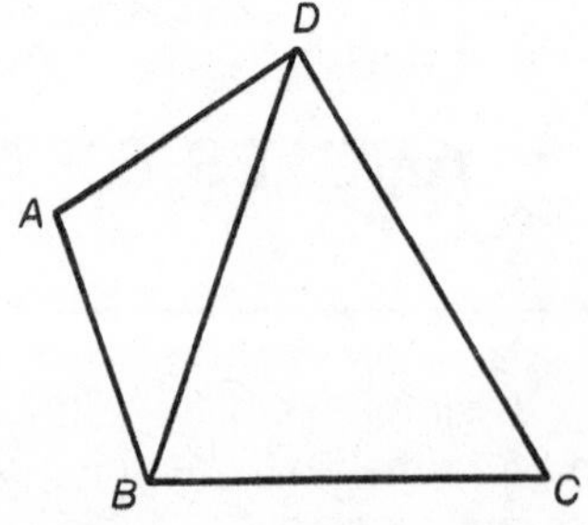

Figure 7.6 Diagonal of a Polygon

THEOREMS: ANGLES OF QUADRILATERALS AND POLYGONS

THEOREM 1: *The sum of the measures of the angles of any quadrilateral is* 360.

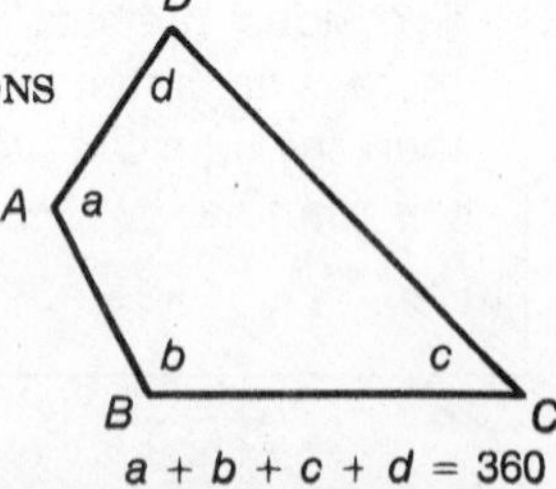

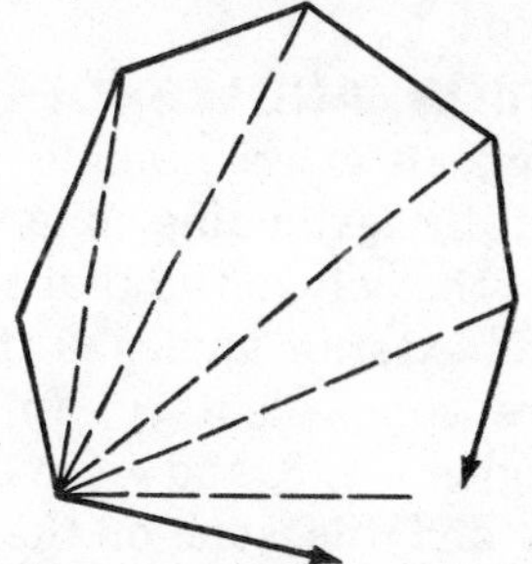

THEOREM 2: *A polygon having* n *sides can be separated into* n − 2 *triangles so that the sum* S *of the measures of the angles of a polygon having* n *sides is given by the formula* **S = (n − 2) · 180.**

Examples

1. The measures of the angles of a quadrilateral are in the ratio of 1 : 2 : 3 : 4. Find the measure of the *largest* angle of the quadrilateral.

Solution: Let x = measure of the *smallest* angle of the quadrilateral.

Then $2x$, $3x$, and $4x$ represent the measures of the remaining angles of the quadrilateral.

$$x + 2x + 3x + 4x = 360$$
$$10x = 360$$
$$x = \frac{360}{10} = 36$$
$$4x = 4(36) = 144$$

The measure of the largest angle of the quadrilateral is **144**.

2. (a) Find the sum of the measures of the angles of a hexagon.

(b) If the hexagon is regular, find the measure of each interior angle.

Solutions: (a) $S = (n - 2) \cdot 180$
$$= (6 - 2) \cdot 180$$
$$= 4 \cdot 180$$
$$= \mathbf{720}$$

(b) A regular polygon is equiangular, so the measure of each angle of a regular hexagon can be found by dividing the sum of the angle measures by 6 (the number of interior angles of the regular hexagon).

$$\text{Measure of each angle} = \frac{\text{Sum of measures of angles}}{\text{Number of angles}}$$

$$= \frac{720}{6} = \mathbf{120}$$

3. If the sum of the measures of the angles of a polygon is 900, determine the number of sides.

Solution: $900 = 180(n - 2)$
$$\frac{900}{180} = n - 2$$
$$5 = n - 2$$
$$n = \mathbf{7}$$

EXTERIOR ANGLES OF A POLYGON. At each vertex of a polygon, an *exterior* angle may be formed by extending one of the sides of the polygon so that the interior and exterior angles at that vertex are supplementary. In Figure 7.7, angles 1, 2, 3, and 4 are exterior angles and the sum of their measures is 360.

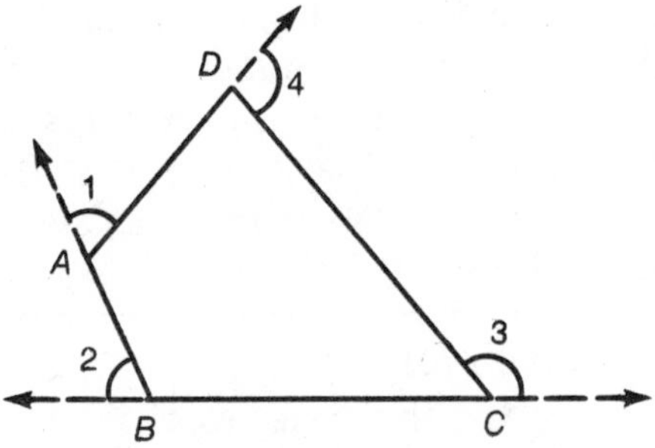

In general:

Figure 7.7 Exterior Angles of a Polygon

THEOREMS: EXTERIOR ANGLES OF POLYGONS

THEOREM 1: *The sum of the measures of the exterior angles of a polygon having any number of sides (one exterior angle at each vertex) is 360.*

THEOREM 2: *If a polygon having* n *sides is regular, then the measure of each exterior angle is* $\frac{360}{n}$.

Examples

4. Find the measure of each interior angle and each exterior angle of a regular decagon.

Solution:

Method 1	Method 2
$\text{Sum} = 180(n-2)$ $= 180(10-2)$ $= 180(8)$ $= 1{,}440$ Since there are 10 interior angles, each of which is identical in measure, $\text{Interior angle} = \frac{1{,}440}{10} = \mathbf{144}$ Since interior and exterior angles are supplementary, $\text{Exterior angle} = 180 - 144 = \mathbf{36}$	First determine the measure of an exterior angle: $\text{Exterior angle} = \frac{360}{10} = \mathbf{36}$ Since an interior angle and an adjacent exterior angle are supplementary, $\text{Interior angle} = 180 - 36 = \mathbf{144}$

5. The measure of each interior angle of a regular polygon is 150. Find the number of sides.

Solution: We use a method similar to the approach illustrated in Method 2 of Example 4 above. Since the measure of each interior angle is 150, the measure of an exterior angle is 180 – 150, or 30. Therefore,

$$30 = \frac{360}{n}$$

$$\text{or } n = \frac{360}{30} = \mathbf{12}$$

EXERCISE SET 7.3

1. Find the sum of the measures of the interior angles of a polygon having:
   (a) 4 sides (b) 6 sides (c) 9 sides (d) 13 sides
2. Find the number of sides of a polygon if the sum of the measures of the interior angles is:
   (a) 1,800 (b) 2,700 (c) 540 (d) 2,160
3. Find the measure of the remaining angle of each of the following figures, given the measures of the other interior angles.
   (a) Quadrilateral: 42, 75, and 118
   (b) Pentagon: 116, 138, 94, 88
   (c) Hexagon: 95, 154, 80, 145, 76
4. Find the measure of each interior angle of a regular polygon having:
   (a) 5 sides (b) 24 sides (c) 8 sides (d) 15 sides
5. Find the number of sides of a regular polygon if the measure of an interior angle is:
   (a) 162 (b) 144 (c) 140 (d) 168
6. Which of the following cannot represent the measure of an exterior angle of a regular polygon?
   (1) 72 (2) 15 (3) 27 (4) 45
7. Find the number of sides in a polygon if the sum of the measures of the interior angles is four times as great as the sum of the measures of the exterior angles.
8. Find the number of sides in a regular polygon if:
   (a) the measure of an interior angle is three times the measure of an exterior angle
   (b) the measure of an interior angle equals the measure of an exterior angle
   (c) the measure of an interior angle exceeds six times the measure of an exterior angle by 12

7.4 INEQUALITY RELATIONSHIPS IN A TRIANGLE

KEY IDEAS

This section illustrates how to apply the following inequality relationships:

- The length of *each* side of a triangle must be less than the sum of the lengths of the remaining two sides:

$AB < BC + CA$ and $BC < AB + CA$ and $CA < AB + BC$.

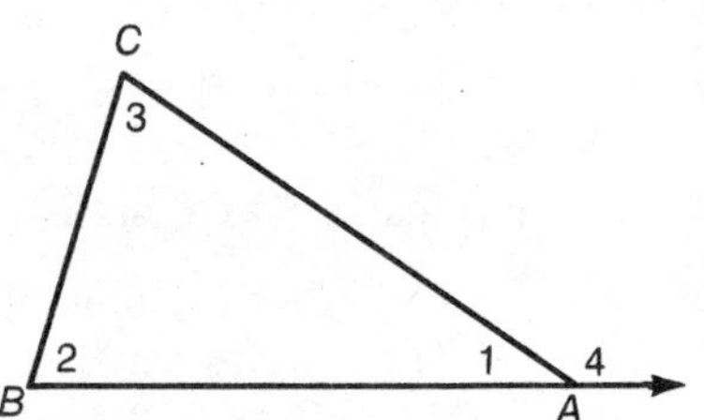

- The measure of an exterior angle of a triangle is greater than either nonadjacent interior angle. For example,

$m\angle 4 > m\angle 2$ and $m\angle 4 > m\angle 3$.

- Noncongruent sides lie opposite noncongruent angles with the longer side opposite the larger angle. For example:

If $m\angle 2 > m\angle 1$, then $AC > BC$.

If $BA > BC$, then $m\angle 3 > m\angle 1$.

TRIANGLE INEQUALITY RELATIONSHIP. To determine whether a set of three positive numbers can represent the lengths of the sides of a triangle, verify that *each* number of the set is less than the sum of the other two.

Examples

1. Which of the following sets of numbers cannot represent the lengths of the sides of a triangle?

(1) $\{9, 40, 41\}$ (2) $\{7, 7, 3\}$ (3) $\{4, 5, 1\}$ (d) $\{6, 6, 6\}$

Solution: The correct answer is **choice (3)**. Although $4 < 5 + 1$ and $1 < 4 + 5$, 5 is *not* less than $4 + 1$.

2. If the lengths of two sides of an isosceles triangle are 3 and 7, what must be the length of the third side?

Solution: Since the triangle must have two sides of equal length, the length of the remaining side is either 3 or 7. The length of this side cannot be 3 since 7 is *not* less than $3+3$. Since $3<7+7$ and $7<3+7$, the length of the third side is **7**.

NONCONGRUENT SIDES IMPLY NONCONGRUENT ANGLES. Here are three examples that illustrate inequality relationships in a triangle.

Examples

3. In $\triangle ABC$, $AB=3$, $BC=5$, and $AC=7$. What is:
(a) the largest angle of the triangle?
(b) the smallest angle of the triangle?

Solutions: (a) $\angle \boldsymbol{B}$ is the largest angle.
(b) $\angle \boldsymbol{C}$ is the smallest angle.

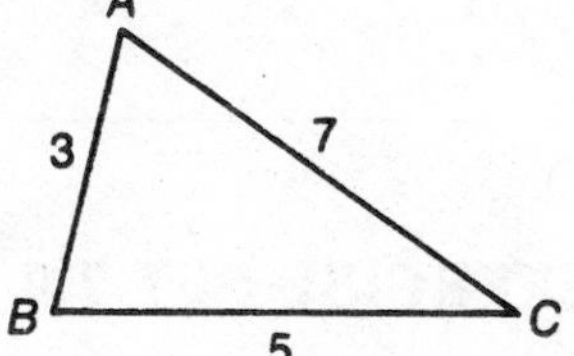

4. The measure of base angle *R* of isosceles triangle *RST* is 50. What is the longest side of the triangle?

Solution: We find the measure of the other two angles of the triangle. $\overline{\boldsymbol{RT}}$ is the longest side since it lies opposite the largest angle ($\angle S$).

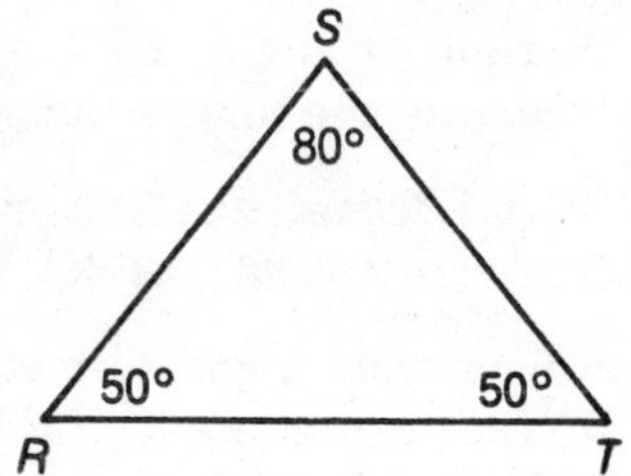

5. Given: $\overline{BD}$ bisects $\angle ABC$.
Prove: $AB > AD$.

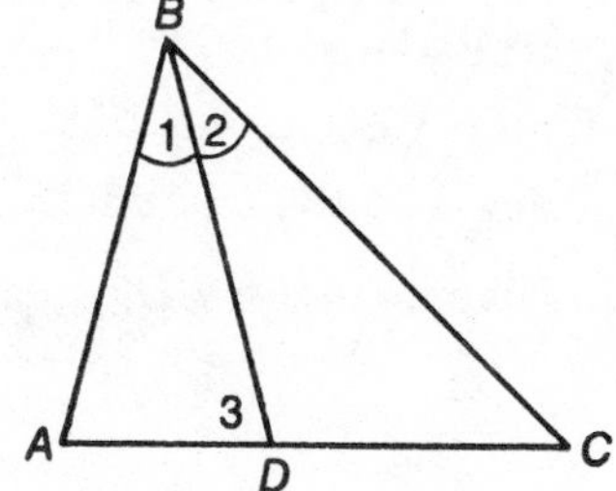

Solution:
PLAN. To prove $AB > AD$, first establish that $m\angle 3$ (the angle opposite $\overline{AB}$) is greater than $m\angle 1$ (the angle opposite $\overline{AD}$).

PROOF

Statement	Reason
1. $m\angle 3 > m\angle 2$	1. The measure of an exterior angle of a triangle is greater than the measure of either nonadjacent interior angle.
2. $\overline{BD}$ bisects $\angle ABC$.	2. Given.
3. $m\angle 1 = m\angle 2$	3. A bisector divides an angle into two angles having the same measure.
4. $m\angle 3 > m\angle 1$	4. Substitution property of inequalities.
5. $AB > AD$	5. If two angles of a triangle are not equal in measure, then the sides opposite are not equal and the longer side is opposite the larger angle.

EXERCISE SET 7.4

1. In $\triangle ABC$, $m\angle A = 50$ and $m\angle B = 60$. Which is the longest side of the triangle?

2. In $\triangle ABC$, $m\angle A = 30$ and the measure of the exterior angle at B is 120. Which is the longest side of the triangle?

3. In right triangle ABC, altitude $\overline{CD}$ is drawn to hypotenuse $\overline{AB}$. Which is the longest side of $\triangle CDB$?

4. In $\triangle ABC$, $m\angle B = 120$, $m\angle A = 55$, and D is the point on $\overline{AC}$ such that $\overline{BD}$ bisects $\angle ABC$. Which is the longest side of $\triangle ABD$?

5. An exterior angle formed at vertex angle J of isosceles triangle JKL by extending leg $\overline{LJ}$ has a degree measure of 115. Which is the longest side of the triangle?

6. In $\triangle ABC$, $BC > AB$ and $AC < AB$. Which is the longest side of the triangle?

7. In $\triangle RST$, $m\angle R < m\angle T$ and $m\angle S > m\angle T$. Which is the largest angle of the triangle?

8. Given: $\overline{AB} \cong \overline{BD}$,
$m\angle 5 > m\angle 6$.

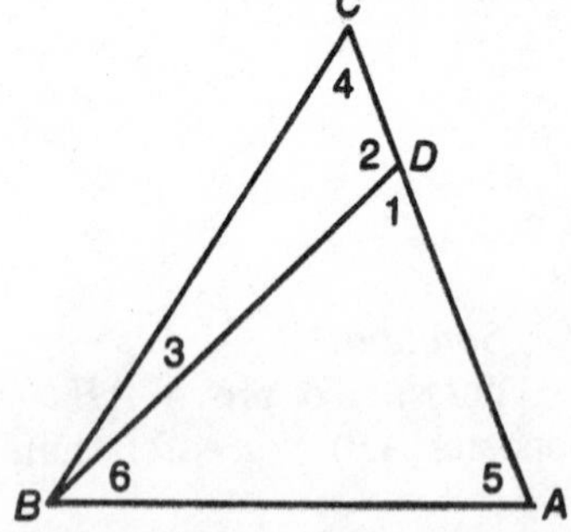

State whether each of the following inequality relationships is true or false:

(a) $m\angle 1 > m\angle 3$ (b) $m\angle 5 > m\angle 1$ (c) $m\angle 3 < m\angle ABC$
(d) $m\angle 2 > m\angle 1$ (e) $AB < AD$ (f) $BC > BA$

9. In $\triangle ABC$, $AB > AC$ and $BC > AC$. Name the smallest angle of $\triangle ABC$.

10. In $\triangle RST$, $ST > RT$ and $RT > RS$.
(a) If one of the angles of the triangle is obtuse, which angle of the triangle must it be?
(b) If the measure of one of the angles of the triangle is 60, which angle of the triangle must it be?

11. Determine whether each of the following sets of numbers can represent the lengths of the sides of a triangle.

(a) $\{8, 17, 15\}$ (b) $\left\{\frac{1}{2}, \frac{1}{3}, \frac{1}{6}\right\}$ (c) $\{1, 1, 3\}$ (d) $\{6, 6, 7\}$

12. Given: $\overline{AB} \cong \overline{CB}$.
Prove: $AB > BD$.

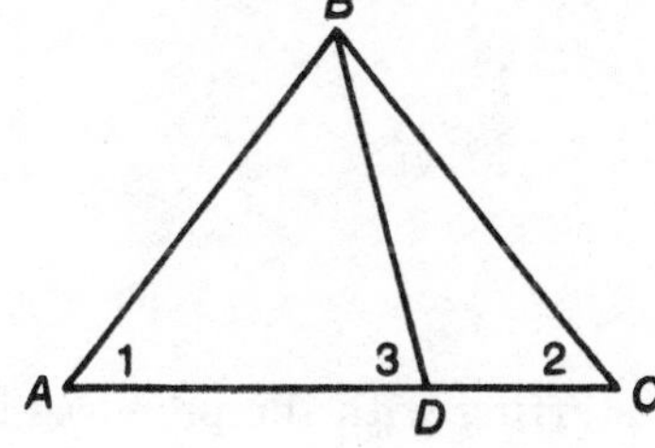

13. Given: $m\angle 1 = m\angle 2$.
Prove: $AD > ED$.

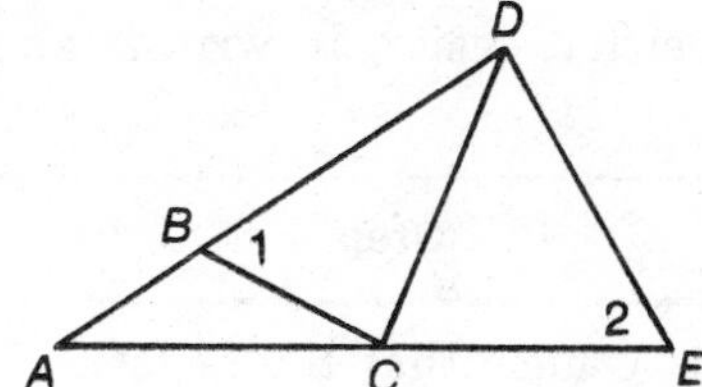

14. Given: Triangles AEC and ABC.
Prove: $m\angle 4 > m\angle AEC$.

15. Given: $AC > BC$.
Prove: $AD > BD$.

16. Given: $AD > BD$,
$\overline{AD}$ bisects $\angle BAC$.
Prove: $AC > DC$.

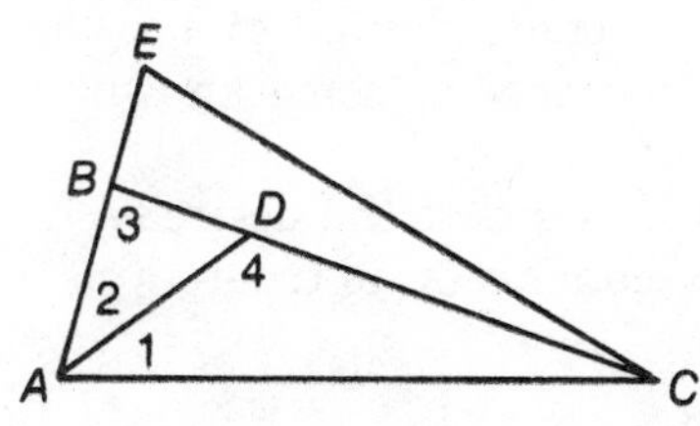

Exercises 14–16

7.5 INDIRECT METHOD OF PROOF

KEY IDEAS

> To prove a conclusion *indirectly*, we begin with the conclusion, assume that it is false, and then demonstrate that this assumption leads to a contradiction of a premise or some known fact. Arriving at a contradiction means that the conclusion with which we started must be true rather than false.

INDIRECT PROOFS. The logic and geometric proofs presented so far have been *direct* proofs. Consider this geometric proof:

Given: $\overline{TW} \perp \overline{RS}$,
$\angle 1 \not\cong \angle 2$.
Prove: $\overline{TW}$ is not the median to side $\overline{RS}$.

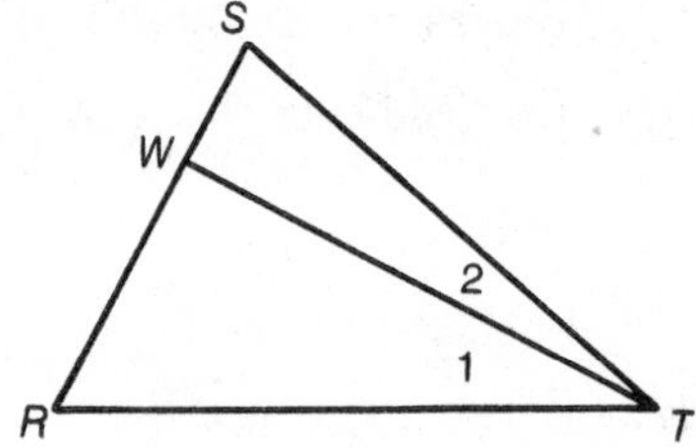

Starting with the premises in the "Given," we could write a direct proof that would lead to the desired conclusion. In this case, however, a proof that begins with the conclusion and uses indirect reasoning is somewhat easier. In writing an indirect proof, follow these steps:

Step	Example
1. Assume that the negation of the desired conclusion is true.	1. Assume that $\overline{TW}$ *is* the median to side $\overline{RS}$.
2. Prove that this assumption is false by showing that it leads to a contradiction of a premise or some known fact.	2. Hence, $\overline{WR} \cong \overline{WS}$ and triangles *TWS* and *TWR* are congruent by SAS $\cong$ SAS. By CPCTC, $< 1 \cong < 2$. But this contradicts the second premise.
3. State that the desired conclusion is true.	3. The conclusion $\overline{TW}$ is *not* the median to side $\overline{RS}$ is true since its negation leads to a contradiction and must therefore be false.

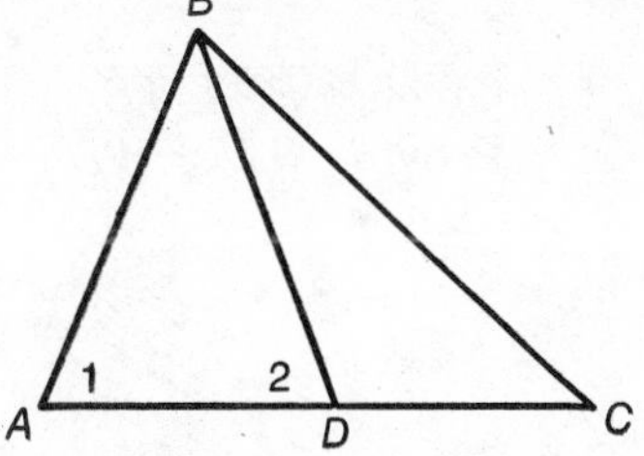

Example

Given: $\overline{AB} \cong \overline{DB}$.
Prove: $\overline{AB} \not\cong \overline{BC}$.

Solution: We use an indirect proof.

Step 1. Assume that $\overline{AB} \cong \overline{BC}$.

Step 2. If $\overline{AB} \cong \overline{BC}$, then $\angle 1 \cong \angle C$. Since $\overline{AB} \cong \overline{DB}$, $\angle 1 \cong \angle 2$. By the transitive property, $\angle 2 \cong \angle C$. But this contradicts the theorem that states that the measure of an exterior angle of a triangle ($\angle 2$) is greater than the measure of either nonadjacent interior angle ($\angle C$).

Step 3. $\overline{AB} \not\cong \overline{BC}$ is true since its negation leads to a contradiction.

INDIRECT REASONING AND LOGIC. Indirect proofs are based on the law of contrapositive inference. In an indirect proof, a statement of the form "If p, then q" is proved by showing that $\sim q \rightarrow \sim p$ is true and then concluding that its contrapositive, $p \rightarrow q$, is also true.

EXERCISE SET 7.5

1. If indirect reasoning is used to prove $a > b$, then which assumption must be proved false?
(1) $a < b$ (2) $a \leq b$ (3) $a = b$ (4) $a \geq b$

2. If indirect reasoning is used to prove the theorem "*If two lines form a pair of congruent corresponding angles, then the lines are parallel*," then which assumption must be proved false?
(1) The corresponding angles are congruent.
(2) The corresponding angles are not congruent.
(3) The lines intersect.
(4) The lines do not intersect.

3. Given: $\angle 1 \cong \angle 3$.
Prove: $\angle 1 \not\cong \angle 2$.

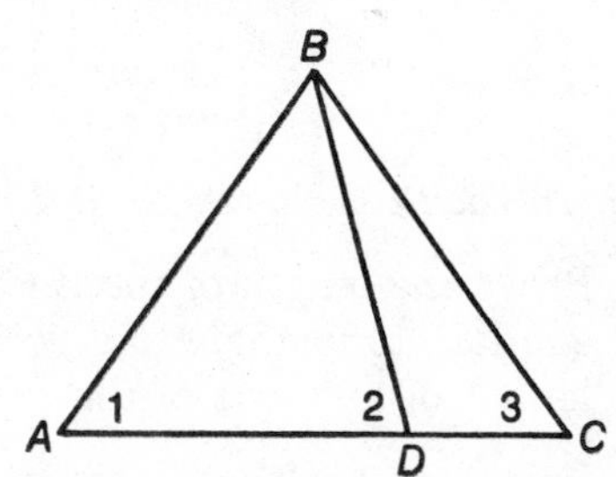

4. Given: $\angle 1 \cong \angle 2$.
Prove: $\overline{AB} \not\cong \overline{BC}$.

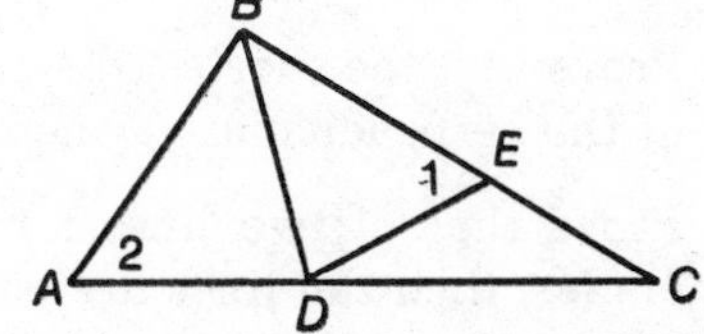

5. Given: $\overline{RS} \cong \overline{TS}$.
Prove: $\overline{RW} \not\cong \overline{WL}$.

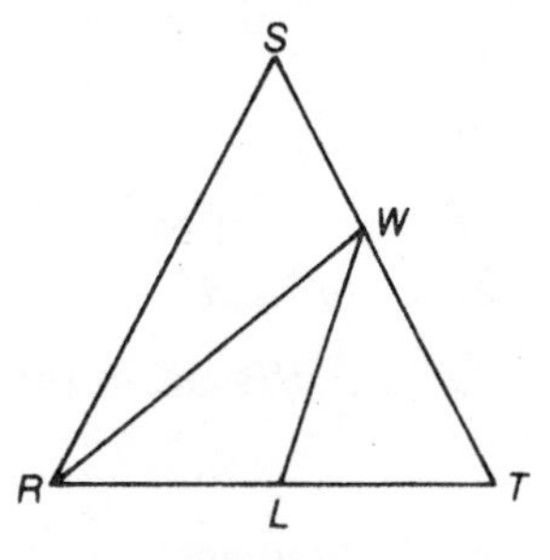

6. Given: $\triangle ABC$ is *not* isosceles,
$\angle ADB \cong \angle CDB$.
Prove: $\triangle ADC$ is *not* isosceles.

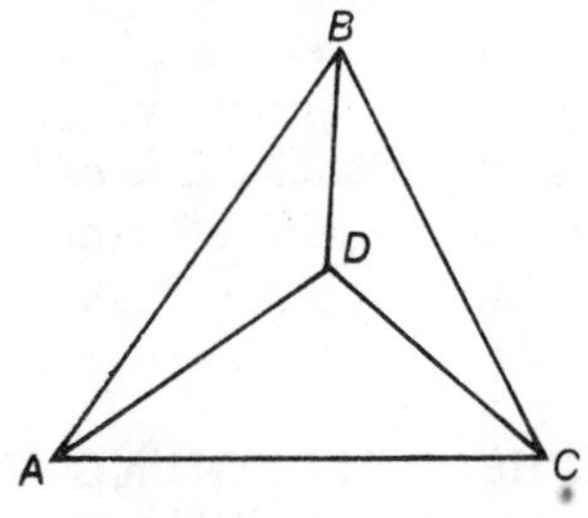

7. Given: $AC > AB$,
$\overline{DE} \cong \overline{CE}$.
Prove: $\overline{AB}$ is *not* parallel to $\overline{DE}$.

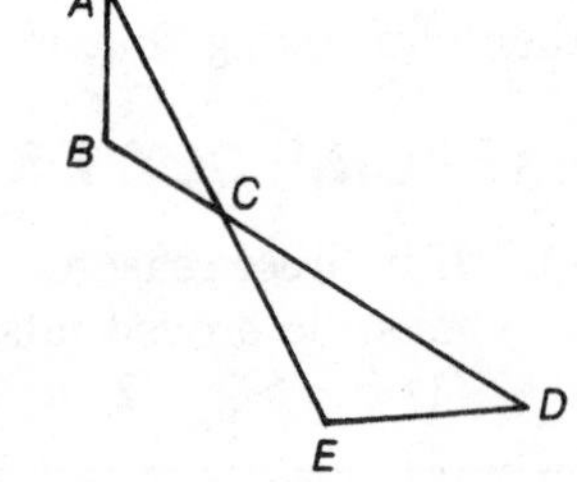

8. Given: $\triangle ABC$ is scalene,
$\overline{BD}$ bisects $\angle ABC$.
Prove: $\overline{BD}$ is *not* perpendicular to $\overline{AC}$.

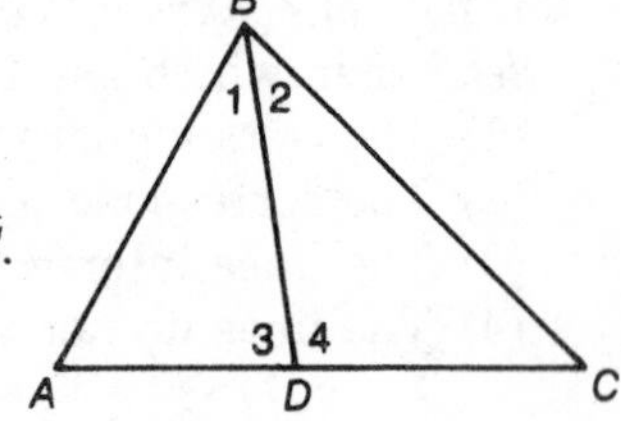

9–12. In each case, write an indirect proof.

9. Prove the length of the line segment drawn from any vertex of an equilateral triangle to a point on the opposite side is less than the length of any side of the triangle.

10. Prove that, if the vertex angle of an isosceles triangle is obtuse, then the base is longer than either leg.

11. Prove that the shortest distance from a point to a line is the length of the perpendicular segment from the point to the line.

12. Prove that, if two lines form a pair of congruent corresponding angles, then the lines are parallel.

7.6 PARALLELOGRAMS AND THEIR PROPERTIES

KEY IDEAS

A **parallelogram** is a quadrilateral having two pairs of parallel sides. A rectangle, a rhombus, and a square are special types of parallelograms. A *rectangle* is a parallelogram having four congruent (right) angles; a *rhombus* is a parallelogram having four congruent sides, a *square* is a parallelogram having four congruent angles *and* four congruent sides.

ANGLES OF A PARALLELOGRAM. The notation $\square ABCD$ is read as "parallelogram *ABCD*." In a parallelogram (Figure 7.8):

- The sum of the measures of the angles is 360.

$a+b+c+d=360$

- Consecutive angles are supplementary.

$a+b=180$ and $b+c=180$
$c+d=180$ and $a+d=180$

- Opposite angles are congruent.

$\angle A \cong \angle C$ and $\angle B \cong \angle D$

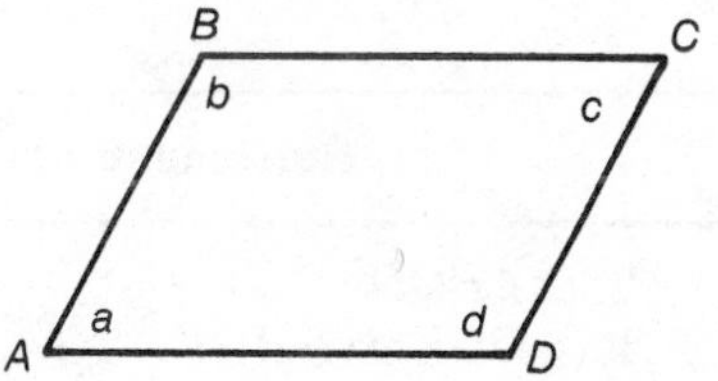

Figure 7.8 Angles of a Parallelogram

SIDES AND DIAGONALS OF A PARALLELOGRAM. In a parallelogram (Figure 7.9):

- Opposite sides are parallel.

  $\overline{AB} \parallel \overline{DC}$ and $\overline{AD} \parallel \overline{BC}$

- Opposite sides are congruent.

  $\overline{AB} \cong \overline{DC}$ and $\overline{AD} \cong \overline{BC}$

- Diagonals bisect each other.

  $\overline{AE} \cong \overline{EC}$ and $\overline{DE} \cong \overline{EB}$

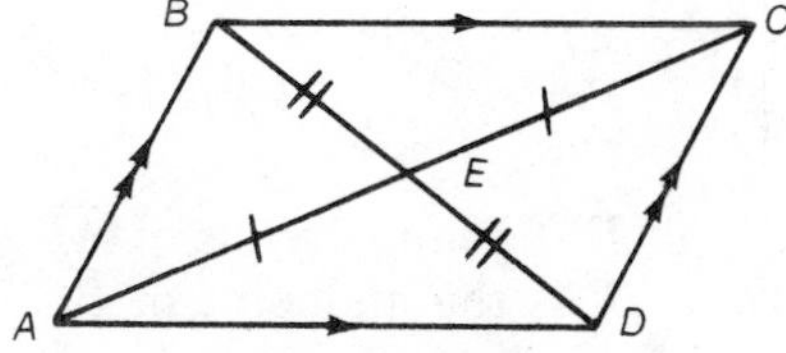

Figure 7.9 Sides and Diagonals of a Parallelogram

Examples

1. In $\square ABCD$, $m\angle B = 5x - 43$ and $m\angle D = 2x - 7$. Find the numerical value of x.

Solution: Angles B and D are opposite angles of a parallelogram and are, therefore, congruent. Hence.

$$
\begin{aligned}
m\angle B &= m\angle D \\
5x - 43 &= 2x - 7 \\
5x &= 2x - 7 + 43 \\
5x &= 2x + 36 \\
5x - 2x &= 36 \\
\frac{3x}{3} &= \frac{36}{3} \\
x &= \mathbf{12}
\end{aligned}
$$

2. Given: $\square ABCD$,
E is the midpoint of $\overline{AD}$,
F is the midpoint of $\overline{BC}$.
Prove: G is the midpoint of $\overline{EF}$.

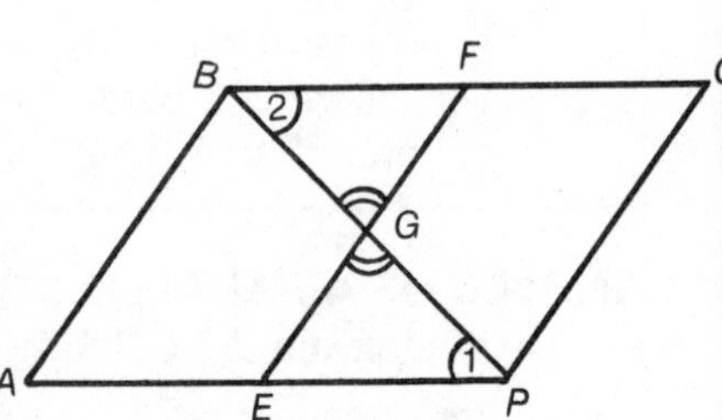

Solution:
PLAN. Show that $\triangle DGE \cong \triangle BGF$ by the AAS theorem.

PROOF

Statement		Reason
1. $\square ABCD$		1. Given.
2. $\angle DGE \cong \angle BGF$	*Angle*	2. Vertical angles are congruent.
3. $\overline{AD} \parallel \overline{BC}$		3. Opposite sides of a parallelogram are parallel.
4. $\angle 1 \cong \angle 2$	*Angle*	4. If two lines are parallel, then alternate interior angles are congruent.
5. $AD = BC$		5. Opposite sides of a parallelogram are equal in length.
6. E is the midpoint of $\overline{AD}$, F is the midpoint of $\overline{BC}$.		6. Given.
7. $DE = \frac{1}{2}AD$, $BF = \frac{1}{2}BC$		7. Definition of midpoint.
8. $DE = BF$		8. Halves of equals are equal.
9. $\overline{DE} \cong \overline{BF}$	*Side*	9. Segments having equal lengths are congruent.
10. $\triangle DGE \cong \triangle BGF$		10. AAS theorem.
11. $\overline{EG} \cong \overline{FG}$		11. CPCTC.
12. G is the midpoint of $\overline{EF}$.		12. Definition of midpoint.

PROVING THAT A QUADRILATERAL IS A PARALLELOGRAM.

To prove that a quadrilateral is a parallelogram, show that any *one* of the following statements is true:

- Opposite angles are congruent.
- Opposite sides are parallel.
- Opposite sides are congruent.
- Diagonals bisect each other.
- One pair of sides is both parallel and congruent.

Examples

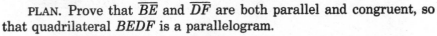

3. Given: ▱*ABCD*,
$\overline{BE} \perp \overline{AC}$, $\overline{DF} \perp \overline{AC}$.
Prove: *BEDF* is a parallelogram.

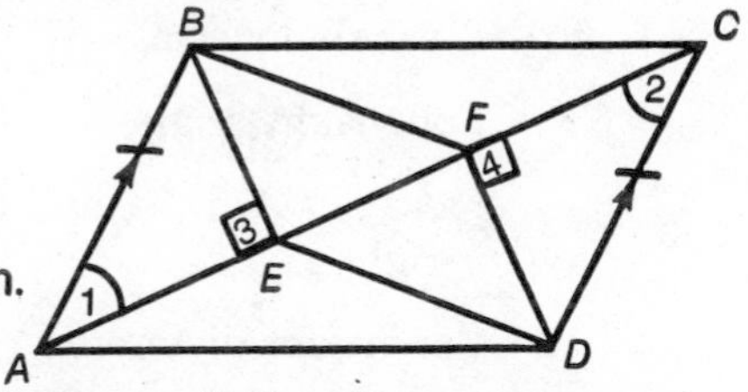

Solution:

PLAN. Prove that $\overline{BE}$ and $\overline{DF}$ are both parallel and congruent, so that quadrilateral *BEDF* is a parallelogram.

PROOF

Statement		Reason
1. $\overline{BE} \perp \overline{AC}$, $\overline{DF} \perp \overline{AC}$		1. Given.
2. $\overline{BE} \parallel \overline{DF}$		2. If two lines are perpendicular to the same line, they are parallel.
3. ▱*ABCD*		3. Given.
4. $\overline{AB} \parallel \overline{DC}$		4. Opposite sides of a parallelogram are parallel.
5. $\angle 1 \cong \angle 2$	*Angle*	5. If two lines are parallel, then alternate interior angles are congruent.
6. Angles 3 and 4 are right angles.		6. Perpendicular lines intersect to form right angles.
7. $\angle 3 \cong \angle 4$	*Angle*	7. All right angles are congruent.
8. $\overline{AB} \cong \overline{DC}$	*Side*	8. Opposite sides of a parallelogram are congruent.
9. $\triangle AEB \cong \triangle CFD$		9. AAS theorem.
10. $\overline{BE} \cong \overline{DF}$		10. CPCTC.
11. Quadrilateral *BEDF* is a parallelogram.		11. If a pair of sides of a quadrilateral are both parallel and congruent, then the quadrilateral is a parallelogram.

RECTANGLE. A rectangle is a parallelogram that is equiangular and may be defined as follows:

DEFINITION OF RECTANGLE

A **rectangle** is a parallelogram whose adjacent sides intersect at right angles.

A rectangle (Figure 7.10) has these special properties:

- All the properties of a parallelogram.
- Four right angles.

  Angles A, B, C, and D are right angles.
- Congruent diagonals.

  $\overline{AC} \cong \overline{DB}$.

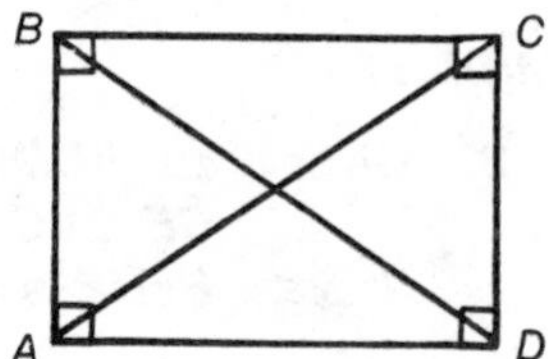

Figure 7.10 Rectangle

Example

4. In rectangle $ABCD$, diagonals $\overline{AC}$ and $\overline{BD}$ intersect at point E. If $AE = 2x - 9$ and $CE = x + 7$, find BD.

Solution: Since the diagonals of a rectangle bisect each other,

$$AE = CE$$
$$2x - 9 = x + 7$$
$$2x = x + 16$$
$$2x - x = 16$$
$$x = 16$$

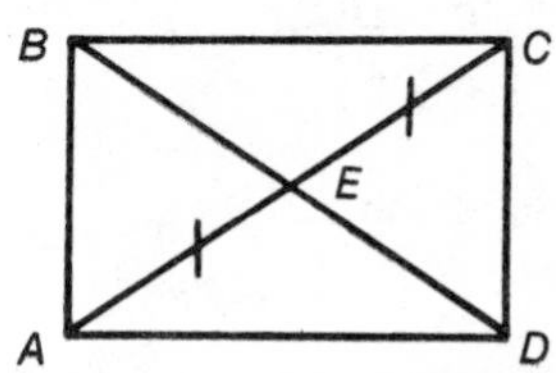

Hence $CE = x + 7 = 16 + 7 = 23$. Also, $AE = 23$. Therefore, $AC = 23 + 23 = 46$. Since the diagonals of a rectangle are congruent,

$$BD = AC = \mathbf{46}.$$

RHOMBUS. A rhombus is a parallelogram that is equilateral and may be defined as follows:

DEFINITION OF RHOMBUS

A **rhombus** is a parallelogram whose adjacent sides are congruent.

A rhombus (Figures 7.11 and 7.12) has these special properties:

- All the properties of a parallelogram.
- Four congruent sides.

  $\overline{AB} \cong \overline{BC} \cong \overline{CD} \cong \overline{DA}$

- Diagonals that bisect opposite pairs of angles.

  $\angle 1 \cong \angle 2$ and $\angle 3 \cong \angle 4$
  $\angle 5 \cong \angle 6$ and $\angle 7 \cong \angle 8$

- Diagonals that are perpendicular to (and bisect) each other.

  $\overline{AC} \perp \overline{BD}$

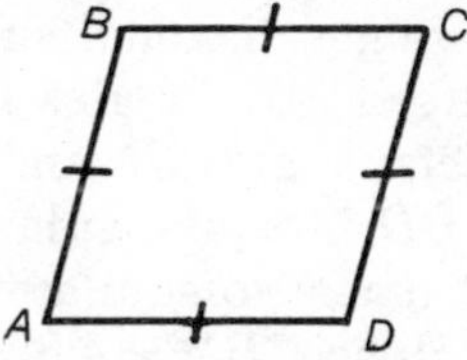

Figure 7.11 Sides of a Rhombus

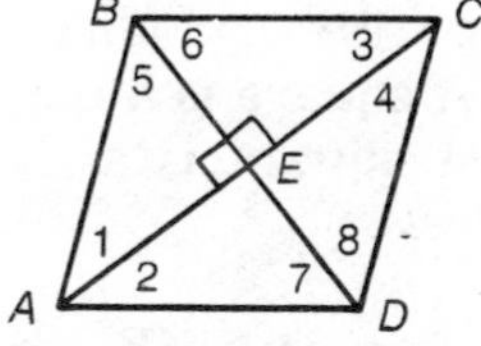

Figure 7.12 Diagonals of a Rhombus

Examples

5. Given that *ABCD* in the accompanying diagram is a rhombus and m$\angle 1 = 40$, find the measure of each of the following angles:
(a) $\angle 2$ (b) $\angle 3$ (c) $\angle ADC$

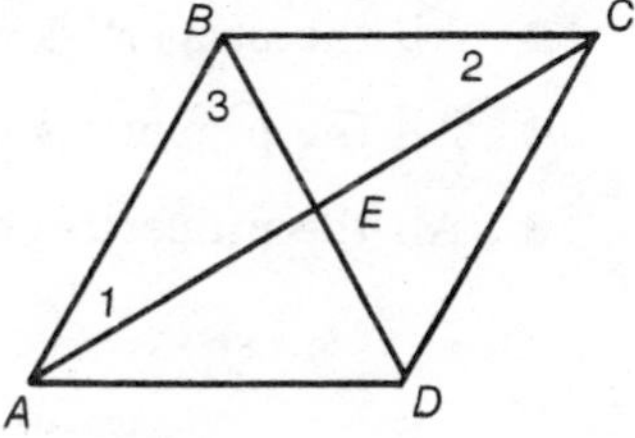

Solutions: (a) Triangle *ABC* is isosceles since $\overline{AB} \cong \overline{BC}$. Hence, the base angles of the triangle must be congruent.

$$m\angle 1 = m\angle 2 = \mathbf{40}$$

(b) In $\triangle AEB$, $\angle AEB$ is a right angle since the diagonals of a rhombus are perpendicular to each other. Since the sum of the measures of the angles of a triangle is 180, the measure of $\angle 3$ must be **50**.

(c) Since the diagonals of a rhombus bisect the angles of the rhombus, if m$\angle 3 = 50$, then m$\angle ABC = 100$. Since opposite angles of a rhombus are equal in measure, m$\angle ADC$ must also equal **100**.

6. The perimeter of rhombus *ABCD* is 20, and the measure of angle *A* is 60. Find the length of the *shorter* diagonal.

Solution: A rhombus is equilateral, so the length of each side is equal to $\frac{20}{4}$ or 5. Since consecutive angles of a rhombus are supplementary, m$\angle B$ = m$\angle D = 120$. The shorter diagonal is $\overline{BD}$ since it lies opposite a 60° angle, while $\overline{AC}$ is

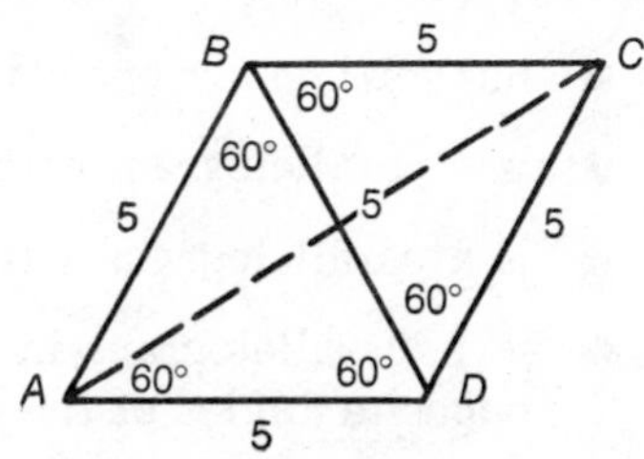

the longer diagonal since it lies opposite a 120° angle. Diagonal $\overline{BD}$ bisects angles B and D, so that $\triangle ABD$ is equiangular; this means that it is also equilateral.

Hence, $BD = AB = AD =$ **5**.

SQUARE. A square is a parallelogram that is both equiangular and equilateral. A square may be defined as a special rectangle or as a special rhombus.

DEFINITION OF SQUARE

A **square** is a rectangle whose adjacent sides are congruent; *or*

A **square** is a rhombus whose adjacent sides intersect at right angles.

A square (Figure 7.13) has these special properties:

- All the properties of a parallelogram.
- All the properties of a rectangle.
- All the properties of a rhombus.

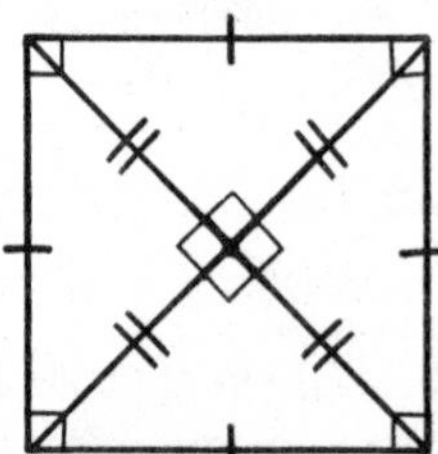

Figure 7.13 Square

PROVING THAT A QUADRILATERAL IS A SPECIAL PARALLELOGRAM. Methods for proving that a quadrilateral is a special type of parallelogram are based on the *converses* of the statements that give the definitions and properties of these figures.

To prove that a quadrilateral is a rectangle, show that the quadrilateral is a parallelogram with *one* of the following properties:

- A right angle.
- Congruent diagonals.

To prove that a quadrilateral is a rhombus, show that the quadrilateral:

- has four congruent sides; *or*
- is a parallelogram with a pair of congruent adjacent sides; *or*
- is a parallelogram with perpendicular diagonals; *or*
- is a parallelogram in which each diagonal bisects a pair of opposite angles of the quadrilateral.

To prove that a quadrilateral is a square, show that the quadrilateral is:

- a rectangle with a pair of congruent adjacent sides; *or*
- a rhombus with a right angle.

APPLICATIONS OF PARALLELOGRAMS. The properties of parallelograms may be used to prove each of the following theorems.

THEOREM: DISTANCE BETWEEN PARALLEL LINES. *Parallel lines are everywhere equidistant.*

In the accompanying figure, $AD = BC = EF, \ldots$. The perpendicular segments drawn from points A and B on line l to line m must be equal since $ABCD$ forms a rectangle. Therefore, $AD = BC$. Similarly, $BC = EF$, and so forth.

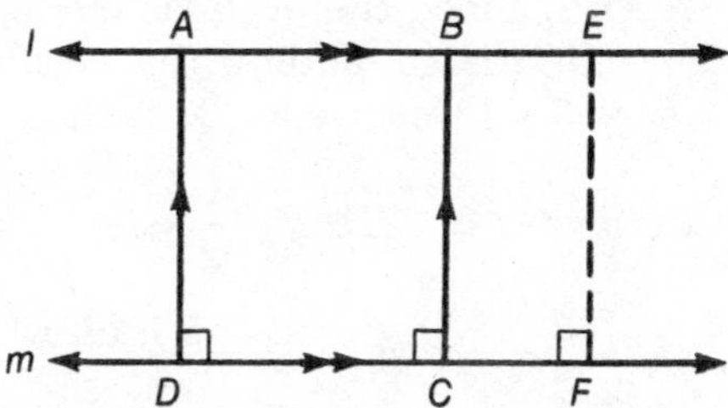

THEOREM: MEDIAN DRAWN TO HYPOTENUSE. *The length of the median drawn to the hypotenuse of a right triangle is one-half the length of the hypotenuse.*

In the accompanying rectangle, since M is the midpoint of $\overline{AC}$, $AM = \frac{1}{2}AC$. Since the diagonals of a rectangle are congruent, $AM = \frac{1}{2}BD$. In right triangle BAD, $\overline{AM}$ represents the median to hypotenuse $\overline{BD}$.

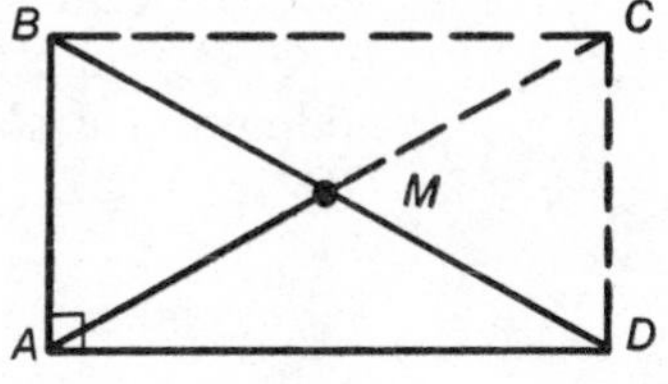

THEOREM: MIDPOINTS OF A TRIANGLE. *The line segment joining the midpoints of two sides of a triangle is parallel to the third side, and its length is one-half the length of the third side.*

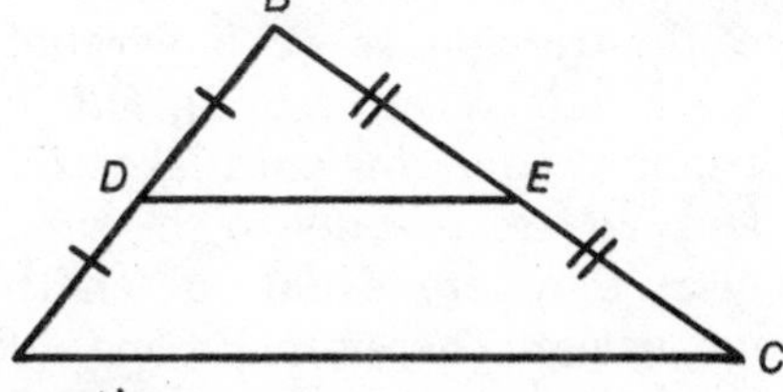

In the accompanying figure, if D and E are midpoints of $\overline{AB}$ and $\overline{BC}$ respectively, then $\overline{DE} \parallel \overline{AC}$ and $DE = \frac{1}{2}AC$.

See Exercise 15 at the end of this section.

COROLLARY TO MIDPOINTS OF A TRIANGLE THEOREM: *The perimeter of the triangle formed by joining the midpoints of the three sides of a triangle is one-half the perimeter of the original triangle.*

Examples

7. The length of each side of an equilateral triangle is 10 inches. If the midpoints of the three sides are joined to form a second triangle, how many inches are there in the perimeter of the second triangle?

Solution: The perimeter of $\triangle DEF$ is one-half the perimeter of $\triangle ABC$. Since the original triangle ($\triangle ABC$) is equilateral, its perimeter is 3×10 inches, or 30 inches.

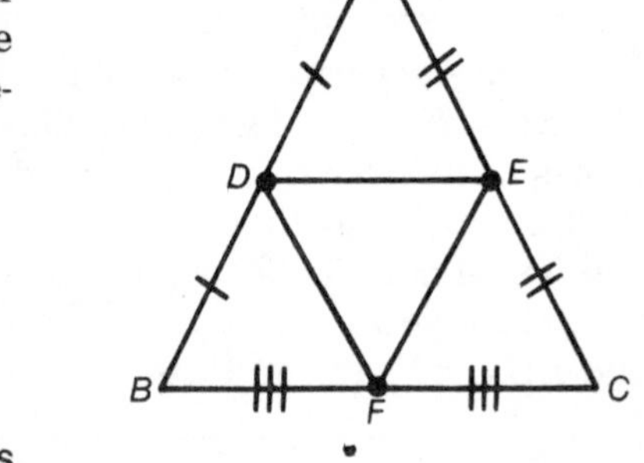

$$\text{Perimeter of } \triangle DEF = \frac{1}{2} \times 30$$

$$= \mathbf{15\ inches}$$

8. Prove that, if the midpoints of a quadrilateral are joined consecutively, a parallelogram is formed.

Solution:

Given: Quadrilateral $ABCD$, P, Q, R, and S are the midpoints of $\overline{AB}$, $\overline{BC}$, $\overline{CD}$, and $\overline{AD}$, respectively.

Prove: $PQRS$ is a parallelogram.

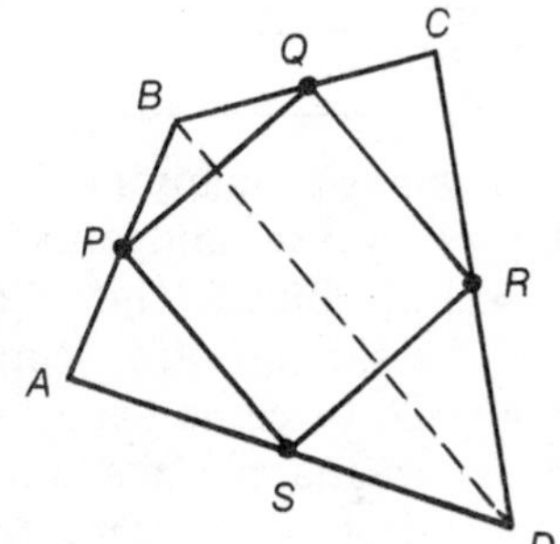

Solution:

PLAN. Show that $\overline{PS} \parallel \overline{QR}$ and $\overline{PS} \cong \overline{QR}$.

PROOF

Draw diagonal $\overline{BD}$. In $\triangle ABD$, $\overline{PS}$ joins the midpoints of two sides of the triangle, so $\overline{PS}$ is parallel to $\overline{BD}$ and equal in length to $\frac{1}{2}BD$.

Similarly, in $\triangle BCD$, $\overline{QR}$ joins the midpoints of two sides of the triangle, so $\overline{QR}$ is parallel to $\overline{BD}$ and equal in length to $\frac{1}{2}BD$. Since $\overline{PS}$ and $\overline{QR}$ are parallel to the same line, they are parallel to each other; also, they are equal in length to the same quantity, so they are congruent. Therefore $\overline{PS}$ and $\overline{QR}$ are both parallel and congruent, and quadrilateral $PQRS$ is a parallelogram.

EXERCISE SET 7.6

1. In rhombus *RSTW*, diagonal $\overline{RT}$ is drawn. If $m\angle RST = 108$, find $m\angle SRT$.

2. In $\square MATH$, the measure of angle *T* exceeds the measure of angle *H* by 30. Find the measure of each angle of the parallelogram.

3. In $\square TRIG$, $m\angle R = 2x + 19$ and $m\angle G = 4x - 17$. Find the measure of each angle of the parallelogram.

4. The length of the median drawn to the hypotenuse of a right triangle is represented by the expression $3x - 7$, while the hypotenuse is represented by $5x - 4$. Find the length of the median.

5. In $\triangle RST$, *E* is the midpoint of $\overline{RS}$ and *F* is the midpoint of $\overline{ST}$. If $EF = 5y - 1$ and $RT = 7y + 10$, find the length of $\overline{EF}$ and of $\overline{RT}$.

6. In $\square RSTW$, diagonals $\overline{RT}$ and $\overline{SW}$ intersect at point *A*. If $SA = x - 13$ and $AW = 2x - 37$, find *SW*.

7. The lengths of the sides of a triangle are 9, 40, and 41. Find the perimeter of the triangle formed by joining the midpoints of the sides.

8. In the accompanying figure, *ABCD* is a parallelogram. If $EB = AB$ and $m\angle CBE = 57$, what is the value of *x*?

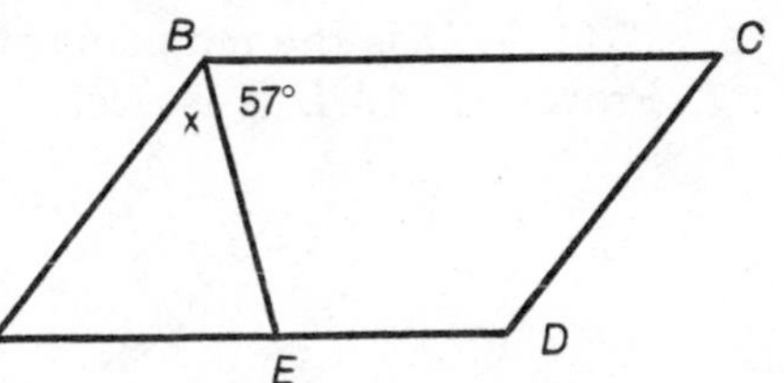

9 and 10. In each case, find the value of x.

9.

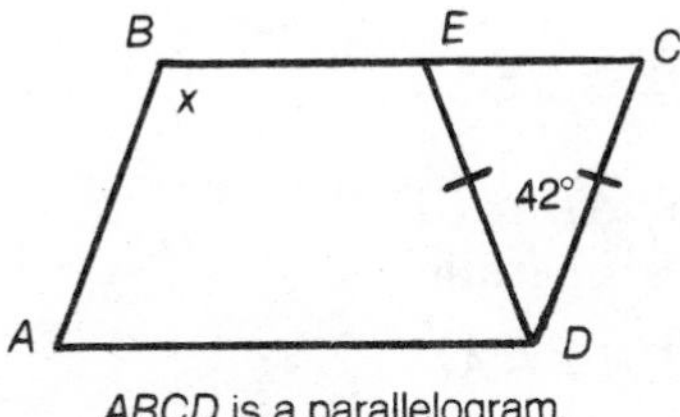

ABCD is a parallelogram.

10.

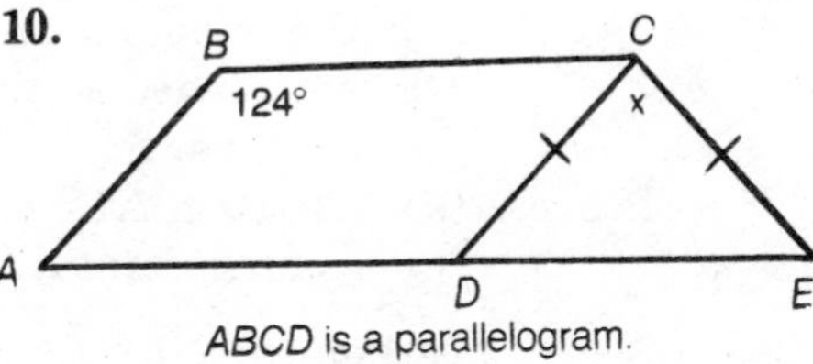

ABCD is a parallelogram.

11. Given: $\square ABCD$,
$\overline{AE} \cong \overline{CF}$.
Prove: $\angle ABE \cong \angle CDF$.

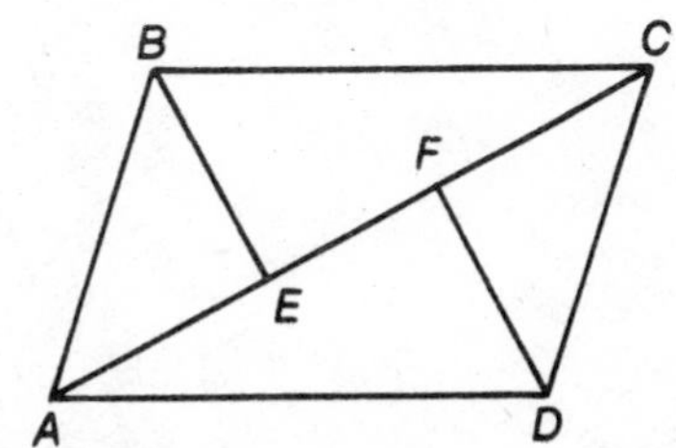

12. Given: $\square ABCD$,
$\overline{EF} \cong \overline{HG}$.
Prove: $\overline{AF} \cong \overline{CG}$.

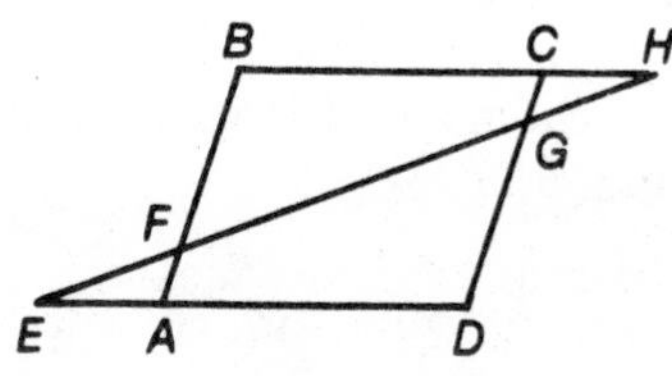

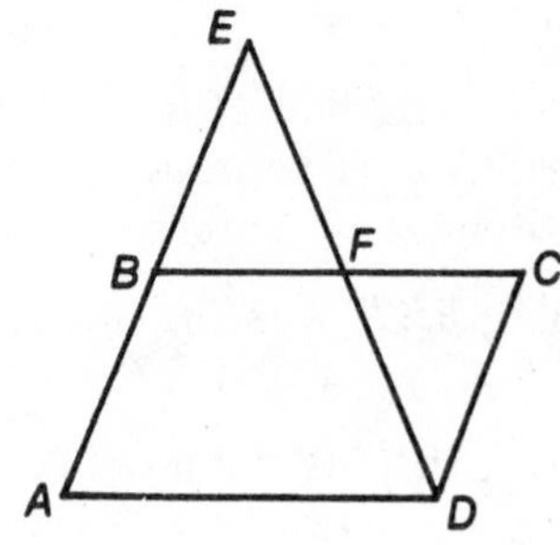

13. Given: $\square ABCD$,
B is the midpoint of $\overline{AE}$.
Prove: $\overline{EF} \cong \overline{FD}$.

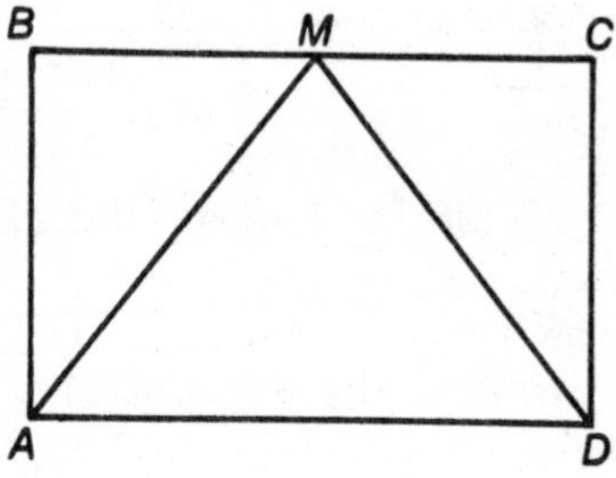

14. Given: Rectangle $ABCD$,
M is the midpoint of $\overline{BC}$.
Prove: $\triangle AMD$ is isosceles.

15. Given: D is the midpoint of $\overline{AB}$,
E is the midpoint of $\overline{BC}$,
$\overline{DE}$ is extended so that $\overline{DE} \cong \overline{EF}$,
$\overline{CF}$ is drawn.
Prove: (a) $\triangle BED \cong \triangle CEF$. (b) $\overline{AD} \cong \overline{CF}$.
(c) Quadrilateral $ADFC$ is a parallelogram. (d) $\overline{DE} \parallel \overline{AC}$ and $DE = \frac{1}{2}AC$.

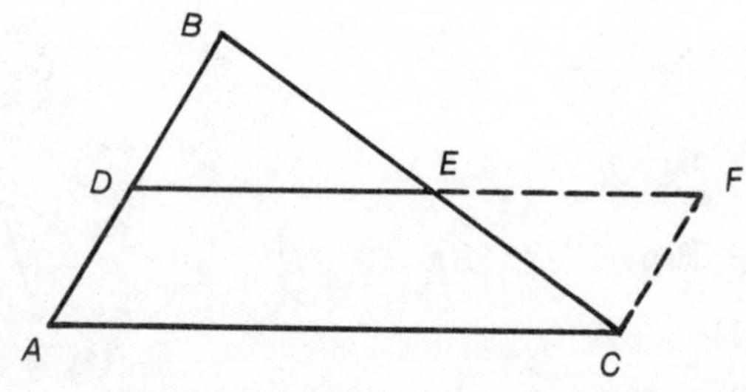

16. Given: Rectangle $ABCD$,
$\overline{BE} \cong \overline{CE}$.
Prove: $\overline{AF} \cong \overline{DG}$.

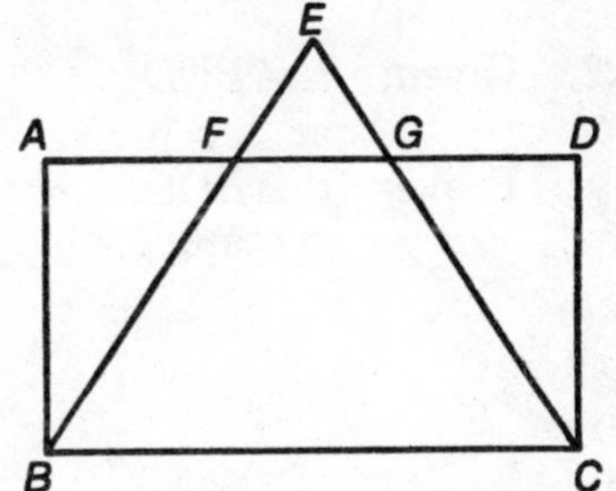

17. Given: $ABCD$ is a parallelogram,
$AD > DC$.
Prove: $m\angle BAC > m\angle DAC$.

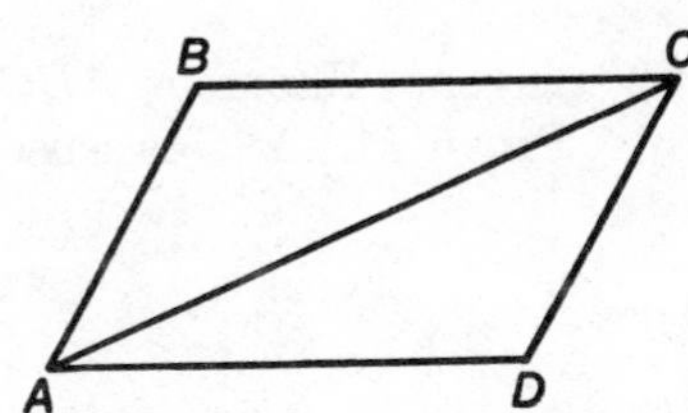

18. Given: $ABCD$ is a parallelogram,
$\overline{BR}$ bisects $\angle ABC$,
$\overline{DS}$ bisects $\angle CDA$.
Prove: $BRDS$ is a parallelogram.

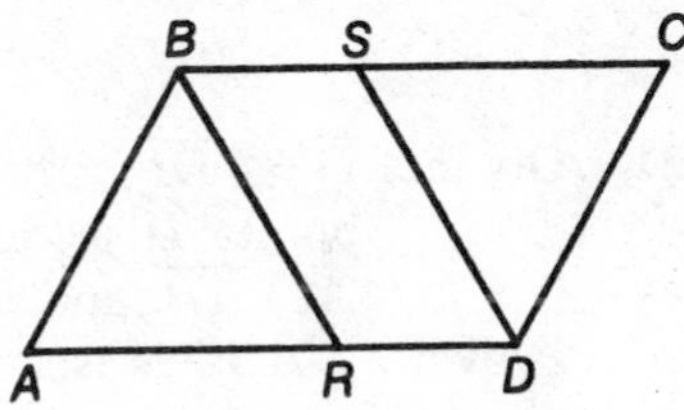

19. Given: $\square BMDL$,
$\overline{AL} \cong \overline{CM}$.
Prove: $ABCD$ is a parallelogram.

20. Given: $\square ABCD$,
$\angle ABL \cong \angle CDM$.
Prove: $BLDM$ is a parallelogram.

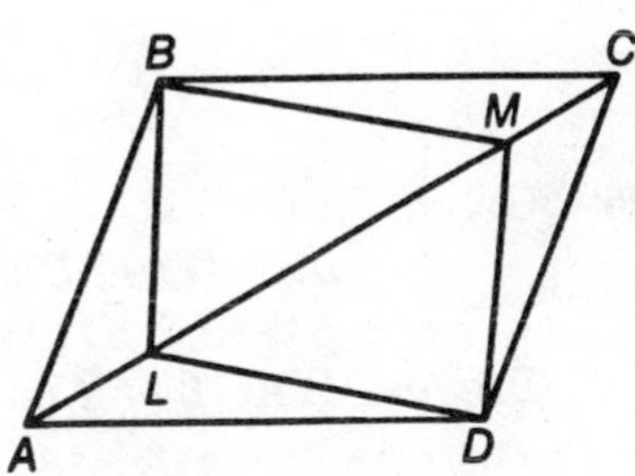

Exercises 19 and 20

21. Given: $ABCD$ is a rhombus,
$\overline{BL} \cong \overline{CM}$, $\overline{AL} \cong \overline{BM}$.
Prove: $ABCD$ is a square.

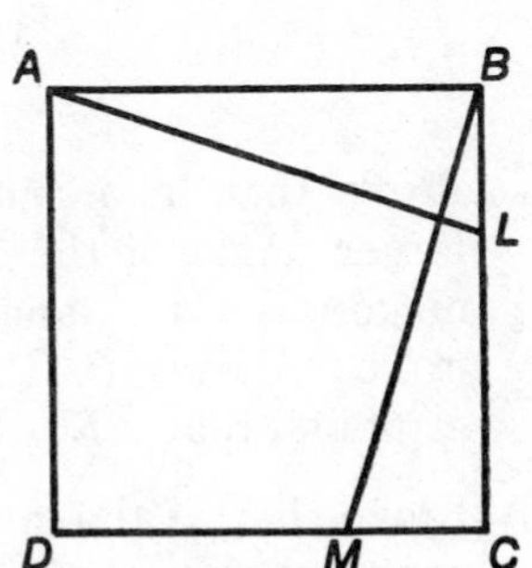

22. Given: $\square ABCD$,
$m\angle 2 > m\angle 1$.
Prove: $\square ABCD$ is *not* a rectangle.

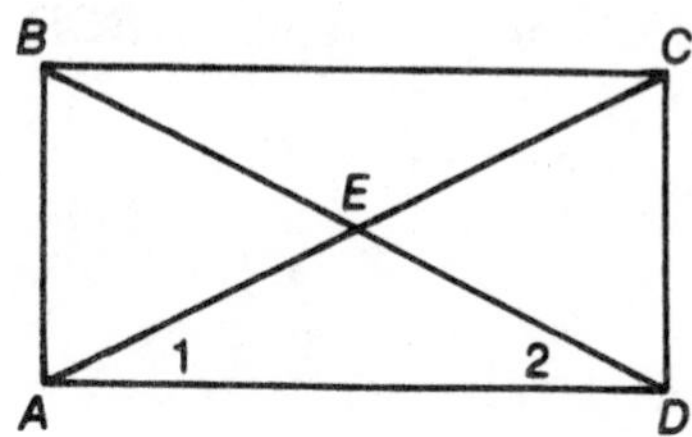

23. Given: Rhombus $ABCD$.
Prove: $\triangle ASC$ is isosceles.

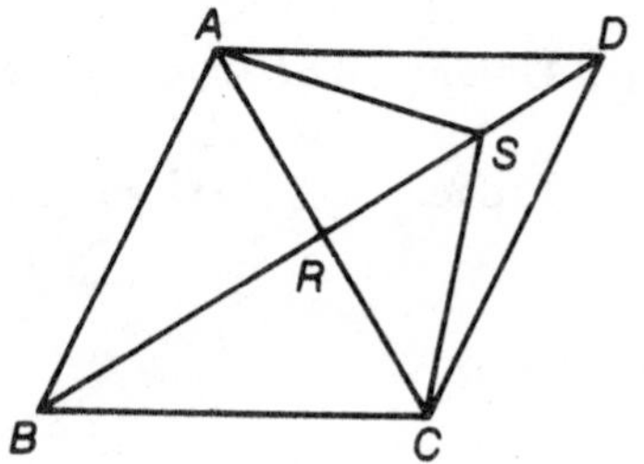

24. Given: $\overline{DE} \cong \overline{DF}$,
points D, E, and F are midpoints of $\overline{AC}$, $\overline{AB}$, and $\overline{BC}$, respectively.
Prove: $\triangle ABC$ is isosceles.

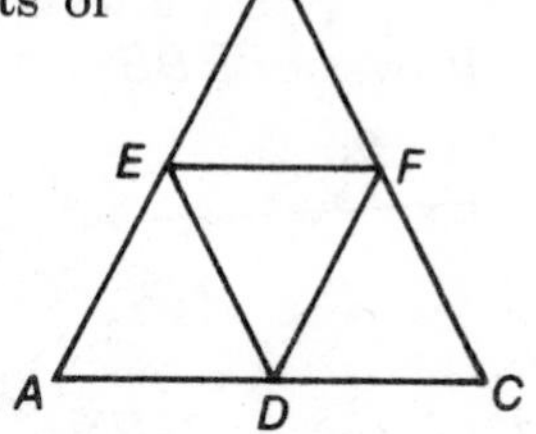

25. Given: $\square RSTW$,
in $\triangle WST$, B and C are midpoints of $\overline{SW}$ and $\overline{ST}$, respectively.
Prove: $WACT$ is a parallelogram.

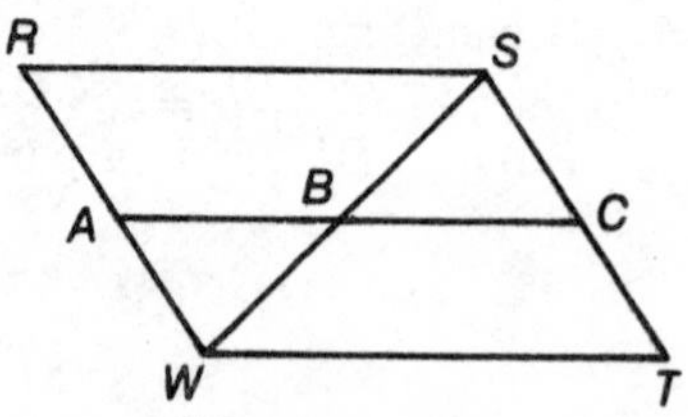

26. Prove that in a rhombus the longer diagonal lies opposite the larger angle of the rhombus. (*Hint*: Given rhombus $ABCD$ with diagonals $\overline{AC}$ and $\overline{DB}$ intersecting at point E, assume $m\angle CDA > m\angle BAD$. Prove $AC > BD$. Work with $\triangle AED$, and first establish that $AE > DE$.)

27. Prove that, if the midpoints of the sides of a rectangle are joined consecutively, the resulting quadrilateral is a rhombus.

7.7 PROPERTIES OF TRAPEZOIDS

KEY IDEAS

A **trapezoid** is *not* a parallelogram since a trapezoid is a quadrilateral that has exactly *one* pair of parallel sides, called *bases*. The nonparallel sides of a trapezoid are called *legs*. If points L and M are the midpoints of legs $\overline{AB}$ and $\overline{DC}$, then $\overline{LM}$ is called the *median* of the trapezoid. If the legs are congruent ($\overline{AB} \cong \overline{DC}$), then the trapezoid is *isosceles*.

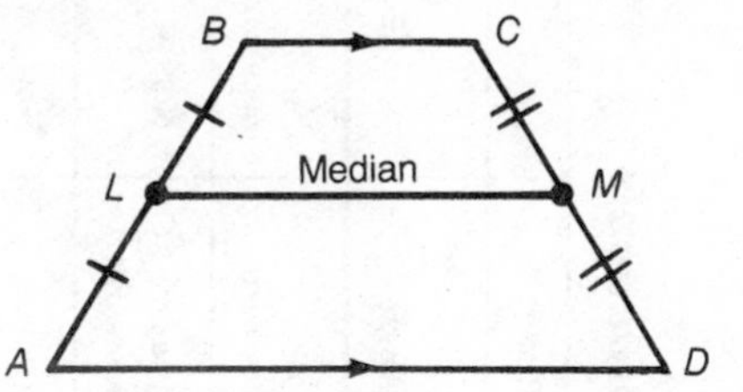

PROPERTIES OF TRAPEZOIDS. A trapezoid has one and only one pair of parallel sides.

An *isosceles trapezoid* (Figure 7.14) has these additional properties:

- The legs are congruent.

  $\overline{AB} \cong \overline{DC}$

- The base angles are congruent.

  $\angle A \cong \angle D$ and $\angle B \cong C$

- The diagonals are congruent.

  $\overline{AC} \cong \overline{DB}$

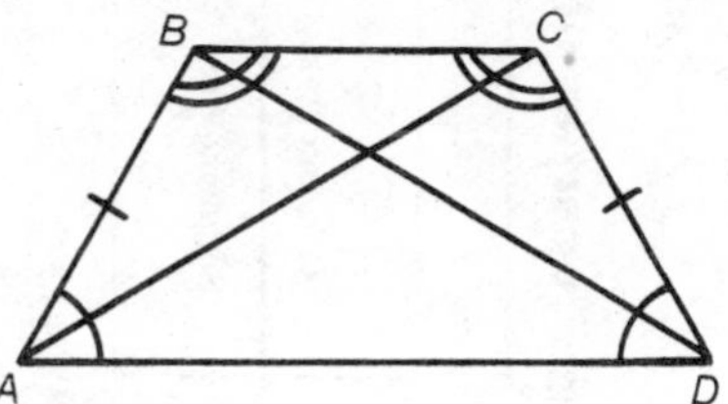

Figure 7.14 Isosceles Trapezoid

Table 7.3 summarizes the properties of special quadrilaterals.

TABLE 7.3 Summary of Properties of Special Quadrilaterals

Special Quadrilateral	Opposite Sides		Angles		Diagonals Bisect		Diagonals	
	$\parallel$	$\cong$	Opposite	Consecutive	Each Other	Angles	$\cong$	$\perp$
Parallelogram	Always	Always	Congruent	Supplementary	Always	Not necessarily	Not necessarily	Not necessarily
Rectangle	Always	Always	4 Right angles			Always necessarily	Not	AlwaysNot necessarily
Rhombus	Always	All sides $\cong$	Congruent	Supplementary	Always	Always	Not necessarily	Always
Square	Always	All sides $\cong$	4 Right angles		Always	Always	Always	Always
Trapezoid	1 Pair (bases)	—	—	Upper and lower base angles are supplementary.	—	—	Not necessarily	—
Isosceles trapezoid	1 Pair (bases)	1 Pair (legs)	Supplementary	Base angles $\cong$	—	—	Always	—

Example

1. In the accompanying diagram of isosceles trapezoid *ABCD*, $\overline{AB} \parallel \overline{DC}$ and diagonals $\overline{DB}$ and $\overline{AC}$ intersect at *E*. Which statement is *not* true?

(1) $\overline{AC} \cong \overline{BD}$
(2) $\angle CDB \cong \angle DBA$
(3) $\triangle ADC \cong \triangle ABC$
(4) $\triangle CBA \cong \triangle DAB$

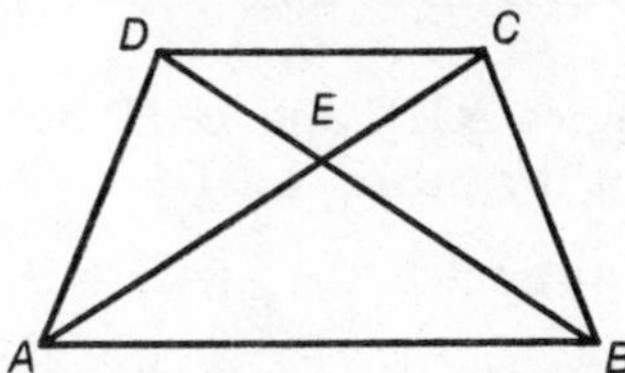

Solution: Choice (1) is true since the diagonals of an isosceles trapezoid are congruent. Choice (2) is true since $\overline{AB} \parallel \overline{DC}$ and therefore alternate interior angles are congruent. Choice (4) is true since $\overline{AD} \cong \overline{BC}$, $\angle DAB \cong \angle CBA$ (base angles of an isosceles trapezoid are congruent), and $\overline{AB} \cong \overline{AB}$, so that the triangles are congruent by the SAS postulate. **Choice (3)** is *not* true.

ALTITUDES AND MEDIAN OF A TRAPEZOID. An **altitude** of a trapezoid is any segment drawn from one base perpendicular to the other base. The **median** of a trapezoid is the segment joining the midpoints of its legs. In Figure 7.15, $\overline{BX}$ and $\overline{CY}$ are altitudes; in Figure 7.16, $\overline{LM}$ is a median. It can be proved that:

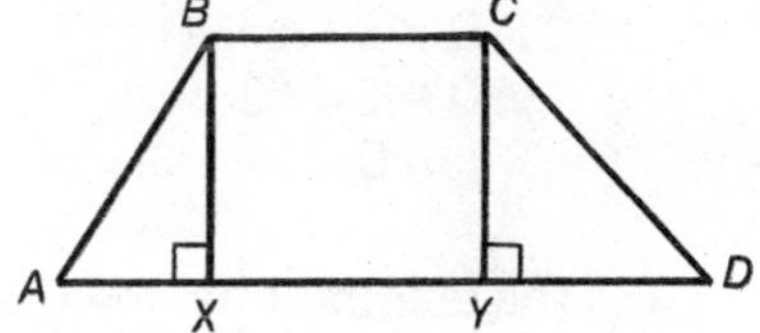

Figure 7.15 Altitudes of a Trapezoid

$$\overline{LM} \parallel \overline{AD} \parallel \overline{BC} \quad \text{and}$$
$$LM = \frac{1}{2}(AD + BC).$$

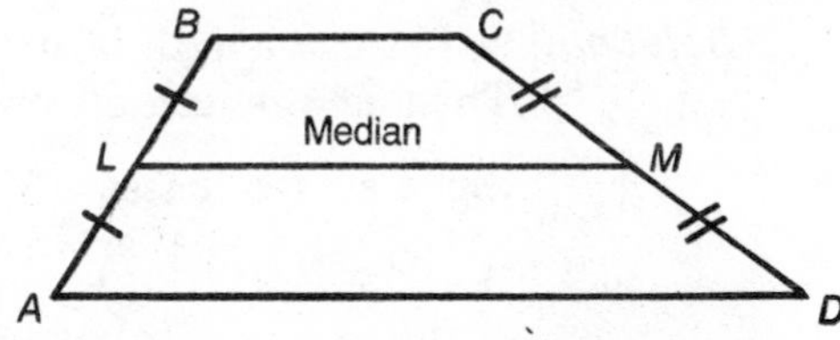

Figure 7.16 Median of a Trapezoid

THEOREM: MEDIAN OF A TRAPEZOID. ***The median of a trapezoid is parallel to the bases and its length is one-half the sum of the lengths of the bases.***

To prove this theorem, it is necessary to draw line segment $\overline{BM}$ and extend it so that it meets the extension of the line segment $\overline{AD}$ (see the accompanying diagram) on page 210).

$\overline{LM}$ is a line segment joining the midpoints of two sides of $\triangle ABE$. Hence,

$\overline{LM} \parallel \overline{ADE}$ and $LM = \frac{1}{2}(AD + DE)$.

Triangles BMC and EMD are congruent by ASA, so $DE = BC$. By substitution, $LM = \frac{1}{2}(AD + BC)$.

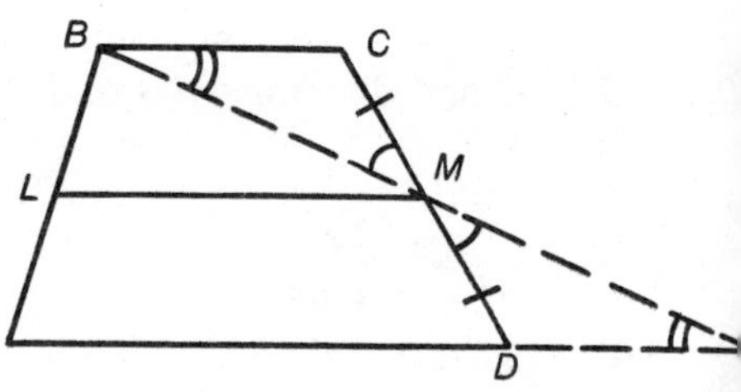

Examples

2. In trapezoid $ABCD$, $\overline{BC} \parallel \overline{AD}$ and $\overline{RS}$ is the median.
(a) If $AD = 13$ and $BC = 7$, find RS.
(b) If $BC = 6$ and $RS = 11$, find AD.

Solutions: (a) $RS = \frac{1}{2}(13 + 7) = \frac{1}{2}(20) = \mathbf{10}$

(b) If median $= \frac{1}{2}$(sum of bases), then

Sum of bases $= 2 \times$ median

$$AD + 6 = 2 \times 11$$
$$AD + 6 = 22$$
$$AD = 22 - 6 = \mathbf{16}$$

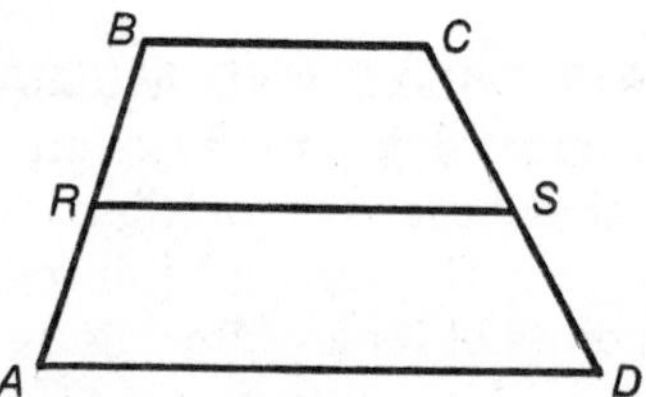

3. The length of the lower base of a trapezoid is three times as long as the length of the upper base. If the median has a 24-inch length, find the lengths of the bases.

Solution: Let $a =$ length of upper base.
Then $3a =$ length of lower base.

Since sum of bases $= 2 \times$ median,

$$a + 3a = 2 \times 24$$
$$4a = 48$$
$$a = \frac{48}{4} = 12$$
$$3a = 36$$

The length of the upper base is **12 inches**, and the length of the lower base is **36 inches**.

PROVING THAT A QUADRILATERAL IS A TRAPEZOID. To prove that a quadrilateral is a trapezoid, show that:

- One pair of sides is parallel; *and*
- One pair of sides is *not* parallel.

PROVING THAT A TRAPEZOID IS ISOSCELES. To prove that a trapezoid is an isosceles trapezoid show that *one* of the following statements is true:

- The legs are congruent.
- The lower (or upper) base angles are congruent.
- The diagonals are congruent.

Example

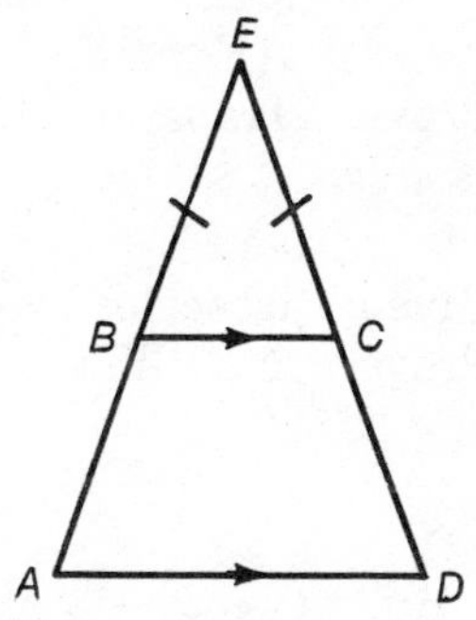

4. Given: Trapezoid $ABCD$, $\overline{BC} \parallel \overline{AD}$, $\overline{EB} \cong \overline{EC}$.

Prove: $ABCD$ is an isosceles trapezoid.

Solution:

PLAN. Show that $\angle A \cong \angle D$.

PROOF

Statement	Reason
1. Trapezoid $ABCD$ with $\overline{BC} \parallel \overline{AD}$, $\overline{EB} \cong \overline{EC}$.	1. Given.
2. $\angle EBC \cong \angle ECB$	2. If two sides of a triangle ($\triangle EBC$) are congruent, then the angles opposite these sides are congruent.
3. $\angle A \cong \angle EBC$ $\angle D \cong \angle ECB$	3. If two lines are parallel, then corresponding angles are congruent.
4. $\angle A \cong \angle D$	4. Transitive property.
5. $ABCD$ is an isosceles trapezoid.	5. If a trapezoid has a pair of congruent base angles, then it is an isosceles trapezoid.

EXERCISE SET 7.7

1. In trapezoid $BYTE$, $\overline{BE} \parallel \overline{YT}$ and median $\overline{LM}$ is drawn.

(a) $LM = 35$. If the length of $\overline{BE}$ exceeds the length of $\overline{YT}$ by 13, find the lengths of the bases.

(b) If $YT = x + 9$, $LM = x + 15$, and $BE = 2x - 5$, find the lengths of $\overline{YT}$, $\overline{LM}$, and $\overline{BE}$.

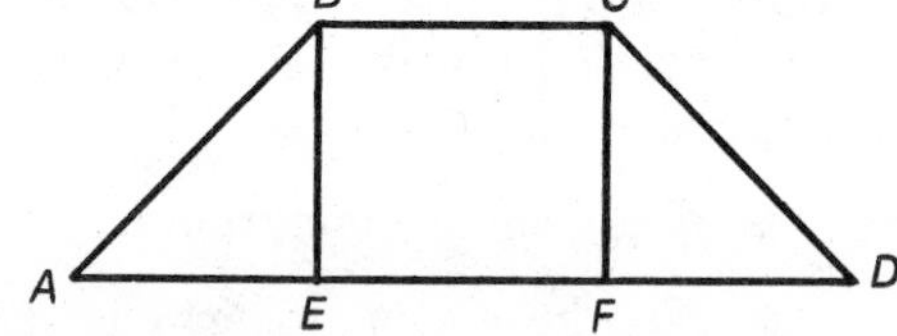

2. Given: Trapezoid $ABCD$,
$\overline{AD} \parallel \overline{BC}$,
$\overline{BE}$ and $\overline{CF}$ are attitudes drawn to $\overline{AD}$,
$\overline{AE} \cong \overline{DF}$.
Prove: Trapezoid $ABCD$ is isosceles.

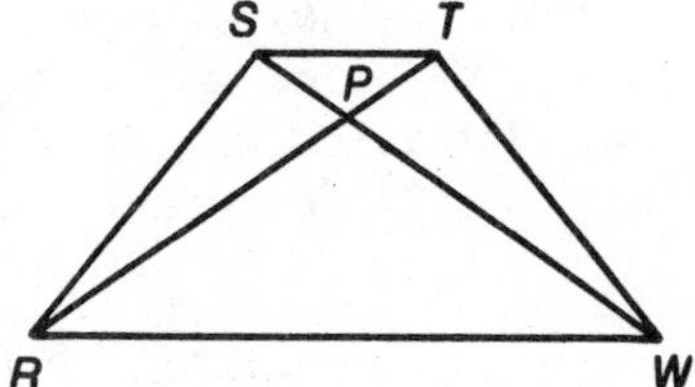

3. Given: Isosceles trapezoid $RSTW$.
Prove: $\triangle RPW$ is isosceles.

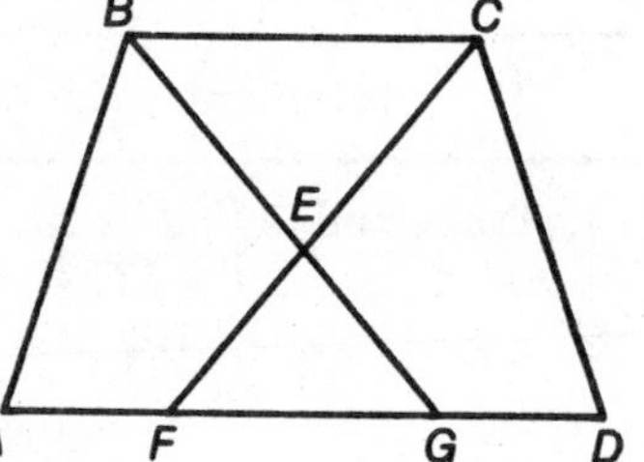

4. Given: Trapezoid $ABCD$,
$\overline{EF} \cong \overline{EG}$, $\overline{AF} \cong \overline{DG}$, $\overline{BG} \cong \overline{CF}$.
Prove: Trapezoid $ABCD$ is isosceles.

5. Given: Trapezoid $ABCD$ with median $\overline{LM}$,
P is the midpoint of $\overline{AD}$,
$\overline{LP} \cong \overline{MP}$.
Prove: Trapezoid $ABCD$ is isosceles.

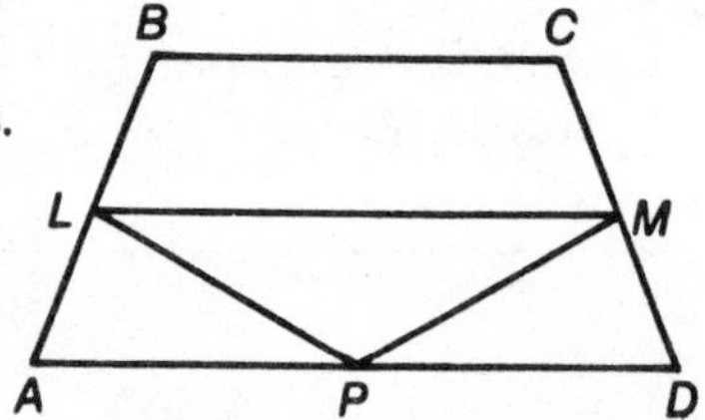

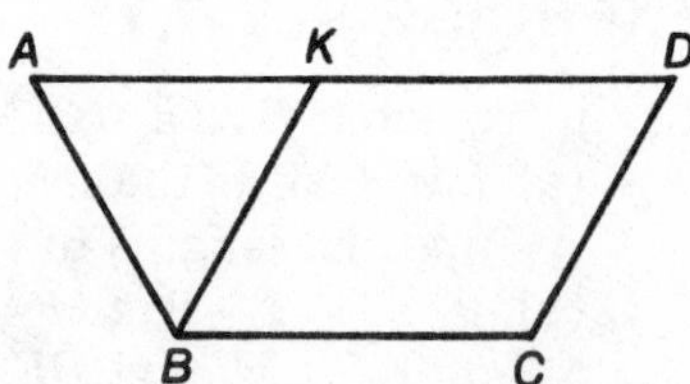

6. Given: Isosceles trapezoid $ABCD$,
$\angle BAK \cong \angle BKA$.
Prove: $BKDC$ is a parallelogram.

CHAPTER 7 REVIEW EXERCISES

1. Two parallel lines are cut by a transversal. The measures of a pair of interior angles on the same side of the transversal are represented by $4x + 50$ and $3x - 10$. Find x.

2. Find the sum of the measures of the interior angles of a polygon with seven sides.

3. In $\square ABCD$, diagonals $\overline{AC}$ and $\overline{BD}$ intersect at E. If $AE = 4x - 3$ and $EC = 17 - x$, find x.

4. How many sides does a regular polygon have if the measure of one of its exterior angles is 45?

5. The length of the line segment joining the midpoints of two sides of an equilateral triangle is 6. Find the perimeter of the *original* triangle.

6. In $\triangle ABC$, $m\angle A = 50$ and $m\angle B = 64$. Which is the *longest* side of the triangle?

7. In rhombus $ABCD$, $AB = 3x + 12$ and $BC = 5x$. What is the value of x?

8. In a right triangle, the measures of the acute angles are x and $5x$. Find x.

9. The length of the longer base of a trapezoid is 18, and the length of the median is 15. Find the length of the *shorter* base of the trapezoid.

10. In isosceles triangle MNQ, $MN = 4$ and $MQ = 11$. Which is the *smallest* angle of the triangle?

11. The measure of one angle of a parallelogram is greater by 24 than the measure of a second angle of the same parallelogram. What is the degree measure of an acute angle of this parallelogram?

12. In $\triangle RST$, $m\angle R = 62$ and the measure of an exterior angle at T is 119. Which is the *longest* side of the triangle?

13. The sum of the measures of the interior angles of a polygon is 1,980. How many sides does this polygon have?

14. The lengths of the sides of $\triangle PQR$ are 7, 10, and 15, respectively. Find the perimeter of the triangle formed by joining the midpoints of the sides of $\triangle PQR$.

15. In $\triangle ABC$, $AB = 12$, $BC = 14$, and $AC = 10$. If D is the midpoint of $\overline{AB}$ and E is the midpoint of $\overline{AC}$, find DE.

16. The length of the median to the hypotenuse of a right triangle is 6. What is the length of the hypotenuse?

17. If the sum of the measures of the interior angles of a polygon equals the sum of the measures of the exterior angles, how many sides does the polygon have?

18. In equilateral triangle ABC, $\overline{AD}$ and $\overline{BE}$, the bisectors of angles A and B, respectively, intersect at point F. What is $m\angle AFB$?

19. If the measures, in degrees, of the three angles of a triangle are x, $2x - 20$, and $3x - 10$, respectively, then the triangle is:
(1) right (2) obtuse (3) isosceles (4) equilateral

20. In an isosceles triangle, the length of one of the legs is always:
(1) equal to the length of the base
(2) equal to one-half of the length of the base
(3) greater than one-half of the length of the base
(4) less than one-half of the length of the base

21. If quadrilateral $ABCD$ is a parallelogram, then which statement is *always* true?
(1) Diagonals $\overline{AC}$ and $\overline{BD}$ are congruent.
(2) Diagonals $\overline{AC}$ and $\overline{BD}$ are perpendicular.
(3) Diagonals $\overline{AC}$ and $\overline{BD}$ bisect each other.
(4) Diagonals $\overline{AC}$ and $\overline{BD}$ bisect the angles through which they pass.

22. If the lengths of two sides of a triangle are 5 and 7, the length of the third side may *not* be:
(1) 12 (2) 7 (3) 3 (4) 5

23. In rhombus $ABCD$, the lines that bisect angles B and C *must* be:
(1) parallel (2) oblique (3) perpendicular (4) congruent

24. In $\triangle ABC$, D is a point on $\overline{AC}$ such that $\overline{BD}$ bisects $\angle ABC$. If $m\angle ABC = 60$ and $m\angle C = 70$, then:
(1) $AD > AB$ (2) $BD > BC$ (3) $AD > BD$ (4) $AB > AD$

25. A quadrilateral *must* be a square if:
(1) its diagonals are congruent
(2) its sides and angles are congruent
(3) its opposite sides and opposite angles are congruent
(4) its diagonals are the perpendicular bisectors of each other

26. In which quadrilateral are the diagonals *not* congruent to each other?
(1) isosceles trapezoid (2) rhombus (3) rectangle (4) square

27. If an exterior angle of a triangle is congruent to the adjacent interior angle, then the remaining two interior angles must be:
(1) supplementary
(2) complementary
(3) congruent
(4) obtuse

28. In $\triangle ABC$, $CB > CA$, D is a point on $\overline{AC}$, and E is a point on $\overline{BC}$ such that $\overline{CE} \cong \overline{CD}$. Which statement is *always* true?
(1) $m\angle CDE > m\angle A$
(2) $m\angle B > m\angle CED$
(3) $EB > AD$
(4) $EB = AD$

29. Which set of numbers could represent the lengths of the sides of a triangle?
(1) $\{1, 2, 3\}$ (2) $\{2, 4, 6\}$ (3) $\{3, 5, 7\}$ (4) $\{5, 10, 20\}$

30. In isosceles triangle ABC, $\overline{AC} \cong \overline{BC}$ and D is a point lying between A and B on base $\overline{AB}$. If $\overline{CD}$ is drawn, then which of the following is true?
(1) $AC > CD$
(2) $CD > AC$
(3) $m\angle A > m\angle ADC$
(4) $m\angle B > m\angle BDC$

31. If both pairs of opposite sides of a quadrilateral are parallel and the diagonals are perpendicular, the quadrilateral *must* be a:
(1) trapezoid (2) rectangle (3) rhombus (4) square

32. In the accompanying diagram, parallel lines $\overleftrightarrow{AB}$ and $\overleftrightarrow{CD}$ are cut by transversal $\overleftrightarrow{AC}$. Segments $\overline{BC}$ and $\overline{AD}$ intersect at E, and $m\angle BAC < m\angle ACD$. If $\overline{CB}$ bisects $\angle ACD$ and $\overline{AD}$ bisects $\angle BAC$, then which statement is true?

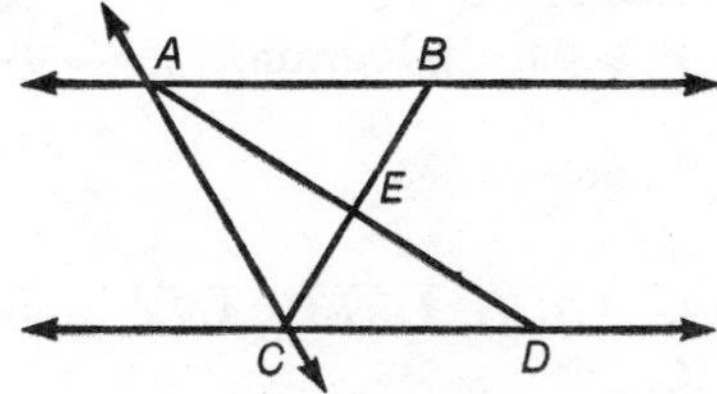

(1) $AC = AB$ (2) $AD < CD$ (3) $AE < CE$
(4) $m\angle DAC > m\angle BCD$

33. Given: $\triangle ABC$, $\overline{BGEA}$, $\overline{BFDC}$,
$\overline{AB} \cong \overline{BC}$,
$\overline{ED} \parallel \overline{AC}$,
$\overline{DG} \perp \overline{AB}$, $\overline{EF} \perp \overline{BC}$.
Prove: $\overline{GE} \cong \overline{FD}$.

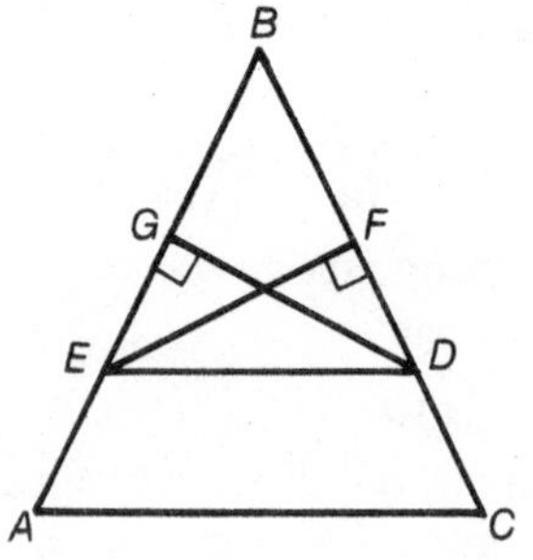

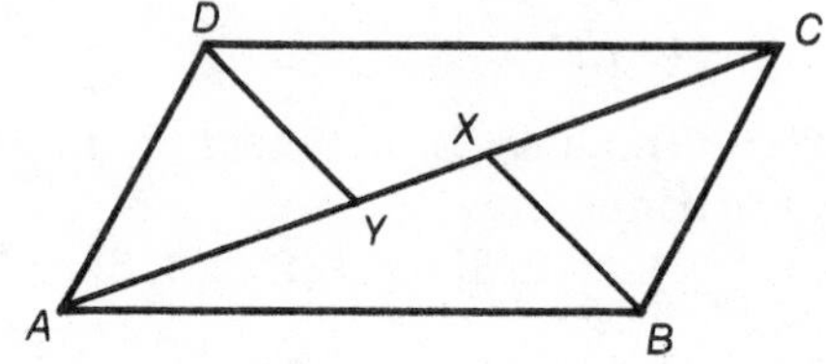

34. Given: $\square ABCD$, $\overline{AYXC}$
$\overline{BX} \parallel \overline{DY}$.
Prove: $\overline{AY} \cong \overline{CX}$.

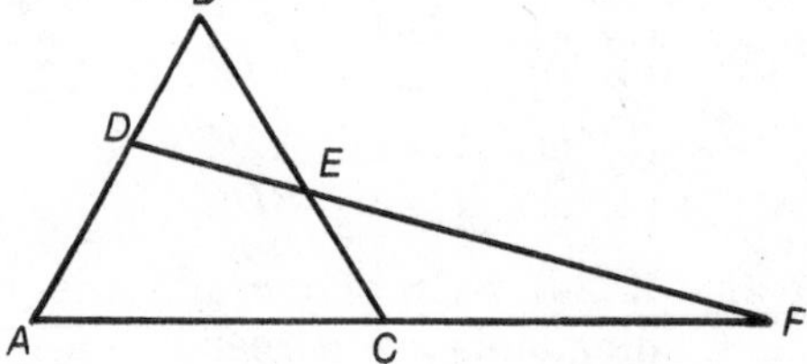

35. Given: $\triangle ABC$, $\overline{ACF}$, $\overline{DEF}$,
$\overline{AB} \cong \overline{BC}$.
Prove: $DF > AD$.

36. Given: quadrilateral $ABCD$,
diagonals $\overline{AC}$ and $\overline{BD}$,
$\angle 1 \cong \angle 2$,
$\overline{BD}$ bisects $\overline{AC}$ at E.
Prove: $ABCD$ is a parallelogram.

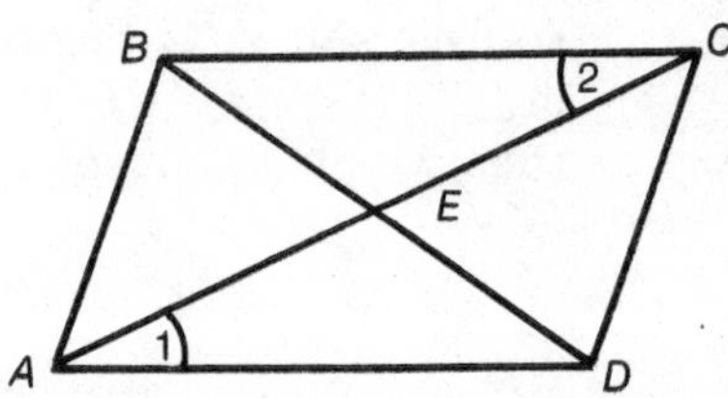

37. Given: rectangle $ABCD$, $\overline{DFEC}$, $\overline{AGE}$, $\overline{BGF}$,
$\overline{DF} \cong \overline{EC}$.
Prove: (a) $\angle 1 \cong \angle 2$.
(b) $\angle 3 \cong \angle 4$.
(c) $AG \cong GB$.

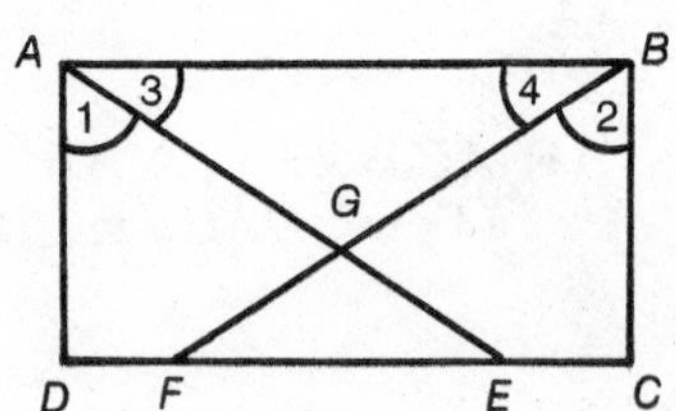

38. Given: $\overline{BC} \parallel \overline{AD}$,
$\triangle ABC$ is *not* isosceles.
Prove: $\overline{AC}$ does *not* bisect $\angle BAD$.

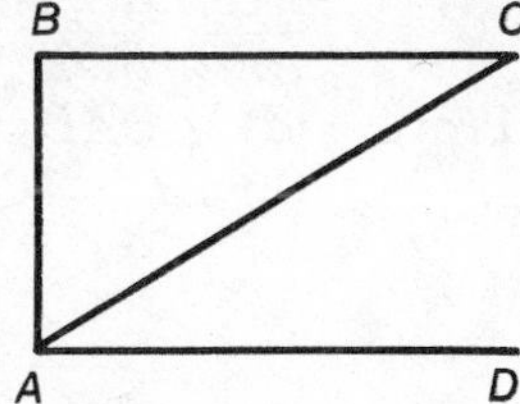

39. Given: $\overline{CM}$ is the median to $\overline{AB}$ of $\triangle ABC$,
$\overline{CEMF}$,
$\overline{BE} \perp \overline{CF}$ at E, $\overline{AF} \perp \overline{CF}$ at F,
$\overline{AE}$ and $\overline{BF}$ are drawn.
Prove: $AFBE$ is a parallelogram.

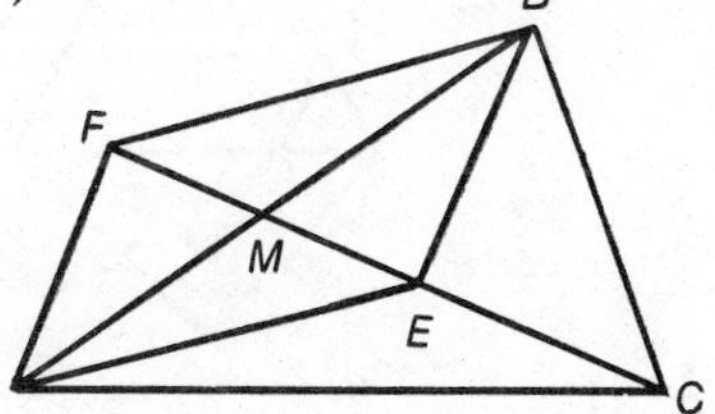

40. Given: trapezoid $ABCD$,
bases $\overline{AB}$ and $\overline{DC}$,
diagonals $\overline{AC}$ and $\overline{BD}$ intersect at E,
$\overline{DE} \cong \overline{EC}$.
Prove: $\triangle ACD \cong \triangle BDC$.

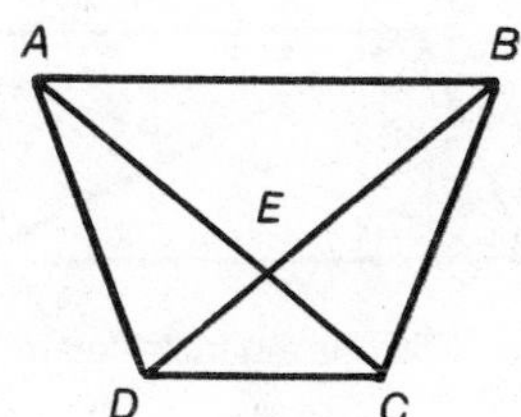

41. Given: trapezoid $ABCD$ with diagonal $\overline{AFC} \cong$ diagonal $\overline{BFD}$;
through C a line is drawn parallel to $\overline{BD}$ and intersecting $\overrightarrow{AB}$ at E.
Prove: (a) $BECD$ is a parallelogram.
(b) $\overline{AC} \cong \overline{CE}$.
(c) $\triangle AFB$ is isosceles.

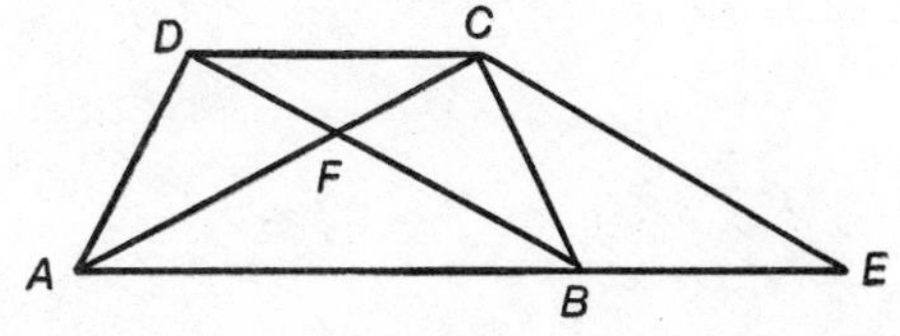

42. Given: $\overline{AC} \cong \overline{BC}$, $\overline{AD} \cong \overline{BD}$,
$\overline{AEC}$, $\overline{BDE}$.
Prove: (a) $\angle CAD \cong \angle CBD$.
(b) $AD > DE$.

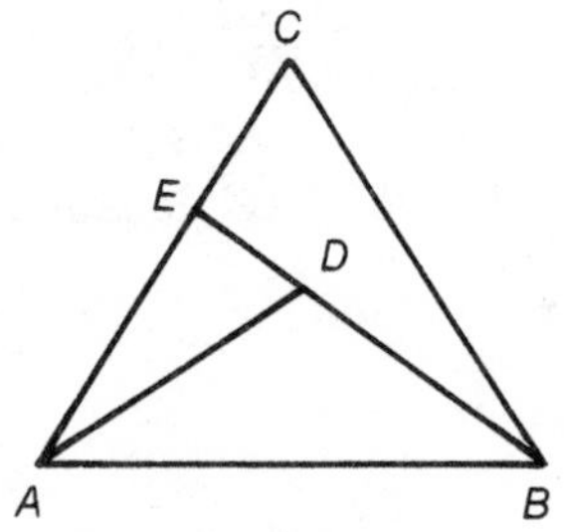

43–46. In each case, find the value of x.

43.

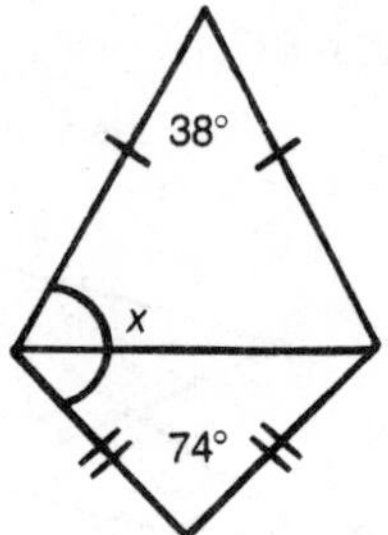

45.

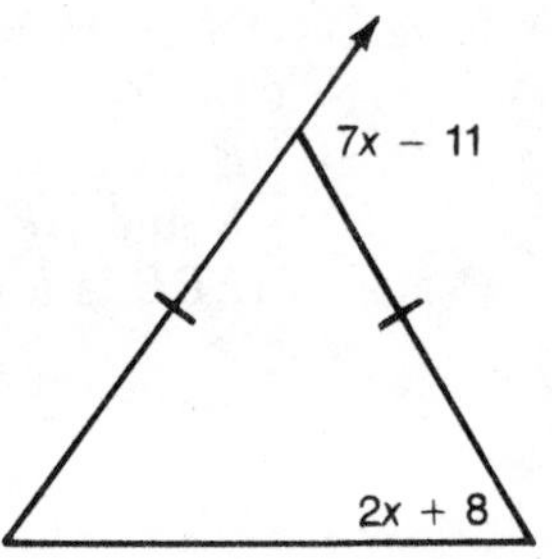

44.

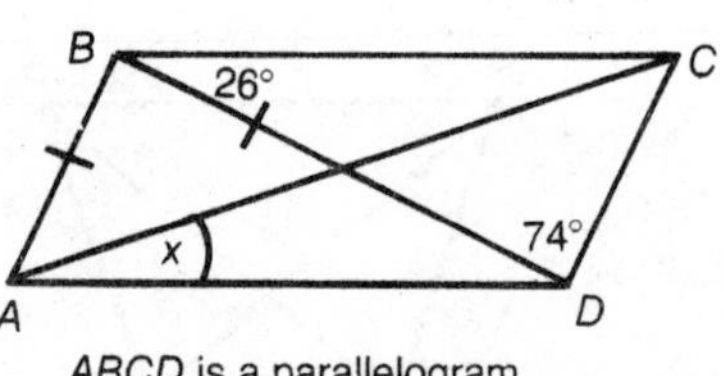

ABCD is a parallelogram.

46.

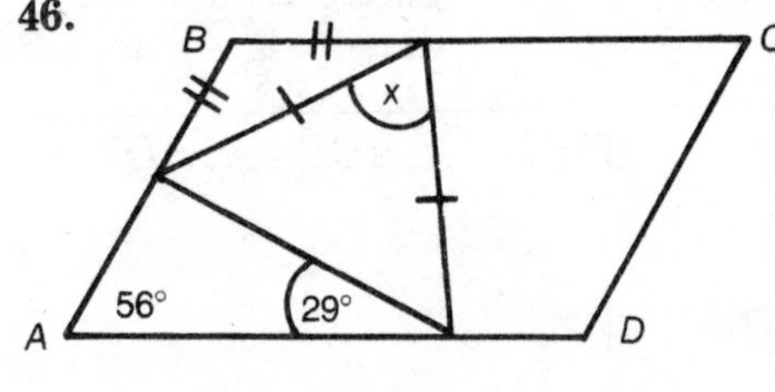

ABCD is a parallelogram.

CHAPTER 8

Similar and Right Triangles

8.1 RATIO AND PROPORTIONS IN A TRIANGLE

KEY IDEAS

The **ratio** of two numbers a and b ($b \neq 0$) is the quotient of the numbers, and is written as

$$\frac{a}{b} \quad \text{or} \quad a:b \text{ (read as ``}a\text{ is to }b\text{'')}.$$

A **proportion** is an equation that states that two ratios are equal. To *solve* a proportion for an unknown member, set the cross-products equal and solve the resulting equation.

$$\text{If} \quad \frac{a}{b}=\frac{c}{d}, \quad \text{then} \quad a \times d = b \times c,$$

where b and d cannot be equal to 0. The terms b and c are called **means**, and the terms a and d are the **extremes**.

THE MEAN PROPORTIONAL. In the proportion $\frac{4}{6}=\frac{6}{9}$, 6 is the *mean proportional* between 4 and 9. Whenever the means of a proportion are identical, the value that appears as the means is referred to as the **mean proportional** or *geometric mean* between the first and last terms (extremes) of the proportion.

Example

1. Find the mean proportional between each pair of extremes.
(a) 3 and 27 (b) 5 and 7

Solutions: (a) Let m = mean proportional between 3 and 27. Then

$$\frac{3}{m} = \frac{m}{27}$$
$$m^2 = 3(27) = 81$$
$$m = \pm\sqrt{81} = \pm 9$$

(b) Let m = mean proportional between 5 and 7. Then

$$\frac{5}{m} = \frac{m}{7}$$
$$m^2 = 35$$
$$m = \pm\sqrt{35}$$

PROPERTIES OF PROPORTIONS. Here are some algebraic properties of proportions that show how the terms of a proportion can be manipulated so that an equivalent proportion results.

Property 1. If the numerators and denominators of both members of a proportion are interchanged, then an equivalent proportion results.

$$\text{If } \frac{a}{b} = \frac{c}{d}, \text{ then } \frac{b}{a} = \frac{d}{c}$$

(provided that a, b, c, and d are nonzero numbers).

Property 2. If either pair of opposite terms of a proportion are interchanged, then an equivalent proportion results.

$$\text{(a) If } \frac{a}{b} = \frac{c}{d}, \text{ then } \frac{d}{b} = \frac{c}{a}.$$
$$\text{(b) If } \frac{a}{b} = \frac{c}{d}, \text{ then } \frac{a}{c} = \frac{b}{d}.$$

Property 3. If the denominator is added to or subtracted from the numerator on each side of the proportion, then an equivalent proportion results.

$$\text{(a) If } \frac{a}{b} = \frac{c}{d}, \text{ then } \frac{a+b}{b} = \frac{c+d}{d}.$$
$$\text{(b) If } \frac{a}{b} = \frac{c}{d}, \text{ then } \frac{a-b}{b} = \frac{c-d}{d}.$$

Property 4. If the product of two nonzero numbers equals the product of another pair of nonzero numbers, then a proportion may be formed by making the factors of one product the extremes, and making the factors of the other product the means. In other words, if $R \times S = T \times W$, then we may:

(a) make R and S the *extremes*: $\frac{R}{T} = \frac{W}{S}$; *or*

(b) make R and S the *means*: $\frac{T}{R} = \frac{S}{W}$.

PROPORTIONS IN A TRIANGLE. In Chapter 7 we saw that a line passing through the midpoints of two sides of a triangle was parallel to the third side (and one-half of its length). Suppose we draw a line parallel to a side of a triangle so that it intersects the other two sides, but not necessarily at their midpoints. Many such lines can be drawn, as shown in Figure 8.1.

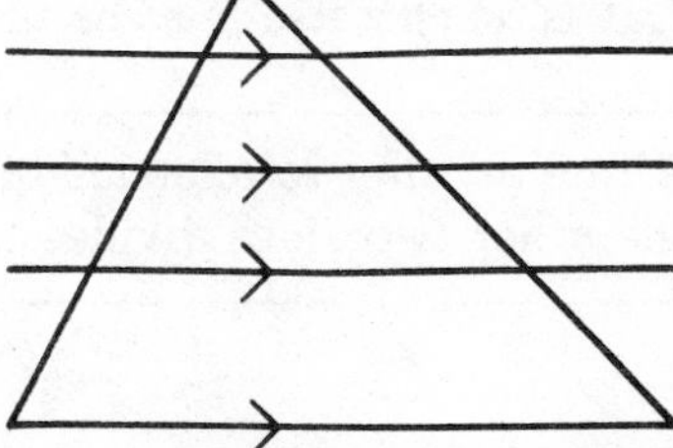

Figure 8.1 Lines Parallel to One Side of a Triangle and Intersecting the Other Two Sides

Consider one of these lines and the segments that it forms on the sides of the triangle, as shown in Figure 8.2.

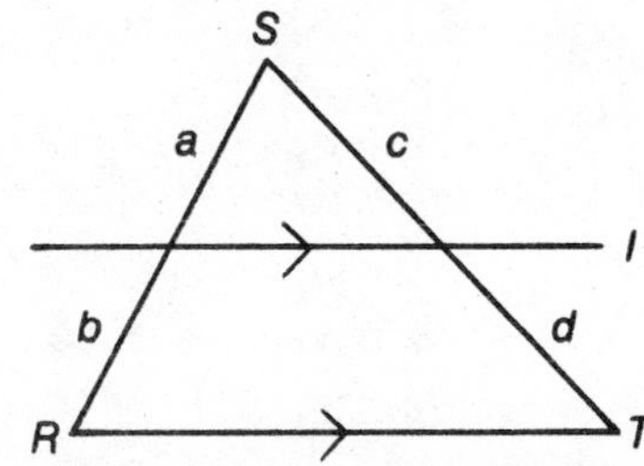

Figure 8.2 Segments Formed by a Line Drawn Parallel to One Side of a Triangle

It will be convenient to postulate that line l divides $\overline{RS}$ and $\overline{ST}$ in such a way that lengths of corresponding segments on each side have the same ratio:

$$\frac{a}{b} = \frac{c}{d} \quad \text{or} \quad \frac{a}{RS} = \frac{c}{ST} \quad \text{or} \quad \frac{b}{RS} = \frac{d}{ST}.$$

If any of the above ratios holds, then the line segments are said to be *divided proportionally*. Notice that these ratios have the form

$$\frac{\text{Upper segment of side } \overline{RS}}{\text{Lower segment of side } \overline{RS}} = \frac{\text{Upper segment of side } \overline{ST}}{\text{Lower segment of side } \overline{ST}},$$

or

$$\frac{\text{Upper segment of side } \overline{RS}}{\text{Whole side } (\overline{RS})} = \frac{\text{Upper segment of side } \overline{ST}}{\text{Whole side } (\overline{ST})},$$

or

$$\frac{\text{Lower segment of side } \overline{RS}}{\text{Whole side } (\overline{RS})} = \frac{\text{Lower segment of side } \overline{ST}}{\text{Whole side } (\overline{ST})}.$$

Keep in mind that the algebraic properties of proportions make it possible to express these three proportions in equivalent forms. For example, the numerator and denominator of each fraction may be interchanged (that is, each ratio may be *inverted*).

Postulate 1: A line parallel to one side of a triangle and intersecting the other two sides divides these sides proportionally.

Example

2. In $\triangle RST$, line segment $\overline{EF}$ is parallel to side $\overline{RT}$, intersecting side $\overline{RS}$ at point E and side $\overline{TS}$ at point F.

(a) If $SE = 8$, $ER = 6$, $FT = 15$, find SF.
(b) If $SF = 4$, $ST = 12$, $SR = 27$, find SE.
(c) If $SE = 6$, $ER = 4$, $ST = 20$, find FT.

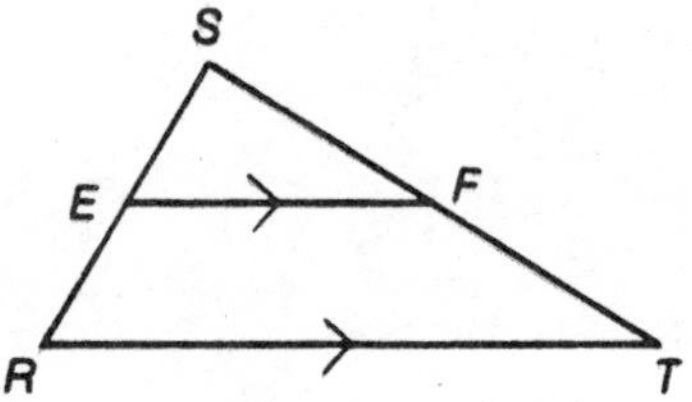

S
E F
R T

Solutions: (a)

$$\frac{SE}{ER} = \frac{SF}{FT}$$

$$\frac{8}{6} = \frac{SF}{15}$$

$$\frac{4}{3} = \frac{SF}{15}$$

$$3 \cdot SF = 60$$

$$SF = \frac{60}{3} = \mathbf{20}$$

(b)

$$\frac{SE}{SR} = \frac{SF}{ST}$$

$$\frac{SE}{27} = \frac{4}{12}$$

$$\frac{SE}{27} = \frac{1}{3}$$

$$3 \cdot SE = 27$$

$$SE = \frac{27}{3} = \mathbf{9}$$

(c)

$$\frac{ER}{RS} = \frac{FT}{ST}$$

$$\frac{4}{4+6} = \frac{FT}{20}$$

$$\frac{4}{10} = \frac{FT}{20}$$

$$\frac{2}{5} = \frac{FT}{20}$$

$$5 \cdot FT = 40$$

$$FT = \frac{40}{5} = 8$$

The converse of Postulate 1 is also true. If a line is drawn so that the ratio of the segment lengths it cuts off on one side of a triangle is equal to the ratio of the segment lengths it cuts off on the second side of a triangle, then the line must be parallel to the third side of the triangle.

***Postulate 2*:** A line that divides two sides of a triangle *proportionally* is parallel to the third side of the triangle.

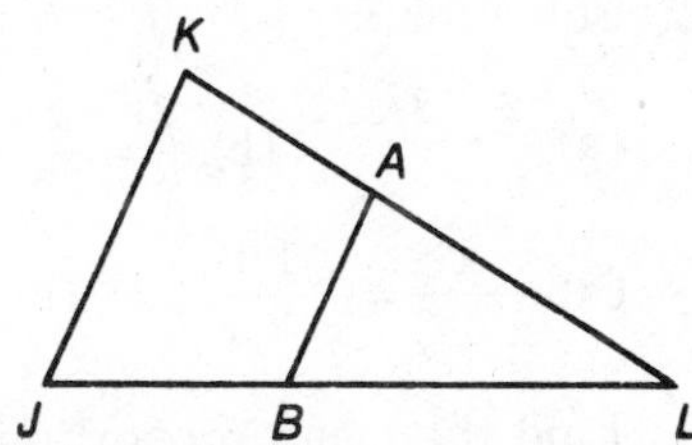

Example

3. Determine whether $\overline{AB} \parallel \overline{KJ}$ if:
(a) $KA = 2$, $AL = 5$, $JB = 6$, and $BL = 15$.
(b) $AL = 3$, $KL = 8$, $JB = 10$, and $JL = 16$.
(c) $AL = 5$, $KA = 9$, $LB = 10$, and $JB = 15$.

Solutions: In each instance, we must write a tentative proportion and then determine whether the proportion is true. If it is true, then $\overline{AB}$ is parallel to $\overline{KJ}$. The proportion written must conform to the meaning of "divides proportionally."

(a) On the basis of the information provided, we use the proportion $\frac{KA}{AL} = \frac{JB}{BL}$ (equivalent proportions may also be formed). We determine whether this proportion is true using the numbers provided:

$$\frac{2}{5} \stackrel{?}{=} \frac{6}{15}$$

$$2 \times 15 = 5 \times 6$$

$$30 = 30$$

Therefore, $\overline{AB}$ **is parallel to** $\overline{KJ}$.

(b) We use the proportion $\frac{AL}{KL} = \frac{BL}{JL}$ (other proportions can also be used). Note that $BL = 16 - 10 = 6$.

$$\frac{3}{8} \stackrel{?}{=} \frac{6}{16}$$

$\overline{AB}$ **is parallel to** $\overline{KJ}$ since $3 \times 16 = 8 \times 6$.

(c) The numbers provided suggest that the proportion $\frac{AL}{KA} = \frac{LB}{JB}$ be used.

$$\frac{5}{9} \stackrel{?}{=} \frac{10}{15}$$

$\overline{AB}$ **is *not* parallel to** $\overline{KJ}$ since $5 \times 15 \neq 9 \times 10$.

EXERCISE SET 8.1

1. Find the measure of the largest angle of a triangle if the measures of its interior angles are in the ratio 3:5:7.

2. Find the measure of the vertex angle of an isosceles triangle if the measures of the vertex angle and a base angle have the ratio 4:3.

3. The measures of a pair of consecutive angles of a parallelogram have the ratio 5:7. Find the measure of each angle of the parallelogram.

4. Solve for x.

   (a) $\frac{2}{6} = \frac{8}{x}$ (b) $\frac{2}{x} = \frac{x}{50}$ (c) $\frac{2x-5}{3} = \frac{9}{4}$

   (d) $\frac{4}{x+3} = \frac{1}{x-3}$ (e) $\frac{3}{x} = \frac{x-4}{7}$

5. Find the mean proportional between each pair of extremes.
   (a) 4 and 16 (b) $3e$ and $12e^3$ (c) $\frac{1}{2}$ and $\frac{1}{8}$ (d) 6 and 9

6. Determine whether each of the following pairs of ratios is in proportion:
   (a) $\frac{1}{2}$ and $\frac{9}{18}$ (b) $\frac{12}{20}$ and $\frac{3}{5}$ (c) $\frac{4}{9}$ and $\frac{12}{36}$ (d) $\frac{15}{25}$ and $\frac{20}{12}$

7. In $\triangle BAG$, $\overline{LM}$ intersects side $\overline{AB}$ at L and side $\overline{AG}$ at M so that $\overline{LM}$ is parallel to $\overline{BG}$. Using this information, write at least three different true proportions. (Do *not* generate equivalent proportions by inverting the numerators and denominators of the ratios.)

8. For each of the following segment lengths, determine whether $\overline{TP} \parallel \overline{BC}$.
   (a) $AT = 5$, $TB = 15$, $AP = 8$, $PC = 24$
   (b) $TB = 9$, $AB = 18$, $AP = 6$, $PC = 6$
   (c) $AT = 4$, $AB = 12$, $AP = 6$, $AC = 15$
   (d) $AT = 3$, $TB = 9$, $PC = 4$, $AC = 12$
   (e) $AT = \frac{1}{3} \cdot AB$ and $PC = 2 \cdot AP$

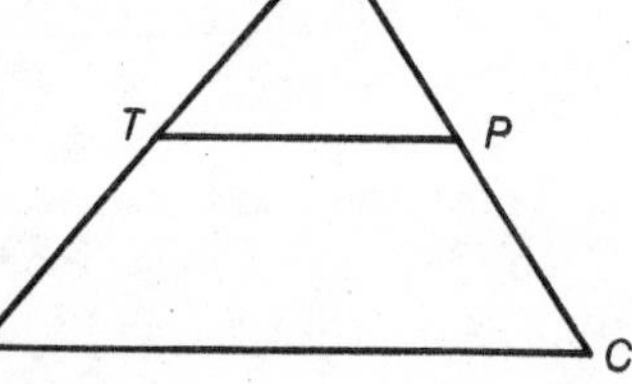

9. If $\overline{KW} \parallel \overline{EG}$, find the lengths of the indicated segments.
   (a) $HE = 20$, $KE = 12$, $WG = 9$, $HG = ?$
   (b) $KH = 7$, $KE = 14$, $HG = 12$, $HW = ?$
   (c) $HW = 4$, $WG = 12$, $HE = 28$, $KH = ?$
   (d) $KH = 9$, $KE = 12$, $HG = 42$, $WG = ?$
   (e) $JH = 2x - 15$, $KE = x$, $HW = 1$, $HG = 4$. Find KH and KE.

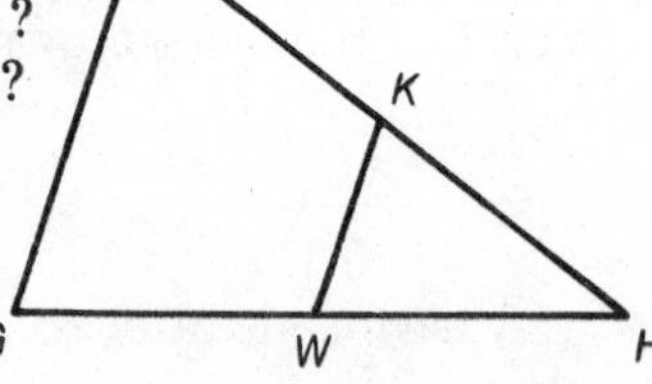

10. In the accompanying diagram of $\triangle ABC$, $\overline{AB} \perp \overline{BC}$ and $\overline{EF} \perp \overline{AB}$ at E. If $BC = 12$, $AB = 16$, and $AE = 8$, find EF.

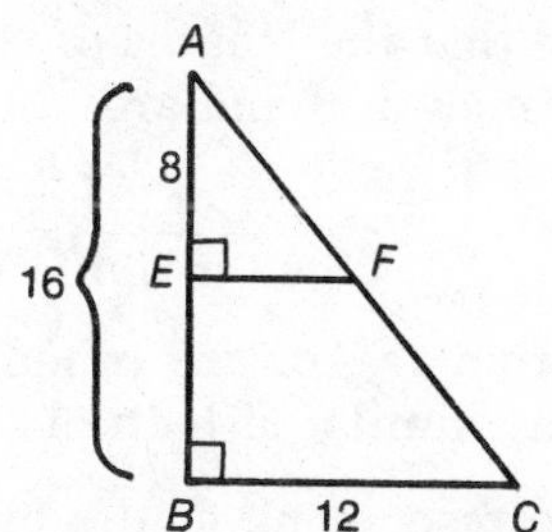

11. In the accompanying diagram of $\triangle ABC$, D is a point on $\overline{AB}$, E is a point on $\overline{BC}$, and $\overline{DE} \parallel \overline{AC}$. If $BD = 5$, $DA = x + 2$, $BE = x + 4$, and $EC = 2x + 4$, find x.

8.2 COMPARING LENGTHS OF SIDES OF SIMILAR POLYGONS

KEY IDEAS

When a photograph is enlarged, the original and enlarged figures are *similar* since they have exactly the same shape. In making a blueprint, every object must be drawn to scale so that the figures in the blueprint are in proportion and are similar to their real-life counterparts.

SIMILAR POLYGONS. Congruent polygons have the same shape *and* the same size, while similar figures have the same shape, but may differ in size. The triangles in Figure 8.3 are similar.

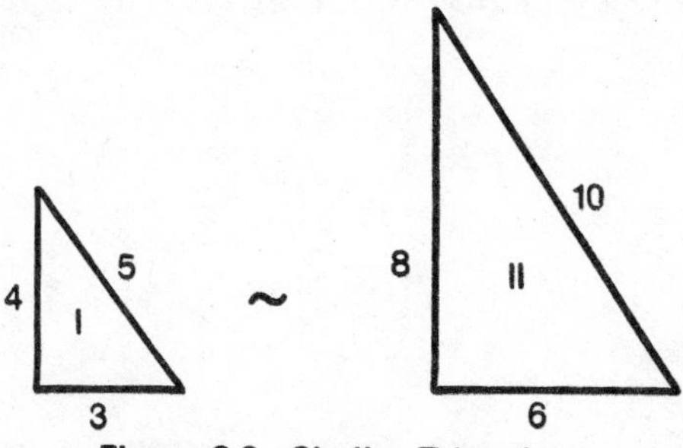

Figure 8.3 Similar Triangles

Notice that the lengths of corresponding sides of triangles I and II have the same ratio and are, therefore, in proportion:

$$\frac{\text{Side in } \triangle \text{I}}{\text{Side in } \triangle \text{II}} = \frac{3}{6} = \frac{4}{8} = \frac{5}{10} = \frac{1}{2} \text{ or } 1:2.$$

Two triangles (or any other two polygons having the same number of sides) are **similar** if both of the following conditions are met:

1. Corresponding angles have the same degree measure.
2. The lengths of corresponding sides are in proportion.

The symbol for similarity is $\sim$. The notation $\triangle ABC \sim \triangle RST$ is read as "triangle *ABC is similar to* triangle *RST*."

Also observe that the perimeters of two similar triangles have the same ratio as the lengths of a pair of corresponding sides:

$$\frac{\text{Perimeter of } \triangle \text{I}}{\text{Perimeter } \triangle \text{II}} = \frac{3+4+5}{6+8+10} = \frac{12}{24} = \frac{1}{2} \text{ or } 1:2.$$

In general:

The perimeters of two similar triangles have the same ratio as the lengths of any pair of corresponding sides.

Examples

1. Quadrilateral *ABCD* is similar to quadrilateral *RSTW*. The lengths of the sides of quadrilateral *ABCD* are 6, 9, 12, and 18. If the length of the longest side of quadrilateral *RSTW* is 24, what is the length of its shortest side?

Solution: Let x = length of the shortest side of quadrilateral *RSTW*. Since the lengths of corresponding sides of similar polygons must have the same ratio, the following proportion is true:

$$\frac{\text{Shortest side of quad } ABCD}{\text{Shortest side of quad } RSTW} = \frac{\text{Longest side of quad } ABCD}{\text{Longest side of quad } RSTW}$$

$$\frac{6}{x} = \frac{18}{24}$$

Write $\frac{18}{24}$ in lowest terms: $\quad \frac{6}{x} = \frac{3}{4}$

Cross-multiply: $\quad 3x = 24$

$$x = \frac{24}{3} = 8$$

The length of the shortest side of quadrilateral *RSTW* is **8**.

2. The longest side of a polygon exceeds twice the length of the longest side of a similar polygon by 3. If the ratio of the perimeters of the two polygons is 4:9, find the length of the longest side of each polygon.

Solution: Let x = length of longest side in smaller polygon. Then $2x + 3$ = length of longest side in larger polygon.

$$\frac{\text{Perimeter of smaller polygon}}{\text{Perimeter of larger polygon}} = \frac{\text{Longest side of smaller polygon}}{\text{Longest side of larger polygon}}$$

$$\frac{4}{9} = \frac{x}{2x+3}$$

Cross-multiply: $9x = 4(2x + 3)$

$$9x = 4(2x + 3)$$
$$9x = 8x + 12$$
$$9x - 8x = 12$$
$$x = 12$$
$$2x + 3 = 2(12) + 3 = 24 + 3 = 27$$

The length of the longest side in the smaller polygon is **12**, and the length of the longest side in the larger polygon is **27**.

ALTITUDES AND MEDIANS OF SIMILAR TRIANGLES. The lengths of corresponding *altitudes*, the lengths of corresponding *medians*, and the *perimeters* of similar triangles (Figure 8.4) have the same ratio as the lengths of any pair of corresponding sides:

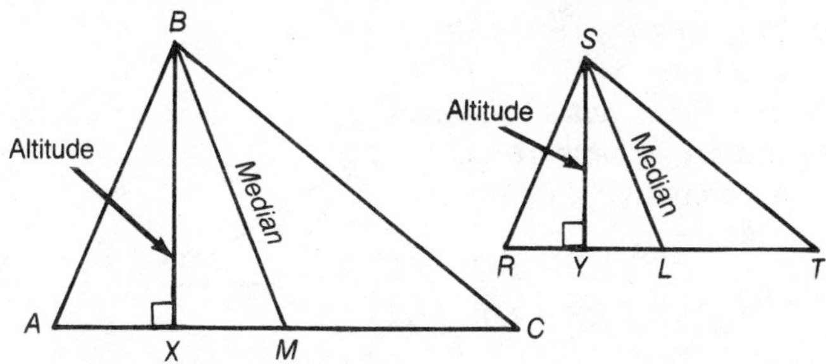

Figure 8.4 Altitudes and Medians in Similar Triangles

$$\frac{\text{Perimeter of } \triangle ABC}{\text{Perimeter of } \triangle RST} = \frac{BX}{SY} = \frac{BM}{SL} = \frac{AB}{RS} = \frac{BC}{ST} = \frac{AC}{RT}.$$

Example

3. Triangle $RST \sim \triangle KLM$. The length of altitude $\overline{SA}$ exceeds the length of altitude $\overline{LB}$ by 5. If $RT = 9$ and $KM = 6$, find the length of each altitude.

Solution: Let x = length of altitude $\overline{LB}$.
Then $x + 5$ = length of altitude $\overline{SA}$.

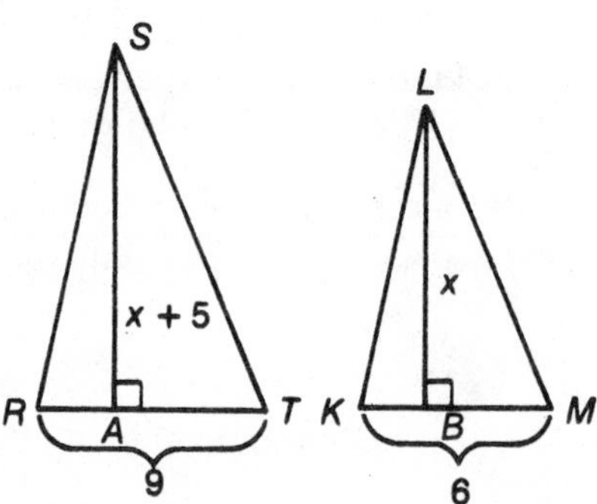

$$\frac{x+5}{x} = \frac{9}{6}$$
$$9x = 6(x+5)$$
$$9x = 6x + 30$$
$$3x = 30$$
$$x = 10$$

Altitude $LB = x =$ **10**
Altitude $SA = x + 5 =$ **15**

METHODS FOR PROVING TRIANGLES SIMILAR. Any one of the following three theorems can be used to prove that two triangles are similar:

- AA THEOREM OF SIMILARITY: *Two triangles are similar if two angles of one triangle are congruent to the corresponding angles in the second triangle.*

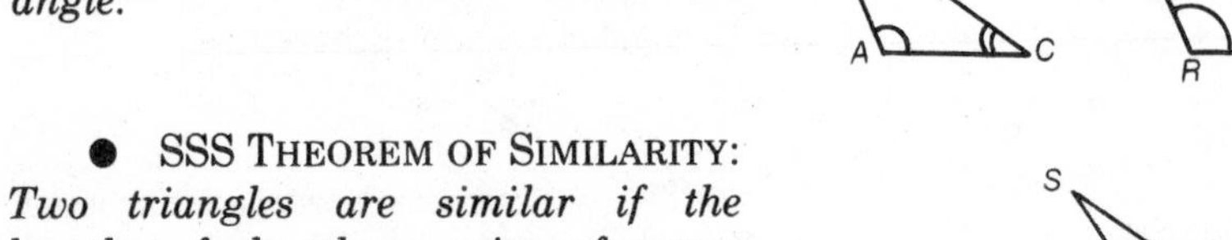

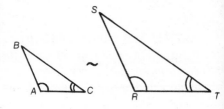

- SSS THEOREM OF SIMILARITY: *Two triangles are similar if the lengths of the three pairs of corresponding sides are in proportion.*

In the accompanying figure,

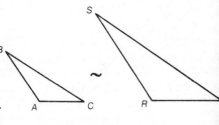

$\triangle ABC \sim \triangle RST$ if $\dfrac{AB}{RS} = \dfrac{BC}{ST} = \dfrac{AC}{RT}$.

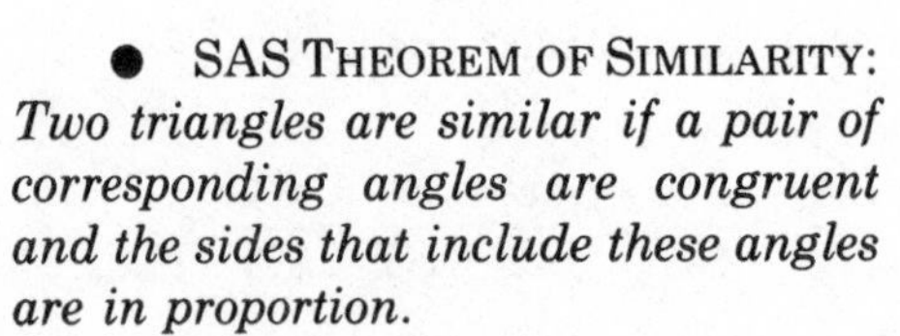

- SAS THEOREM OF SIMILARITY: *Two triangles are similar if a pair of corresponding angles are congruent and the sides that include these angles are in proportion.*

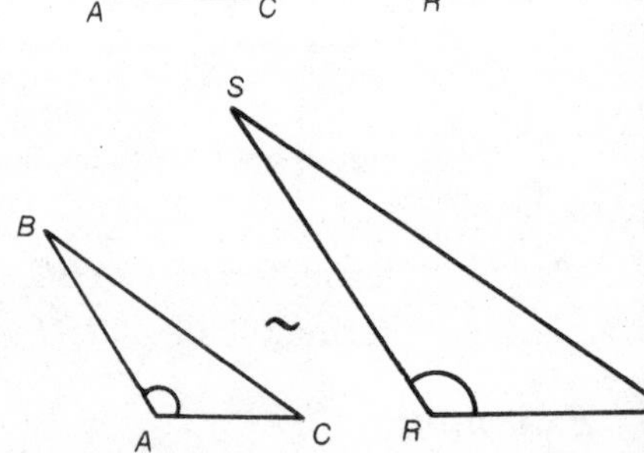

In the accompanying figure,

$\triangle ABC \sim \triangle RST$ if $\angle A \cong \angle R$ and $\dfrac{AB}{RS} = \dfrac{AC}{RT}$.

SOME ADDITIONAL NOTES ON SIMILARITY. Here are some additional facts about similar polygons that you should know:

1. If two polygons are similar, then the ratio of the lengths of any pair of corresponding sides is sometimes referred to as the *ratio of similitude.*

2. If two polygons are congruent, then they are also similar and their ratio of similitude is 1. Two similar polygons are *not* necessarily congruent.

3. Regular polygons having the same number of sides are similar.

4. Since similarity is an equivalence relation, the reflexive, symmetric, and transitive properties may be applied to similarity of polygons. For example, if $\triangle I \sim \triangle II$ and $\triangle II \sim \triangle III$, then $\triangle I \sim \triangle III$.

Examples

4. In the accompanying figure, $\overline{DE} \parallel \overline{AB}$.

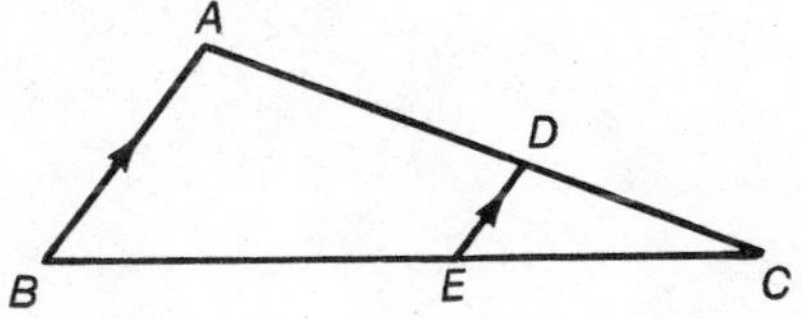

(a) Is $\triangle DEC \sim \triangle ABC$? Give a reason for your answer.

(b) If $CD = 6$, $CA = 18$, and $DE = 4$, what is the length of $\overline{AB}$?

Solution: (a) $\overline{DE} \parallel \overline{AB}$, so transversal $\overline{AC}$ forms congruent corresponding angles, making $m\angle A = m\angle CDE$. Also, transversal $\overline{BC}$ forms congruent corresponding angles, making $m\angle B = m\angle CED$. **Triangle** $\boldsymbol{DEC \sim \triangle ABC}$ since two angles of $\triangle DEC$ are equal in degree measure to two angles of $\triangle ABC$.

(b) Let x = length of $\overline{AB}$. Since the lengths of corresponding sides of similar triangles are in proportion,

$$\frac{CD}{CA} = \frac{DE}{AB}$$

$$\frac{6}{18} = \frac{4}{x}$$

$$\frac{1}{3} = \frac{4}{x}$$

$$x = 3 \cdot 4 = 12$$

The length of $\overline{AB}$ is **12**.

5. A pole 10 feet high casts a 15-foot-long shadow on level ground. At the same time a man casts a shadow that is 9 feet in length. How tall is the man?

Solution: By assuming that the shadows are perpendicular to the pole and the man, we may use right triangles to represent these situations, where x represents the height of the man.

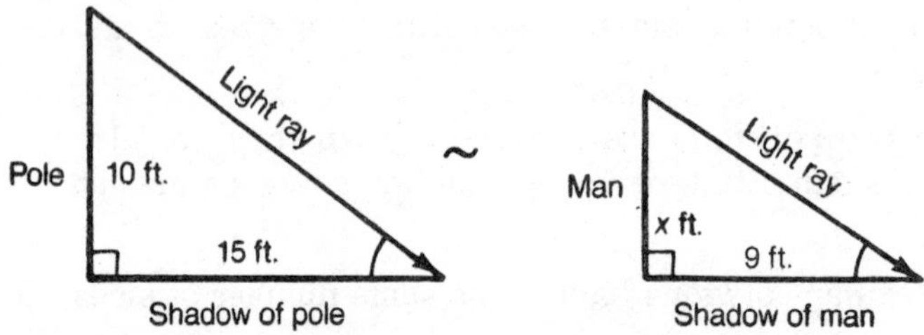

We also assume that in each triangle the light rays make angles with the ground that have the same degree measure. Since all right angles have the same degree measure, the two right triangles are similar and the lengths of their sides are in proportion.

$$\frac{\text{Height of pole}}{\text{Height of man}} = \frac{\text{Shadow of pole}}{\text{Shadow of man}}$$

$$\frac{10}{x} = \frac{15}{9}$$

$$\frac{10}{x} = \frac{5}{3}$$

$$5x = 3 \cdot 10$$

$$x = \frac{30}{5} = 6$$

The man is **6 feet** tall.

EXERCISE SET 8.2

1. Triangle $GAL \sim \triangle SHE$. Name three pairs of congruent angles and three equal ratios.
2. The ratio of perimeters of two similar polygons is 3:5. If the length of the shortest side of the smaller polygon is 24, find the length of the shortest side of the larger polygon.
3. Triangle $ZAP \sim \triangle MYX$. If $ZA = 3$, $AP = 12$, $ZP = 21$, and $YX = 20$, find the lengths of the remaining sides of $\triangle MYX$.
4. Quadrilateral $ABCD \sim$ quadrilateral $RSTW$. The lengths of the sides of quadrilateral $ABCD$ are 3, 6, 9, and 15. If the length of the longest side of quadrilateral $RSTW$ is 20, find the perimeter of $RSTW$.

5. The longest side of a polygon exceeds twice the length of the longest side of a similar polygon by 3. If the ratio of similitude of the polygons is 4:9, find the length of the longest side of each polygon.

6. Triangle $RST \sim \triangle JKL$.
   (a) $\overline{RA}$ and $\overline{JB}$ are medians to sides $\overline{ST}$ and $\overline{KL}$, respectively. $RS = 10$ and $JK = 15$. If the length of $\overline{JB}$ exceeds the length of $\overline{RA}$ by 4, find the lengths of medians $\overline{JB}$ and $\overline{RA}$.
   (b) $\overline{SH}$ and $\overline{KO}$ are altitudes to sides $\overline{RT}$ and $\overline{JL}$, respectively. If $SH = 12$, $KO = 15$, $LK = 3x - 2$, and $TS = 2x + 1$, find the lengths of $\overline{LK}$ and $\overline{TS}$.
   (c) The perimeter of $\triangle RST$ is 25 and the perimeter of $\triangle JKL$ is 40. If $ST = 3x + 1$ and $KL = 4x + 4$, find the lengths of $\overline{ST}$ and $\overline{KL}$.
   (d) The ratio of perimeters of $\triangle RST$ and $\triangle JKL$ is $3:x$. The length of altitude $\overline{SU}$ is $x - 4$, and the length of altitude $\overline{KV}$ is 15. Find the length of altitude $\overline{SU}$.

7. State whether each of the given pairs of figures are *always*, *sometimes*, or *never* similar.
   (a) Two right triangles
   (b) Two equilateral triangles
   (c) Two rhombuses
   (d) Two squares
   (e) Two isosceles triangles having congruent vertex angles
   (f) A trapezoid and a parallelogram

8–10. In the accompanying figure, $\overline{AB} \perp \overline{BE}$ *and* $\overline{DE} \perp \overline{BE}$.

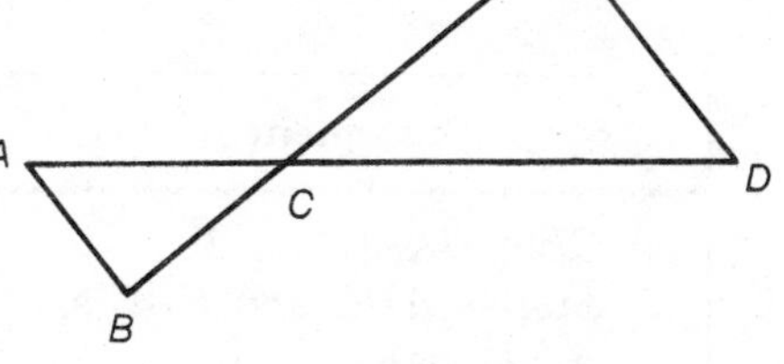

8. If $AB = 4$, $BC = 2$, and $DE = 6$, find EC.

9. If $BC = 5$, $BE = 15$, and $AC = 3$, find DC.

10. If $AC = 8$, $DC = 12$, and $BE = 15$, find BC.

11. Given $\triangle ABC$, $\triangle DEF$, $\angle A \cong \angle D$, $\angle B \cong \angle E$, $AB = 4$, $DF = 6$, $DE = x$, and $AC = x + 5$.
   (a) Write an equation in terms of x that can be used to find DE.
   (b) Find DE.

8.3 PROOFS INVOLVING SIMILAR TRIANGLES

KEY IDEAS

The method most often used to prove that two triangles are similar is the AA theorem of similarity. To prove that two triangles are similar using this method, simply show that two angles of one triangle are congruent to two angles of the second triangle.

PROVING THAT TRIANGLES ARE SIMILAR. Unlike congruence proofs, proofs that use the AA theorem of similarity to prove that two triangles are similar require that *two*, rather than three, pairs of parts (in this case, angles) be proved congruent.

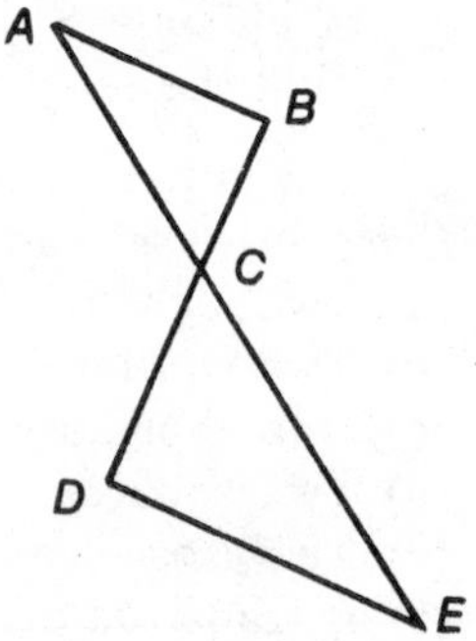

A
B
C
D
E

Example

1. Given: $\overline{CB} \perp \overline{BA}$, $\overline{CD} \perp \overline{DE}$.
   Prove: $\triangle ABC \sim \triangle EDC$.

Solution:

PLAN. Use the AA theorem. The two triangles include right and vertical angles that yield two pairs of congruent angles.

PROOF

Statement	Reason
1. $\overline{CB} \perp \overline{BA}$, $\overline{CD} \perp \overline{DE}$	1. Given.
2. Angles *ABC* and *EDC* are right angles.	2. Perpendicular lines intersect to form right angles.
3. $\angle ABC \cong \angle EDC$ *Angle*	3. All right angles are congruent.
4. $\angle ACB \cong \angle ECD$ *Angle*	4. Vertical angles are congruent.
5. $\triangle ABC \sim \triangle EDC$	5. AA theorem.

USING SIMILAR TRIANGLES TO PROVE THAT LENGTHS OF SEGMENTS ARE IN PROPORTION. To establish that the lengths of four segments are in proportion, we first show that a pair of triangles that contain these segments as corresponding sides are similar. We may then apply the reverse of the definition of similarity and conclude that the lengths of the segments are in proportion, using, as a reason: *The lengths of corresponding sides of similar triangles are in proportion.*

Examples

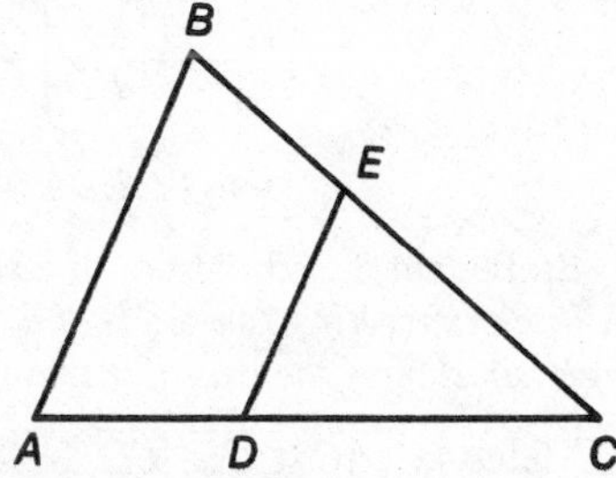

2. Given: $\overline{AB} \parallel \overline{DE}$.

Prove: $\dfrac{EC}{BC} = \dfrac{ED}{AB}$.

Solution:

PLAN. *Step 1.* Select the triangles that contain these segments as sides. Read across the proportion:

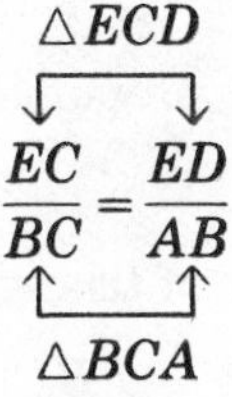

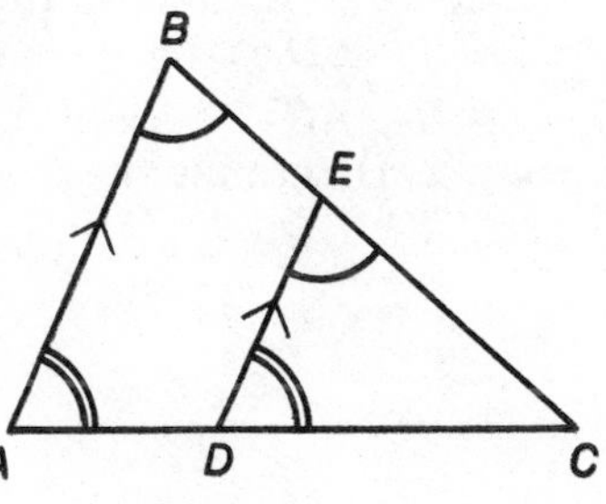

Step 2. Mark off the diagram with the "Given" and all pairs of corresponding congruent angles.

Step 3. Write the proof.

PROOF

Statement	Reason
1. $\overline{AB} \parallel \overline{DE}$	1. Given.
2. $\angle CED \cong \angle CBA$, *Angle* $\angle CDE \cong \angle CAB$ *Angle*	2. If two lines are parallel, then their corresponding angles are congruent.
3. $\triangle ECD \sim \triangle BCA$	3. AA theorem.
4. $\dfrac{EC}{BC} = \dfrac{ED}{AB}$	4. The lengths of corresponding sides of similar triangles are in proportion.

Notes

● If the original proportion in the "Prove" of Example 2 had been written as $\dfrac{EC}{ED} = \dfrac{BC}{AB}$, then reading *across* the proportion would *not* yield the vertices of the triangles to be proved similar. In that case, reading down each ratio would give the required triangles:

$$\frac{EC}{ED} = \frac{BC}{AB}$$

$$\triangle EDC \sim \triangle BCA$$

● Statement 3 of the proof of Example 2 establishes the following relationship: *A line intersecting two sides of a triangle and parallel to the third side forms a triangle similar to the original triangle.*

3. Given: $\overline{AC} \perp \overline{CB}$, $\overline{ED} \perp \overline{AB}$.
Prove: $EB : AB = ED : AC$.

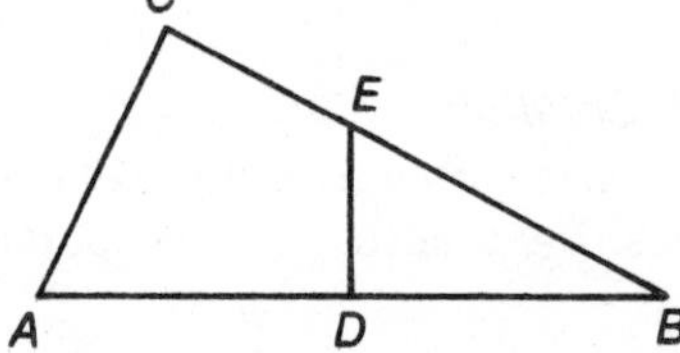

Solution:

PLAN. *Step 1.* Rewrite the proportion in the "Prove" in fractional form: $\frac{EB}{AB} = \frac{ED}{AC}$.

Step 2. Determine the pair of triangles that must be proved similar.

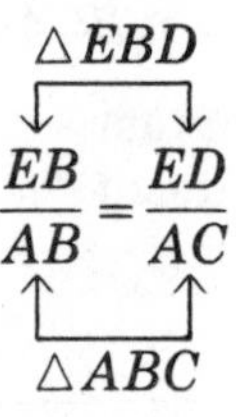

Step 3. Mark off the diagram with the "Given" and all pairs of corresponding congruent angle pairs.

Step 4. Write the proof.

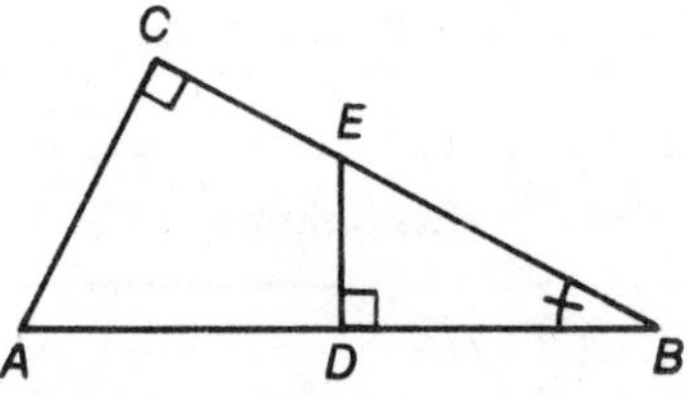

PROOF

Statement		Reason
1. $\overline{AC} \perp \overline{CB}$, $\overline{ED} \perp \overline{AB}$		1. Given.
2. Angles C and EDB are right angles.		2. Perpendicular lines intersect to form right angles.
3. $\angle C \cong \angle EDB$	*Angle*	3. All right angles are congruent.
4. $\angle B \cong \angle B$	*Angle*	4. Reflexive property of congruence.
5. $\triangle EBD \sim \triangle ABC$		5. AA theorem.
6. $EB{:}AB = ED{:}AC$		6. The lengths of corresponding sides of similar triangles are in proportion.

USING SIMILAR TRIANGLES TO PROVE THAT PRODUCTS OF SEGMENT LENGTHS ARE EQUAL. If $\frac{A}{B} = \frac{C}{D}$, then $A \times D = B \times C$. The reason is that in a proportion the product of the means equals the product of the extremes. Instead of generating a product from a proportion, it is sometimes necessary to be able to take a product and determine the related proportion that would yield that product.

Suppose that the lengths of four segments are related in such a way that

$$KM \times LB = LM \times KD.$$

What proportion gives this result when the products of its means and extremes are set equal to each other? A true proportion may be derived from the product by making a pair of terms appearing on the same side of the equal sign (say, KM and LB) the extremes. The pair of terms on the opposite side of the equal sign (LM and KD) then become the means:

$$\frac{KM}{LM} = \frac{KD}{LB}.$$

***Note*:** An equivalent proportion results if KM and LB are made the means rather than the extremes.

This analysis will be useful in the example that follows.

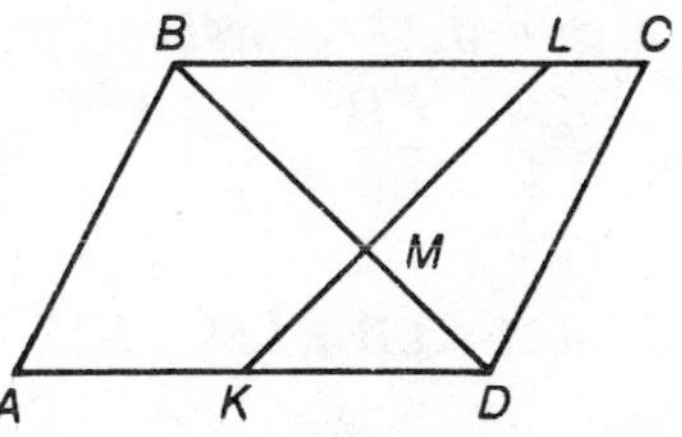

Example

4. Given: ▱$ABCD$.
Prove: $KM \times LB = LM \times KD$.

Solution:

PLAN. Analyze the solution to this type of problem by *working backward*, beginning with the "Prove."

Step 1. Express the product as an equivalent proportion (this was accomplished in the discussion that preceded this example).

$$\frac{KM}{LM} = \frac{KD}{LB}$$

Step 2. From the proportion (and, in some problems, in conjunction with the "Given"), determine the pair of triangles to be proved similar.

$$\triangle KMD$$

$$\frac{KM}{LM} = \frac{KD}{LB}$$

$$\triangle LMB$$

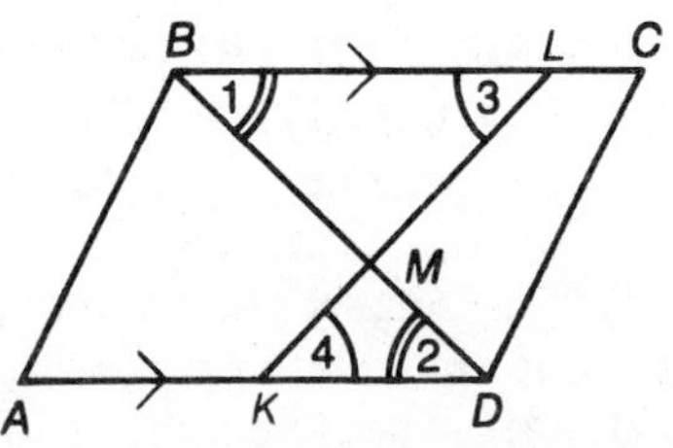

Step 3. Mark off the diagram with the "Given" and decide how to show the triangles similar. STRATEGY: Use the AA theorem.

Step 4. Write the formal two-column proof. The steps in the proof should reflect the logic of this analysis, proceeding from step 4 back to step 1 (proving the triangles similar, forming the appropriate proportion, and, lastly, writing the product):

PROOF

Statement	Reason
1. $\square ABCD$	1. Given.
2. $\overline{AD} \parallel \overline{BC}$	2. Opposite sides of a parallelogram are parallel.
3. $\angle 1 \cong \angle 2$, $\angle 3 \cong \angle 4$	3. If two lines are parallel, then their alternate interior angles are congruent.
4. $\triangle KMD \sim \triangle LMB$	4. AA theorem.
5. $\frac{KM}{LM} = \frac{KD}{LB}$	5. The lengths of corresponding sides of similar triangles are in proportion.
6. $KM \times LB = LM \times KD$	6. In a proportion, the product of the means equals the product of the extremes.

***Note*:** In our analysis of the product $KM \times LB = LM \times KD$, suppose we formed the proportion

$$\frac{KM}{KD} = \frac{LM}{LB}.$$

Reading across the top (*K-M-L*), we do not find a set of letters that correspond to the vertices of a triangle. Reading down the first ratio (*K-M-D*), however, gives us the vertices of one of the desired triangles, and reading down the second ratio give the vertices of the other triangle.

SUMMARY

To prove that the products of the lengths of the sides of two triangles are equal, follow these steps:

Step 1. From the product, write a proportion by making one pair of factors the means and the other pair of factors the extremes.

Step 2. Use the proportion to identify the triangles that contain the desired segments as sides.

Step 3. Mark off the diagram with the "Given," as well as with any additional congruent parts resulting from vertical angles, right angles, congruent alternate interior angles, and so forth.

Step 4. Determine the similarity method of proof to be used. In the overwhelming majority of cases, look to apply the AA theorem of similarity.

Step 5. Write the formal proof:

A. Prove that the triangles are similar.

B. Write the desired proportion, using, as a reason: *Lengths of corresponding sides of similar triangles are in proportion.*

C. Write the desired product, using, as a reason: *In a proportion the product of the means equals the product of the extremes.*

EXERCISE SET 8.3

1. Given: $\overline{XW} \cong \overline{XY}$,
$\overline{HA} \perp \overline{WY}$, $\overline{KB} \perp \overline{WY}$.
Prove: $\triangle HWA \sim \triangle KYB$.

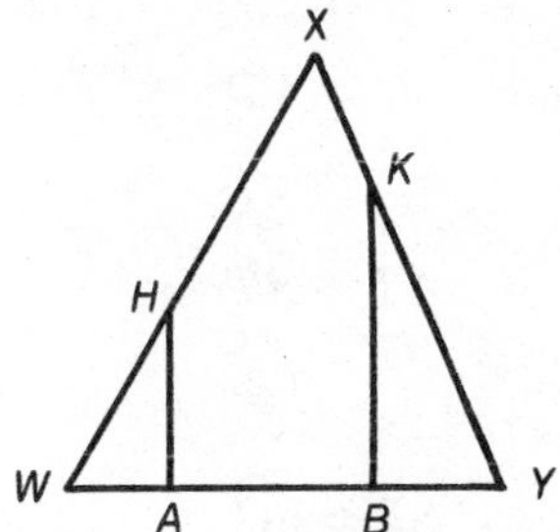

2. Given: $\triangle ABC \sim \triangle RST$,
$\overline{BX}$ bisects $\angle ABC$,
$\overline{SY}$ bisects $\angle RST$.
Prove: $\triangle BXC \sim \triangle SYT$.

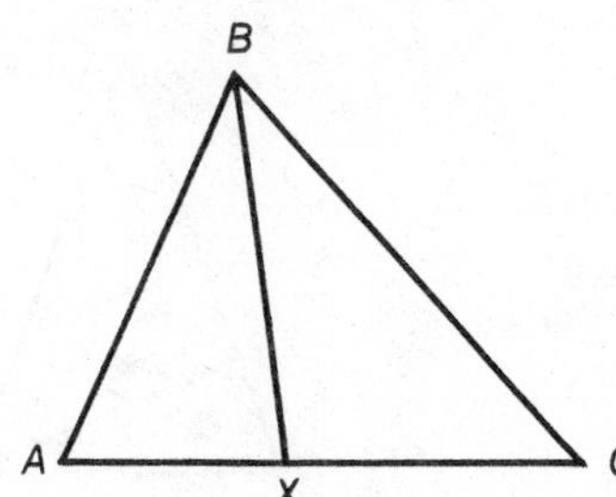

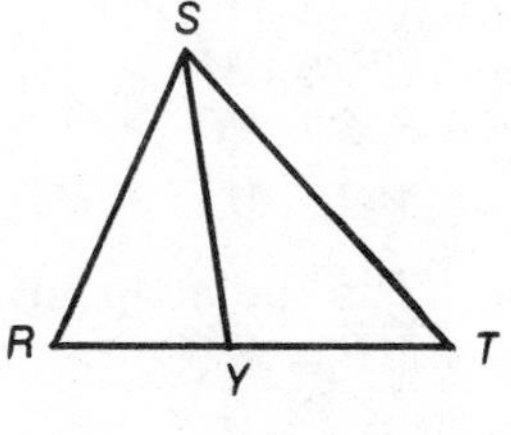

3. Given: Rectangle $DEFG$,
$\triangle ABC$ with a right angle at B,
$\overline{AEB}$, $\overline{BDC}$, $\overline{AFGC}$.
Prove: $\frac{AE}{ED} = \frac{EF}{BD}$.

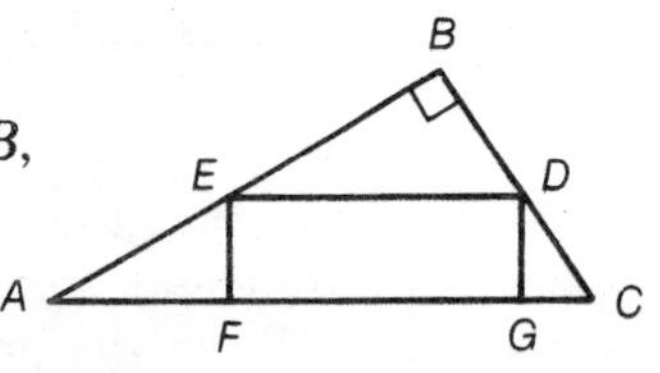

4. Given: $\triangle MCT \sim \triangle BAW$,
$\overline{SW}$ bisects $\angle AWM$.
Prove: $\triangle MCT \sim \triangle BCW$.

5. Given: $\overline{AW} \parallel \overline{ST}$,
$\overline{MS} \cong \overline{MW}$, $\overline{WC} \cong \overline{WA}$.
Prove: $\triangle BCW \sim \triangle BTS$.

6. Given: $\overline{WB} \cong \overline{WC}$,
$\overline{ST} \parallel \overline{AW}$,
$\overline{AT}$ bisects $\angle STW$.
Prove: $\triangle ABW \sim TCW$.

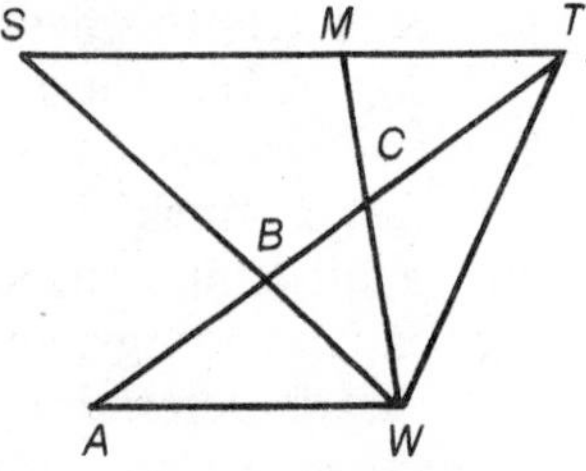

Exercises 4–6

7. Given: $\overline{HW} \parallel \overline{TA}$, $\overline{HY} \parallel \overline{AX}$.
Prove: $\frac{AX}{HY} = \frac{AT}{HW}$.

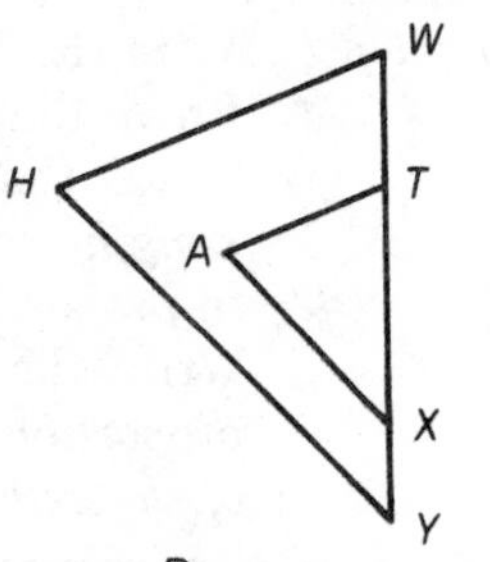

8. Given: $\overline{MN} \parallel \overline{AT}$,
$\angle 1 \cong \angle 2$.
Prove: $\frac{NT}{AT} = \frac{RN}{RT}$.

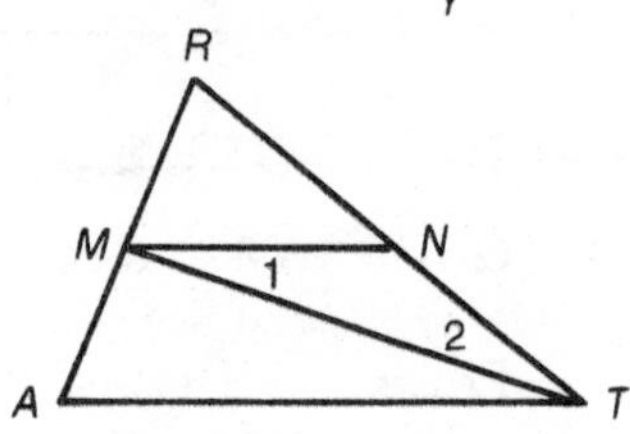

9. Given: $\overline{SR} \cong \overline{SQ}$,
$\overline{RQ}$ bisects $\angle SRW$.
Prove: $\frac{SQ}{RW} = \frac{SP}{PW}$.

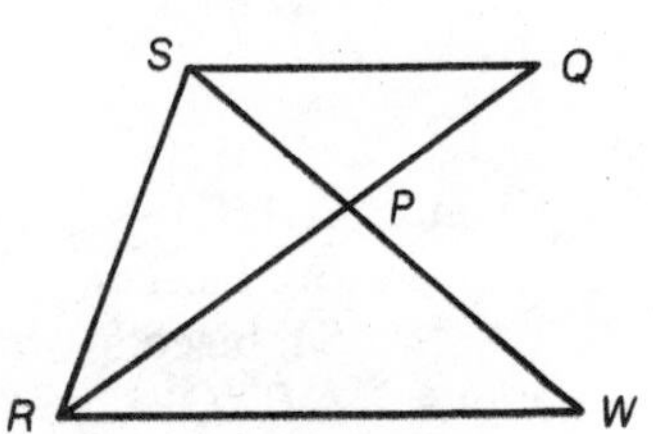

10. Given: $\overline{MC} \perp \overline{JK}$, $\overline{PM} \perp \overline{MQ}$,
$\overline{TP} \cong \overline{TM}$.
Prove: $\frac{PM}{MC} = \frac{PQ}{MK}$.

11. Given: T is the midpoint of $\overline{PQ}$,
$\overline{MP} \cong \overline{MQ}$,
$\overline{JK} \parallel \overline{MQ}$.
Prove: $\frac{PM}{JK} = \frac{TQ}{JT}$.

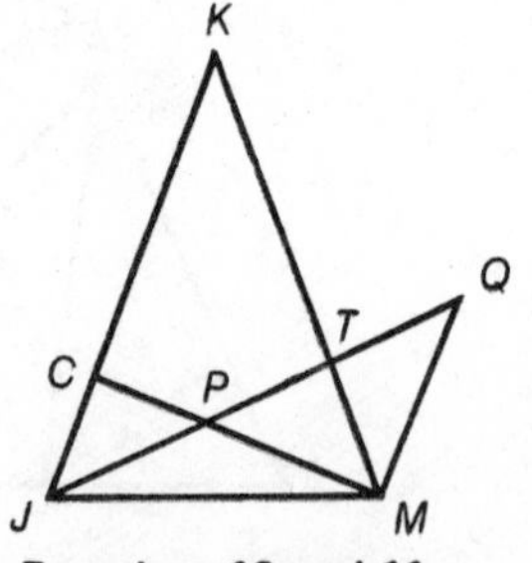

Exercises 10 and 11

12. Given: $\overline{EF}$ is the median of trapezoid $ABCD$.
Prove: $EI \times GH = IH \times EF$.

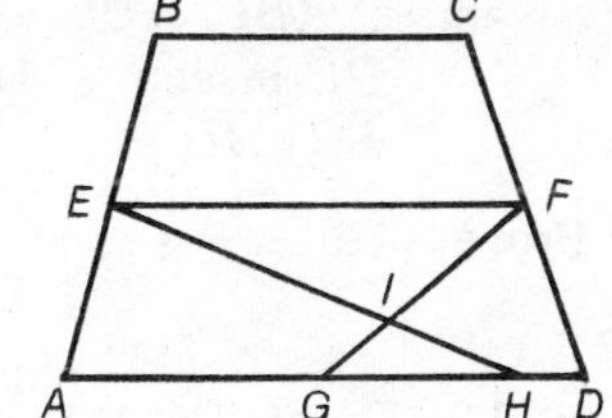

13. Given: $\overline{RS} \perp \overline{ST}$, $\overline{SW} \perp \overline{RT}$.
Prove: $(ST)^2 = TW \times RT$.

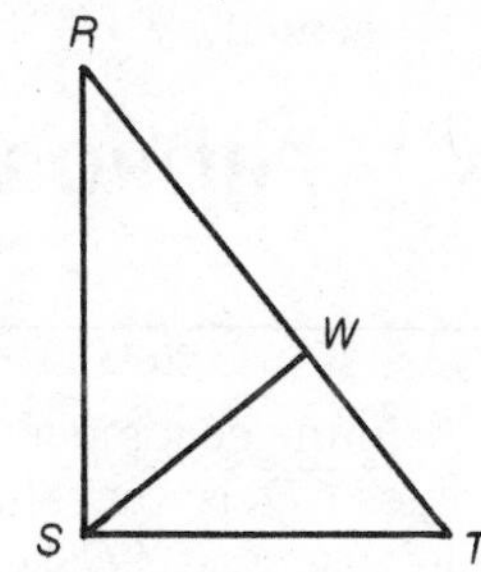

14. Given: $\overline{BH} \perp \overline{AC}$, $\overline{AF} \perp \overline{BC}$.
Prove: $BH : AF = BC : AC$.

15. Given: $\overline{AF}$ bisects $\angle BAC$,
$\overline{BH}$ bisects $\angle ABC$,
$\overline{BC} \cong \overline{AC}$.
Prove: $AH \times EF = BF \times EH$.

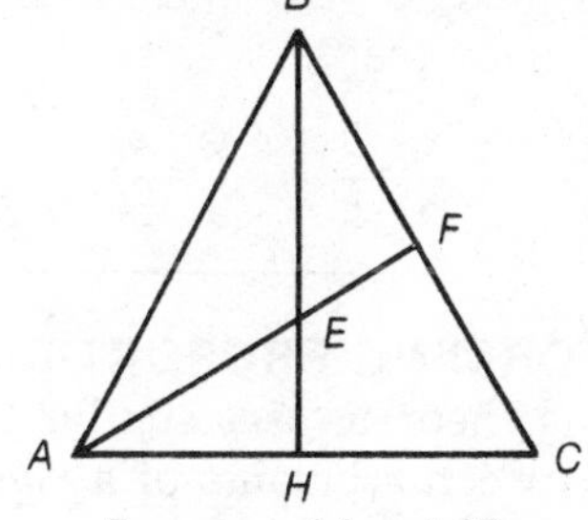

Exercises 14 and 15

16. Given: $\overline{XY} \parallel \overline{LK}$, $\overline{XZ} \parallel \overline{JK}$.
Prove: $JY \times ZL = XZ \times KZ$.

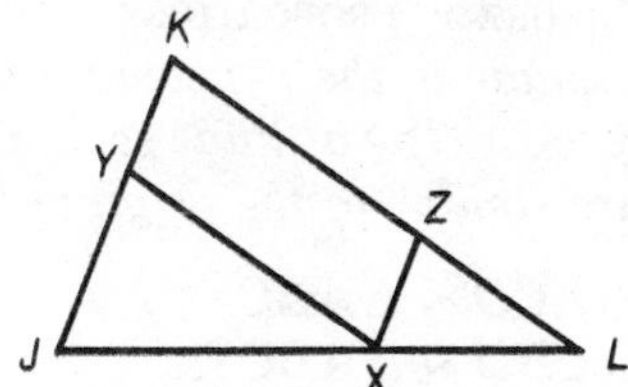

17. Given: $\square ABCD$,
$\overline{CB}$ is extended through B to E,
$\overline{DE}$ is drawn intersecting $\overline{AB}$ at F.

(a) Prove: $\dfrac{EB}{DA} = \dfrac{BF}{AF}$.

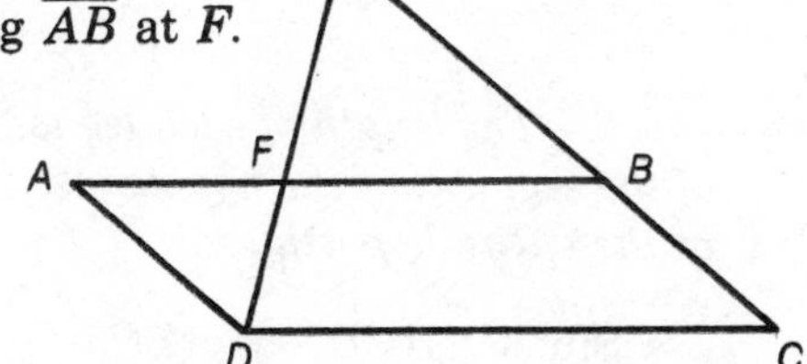

(b) If $DF = 4$, $EF = 6$, EB is represented by x, and AD is represented by $x - 3$, find x.

18. Given: $\triangle ABC$, $\overline{ABE}$, $\overline{ADC}$, $\overline{BD}$ bisects $\angle ABC$, $\overline{EC} \parallel \overline{BD}$.

Prove: (a) $\dfrac{AD}{DC} = \dfrac{AB}{BE}$.

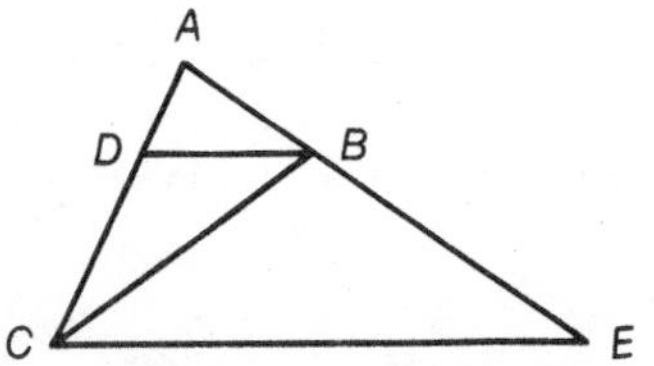

(b) If $AD = 6$, $DC = 9$, AB is represented by x, and BE is represented by $x + 4$, find x.

8.4 PROPORTIONS IN A RIGHT TRIANGLE

KEY IDEAS

In the accompanying right triangle, $\overline{CD}$ is the altitude to hypotenuse $\overline{AB}$. $\overline{CD}$ separates right triangle ABC into similar triangles such that:

$$\frac{x}{b} = \frac{b}{c}, \quad \frac{y}{a} = \frac{a}{c}, \quad \frac{x}{h} = \frac{h}{y}.$$

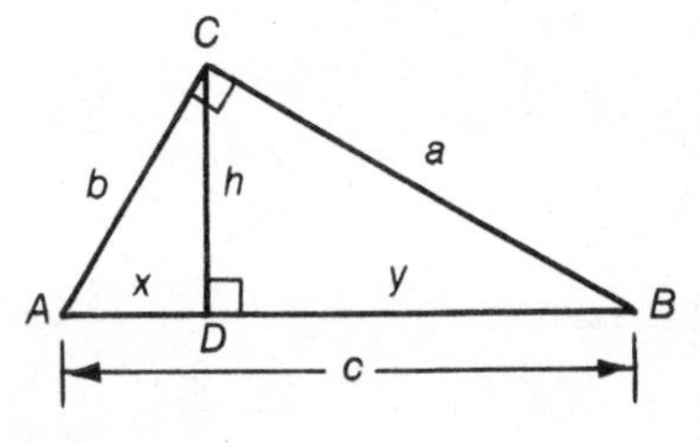

FORMING PROPORTIONS IN A RIGHT TRIANGLE. The following theorem uses similar triangles to establish important relationships between segments of a right triangle.

THEOREM: PROPORTIONS IN A RIGHT TRIANGLE. *If in a right triangle the altitude to the hypotenuse is drawn, then:*

1. *The altitude separates the original triangle into two triangles that are similar to the original triangle and to each other.*

$\triangle ACD \sim \triangle ABC$
$\triangle CBD \sim \triangle ABC$
$\triangle ACD \sim \triangle CBD$

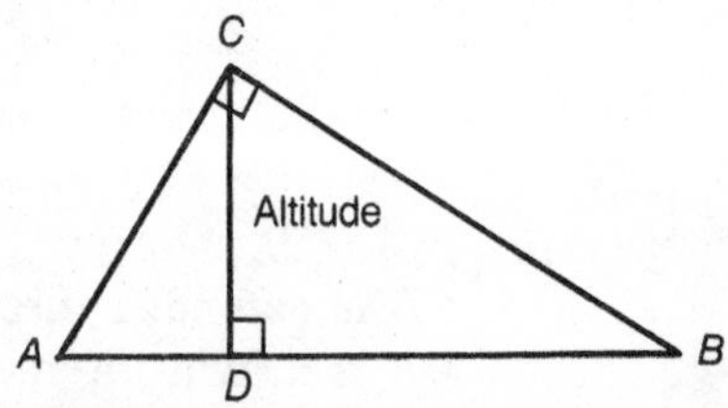

2. *The length of each leg is the mean proportional between the length of the segment of the hypotenuse that is adjacent to the leg and the length of the entire hypotenuse.*

Since $\triangle ACD \sim \triangle ABC$, $\dfrac{AD}{AC} = \dfrac{AC}{AB}$.

Since $\triangle CBD \sim \triangle ABC$, $\dfrac{BD}{BC} = \dfrac{BC}{AB}$.

3. The length of the altitude is the mean proportional between the lengths of the segments that it forms on the hypotenuse.

Since $\triangle ACD \sim \triangle CBD$, $\dfrac{AD}{CD} = \dfrac{CD}{DB}$.

Examples

1. In each case, find the value of *x*.

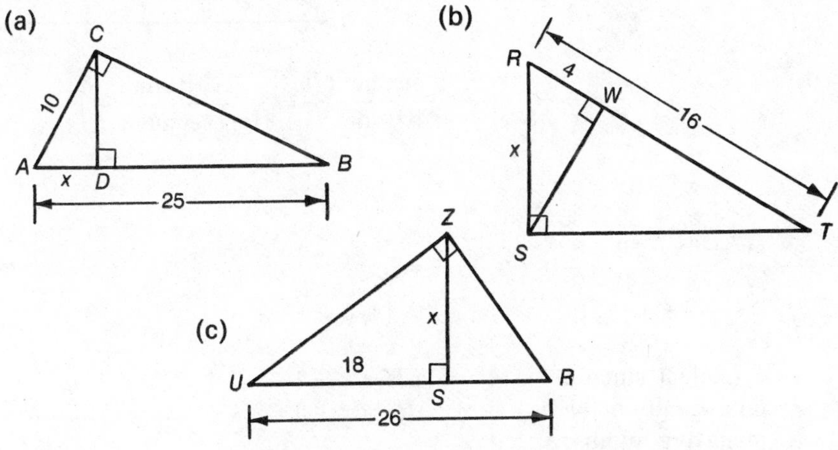

Solutions: (a) $\dfrac{AD}{AC} = \dfrac{AC}{AB}$ $\left(\dfrac{\text{Hyp segment}}{\text{Leg}} = \dfrac{\text{Leg}}{\text{Hyp}}\right)$

$$\frac{x}{10} = \frac{10}{25}$$

$$25x = 100$$

$$x = \frac{100}{25} = \mathbf{4}$$

(b) $\dfrac{RW}{RS} = \dfrac{RS}{RT}$ $\left(\dfrac{\text{Hyp segment}}{\text{Leg}} = \dfrac{\text{Leg}}{\text{Hyp}}\right)$

$$\frac{4}{x} = \frac{x}{16}$$

$$x^2 = 64$$

$$x = \sqrt{64} = \mathbf{8}$$

(c) $\dfrac{US}{ZS} = \dfrac{ZS}{RS}$ $\left(\dfrac{\text{Hyp segment 1}}{\text{Altitude}} = \dfrac{\text{Altitude}}{\text{Hyp segment 2}}\right)$

$$\frac{18}{x} = \frac{x}{8}$$ NOTE: $RS = 26 - 18 = 8$

$$x^2 = 144$$

$$x = \sqrt{144} = \mathbf{12}$$

2. In right triangle JKL, $\angle K$ is the right angle. Altitude $\overline{KH}$ is drawn in such a way that the length of $\overline{JH}$ exceeds the length of $\overline{HL}$ by 5. If $KH = 6$, find the length of the hypotenuse.

Solution:

Let $x =$ length of $\overline{LH}$.
Then $x + 5 =$ length of $\overline{JH}$.

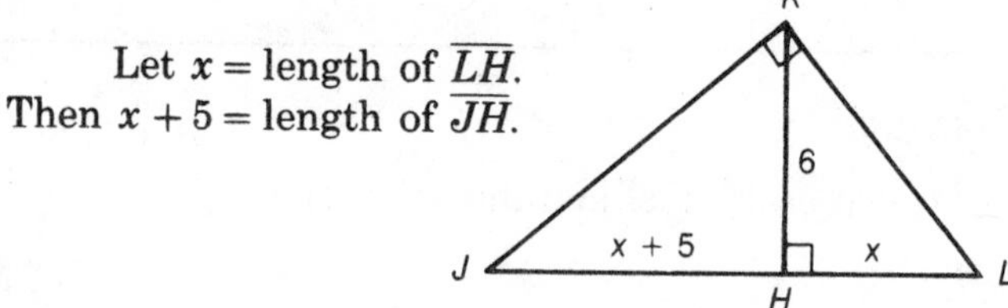

$$\frac{x}{6} = \frac{6}{x+5} \quad \left(\frac{\text{Hyp segment 1}}{\text{Altitude}} = \frac{\text{Altitude}}{\text{Hyp segment 2}}\right)$$

$$x(x+5) = 36$$
$$x^2 + 5x = 36$$
$$x^2 + 5x - 36 = 36 - 36$$
$$x^2 + 5x - 36 = 0$$
$$(x+9)(x-4) = 0$$

$x + 9 = 0$ or $x - 4 = 0$

$x = -9$	$x = 4$
(Reject since a length cannot be a negative number.)	$LH = x = 4$ $JH = x + 5 = 9$

JL (hypotenuse length) $= 4 + 9 = \mathbf{13}$.

EXERCISE SET 8.4

1–3. In each case, find the value of r, s, *and* t.

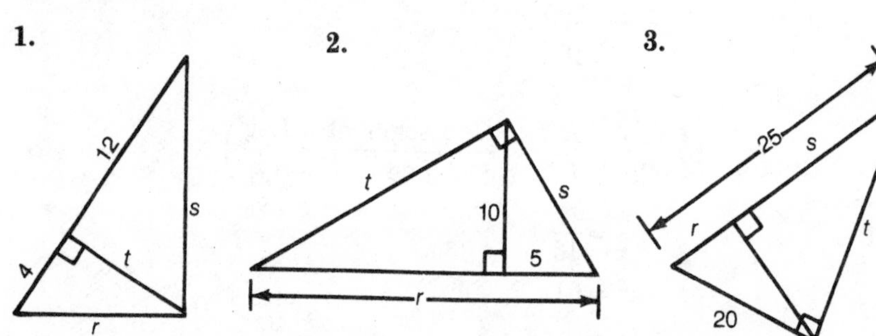

4–9. In right triangle JKL, $\angle$JKL *is a right angle and* $\overline{\text{KH}} \perp \overline{\text{JL}}$.

4. If $JH = 4$ and $HL = 16$, find KH.

5. If $JH = 5$ and $HL = 4$, find KL.

6. If $JH = 8$, $JL = 20$, find KH.

7. If $KL = 18$, $JL = 27$, find JK.

8. If $JK = 14$, $HL = 21$, find JH.

9. If $KH = 12$, $JL = 40$, find JK (assume $\overline{JK}$ is the shorter leg of right triangle JKL).

10. The altitude drawn to the hypotenuse of a right triangle divides the hypotenuse into segments such that their lengths are in the ratio of 1:4. If the length of the altitude is 8, find the length of:
(a) each segment of the hypotenuse
(b) the longer leg of the triangle

11. The altitude drawn to the hypotenuse of a right triangle divides the hypotenuse into segments of lengths 2 and 8. Find the length of the altitude.

12. In right triangle ABC, altitude $\overline{CD}$ is drawn to the hypotenuse. If $AD = 4$ and $DB = 5$, find AC.

13. If the altitude drawn to the hypotenuse of a right triangle has length 8, the lengths of the segments of the hypotenuse may be
(1) 4 and 16 (2) 2 and 4 (3) 3 and 5 (4) 32 and 32

14–18. In each case, solve algebraically.

14. In right triangle ABC, $\overline{CD}$ is the altitude to hypotenuse $\overline{AB}$. If $CD = 12$, $AD = x$, and BD is 7 more than AD, find x.

15. In right triangle ABC, altitude $\overline{CD}$ is drawn to hypotenuse $\overline{AB}$. If $AC = 4$ and DB is 4 more than the length of $\overline{AD}$, find AD.

16. In right triangle RST, altitude $\overline{TP}$ is drawn to hypotenuse $\overline{RS}$. If $TP = 6$ and RP is less than PS, find the length of hypotenuse $\overline{RS}$.

17. In right triangle ABC, altitude $\overline{CD}$ is drawn to hypotenuse $\overline{AB}$. If AB is four times as large as AD and AC is 3 more than AD, find the length of $\overline{AD}$.

18. In right triangle ABC, altitude $\overline{CD}$ is drawn to hypotenuse $\overline{AB}$. If AD is 12 and DB is 3 less than the altitude, find the length of $\overline{CD}$.

8.5 THE PYTHAGOREAN THEOREM

KEY IDEAS

The Pythagorean theorem states that the lengths of the sides of a *right* triangle satisfy the following relationship:

$$(\text{Leg}_1)^2 + (\text{Leg}_2)^2 = (\text{Hypotenuse})^2$$
$$a^2 + b^2 = c^2$$

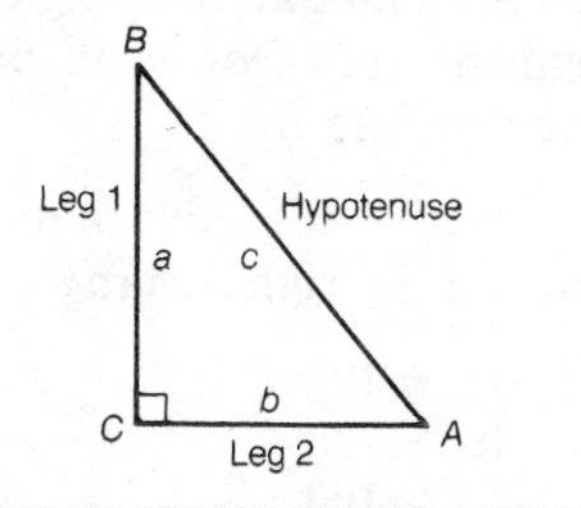

APPLYING THE PYTHAGOREAN THEOREM. When the lengths of any two sides of a right triangle are known, the Pythagorean relationship provides a way of determining the length of the remaining side.

Examples

1. For each triangle, find the value of x.

(a)

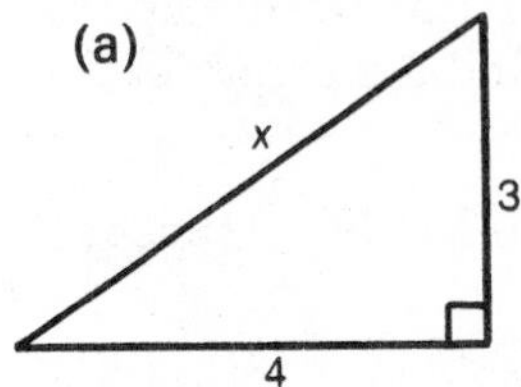

(b)

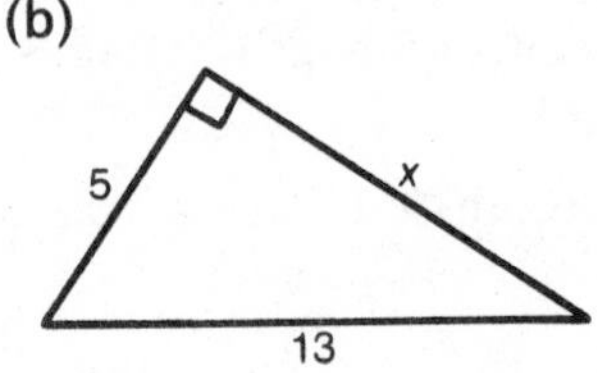

(c)

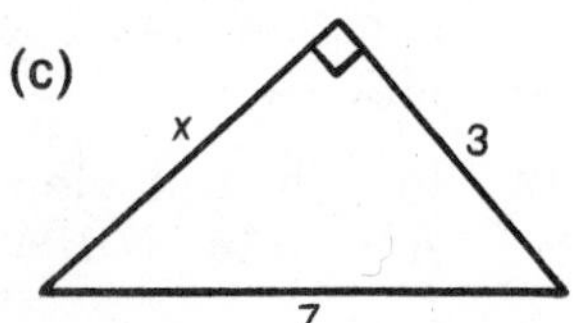

Solutions:

(a)
$$3^2 + 4^2 = x^2$$
$$9 + 16 = x^2$$
$$25 = x^2$$
$$\sqrt{x^2} = \pm\sqrt{25}$$
$$x = \mathbf{5}$$

(b)
$$x^2 + 5^2 = 13^2$$
$$x^2 + 25 = 169$$
$$x^2 = 169 - 25$$
$$\sqrt{x^2} = \pm\sqrt{144}$$
$$x = \mathbf{12}$$

(c)
$$x^2 + 3^2 = 7^2$$
$$x^2 + 9 = 49$$
$$x^2 = 49 - 9$$
$$\sqrt{x^2} = \pm\sqrt{40}$$
$$x = \sqrt{40} = \sqrt{4} \cdot \sqrt{10} = \mathbf{2\sqrt{10}}$$

***Note*:** In each of these examples, the negative value of x is discarded since the length of a side of a triangle cannot be negative.

2. If the length of a diagonal of a square is 10, what is the length of a side of the square?

Solution: Let $x =$ length of a side of the square.

A diagonal of a figure such as a square (or rectangle) is a line segment that connects any two nonconsecutive corners (called **vertices**) of the figure.

Apply the Pythagorean relationship in right triangle ABC.

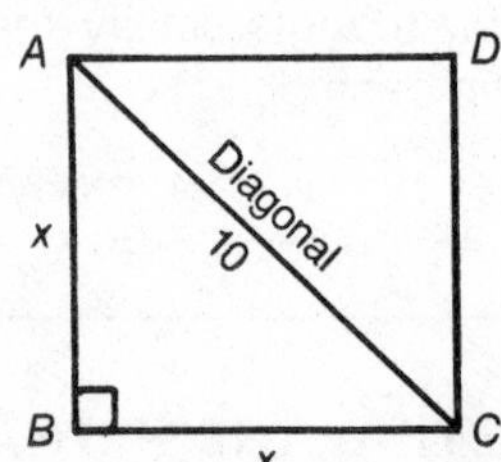

$$x^2 + x^2 = 10^2$$
$$2x^2 = 100$$
$$x^2 = \frac{100}{2}$$
$$x = \sqrt{50} = \sqrt{25}\sqrt{2} = \mathbf{5\sqrt{2}}$$

3. The perimeter of a right triangle is 60. If the length of the hypotenuse is 26, find the length of the shorter leg of the triangle.

Solution: The sum of the lengths of the legs = 60 − 26 = 34.

Let x = length of the shorter leg.
Then $34 - x$ = length of the remaining leg.

$$x^2 + (34 - x)^2 = 26^2$$
$$x^2 + x^2 - 68x + 1156 = 676$$
$$2x^2 - 68x + 1156 - 676 = 0$$
$$2x^2 - 68x + 480 = 0$$
$$\frac{2x^2}{2} - \frac{68x}{2} + \frac{480}{2} = \frac{0}{2}$$
$$x^2 - 34x + 240 = 0$$
$$(x - 10)(x - 24) = 0$$
$$(x - 10 = 0) \vee (x - 24 = 0)$$
$$x = 10 \vee \quad x = 24$$

The length of the shorter leg of the right triangle is **10**.

PYTHAGOREAN TRIPLES. A **Pythagorean triple** is a set of positive integers $\{a, b, c\}$ that satisfy the equation $a^2 + b^2 = c^2$. There are many Pythagorean triples. The sets

$$\{3, 4, 5\} \quad \text{and} \quad \{5, 12, 13\} \quad \text{and} \quad \{8, 15, 17\}$$

are commonly encountered Pythagorean triples. Observe that:

$3^2 + 4^2 = 5^2$		$5^2 + 12^2 = 13^2$		$8^2 + 15^2 = 17^2$	
$9 + 16$	25	$25 + 144$	169	$64 + 225$	289
$25 = 25$ ✓		$169 = 169$ ✓		$289 = 289$ ✓	

Whole-number multiples of any Pythagorean triple also comprise a Pythagorean triple. For example, if each member of the set $\{3, 4, 5\}$ is multiplied by 2, the set $\{6, 8, 10\}$ that is obtained is also a Pythagorean triple since $6^2 + 8^2 = 10^2$ $(36 + 64 = 100)$. If each member of the set $\{3, 4, 5\}$ is multiplied by 3, the set $\{9, 12, 15\}$, which is also a Pythagorean triple since $9^2 + 12^2 = 15^2$ $(81 + 144 = 225)$, is obtained.

Examples

4. Which of the following is *not* a Pythagorean triple?
(1) $\{9, 40, 41\}$ (2) $\{15, 20, 25\}$ (3) $\{8, 12, 17\}$ (4) $\{10, 24, 26\}$

Solution: Choice (1) represents a Pythagorean triple since

$9^2 + 40^2 = 41^2$	
$81 + 1600$	1681
$1681 = 1681$ ✓	

Choice (2) is a multiple of $\{3, 4, 5\}$ where each element is multiplied by 5. Choice (4) is a multiple of $\{5, 12, 13\}$ where each element is multiplied by 2. In choice (3) you can easily verify that $8^2 + 12^2 \neq 17^2$, so $\{8, 12, 17\}$ is *not* a Pythagorean triple.

The correct answer is **choice (3).**

5. The perimeter of an isosceles triangle is 36, and the length of the base is 10. Find the length of the altitude to the base.

Solution: The sum of the lengths of the legs $= 36 - 10$, or 26, so that the length of each leg is 13. The altitude drawn to the base of an isosceles triangle bisects the base so that a 5-12-13 right triangle is formed.

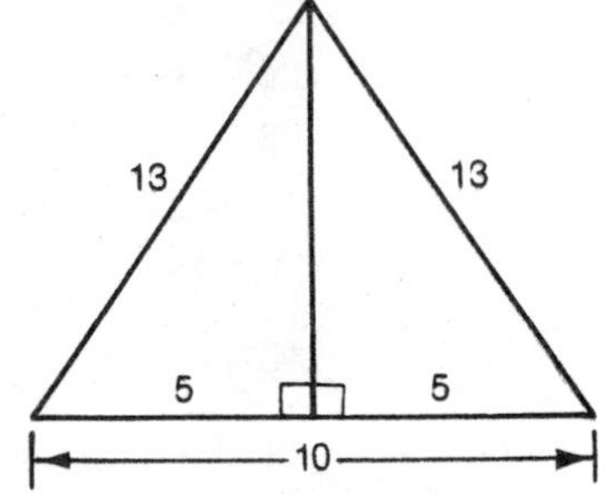

The length of the altitude drawn to the base is **12**.

6. The lengths of the diagonals of a rhombus are 18 and 24. Find the length of a side of the rhombus.

Solution: Recall that the diagonals of a rhombus bisect each other and intersect at right angles.

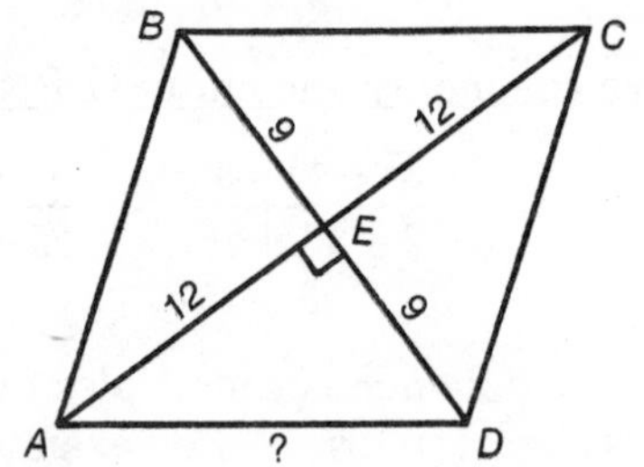

The lengths of the sides of $\triangle AED$ form a multiple of a 3-4-5 right triangle. Each member of the triple is multiplied by 3. Hence, the length of the hypotenuse is $3 \cdot 5$, or 15.

Side $\overline{AD}$ of the rhombus has a length of **15**.

7. Each leg of an isosceles trapezoid has a length of 17. The lengths of its bases are 9 and 39. Find the length of an altitude.

Solution: Drop two altitudes, one from each of the upper vertices.

Quadrilateral $BEFC$ is a rectangle (since $\overline{BE}$ and $\overline{CF}$ are congruent, are parallel, and intersect $\overline{AD}$ at right angles). Hence $BC = EF = 9$. Since right triangle $AEB \cong$ right triangle DFC, $AE = DF = 15$. Triangle AEB is an 8-15-17 right triangle, where 17 is the length of the hypotenuse. The length of an altitude is **8**.

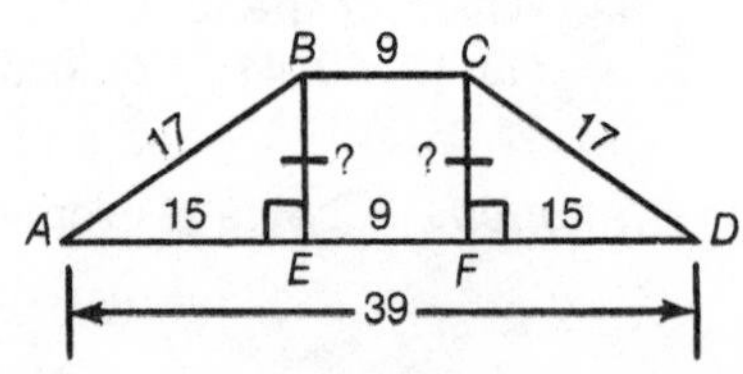

CONVERSE OF THE PYTHAGOREAN RELATIONSHIP. If the lengths of the sides of a triangle satisfy the Pythagorean relationship, then the triangle is a right triangle.

Example

8. The lengths of the sides of a triangle are 2, $\sqrt{5}$, and 3. Determine whether the triangle is a right triangle.

Solution:
$$2^2 + (\sqrt{5})^2 \stackrel{?}{=} 3^2$$
$$4 + 5 \mid 9$$
$$9 = 9 \checkmark$$

Therefore the **triangle is a right triangle.**

EXERCISE SET 8.5

1–4. In each case, find the value of x.

1.

2.

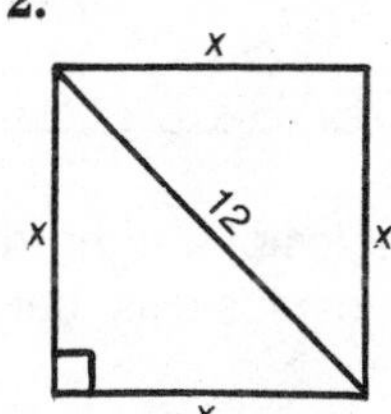

3.

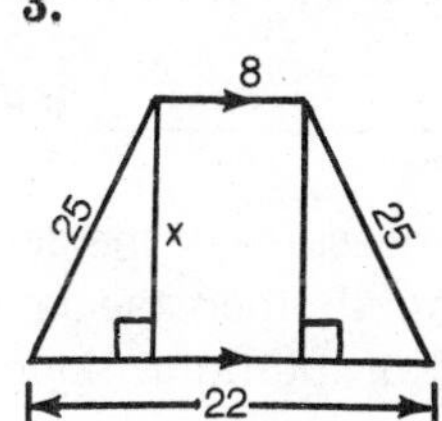

4.

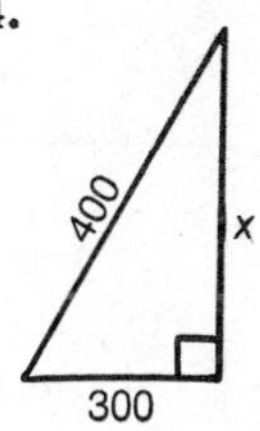

5. If the lengths of the diagonals of a rhombus are 12 and 16, find the perimeter of the rhombus.
6. If the perimeter of a rhombus is 164 and the length of the longer diagonal is 80, find the length of the shorter diagonal.
7. Find the length of the altitude drawn to a side of an equilateral triangle whose perimeter is 30.
8. The length of the base of an isosceles triangle is 14. If the length of the altitude drawn to the base is 5, find the length of each of the legs of the triangle.
9. In a right triangle, one leg has length 7 and the hypotenuse has length 10. What is the length, in radical form, of the other leg?
10. Find the length of a diagonal of a rectangle whose length is 9 and whose width is 12.
11. The length of the hypotenuse of a right triangle exceeds the length of the longer leg by 1. If the length of the shorter leg is 7, find the length of the longer leg.
12. Find the length of a diagonal of a square whose perimeter is 16.

13. If the hypotenuse of an isosceles right triangle is 12, find the length of each leg.

14. Determine whether the set $\{3, \sqrt{7}, \text{and } 4\}$ can represent the lengths of the sides of a right triangle.

15. In isosceles triangle ABC, $\overline{AB} \cong \overline{BC}$, and $\overline{BD}$ is the altitude to base $\overline{AC}$. If $BD = x$, $AB = 2x - 1$, and $AC = 2x + 2$, find the length of $\overline{BD}$.

16. In right triangle ABC, $BC = x$, $AC = 8 - x$, and hypotenuse $AB = 6$. Find x. (*Answer may be left in radical form.*)

8.6 SPECIAL RIGHT TRIANGLE RELATIONSHIPS

KEY IDEAS

If the measures of the acute angles of a right triangle are 30 and 60, or 45 and 45, then special relationships exist between the lengths of the sides of the right triangle.

30-60 RIGHT TRIANGLE RELATIONSHIPS. In a 30-60 right triangle (Figure 8.5) the following relationships hold:

- The length of the *shorter* leg (the side opposite the 30° angle) is one-half the length of the hypotenuse:

$$AD = \frac{1}{2}AB.$$

- The length of the *longer* leg (the side opposite the 60° angle) is one-half the length of the hypotenuse multiplied by $\sqrt{3}$:

$$BD = \frac{1}{2}AB \cdot \sqrt{3}.$$

- The length of the *longer* leg is equal to the length of the shorter leg multiplied by $\sqrt{3}$:

$$BD = AD \cdot \sqrt{3}.$$

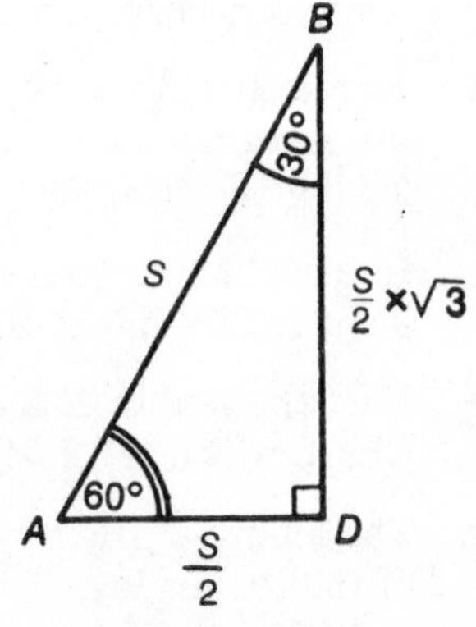

Figure 8.5 30-60 Right Triangle

Examples

1. Fill in the missing values in the following table which is based on right triangle *TSR*:

	RS	*ST*	*RT*
(a)	?	?	12
(b)	4	?	?
(c)	?	$7\sqrt{3}$	?

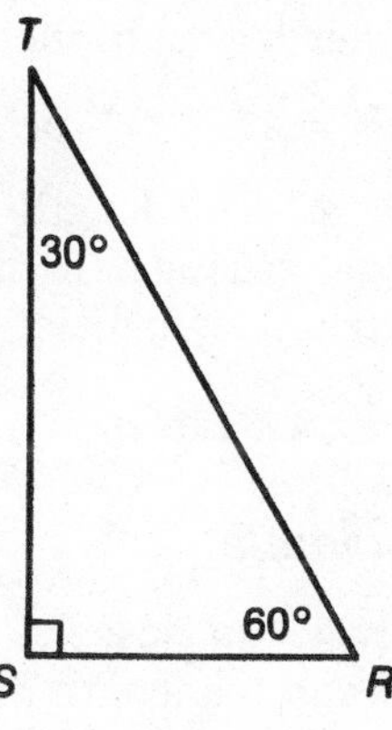

Solutions: (a) $RS = \frac{1}{2}(RT) = \frac{1}{2}(12) = \mathbf{6}$

$ST = RS \cdot \sqrt{3} = \mathbf{6\sqrt{3}}$

(b) $ST = RS \cdot \sqrt{3} = \mathbf{4\sqrt{3}}$

$RT = 2(RS) = 2(4) = \mathbf{8}$

(c) $RS = \mathbf{7}$

$RT = 2(RS) = 2(7) = \mathbf{14}$

2. In ▱*RSTW*, m∠*R* = 30 and *RS* = 12. Find the length of an altitude.

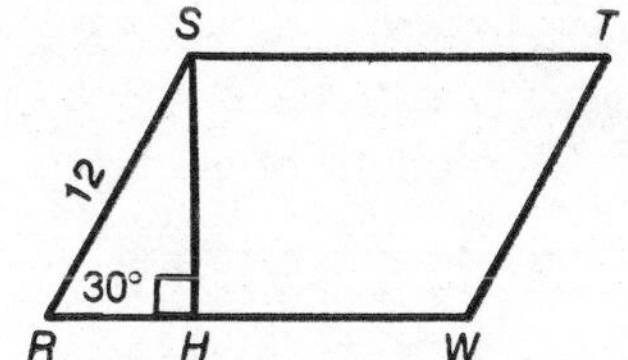

Solution: Altitude SH $= \frac{1}{2}(12) = \mathbf{6}$.

3. In △*JKL*, m∠*K* = 120 and *JK* = 10. Find the length of the altitude drawn from vertex *J* to side $\overline{LK}$ (extended if necessary).

Solution: Since ∠*K* is obtuse, the altitude falls in the exterior of the triangle, intersecting the extension of $\overline{LK}$, say at point *H*. Triangle *JHK* is a 30-60 right triangle. Hence,

$$JH = \frac{1}{2} \cdot 10 \cdot \sqrt{3} = 5\sqrt{3}$$

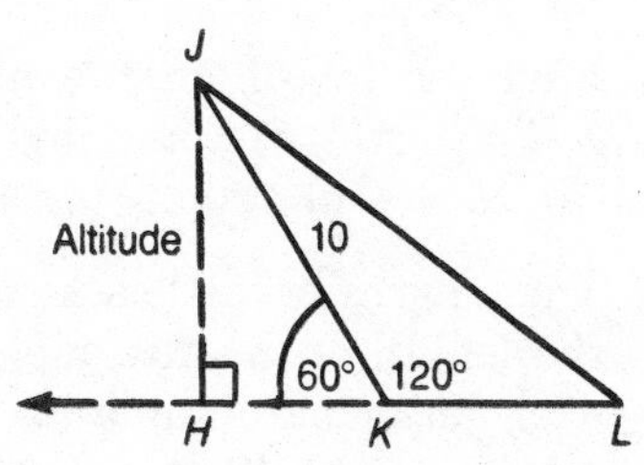

The length of the altitude from vertex *J* to side $\overline{LK}$ is $\mathbf{5\sqrt{3}}$.

45-45 RIGHT TRIANGLE RELATIONSHIPS. In a 45-45 (isosceles) right triangle (Figure 8.6) the following relationships hold:

- The lengths of the *legs* are equal:

$$AC = BC.$$

- The length of the *hypotenuse* is equal to the length of either leg multiplied by $\sqrt{2}$:

$$AB = AC \cdot \sqrt{2} \quad \text{or} \quad AB = BC \cdot \sqrt{2}$$

- The length of either *leg* is equal to one-half the length of the hypotenuse multiplied by $\sqrt{2}$:

$$AC \text{ (or } BC) = \frac{1}{2} AB \cdot \sqrt{2}.$$

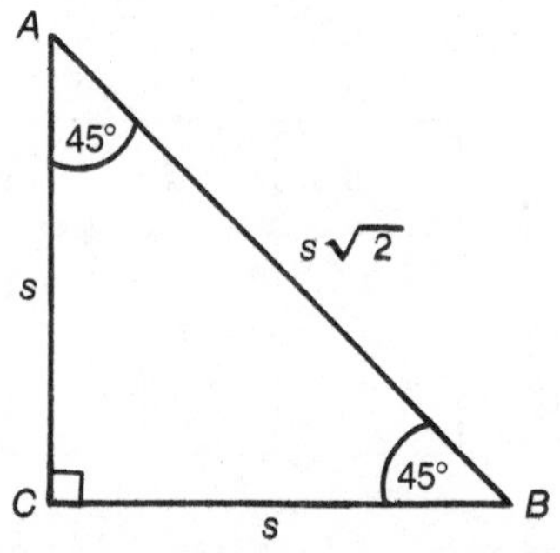

Figure 8.6 45-45 Right Triangle

Example

4. In isosceles trapezoid $ABCD$, the measure of a lower base angle is 45 and the length of upper base $\overline{BC}$ is 5. If the length of an altitude is 7, find the lengths of the legs, $\overline{AB}$ and $\overline{DC}$.

Solution: Drop altitudes from B and C, forming two congruent 45-45 right triangles. $AE = BE = 7$. Also, $FD = 7$. $AB = AE \cdot \sqrt{2} = 7\sqrt{2}$. The length of each leg is $\mathbf{7\sqrt{2}}$.

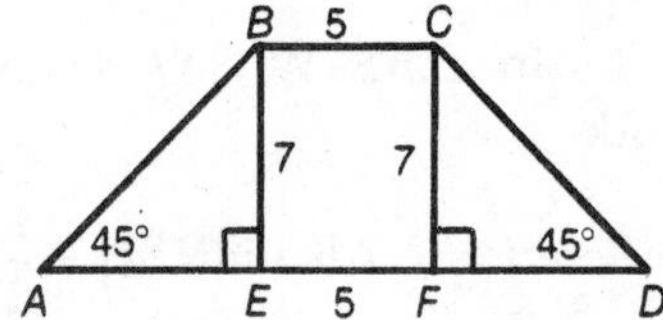

EXERCISE SET 8.6

1. The measure of the vertex angle of an isosceles triangle is 120, and the length of each leg is 8. Find the length of:
   (a) the altitude drawn to the base (b) the base

2. If the perimeter of a square is 24, find the length of a diagonal.

3. If the length of a diagonal of a square is 18, find the perimeter of the square.

4. Find the length of the altitude drawn to side $\overline{AC}$ of $\triangle ABC$ if $AB = 8$, $AC = 14$, and $m\angle A$ equals:
   (a) 30 (b) 120 (c) 135

5. The lengths of the bases of an isosceles trapezoid are 9 and 25. Find the length of the altitude and of each of the legs if the measure of each lower base angle is:
   (a) 30 (b) 45 (c) 60

6. Find the values of x, y, and z.

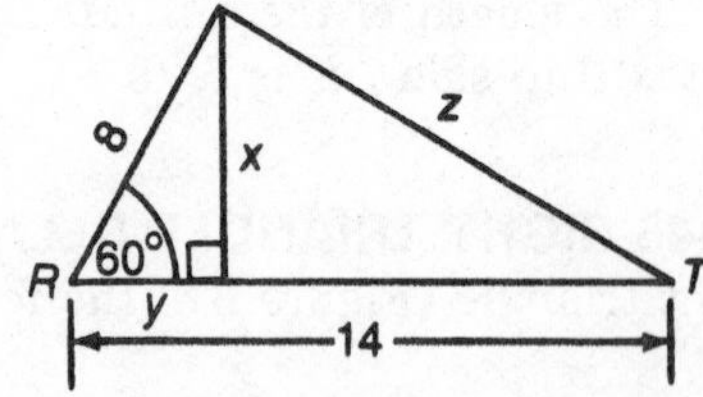

7. The lengths of two adjacent sides of a parallelogram are 6 and 14. If the measure of an included angle is 60, find the length of the shorter diagonal of the parallelogram.

8. The length of each side of a rhombus is 10, and the measure of an angle of the rhombus is 60. Find the length of the longer diagonal of the rhombus.

9. Find the length of a diagonal of a square whose area is:
(a) 16 (b) 25 (c) $9x^2$

10. Find the area of an equilateral triangle whose perimeter is:
(a) 24 (b) 30 (c) equal to the perimeter of a square of area 81

11. For the accompanying figure, fill in the missing values in the table.

	a	b	c	d
(a)	8	?	?	?
(b)	?	?	6	?
(c)	?	$5\sqrt{3}$	?	?
(d)	?	?	?	$10\sqrt{2}$

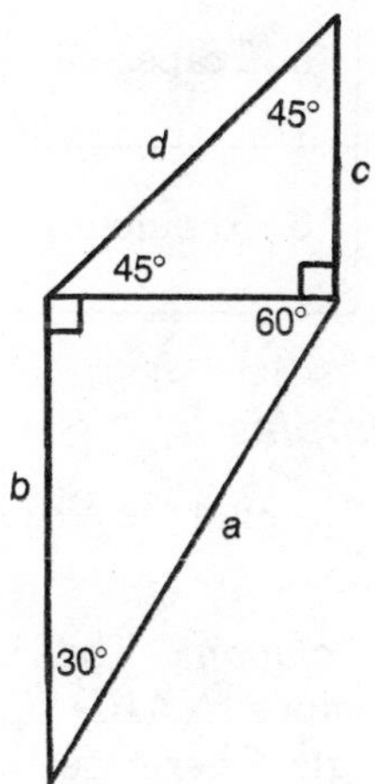

8.7 AREA FORMULAS

KEY IDEAS

Sometimes it is necessary to apply a special right triangle relationship in order to find the length of a segment needed to determine the area of a special quadrilateral or triangle.

SOME AREA FORMULAS. Table 8.1 summarizes area formulas for some special quadrilaterals and triangles.

TABLE 8.1 Area Formulas

Figure	Area =
1. Rectangle	Base × Height
2. Square	Side × Side
3. Parallelogram	Base × Height
4. Triangle	$\frac{1}{2}$ × Base × Height
5. Equilateral Triangle	$\frac{1}{4} \times (\text{Side})^2 \times \sqrt{3}$
5. Trapezoid	$\frac{1}{2}$ × Altitude × (Sum of bases)
6. Rhombus	$\frac{1}{2} \times (\text{Diagonal}_1) \times (\text{Diagonal}_2)$

Examples

1. Find the area of a rectangle whose diagonal is 10 and whose base is 8.

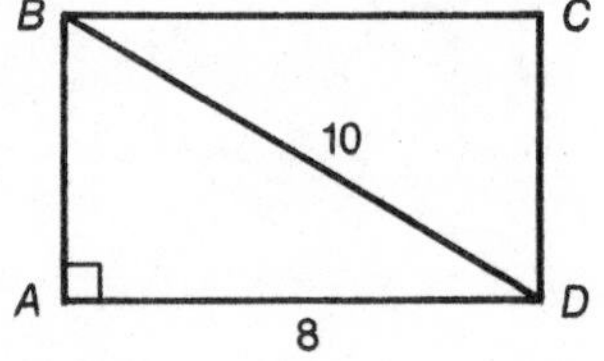

Solution: The height (width) is 6 since $\triangle ABD$ is a 6-8-10 right triangle. Therefore,

$$\text{Area} = \text{Base} \times \text{Height}$$
$$= 8 \times 6 = \mathbf{48}$$

2. The lengths of a pair of adjacent sides of a parallelogram are 6 and 10 centimeters. If the measure of their included angle is 30, find the area of the parallelogram.

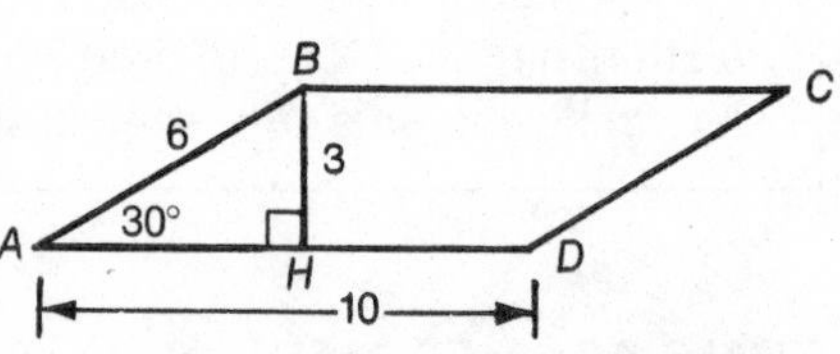

Solution: In $\square ABCD$, altitude $BH = 3$ since the length of the side opposite a 30° angle in a 30-60 right triangle is one-half the length of the hypotenuse (side $\overline{AB}$).

$$\text{Area } \square ABCD = BH$$
$$= AD \cdot BH$$
$$= 10 \cdot 3 = \mathbf{30\ cm^2}$$

3. In the accompanying figure, $ABCD$ is a trapezoid with $\overline{AB} \parallel \overline{CD}$, $\overline{BA} \perp \overline{AD}$, $\overline{CD} \perp \overline{AD}$, $AB = 10$, $BC = 17$, and $AD = 15$. Find the area of trapezoid $ABCD$.

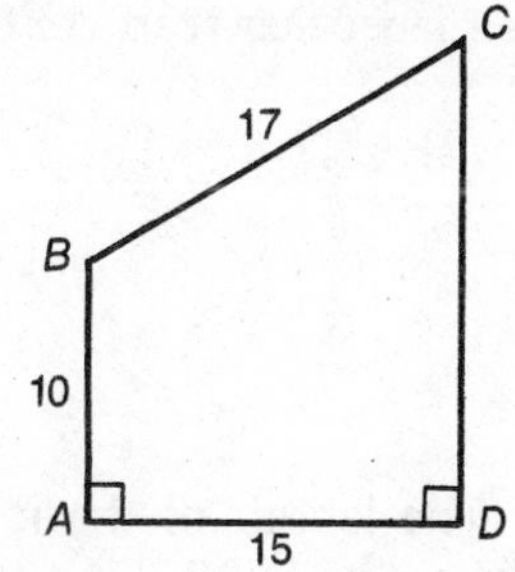

Solution: First find the length of base $\overline{CD}$ by drawing altitude $\overline{BH}$ to $\overline{CD}$. Since parallel lines are everywhere equidistant, $BH = AD = 15$. The lengths of the sides of right triangle BHC form an 8-15-17 Pythagorean triple, where $CH = 8$. Thus, $CD = CH + HD = 8 + 10 = 18$.

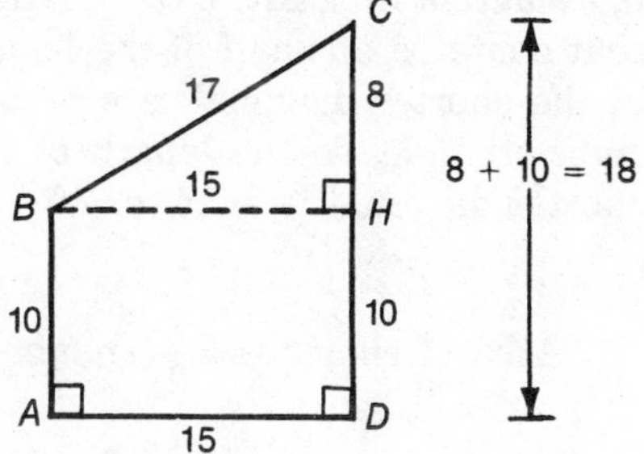

$$\text{Area trap } ABCD = \frac{1}{2} \times \text{Altitude} \times (\text{Sum of bases})$$

$$= \frac{1}{2}(BH)(AB + CD)$$

$$= \frac{1}{2}(15)(10 + 18)$$

$$= \frac{1}{2}(15)(28)$$

$$= \frac{1}{2}(420) = \mathbf{210}$$

4. Find the area of an isosceles trapezoid the length of whose bases are 8 and 20 and whose lower base angle has a measure of 45.

Solution: Since $\overline{BH}$ is the side opposite the 45° angle in a 45-45 right triangle, $BH = AH = 6$.

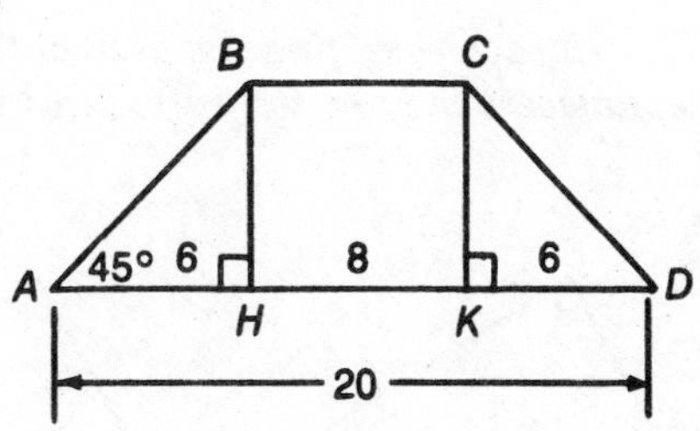

$$\text{Area isosceles trap } ABCD = \frac{1}{2} \times \text{Altitude} \times (\text{Sum of bases})$$
$$= \frac{1}{2} BH(AD + BC)$$
$$= \frac{1}{2}(6)(20 + 8)$$
$$= 3(28) = \mathbf{84}$$

5. The length of the longer diagonal of a rhombus is 24, and the length of a side is 13. Find the area of the rhombus.

Solution: As the accompanying diagram illustrates, the triangle that contains one-half of the length of the shorter diagonal is a 5-12-13 right triangle, so the length of the shorter diagonal is 2×5, or 10.

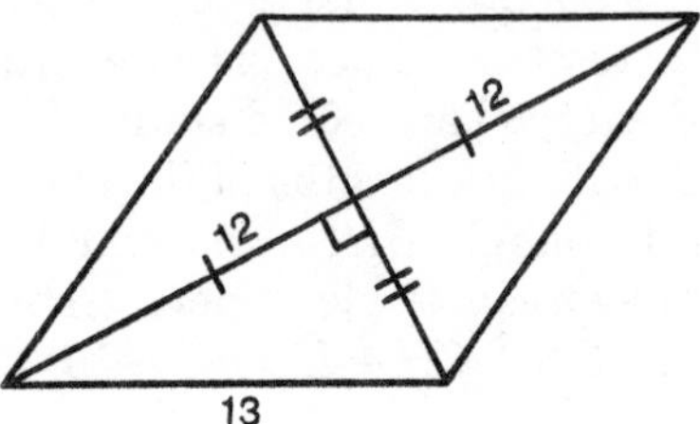

$$\text{Area of rhombus} = \frac{1}{2}(\text{Diagonal}_1) \times (\text{Diagonal}_2)$$
$$= \frac{1}{2}(10 \times 24)$$
$$= \frac{1}{2}(240) = \mathbf{120}$$

EXERCISE SET 8.7

1. Find the area of each of the following:
   (a) a rectangle whose base is 6 and whose diagonal is 10
   (b) a square whose diagonal is 8
   (c) a parallelogram having two adjacent sides of 12 and 15 centimeters and an included angle of measure 60

2. Find the dimensions of each of the following:
   (a) a rectangle whose area is 75 and whose base and altitude are in the ratio of 3 : 1
   (b) a rectangle that has an area of 135 and whose base is represented by $x + 2$ and whose altitude is represented by $2x + 1$

3–11. In each case, find the area of the figure shown. (Whenever appropriate, answers may be left in radical form.)

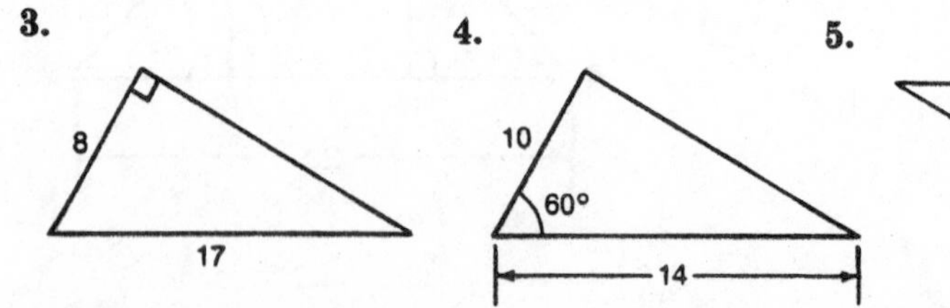

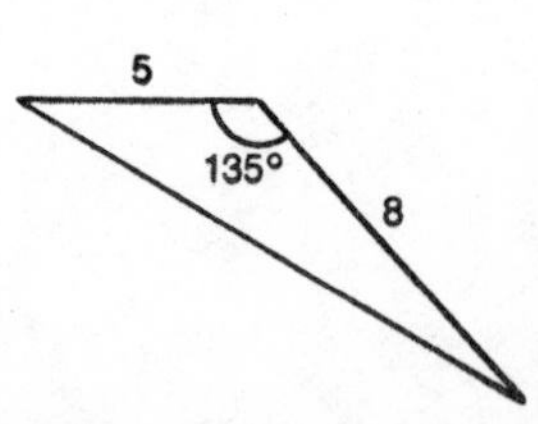

6.

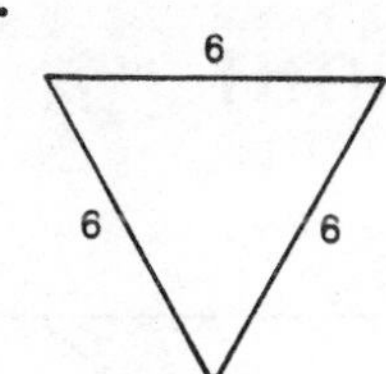

7.

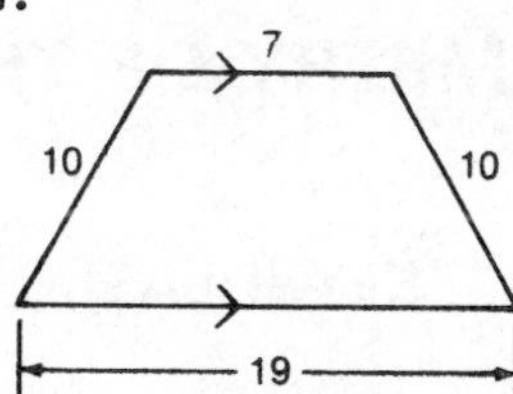

8.

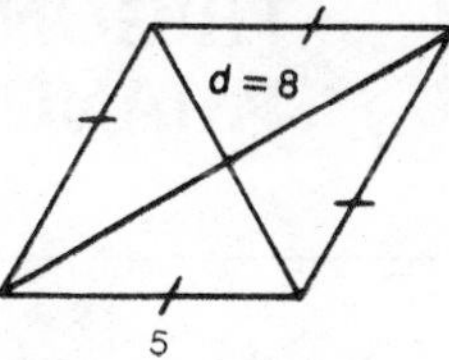

9.

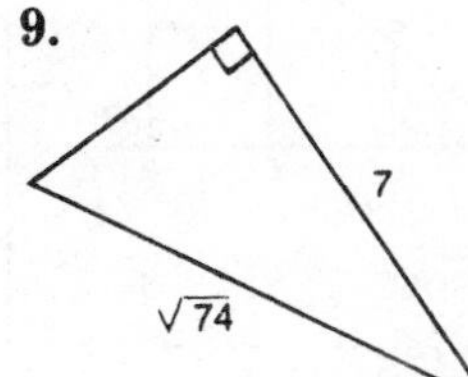

10.

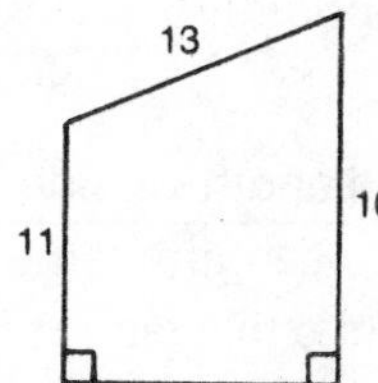

11.

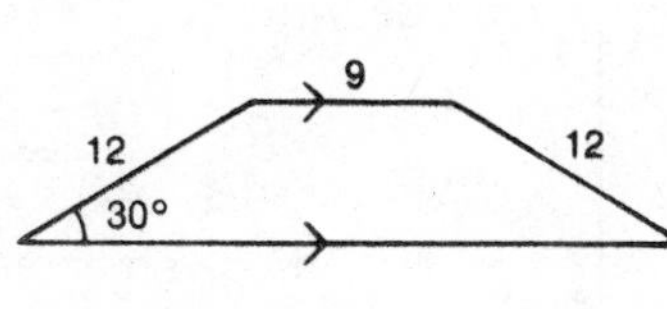

12. Find the length of the shorter diagonal of a rhombus if:
(a) the length of the longer diagonal is 15 and the area is 90
(b) the lengths of the diagonals are in the ratio of 2 : 3, and the area of the rhombus is 147

13. Find the length of an altitude of a trapezoid if:
(a) its area is 72 and the sum of the lengths of the bases is 36
(b) its area is 80 and the length of its median is 16
(c) the sum of the lengths of the bases is numerically equal to one-third of the area of the trapezoid

14. Find the area of a rhombus if its perimeter is 68 and the length of one of its diagonals is 16.

15. Find the area of a triangle if the lengths of a pair of adjacent sides are 6 and 14 and the measure of the included angle is:
(a) 90 (b) 30 (c) 120 (*Leave answer in radical form.*)

16. In a trapezoid, the length of one base is five times the length of the other base. The height of the trapezoid is 1 less than the length of the shorter base. If the area of the trapezoid is 90, find the length of the *shorter* base. [*Only an algebraic solution will be accepted.*]

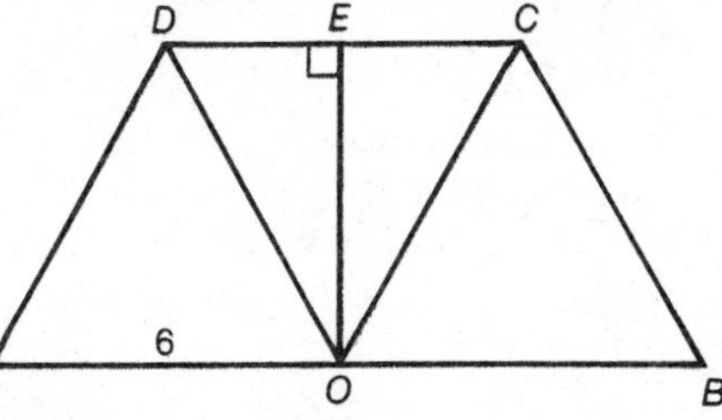

17. In the accompanying figure, $ABCD$ is a trapezoid with $\overline{CD} \parallel \overline{AB}$. Triangles AOD, DOC, and COB are equilateral, $\overline{OE} \perp \overline{CD}$, and $OA = 6$.
(a) Find OE.
(b) Find the area of $\triangle DOC$.
(c) Find the area of trapezoid $ABCD$.
(d) Find the area of trapezoid $OECB$.
(*Answers to parts (b), (c), and (d) may be left in radical form.*)

8.8 TRIGONOMETRY OF THE RIGHT TRIANGLE

KEY IDEAS

For the accompanying figure, since $\triangle ABC \sim \triangle ADE \sim \triangle AFG$, the following extended proportion may be written:

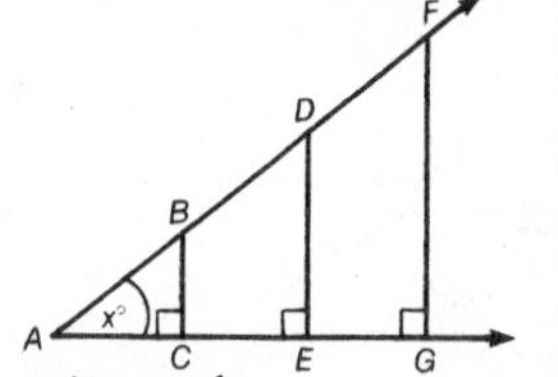

$$\frac{BC}{AB} = \frac{DE}{AD} = \frac{FG}{AF} = \frac{\text{Length of side opposite } \angle A}{\text{Length of the hypotenuse}}.$$

For any given acute angle, this ratio is the same regardless of the size of the right triangle. It will be convenient to refer to this type of ratio by a special name—the *sine ratio*. The **sine** of an acute angle of a right triangle is the ratio formed by taking the length of the side opposite the angle and dividing it by the length of the hypotenuse. In a similar fashion, other ratios may also be formed, with each being given a special name.

DEFINITIONS OF TRIGONOMETRIC RATIOS. A **trigonometric ratio** is the ratio of the lengths of a selected pair of sides of a right triangle (Figure 8.7). Three commonly formed trigonometric ratios are called *sine*, *cosine*, and *tangent*:

$$\text{Sine of } \angle A = \frac{\text{Length of side opposite } \angle A}{\text{Length of the hypotenuse}} = \frac{BC}{AB};$$

$$\text{Cosine of } \angle A = \frac{\text{Length of side adjacent to } \angle A}{\text{Length of the hypotenuse}} = \frac{AC}{AB};$$

$$\text{Tangent of } \angle A = \frac{\text{Length of side opposite } \angle A}{\text{Length of side adjacent to } \angle A} = \frac{BC}{AC}$$

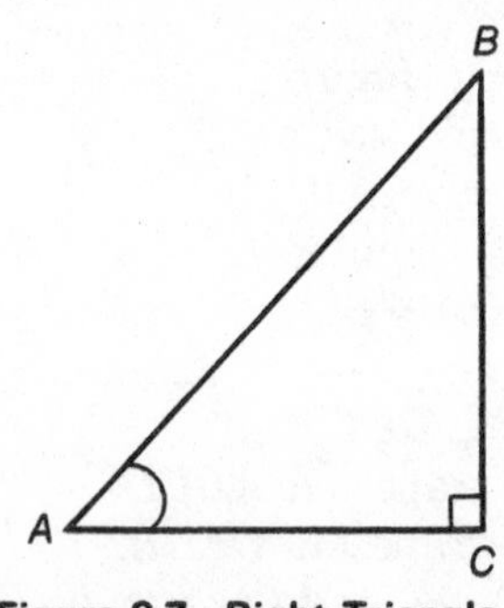

Figure 8.7 Right Triangle

***Notes*:**

- The trigonometric ratios of sine, cosine, and tangent may be abbreviated as *sin*, *cos*, and *tan*, respectively.

- In a right triangle, the trigonometric ratios may be taken with respect to either of the acute angles of the triangle:

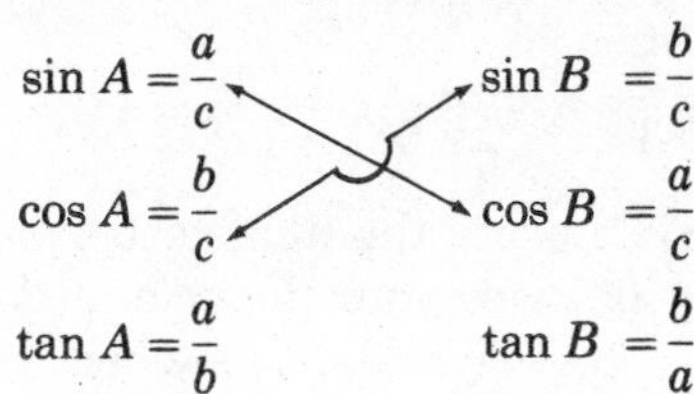

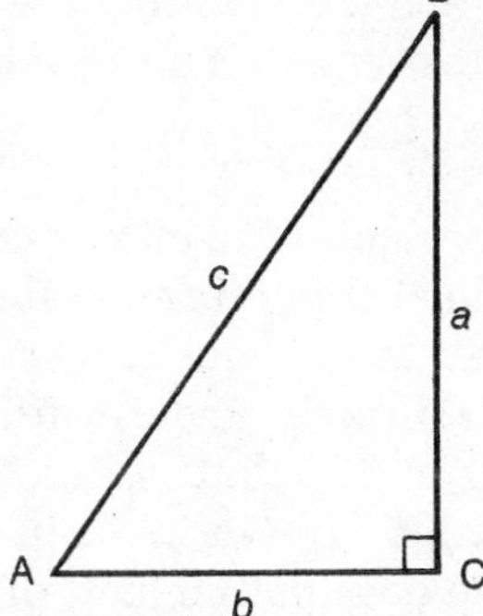

- Since sin A and cos B are both equal to $\frac{a}{c}$, they are equal to each other. Similarly, cos A and sin B are equal since both are equal to $\frac{b}{c}$. Angles A and B are complementary. In general, the sine of an angle is equal to the cosine of the angle's complement. For example, $\sin 60° = \cos 30°$.

- The definitions of the three trigonometric ratios should be memorized.

Examples

1. In right triangle ABC, $\angle C$ is the right angle. If $AC = 4$ and $BC = 3$, find the values of tan A, sin A, and cos A.

Solution: Since $\triangle ABC$ is a 3-4-5 right triangle, $AB = 5$.

$$\tan A = \frac{\text{Side opposite } \angle A}{\text{Side adjacent } \angle A} = \frac{3}{4} = \mathbf{0.75}$$

$$\sin A = \frac{\text{Side opposite } \angle A}{\text{Hypotenuse}} = \frac{3}{5} = \mathbf{0.6}$$

$$\cos A = \frac{\text{Side adjacent } \angle A}{\text{Hypotenuse}} = \frac{4}{5} = \mathbf{0.8}$$

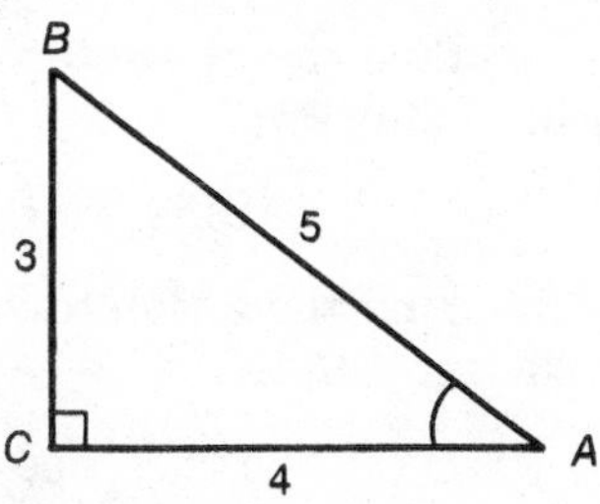

2. Find the value of x in the equality $\sin 2x = \cos 3x$.

Solution: If angles A and B are complementary, $\sin A = \cos B$. Therefore,

$$2x + 3x = 90$$
$$5x = 90$$
$$x = \frac{90}{5} = \mathbf{18}$$

USING TRIGONOMETRIC TABLES. A table of Trigonometric Values located on page 400 lists the values of the three trigonometric ratios for angles between 0° and 90°, correct to the nearest ten thousandth (fourth decimal-place position). For example, the values of the trigonometric functions of 38° can be found by locating 38° in the "Angle" column and then reading the decimal values under the column headings from left to right. You should verify that:

$$\sin 38° = 0.6157 \qquad \cos 38° = 0.7880 \qquad \tan 38° = 0.7813.$$

For what value of x is $\cos x = 0.8387$¶ Locate the number 0.8387 in the cosine column in the table on page 400. Place your finger on 0.8387, and then move your finger horizontally to the left. Stop when your finger is directly underneath the "Angle" column. You should now be pointing at 33°, that is, cos **33°** = 0.8387.

For what value of x is $\sin x = 0.6751$? Notice that the sine column of the table does *not* include this number. The sine column however, does include a pair of consecutive decimal numbers, one less than 0.6751 (sin 42°) and one greater than 0.6751 (sin 43°).

Angle	Sin
42°	0.6691
x	0.6751
43°	0.6820

difference = 0.0060

difference = 0.0069

To find out whether x is closer to 42° or to 43°, proceed as follows:

1. Find the absolute value of the difference between the value of sin 42° and 0.6751.

$$\text{Difference} = |0.6691 - 0.6751| = 0.0060.$$

2. Find the absolute value of the difference between the value of sin 43° and 0.6751.

$$\text{Difference} = |0.6820 - 0.6751| = 0.0069.$$

3. Compare the calculated differences. The measure of angle x, *correct to the nearest degree*, is the angle that gives the *smallest* calculated difference. Since 0.0060 is smaller than 0.0069, the value of x, *correct to the nearest degree*, is **42°**.

Examples

3. Find the value of each of the following:
(a) tan 27° (b) sin 64° (c) cos 35°

Solutions: (a) **0.5095** (b) **0.8988** (c) **0.8192**

4. Find the degree measure of angle x.
(a) $\sin x = 0.7071$ (b) $\tan x = 0.6009$ (c) $\cos x = 0.7986$

Solutions: (a) $x = 45°$ (b) $x = 31°$ (c) $x = 37°$

5. Find the degree measure of $\angle A$ correct to the nearest degree.
(a) $\tan A = 0.7413$ (c) $\sin A = 0.6599$
(b) $\cos A = 0.8854$ (d) $\tan A = 2.2500$

Solution: (a) $A = 37°$ (b) $A = 28°$ (c) $A = 41°$ (d) $A = 66°$.

6. Find the value of x correct to the nearest tenth.

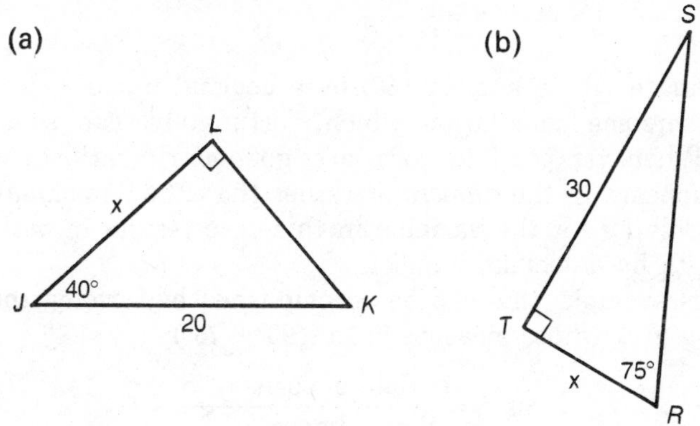

Solutions: (a) *Step 1.* Decide which trigonometric ratio to use, and then write the corresponding equation:

$$\cos J = \frac{\text{Side adjacent to } \angle J}{\text{Hypotenuse}}$$

$$\cos 40° = \frac{x}{20}.$$

Step 2. Consult the Table of Trigonometric Values in order to evaluate the trigonometric ratio. Since $\cos 40° = 0.7660$,

$$0.7660 = \frac{x}{20}.$$

Step 3. Solve the equation. It may help to think of the equation as a proportion:

$$\frac{0.7660}{1} = \frac{x}{20}.$$

Solve by cross-multiplying:

$$x = 20(0.7660) = 15.32.$$

Step 4. Round off your answer to the desired accuracy. In this example we are asked to express the answer correct to the nearest tenth. Hence,

$$x = \mathbf{15.3}$$

(b) Since the problem involves both legs of the right triangle, use the tangent ratio.

$$\tan R = \frac{\text{Side opposite } \angle R}{\text{Side adjacent to } \angle R}$$

$$\tan 75^\circ = \frac{30}{x}$$

$$x = \frac{30}{\tan 75^\circ}$$

Dividing a whole number (30) by a decimal number (tan 75°) is a very cumbersome calculation which, fortunately, can sometimes be avoided. Whenever possible, form a trigonometric ratio in which the variable appears in the *numerator* rather than the denominator of the fraction. Solving for the variable in this case results in *multiplying* a given length by a decimal number.

In this example, this can be accomplished by forming the tangent ratio using $\angle S$, whose measure is 15° (90° – 75°).

$$\tan S = \frac{\text{Side opposite } \angle S}{\text{Side adjacent to } \angle S}$$

$$\tan 15^\circ = 0.2679 = \frac{x}{30}$$

$$x = 30(0.2679) = 8.037$$

$$= \mathbf{8.0} \text{ (correct to the nearest tenth)}$$

INDIRECT MEASUREMENT AND TRIGONOMETRY. A trigonometric ratio may be used to arrive at the measure of a part of a right triangle that may be difficult, if not impossible, to calculate by direct measurement.

Example

7. A plane takes off from a runway, and climbs while maintaining a constant angle with the ground. When the plane has traveled 1,000 meters, its altitude is 290 meters. Find, correct to the nearest degree, the angle at which the plane has risen with respect to the horizontal ground.

Solution:

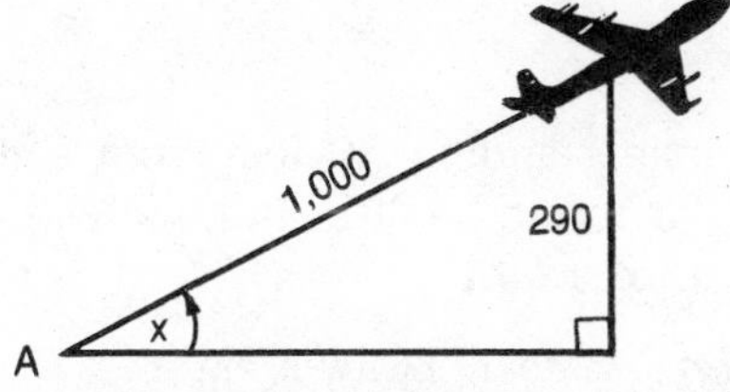

To find the value of $\angle x$, first determine the appropriate trigonometric ratio. The sine ratio relates the three quantities under consideration:

$$\sin x = \frac{\text{Side opposite } \angle A}{\text{Hypotenuse}},$$

$$= \frac{290}{1{,}000} = 0.2900.$$

To find the value of x to the nearest degree, consult the Table of Trigonometric Values. A search for the decimal number in the sine column that is closest to 0.2900 leads to the table value of 0.2924, which corresponds to 17°.

Hence, $\angle x = \mathbf{17°}$, which is correct to the nearest degree.

ANGLES OF ELEVATION AND DEPRESSION. The angles formed by an observer's line of vision and a horizontal line are sometimes referred to by special names. The **angle of elevation** represents the angle through which an observer must *raise* his or her line of sight with respect to a horizontal line in order to see an object. For example, if to see a bird in flight John must raise his line of sight 35° with respect to the horizontal ground, then the angle of elevation of the bird is 35°.

If in order to view an object the observer must *lower* his or her line of sight with respect to a horizontal line, the angle thus formed is called the **angle of depression**. For example, if a pilot of an airplane in flight must lower her line of sight 23° to spot a landmark on the ground, then the angle of depression of the landmark is 23°. As Figure 8.8 illustrates:

- The angle of elevation e is an angle of a right triangle that has the line of sight as its hypotenuse, while the angle of depression d falls outside this triangle.

- The angle of elevation and the angle of depression are numerically equal ($e = d$) since they are alternate interior angles formed by parallel (horizontal) lines and a transversal (the line of sight).

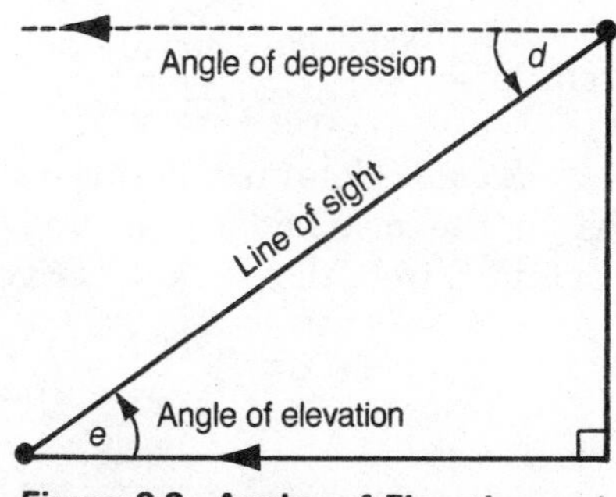

Figure 8.8 Angles of Elevation and Depression

Examples

8. A man standing 30 feet from a flagpole observes the angle of elevation of its top to be 48°. Find the height of the flagpole, correct to the nearest tenth of a foot.

Solution: *Step 1.* Draw a right triangle and label it with the given information.

Step 2. Decide which trigonometric ratio to use, and then write the corresponding equation:

$$\tan 48^\circ = \frac{x \text{ (side opposite angle)}}{30 \text{ (side adjacent to angle)}}.$$

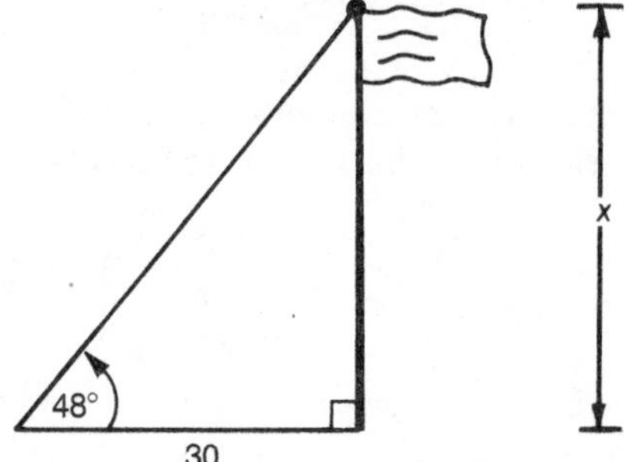

Step 3. Replace $\tan 48^\circ$ by its value, obtained from the Table of Trigonometric Values, and then solve for x:

$$1.1106 = \frac{x}{30}$$

$$x = 30(1.1106) = 33.318$$

$$= \mathbf{33.3} \text{ (correct to the nearest tenth of a foot)}$$

9. An airplane pilot observes the angle of depression of a point on a landing field to be 28°. If the plane's altitude at this moment is 900 meters, find the distance from the pilot to the observed point on the landing field, correct to the nearest meter.

Solution: *Step 1.* Draw a right triangle, and label it with the given information. Use the fact that the angle of elevation and the angle of depression are numerically equal.

Step 2. Decide which trigonometric ratio to use and then write the corresponding equation:

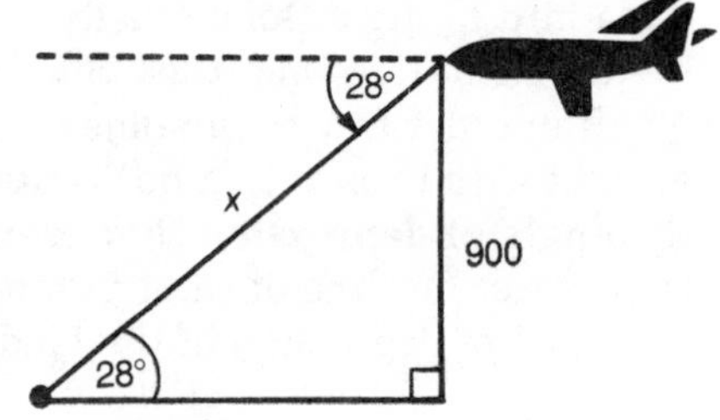

$$\sin 28^\circ = \frac{900 \text{ (side opposite angle)}}{x \text{ (hypotenuse)}}.$$

Step 3. Note that in this case forming a ratio in which the variable is in the denominator is unavoidable. Replace $\sin 28^\circ$ by its value obtained from the Table of Trigonometric Values, and then solve for x:

$$0.4695 = \frac{900}{x}$$

$$x = \frac{900}{0.4695} = 1916.93$$

$$= \mathbf{1917} \text{ (correct to the nearest meter)}$$

USING TRIGONOMETRY TO SOLVE GEOMETRY PROBLEMS. Sometimes trigonometry is needed to calculate the measures of parts of geometric figures.

Examples

10. The lengths of two adjacent sides of a triangle are 10 and 16. If the measure of their included angle is 39°, find the area of the triangle, correct to the nearest square unit.

Solution: Draw the altitude to the side whose length is 16. In the right triangle formed, use the sine ratio to find the length of the altitude.

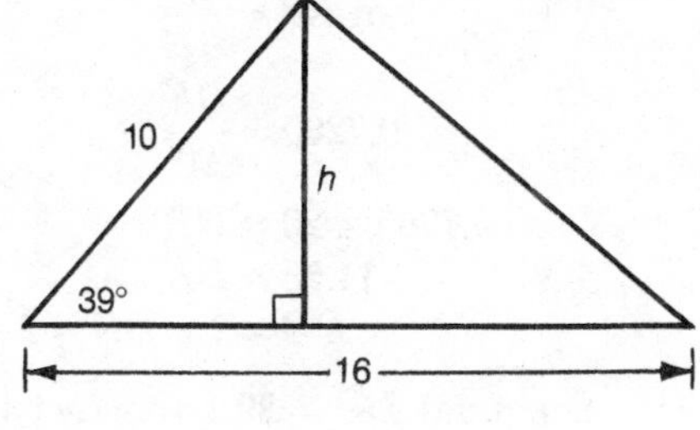

$$\sin 39^\circ = \frac{h}{10}$$

$$0.6293 = \frac{h}{10}$$

$$h = 10(0.6293) = 6.293 \text{ or } 6.3 \text{ to the nearest tenth}$$

$$\text{Area} = \frac{1}{2}\text{Base} \times \text{Height}$$

$$= \frac{1}{2}(16)(6.3)$$

$$= \frac{1}{2}(100.8)$$

$$= 50.4$$

$= \mathbf{50}$ (correct to the nearest square unit)

11. In rhombus $ABCD$, $AC = 40$ and $m\angle DAB = 72$.
(a) Find diagonal DB to the nearest tenth.
(b) Find the area of rhombus $ABCD$ to the nearest integer.
(c) Find the length of a side of rhombus $ABCD$ to the nearest integer.

Solution: The following properties of a rhombus are needed in order to make use of right triangle trigonometry:

- The diagonals of a rhombus bisect the angles at the vertices they connect, so

$$m\angle EAD = \frac{1}{2}m\angle DAB = \frac{1}{2}(72) = 36.$$

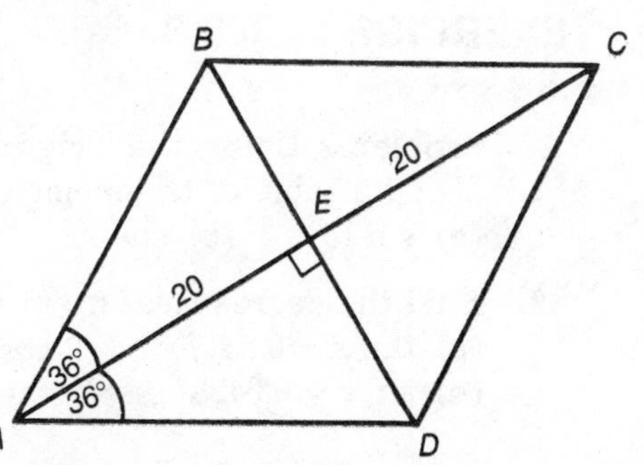

- The diagonals of a rhombus are the perpendicular bisectors of each other. Therefore, if diagonals $\overline{AC}$ and $\overline{BD}$ intersect at point E, then

(1) $\triangle AED$ is a right triangle, and

(2) $AE = \frac{1}{2}AC = \frac{1}{2}(40) = 20.$

(a) To find DB, first find DE, using the tangent ratio in right triangle AED.

$$\tan \angle EAD = \frac{\text{Leg opposite } \angle EAD}{\text{Leg adjacent to } \angle EAD}$$

$$\tan 36^\circ = \frac{DE}{AE}$$

$$0.7265 = \frac{DE}{20}$$

$$0.7265 \times 20 = DE$$

$$14.53 = DE$$

$$DB = 2 \times DE = 2 \times 14.53 = 29.06$$

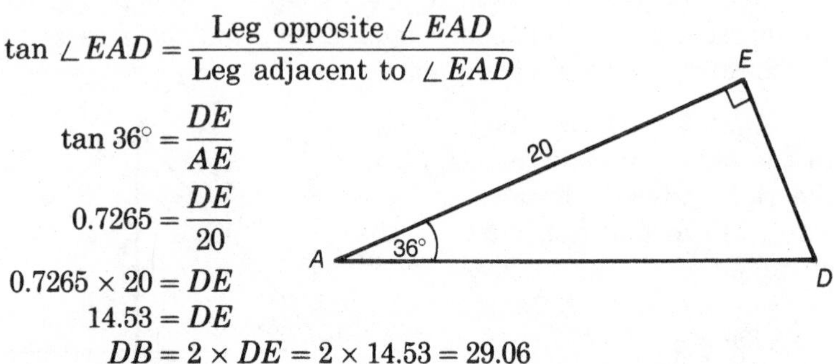

Diagonal $DB = \mathbf{29.1}$ (correct to the nearest tenth).

(b) The area of a rhombus is equal to one-half the product of the lengths of its diagonals.

$$\text{Area rhombus } ABCD = \frac{1}{2}(AC)(DB) = \frac{1}{2}(40)(29.1)$$
$$= (20)(29.1)$$

Area of rhombus $ABCD = \mathbf{582}$ (correct to the nearest integer).

(c) To find AD, use the cosine ratio:

$$\cos \angle EAD = \frac{\text{Leg adjacent to } \angle EAD}{\text{Hypotenuse}} = \frac{AE}{AD}$$

$$\cos 36^\circ = \frac{20}{AD}$$

$$AD = \frac{20}{\cos 36^\circ} = \frac{20}{0.8090} = \frac{20}{0.81} = 24.69$$

Side $AD = \mathbf{25}$ (correct to the nearest integer).

EXERCISE SET 8.8

A. Problems Using the Trigonometric Tables

1. Using a table of trigonometric values, find:
(a) sin 15° (b) cos 59° (c) tan 73° (d) sin 82°

2. Find the degree measure of angle x if:
(a) $\tan x = 0.5317$ (c) $\cos x = 0.3584$ (e) $\sin x = 0.9744$
(b) $\sin x = 0.6428$ (d) $\tan x = 1.2349$ (f) $\cos x = 0.8988$

3. In each case, find the degree measure of angle x, correct to the *nearest degree.*
   (a) $\cos x = 0.8700$
   (b) $\sin x = 0.9298$
   (c) $\tan x = 1.0110$
   (d) $\cos x = \frac{3}{5}$
   (e) $\sin x = \frac{5}{13}$
   (f) $\tan x = 0.5036$

B. Problems Involving Trigonometric Definitions and Relationships

4. In each case, find the degree measure of angle x.
   (a) $\sin x = \cos (x + 28)$
   (b) $\cos 4x = \sin 18^\circ$
   (c) $\sin (2x + 17) = \cos (x + 13)$
   (d) $\cos (5x + 12) = \sin (2x - 13)$

5. Express the value of each of the following *in radical form*:
   (a) $\sin 45^\circ$
   (b) $\cos 45^\circ$
   (c) $\sin 60^\circ$
   (d) $\cos 30^\circ$
   (e) $\tan 30^\circ$
   (f) $\tan 60^\circ$

6. In right triangle ABC, $\angle C$ is the right angle, $AB = 25$, and $AC = 24$. Find the values of the sine, cosine, and tangent of the *smallest* angle of the triangle.

7. In right triangle RST, $\angle T$ is the right angle. If $\sin R = \frac{9}{41}$, find the values of $\cos R$ and $\tan R$.

8. In right triangle ABC, $\angle C$ is the right angle, $BC = 6$, and $AB = 10$. Express each of the following as a single fraction in lowest terms:
   (a) $\sin A + \cos A$
   (b) $\tan A + \tan B$
   (c) $\sin A + \cos B$
   (d) $\frac{\sin A}{\cos A}$

9. In right triangle ABC, $\angle C$ is the right angle. The ratio of leg AC to hypotenuse AB is 4 to 5. Find, to the *nearest degree*, the measure of $\angle B$.

10. As shown in the accompanying diagram, a 25-foot ladder leans against the side of a house. The base of the ladder is 12 feet from the house on level ground.
    (a) Find, to the *nearest degree*, the measure of the angle that the ladder makes with the ground.
    (b) Find, to the *nearest foot*, the distance from the top of the ladder to the ground.

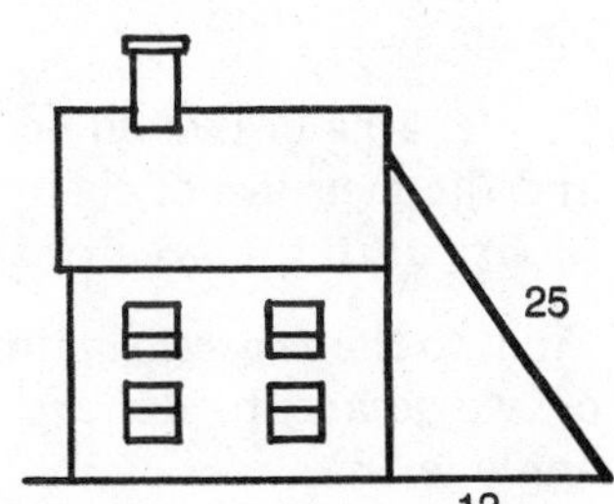

11. As shown in the accompanying diagram, a kite is flying at the end of a 200-meter straight string. If the string makes an angle of 68° with the ground, how high, to the *nearest meter*, is the kite?

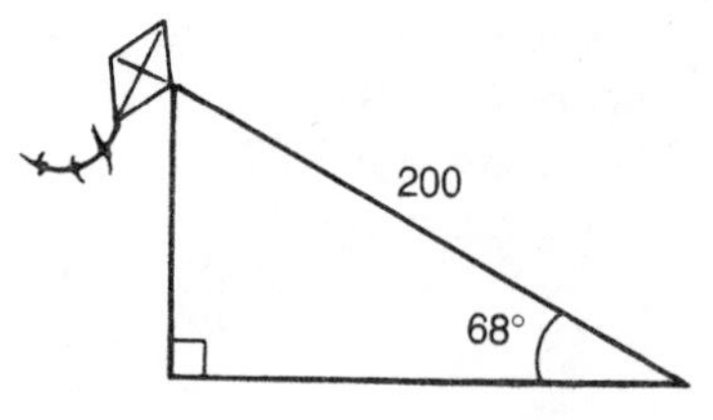

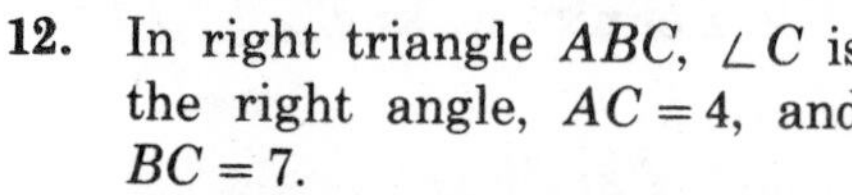

12. In right triangle ABC, $\angle C$ is the right angle, $AC = 4$, and $BC = 7$.
    (a) Find, to the *nearest degree*, the measure of $\angle B$.
    (c) Find, to the *nearest integer*, the length of $\overline{AB}$.

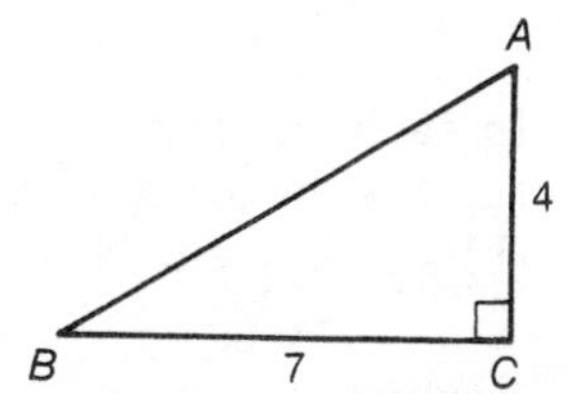

C. Problems Involving Angles of Elevation and Depression

13. At noon, a tree having a height of 10 feet casts a shadow 15 feet in length. Find, to the *nearest degree*, the angle of elevation of the sun at this time.

14. Find, to the *nearest foot*, the height of a building that casts a shadow of 80 feet when the angle of elevation of the sun is 42°.

15. A man observes the angle of depression from the top of a cliff overlooking the ocean to a ship to be 37°. If at this moment the ship is 1,000 meters from the foot of the cliff, find, to the *nearest meter*, the height of the cliff.

16. When the altitude of a plane is 800 meters, the pilot spots a target at a distance of 1,200 meters. At what angle of depression does the pilot observe the target?

D. Problems Involving Geometric Figures

17. The lengths of a pair of adjacent sides of a rectangle are 10 and 16. Find, to the *nearest degree*, the angle a diagonal makes with the longer side.

18. The lengths of two adjacent sides of a parallelogram are 8 and 12, and the measure of their included angle is 42. Find, to the *nearest square unit*, the area of the parallelogram.

19. Find, to the *nearest square unit*, the area of a triangle if the lengths of two adjacent sides are 7 and 12 and the measure of the included angle is 58°.

20. The lengths of diagonals of a rhombus are 12 and 16. Find, to the *nearest degree*, the measures of the four angles of the rhombus.

21. The length of each leg of an isosceles triangle is 20. If each leg makes an angle of 50° with the base, find:
(a) the length of the base, to the *nearest tenth*
(b) the area of the triangle, to the *nearest square unit*

22. The measure of the vertex angle of an isosceles triangle is 72. If the length of the altitude drawn to the base is 10, find:
(a) the length of the base, to the *nearest tenth*
(b) the area of the triangle, to the *nearest square unit*
(c) the length of a leg, to the *nearest tenth*

23. In the accompanying diagram, quadrilateral $ABCD$ is a trapezoid with $\overline{AB} \parallel \overline{CD}$, $m\angle A = 67$, $m\angle B = 90$, $DC = 12$, and $AD = 8$.
(a) Find, to the *nearest tenth*, the length of an altitude of the trapezoid.
(b) Find, to the *nearest integer*, the length of $\overline{AB}$.
(c) Find, to the *nearest integer*, the area of the trapezoid.

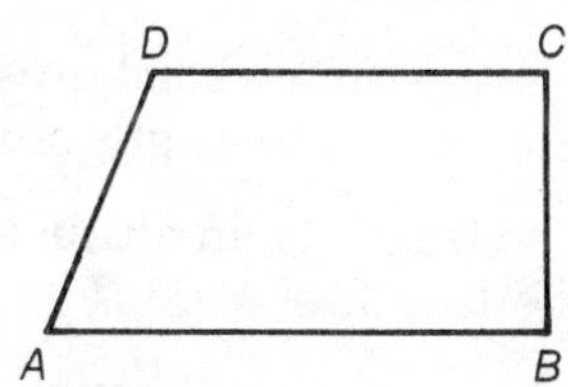

24. In the accompanying figure, $ABCD$ is a trapezoid. If $AB = 14$, $BC = 10$, and $m\angle BCD = 38$, find:
(a) AD, to the *nearest tenth*
(b) CD, to the *nearest tenth*
(c) the area of trapezoid $ABCD$, to the *nearest square unit*

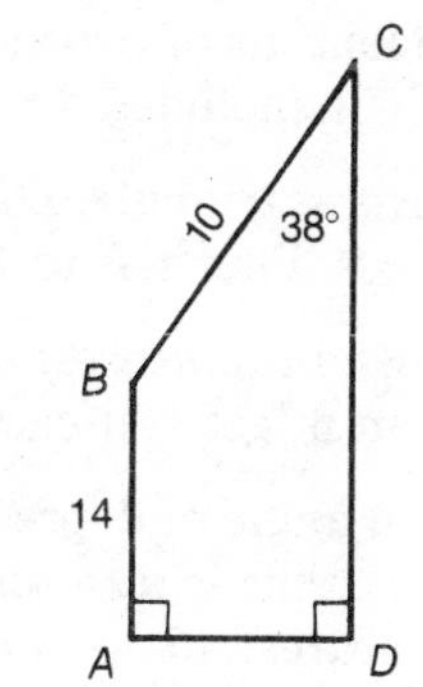

CHAPTER 8 REVIEW EXERCISES

1. A person 5 feet tall is standing near a tree 30 feet high. If the person's shadow is 4 feet long, how many feet long is the shadow of the tree?

2. If $\tan A = 0.4000$, find the measure of $\angle A$ to the *nearest degree.*

3. If the lengths of the legs of a right triangle are 4 and 7, find the length of the hypotenuse in radical form.

4. If $\cos A = 0.8155$, find $\angle A$ to the *nearest degree.*

5. In $\triangle ABC$, $m\angle C = 90$ and $\overline{CD}$ is the altitude to hypotenuse $\overline{AB}$.
(a) If $CD = 12$ and $AD = 6$, find the length of $\overline{DB}$.
(b) If $AD = 5$ and $DB = 15$, find the length of $\overline{AC}$.

6. What is the area of an equilateral triangle whose side has a length of 4 inches?

7. The lengths of the bases of an isosceles trapezoid are 6 and 12, and the length of a leg is 5. Find:
(a) the length of an altitude (b) the area of the trapezoid

8. If the length of a diagonal of a square is $6\sqrt{2}$, what is the area of the square?

9. The area of a rhombus is 40, and the length of one of its diagonals is 8. Find the length of the other diagonal.

10. The length of an altitude of an equilateral triangle is $6\sqrt{3}$. What is the length of a side?

11. In two similar triangles, the ratio of the lengths of a pair of corresponding sides is 5:8. If the perimeter of the larger triangle is 32, find the perimeter of the *smaller* triangle.

12. The lengths of two sides of a triangle are 8 and 15, and the measure of their included angle is 30. Find the area of the triangle.

13. In right triangle DEF, $\angle F$ is the right angle, $m\angle D = 50$, and $EF = 8$. Find DF to the *nearest integer.*

14. Find, to the *nearest degree*, the angle of elevation of the sun when a man 6 feet tall casts a shadow 4 feet long.

15. At an angle of depression of 42°, an airplane pilot is able to view a target that is at a distance of 1,000 meters from the pilot. Find, to the *nearest 10 meters*, the altitude of the plane.

16. The length of each leg of an isosceles triangle is 10, and the length of the base is 16.
(a) What is the length of the altitude drawn to the base?
(b) What is the measure of the acute angle formed by the base and a leg?

17. (a) As shown in the accompanying diagram, a ship at sea is sighted from the top of a 60-foot lighthouse. If the angle of depression of the ship from the top of the lighthouse measures 15°, find, to the *nearest foot*, how far the ship is from the base of the lighthouse.

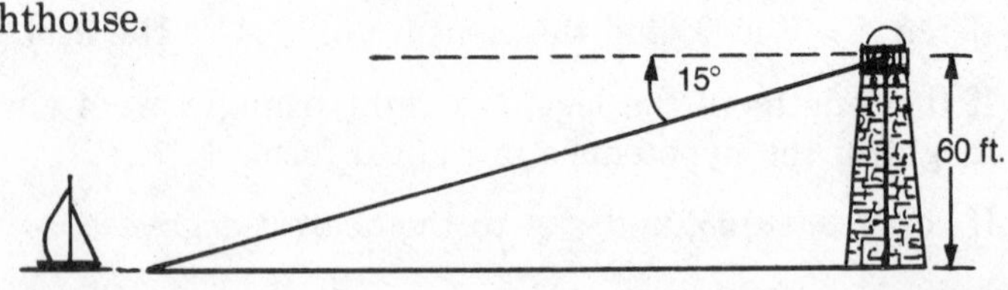

(b) In right triangle *DEF*, the measure of $\angle E$ is 90°, the length of side $\overline{EF}$ is 7 centimeters, and the length of side $\overline{DF}$ is 12 centimeters. Find the measure of $\angle F$ to the *nearest degree.*

18. In $\triangle PRT$, K is a point on $\overline{TP}$ and G is a point on $\overline{TR}$ such that $\overline{KG} \parallel \overline{PR}$. If $TP = 20$, $KP = 4$, and $GR = 7$, find TG.

19. If the lengths of the bases of an isosceles trapezoid are 16 and 10, and each leg makes an angle of 45° with the longer base, find the area of the trapezoid.

20. The length of base $\overline{AB}$ of $\triangle ABC$ and the length of a side of a square are each 8. If the area of the triangle is equal to the area of the square, find the length of the altitude drawn to $\overline{AB}$.

21. In $\triangle ABC$, $m\angle C = 90$, $m\angle A = 30$, and $AB = 12$. What is the length of $\overline{AC}$?

22. In right triangle ABC, $\overline{BD}$ is the altitude to hypotenuse $\overline{AC}$. If $AD = 4$ and $AC = 20$, what is the length of $\overline{BD}$?

23. The lengths of two adjacent sides of a parallelogram are 10 and 8, and the measure of the included angle is 30. What is the area of the parallelogram?

24. In right triangle ABC, altitude $\overline{CD}$ is drawn to hypotenuse $\overline{AB}$. If $AC = 10$ and $AB = 25$, what is the length of $\overline{AD}$?

25. In right triangle ABC, $m\angle C = 90$, $BC = 7$, and $AC = 10$. Find, to the *nearest degree*, the measure of $\angle A$.

26. The altitude drawn to the hypotenuse of a right triangle divides the hypotenuse into segments whose lengths are 4 and 16. What is the length of the altitude to the hypotenuse?

27. If the perimeter of a square is 20, find the length of a diagonal of the square.

28. *Solve algebraically*: The altitude drawn to the hypotenuse of a right triangle divides the hypotenuse into two segments such that the length of the longer segment exceeds twice the length of the shorter segment by 1. If the length of the altitude is 6, find:
(a) the length of the shorter segment of the hypotenuse
(b) the length of the shorter leg of the triangle (*Answer may be left in radical form.*)
(c) the measure of the smallest acute angle of the right triangle, to the *nearest degree.*

29. Given: Rhombus $ABCD$ with diagonals $\overline{BD}$ and $\overline{AC}$ intersecting at E, $AB = 13$, $BD = 10$.

Find:
(a) AC
(b) area of rhombus $ABCD$
(c) $m\angle EAB$, to the *nearest degree*

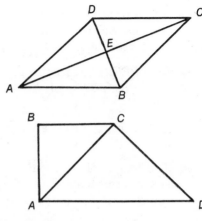

30. Given: Quadrilateral $ABCD$, $\overline{AB} \perp \overline{BC}$, $\overline{AB} \perp \overline{AD}$, $\overline{AC} \perp \overline{CD}$.

Prove: $BC \times AD = (AC)^2$.

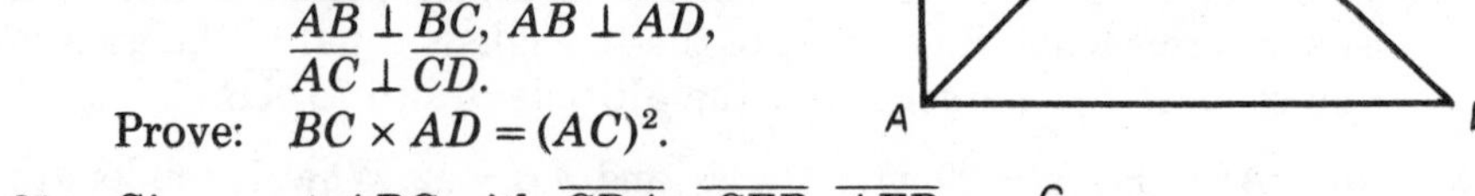

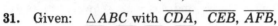

31. Given: $\triangle ABC$ with $\overline{CDA}$, $\overline{CEB}$, $\overline{AFB}$, $\overline{DE} \parallel \overline{AB}$, $\overline{EF} \parallel \overline{AC}$, $\overline{CF}$ intersects $\overline{DE}$ at G.

Prove: (a) $\triangle CAF \sim \triangle FEG$.
(b) $DG \times GF = EG \times GC$.

REGENTS TUNE-UP: CHAPTERS 6–8

Here is an opportunity for you to review Chapters 6–8 and, at the same time, prepare for the Course II Regents Examination. Problems included in this section are similar in form and difficulty to those found on the New York State Regents Examination for Course II of the Three-Year Sequence for High School Mathematics. Problems preceded by an asterisk have actually appeared on a previous Course II Regents Examination.

***1.** In rectangle $ABCD$ with diagonals $\overline{AC}$ and $\overline{BD}$, $AC = 3x - 15$ and $BD = 7x - 55$. Find x.

***2.** In the accompanying diagram, $\overleftrightarrow{WX} \parallel \overleftrightarrow{YZ}$; $\overleftrightarrow{AB}$ and $\overleftrightarrow{CD}$ intersect $\overleftrightarrow{WX}$ at E and $\overleftrightarrow{YZ}$ at F and G, respectively. If $m\angle CEW = m\angle BEX = 50$, find $m\angle EGF$.

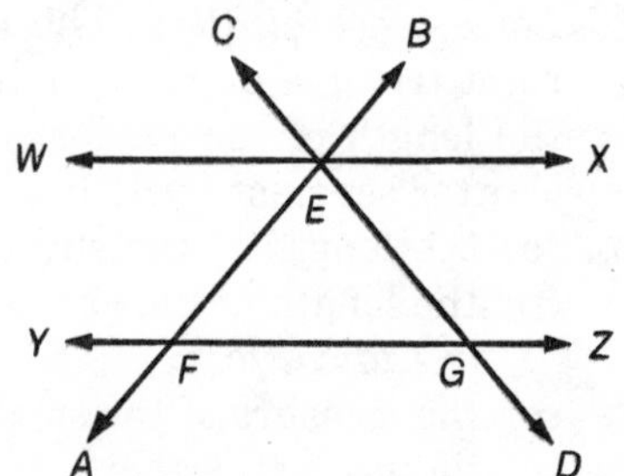

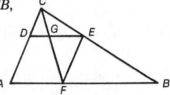

***3.** Find, in radical form, the length of a diagonal of a square if the perimeter of the square is 20.

***4.** In $\triangle ABC$, D is a point on $\overline{AB}$ and E is a point on $\overline{AC}$ such that $\overline{DE} \parallel \overline{BC}$. If $AD = 4$, $DB = 2$, and $AC = 9$, find AE.

***5.** What is the length of a side of a square whose diagonal measures $3\sqrt{2}$?

***6.** In right triangle *ABC*, hypotenuse $AB = 10$. The altitude drawn from *C* to $\overline{AB}$ intersects $\overline{AB}$ at *D*. If $AD = 2$, find *CD*.

***7.** In $\triangle ABC$, $m\angle C = 55$ and $m\angle C > m\angle B$. What is the *longest* side of the triangle?

***8.** A rhombus has a side of length 10 and one diagonal of length 16. Find the length of the other diagonal.

***9.** In the accompanying diagram, $\overleftrightarrow{AB}$ intersects $\overleftrightarrow{PQ}$ and $\overleftrightarrow{RS}$ at *C* and *D*, respectively. If $\overleftrightarrow{PQ} \parallel \overleftrightarrow{RS}$, $m\angle RDB = 2x - 10$, and $m\angle QCA = 3x - 65$, find *x*.

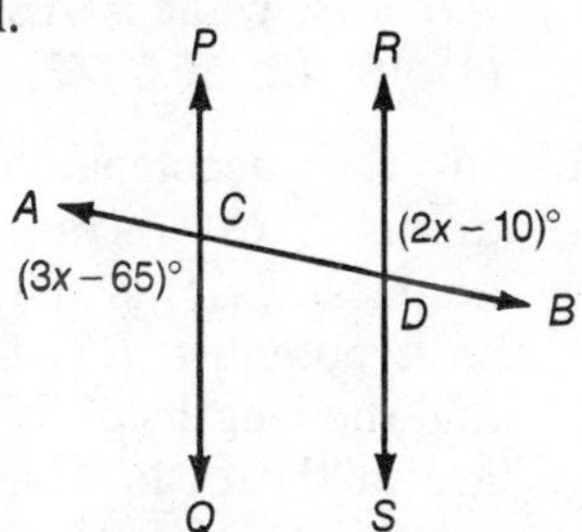

***10.** Two triangles are similar. The lengths of the sides of the smaller triangle are 4, 6, and 8, and the perimeter of the larger triangle is 27. Find the length of the *shortest* side of the larger triangle.

***11.** In $\triangle ABC$, point *D* is on $\overline{AC}$ and point *E* is on $\overline{BC}$ such that $\overline{DE} \parallel \overline{AB}$, $DE = 4$, $CD = 6$, and $DA = 3$. Find *AB*.

***12.** The length of a side of an equilateral triangle is 10. What is the length, in radical form, of an altitude of the triangle?

***13.** In $\square ABCD$, $m\angle A = 3x - 40$ and $m\angle C = 7x - 100$. Find the numerical value of *x*.

***14.** In equilateral triangle *ABC*, $AB = 16$. Find the perimeter of the triangle formed by connecting the midpoints of the sides of $\triangle ABC$.

***15.** In rhombus *ABCD*, $m\angle BCD = 80$. Find $m\angle BDA$.

***16.** In right triangle *ABC*, altitude $\overline{CD}$ is drawn to hypotenuse $\overline{AB}$. If $AD = 4$ and $DB = 5$, find *AC*.

17. Find, to the nearest foot, the height of a tree that casts a 12-foot shadow when the angle of elevation of the sun is 38°.

***18.** In the accompanying diagram of trapezoid *ABCD*, $CB = 6$, $m\angle A = 45$, $m\angle B = 90$, and base $DC = 2$. Find the length of base $\overline{AB}$.

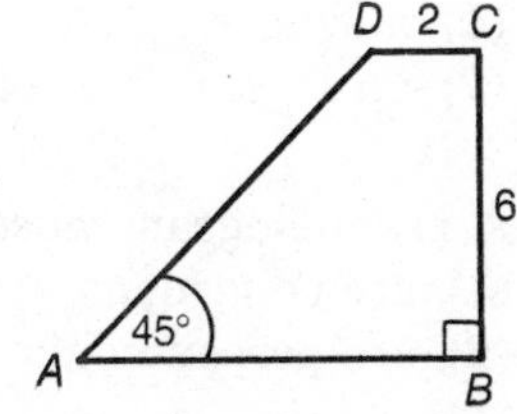

***19.** If the diagonals of a quadrilateral are perpendicular and *not* congruent, the quadrilateral may be:
(1) a rhombus (3) a rectangle
(2) an isosceles trapezoid (4) a square

***20.** In the accompanying diagram, $\overline{AB}$ and $\overline{CD}$ intersect at point E so that $\overline{AC}$ is parallel to $\overline{DB}$. If $AC = 3$, $DB = 4$, and $AB = 14$, what is AE?
(1) 19 (2) 10.5 (3) 8 (4) 6

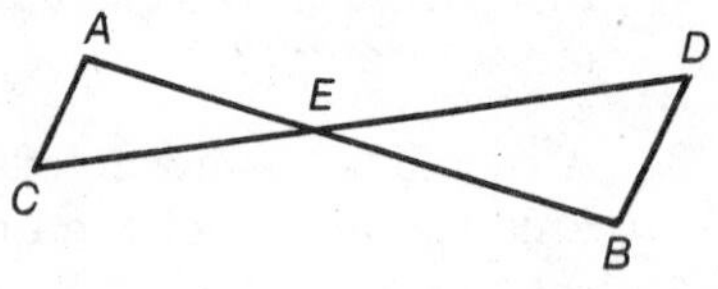

***21.** In the accompanying diagram, $\triangle RST$ is a right triangle and $\overline{SP}$ is the altitude to hypotenuse $\overline{RT}$. If $SP = 6$ and the lengths of $\overline{RP}$ and $\overline{PT}$ are in the ratio 1:4, what is the length of $\overline{RP}$?
(1) 12 (2) 15 (3) 3 (4) 9

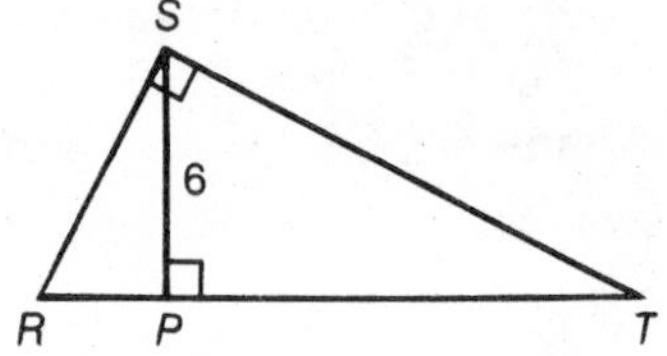

***22.** If the altitude drawn to the hypotenuse of a right triangle has length 10, the lengths of the segments of the hypotenuse may be:
(1) 5 and 20 (2) 2 and 5 (3) 3 and 7 (4) 50 and 50

***23.** Triangle ABC is equilateral, and D is any point on $\overline{AB}$. Which of the following is *always* a correct conclusion?
(1) $m\angle A > m\angle ADC$ (3) $CD > DB$
(2) $m\angle B > m\angle BDC$ (4) $DB > CD$

***24.** In the accompanying diagram, $\overline{ADB} \cong \overline{AEC}$. It can be proved that $\overline{CD} \cong \overline{BE}$ if what else is also known?
(1) $\angle 1 \cong \angle 2$
(2) $\angle 3 \cong \angle 4$
(3) $\angle 3 \cong \angle 5$
(4) $\angle 4 \cong \angle 6$

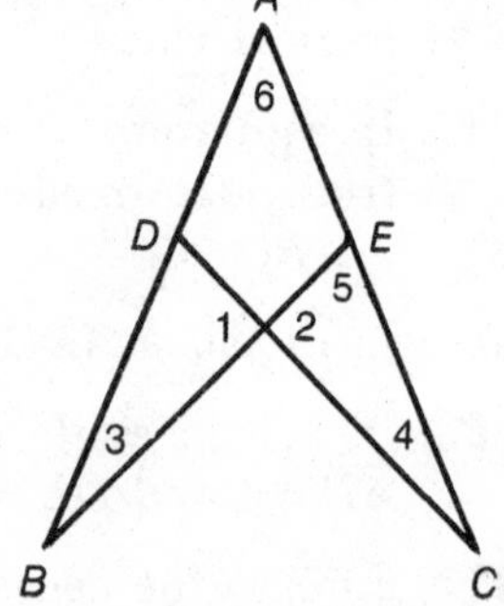

***25.** An example of a quadrilateral whose diagonals are congruent but do not bisect each other is
(1) a square (3) a rhombus
(2) an isosceles trapezoid (4) a rectangle

***26.** A parallelogram *must* be a rectangle if the diagonals
(1) are congruent (3) bisect the angles
(2) are perpendicular (4) bisect each other

***27.** If the midpoints of the sides of a quadrilateral are joined consecutively, the resulting figure will *always* be a
(1) rhombus (2) square (3) rectangle (4) parallelogram

***28.** In $\triangle ABC$, $m\angle B = 120$, $m\angle A = 55$, and D is a point on $\overline{AC}$ such that $\overline{BD}$ bisects $\angle ABC$. Which is the longest side of $\triangle ABD$?
(1) $\overline{AB}$ (2) $\overline{AD}$ (3) $\overline{BD}$ (4) $\overline{DC}$

***29.** Which set of numbers *cannot* be the lengths of the sides of a triangle?
(1) $\{2, 3, 4\}$ (2) $\{3, 5, 8\}$ (3) $\{6, 9, 10\}$ (4) $\{7, 8, 9\}$

30–32. Solve each problem algebraically.

***30.** In right triangle ABC, $\overline{CD}$ is the altitude drawn to hypotenuse $\overline{AB}$. The length of $\overline{DB}$ is 5 units longer than the length of $\overline{AD}$. If $CD = 3$, find the length of $\overline{AD}$ in radical form.

***31.** In $\triangle ABC$, D is a point on $\overline{AB}$ and E is a point on $\overline{AC}$ such that $\overline{DE} \parallel \overline{BC}$. If $AD = 2$, $DB = x - 1$, $AE = x$, and $EC = x + 2$, find AE.

***32.** In right triangle ABC, $\overline{CD}$ is the altitude to hypotenuse $\overline{AB}$. If $CD = 8$, and $AD : AB = 1:5$, find AD.

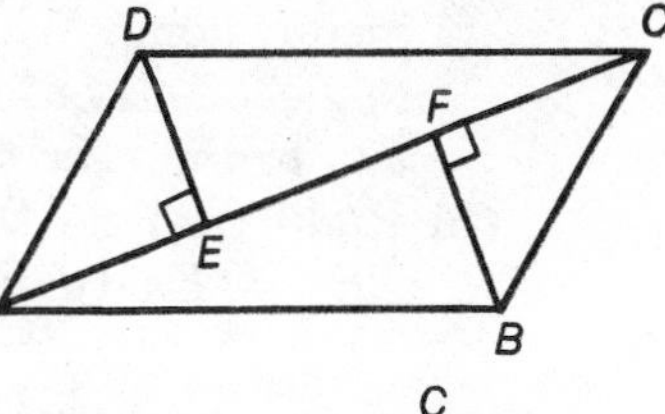

***33.** Given: Quadrilateral $ABCD$, diagonal $\overline{AEFC}$, $\overline{DE} \perp \overline{AC}$, $\overline{BF} \perp \overline{AC}$, $\overline{AE} \cong \overline{CF}$, $\overline{DE} \cong \overline{BF}$.
Prove: $ABCD$ is a parallelogram.

***34.** Given: $\overline{AC} \cong \overline{CB}$, $\overline{CDE} \perp \overline{ADB}$, $\overline{EB} \parallel \overline{AC}$.
Prove: (a) $\angle ACD \cong \angle BCD$. (b) $\overline{CD} \cong \overline{DE}$.

***35.** Given: V is a point on $\overline{ST}$ such that $\overline{RVW}$ bisects $\angle SRT$, $\overline{TW} \cong \overline{TV}$.
Prove: $RW \times SV = RV \times TW$.

36. Given: $ABCD$ is a parallelogram, diagonals $\overline{AC}$ and $\overline{BD}$ are drawn, $AB > BD$.
Prove: $m\angle ADC > m\angle DCB$.

***37.** Given: $\overline{AEC}$, $\overline{BFC}$, $\overline{EGB}$, $\overline{FGA}$, $\overline{FG} \cong \overline{EG}$, $\angle EGC \cong \angle FGC$.
Prove: $\overline{AC} \cong \overline{BC}$.

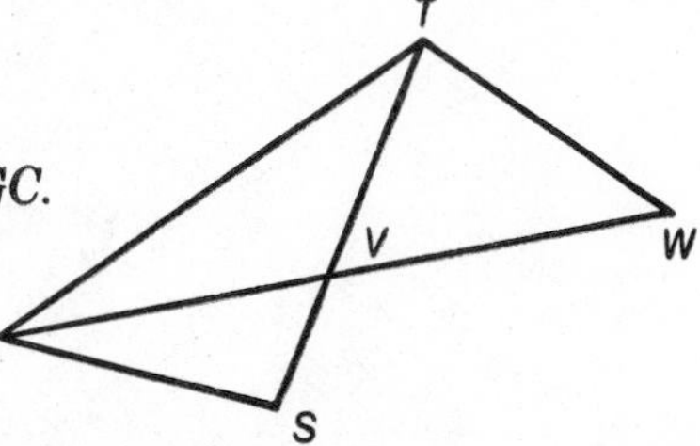

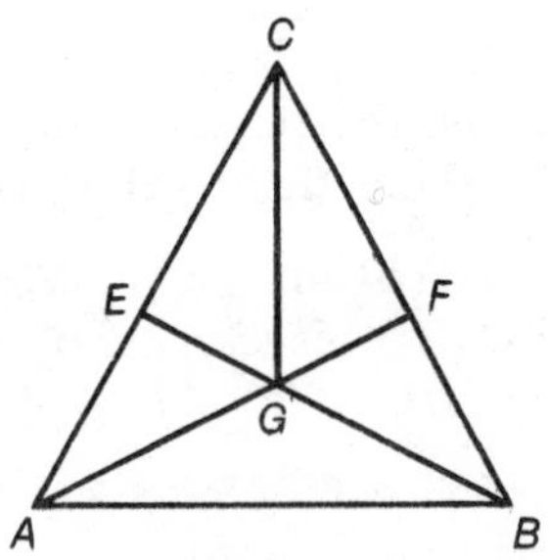

38. (a) In $\triangle ABC$, $\angle C$ is a right angle, $AC = 12$, and $\angle A$ is $35°$. Find BC to the *nearest integer.*

(b) The width and length of a rectangle are 5 and 12 inches, respectively. Find, to the *nearest degree*, the angle formed by a diagonal and a longer side.

39. (a) In the accompanying diagram, a tree 15 meters high casts a shadow 10 meters long. What is the angle of elevation of the sun to the *nearest degree*?

(b) Right triangle DEF has the right angle at F, $\angle D = 50°$, and $EF = 8$. Find DF to the *nearest integer*.

UNIT IV: ANALYTIC GEOMETRY

CHAPTER 9

Coordinate Geometry

9.1 COORDINATES AND AREA

KEY IDEAS

A **coordinate plane** may be created by drawing a horizontal number line called the ***x*-axis** and a vertical number line called the ***y*-axis**. The *x*-axis and the *y*-axis are sometimes referred to as the **coordinate axes**, and their point of intersection is called the **origin**. The process of locating a point or a series of points in the coordinate plane is called **graphing**.

Coordinates can be used to find the area of a triangle or a quadrilateral whose vertices are given.

GRAPHING ORDERED PAIRS. Each point in the coordinate plane is located (see Figure 9.1) using an ordered pair of numbers of the form (x, y), in which the first number of the pair is the x-coordinate, and the second number is the y-coordinate. The x-coordinate, sometimes called the **abscissa**, tells the number of units the point is located to the right ($x > 0$) or to the left ($x < 0$) of the origin. The y-coordinate, sometimes called the **ordinate**, gives the number of units the point is located

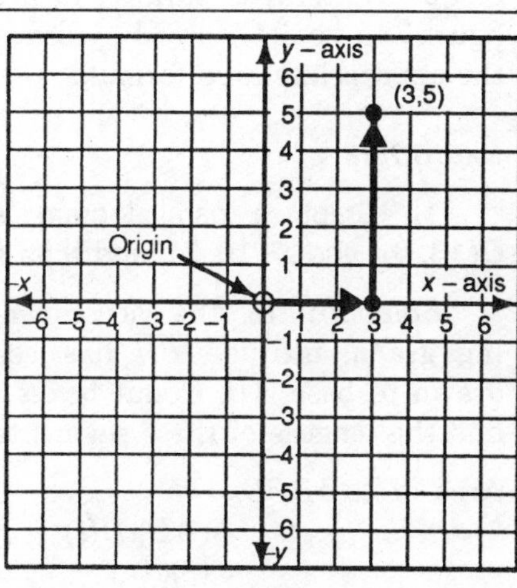

Figure 9.1 Graphing Ordered Pairs

above ($y > 0$) or below ($y < 0$) the origin. For example, to graph point (3, 5), start at the origin, move 3 units to the right, and then 5 units up.

THE FOUR QUADRANTS. The coordinate axes divide the plane into four regions called **quadrants**. As shown in Figure 9.2, the quadrants are numbered in counterclockwise order, beginning at the upper right and using Roman numerals. Notice that points $A(2, 3)$, $B(-4, 5)$, $C(-3, -6)$, and $D(3, -3)$ lie in different quadrants. The signs of the x- and y-coordinates of a point determine the quadrant in which the point lies.

Coordinates	Location of Point
(+, +)	Quadrant I
(−, +)	Quadrant II
(−, −)	Quadrant III
(+, −)	Quadrant IV

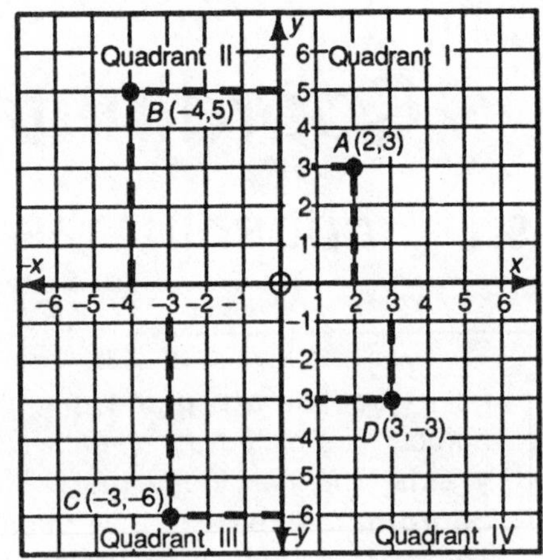

Figure 9.2 The Four Quadrants

FINDING AREA USING COORDINATES. If one side of a triangle or quadrilateral is parallel to a coordinate axis, then the area of the figure can be determined by drawing an altitude to this side and using the appropriate area formula.

Examples

1. Graph a parallelogram whose vertices are $A(2, 2)$, $B(5, 6)$, $C(13, 6)$, and $D(10, 2)$, and then find its area.

Solution: In the accompanying graph, altitude $\overline{BH}$ has been drawn to base $\overline{AD}$. Count boxes to find the lengths of these segments.

$$\begin{aligned}\text{Area of } \square ABCD &= bh \\ &= (AD)(BH) \\ &= (8)(4) \\ &= \mathbf{32 \text{ square units}}\end{aligned}$$

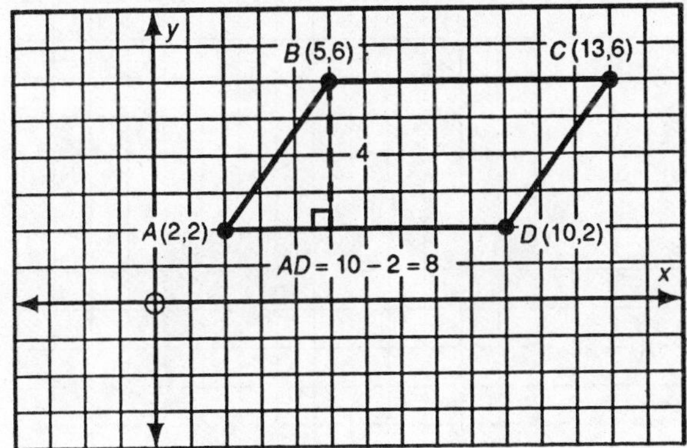

2. Graph a trapezoid whose vertices are $A(-4, 0)$, $B(-4, 3)$, $C(0, 6)$, and $D(0, 0)$, and then find its area.

Solution: In the accompanying graph, $\overline{AB}$ and $\overline{CD}$ are the bases of trapezoid $ABCD$. $\overline{AD}$ may be considered an altitude since it is perpendicular to both bases.

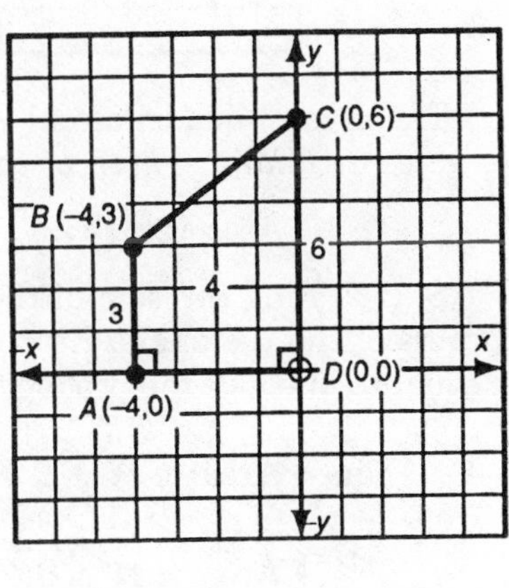

$$\begin{aligned}\text{Area of trap } ABCD &= \frac{1}{2}h(b_1 + b_2)\\ &= \frac{1}{2}(AD)(AB + CD)\\ &= \frac{1}{2}(4)(3 + 6)\\ &= 2(9)\\ &= \textbf{18 square units}\end{aligned}$$

USING SUBTRACTION TO FIND AREA. Example 3 illustrates a method that can be used to find the area of a quadrilateral or the area of a triangle that does not have a vertical or horizontal side.

Example

3. Find the area of the quadrilateral whose vertices are $A(-2, 2)$, $B(2, 5)$, $C(8, 1)$, and $D(-1, -2)$.

Solution: Circumscribe a rectangle about quadrilateral $ABCD$ by drawing intersecting horizontal and vertical segments through the vertices of the quadrilateral as shown in the accompanying diagram.

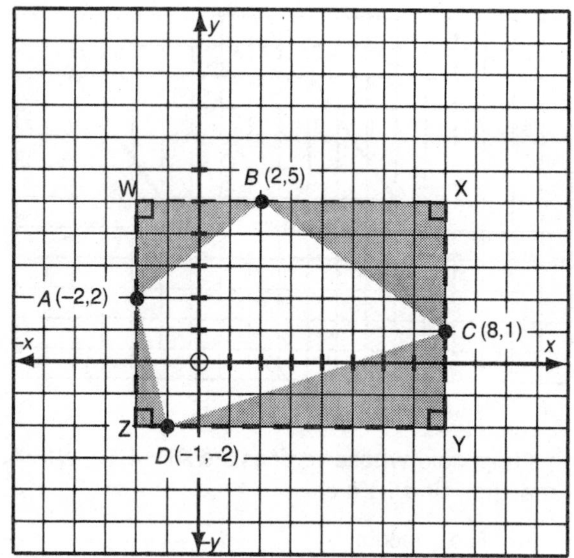

The area of quadrilateral *ABCD* is calculated *indirectly* as follows:
Step 1. Find the area of the rectangle.

$$\text{Area rect } WXYZ = (ZY)(YX) = (10)(7) = 70$$

Step 2. Find the sum of the areas of the right triangles in the four corners of the rectangle. Keep in mind that the area of a right triangle is equal to one-half the product of the lengths of the legs of the triangle.

$$\begin{aligned}
\text{Sum} = \text{Area of rt } \triangle BWA &= \frac{1}{2}(BW)(WA) = \frac{1}{2}(4)(3) = 6 \\
+\text{Area of rt } \triangle BXC &= \frac{1}{2}(BX)(XC) = \frac{1}{2}(6)(4) = 12 \\
+\text{Area of rt } \triangle DYC &= \frac{1}{2}(DY)(YC) = \frac{1}{2}(9)(3) = 13.5 \\
+\text{Area of rt } \triangle DZA &= \frac{1}{2}(DZ)(ZA) = \frac{1}{2}(1)(4) = 2 \\
\text{Sum of } \triangle \text{ areas} &= 33.5
\end{aligned}$$

Step 3. Subtract the sum of the areas of the right triangles from the area of the rectangle.

$$\begin{aligned}
\text{Area quad } ABCD &= \text{Area rect } WXYZ - \text{Sum of areas of right triangles} \\
&= 70 - 33.5 \\
&= \mathbf{36.5}
\end{aligned}$$

EXERCISE SET 9.1

1. Find the areas of the triangles whose vertices are:
(a) $A(0, 5)$, $B(6, 0)$, $C(0, 0)$
(b) $A(-4, 0)$, $B(0, 0)$, $C(0, -9)$
(c) $A(2, 2)$, $B(2, 7)$, $C(5, 2)$
(d) $X(-3, 0)$, $Y(0, 8)$, $Z(3, 0)$
(e) $R(-3, 3)$, $S(4, 9)$, $T(9, 5)$
(f) $J(-5, 2)$, $K(-3, 6)$, $L(3, 1)$
(g) $D(2, -7)$, $E(11, 7)$, $F(8, -3)$
(h) $K(-10, -6)$, $L(0, 6)$, $M(6, 1)$

2. Given the triangle determined by points $A(1, 4)$, $B(1, 1)$ and $C(x, 1)$, find x if the area of $\triangle ABC$ is 6.

3. The rectangle whose vertices are $A(0, 0)$, $B(0, 5)$, $C(h, k)$, and $D(8, 0)$ lies in the first quadrant.
(a) What are the values of h and k?
(b) What is the area of rectangle $ABCD$?

4. Find the area of the parallelogram whose vertices are:
(a) $A(2, 3)$, $B(5, 9)$, $C(13, 9)$, $D(10, 3)$
(b) $A(-4, -2)$, $B(-2, 6)$, $C(10, 6)$, $D(8, -2)$

5. Find the area of the trapezoid whose vertices are:
(a) $A(0, 0)$, $B(0, 5)$, $C(7, 11)$, $D(7, 0)$
(b) $A(-3, 0)$, $B(-3, 2)$, $C(5, 6)$, $D(5, 0)$
(c) $A(0, 0)$, $B(-2, -6)$, $C(9, -6)$, $D(7, 0)$
(d) $T(-4, -4)$, $R(-1, 5)$, $A(6, 5)$, $P(9, -4)$

6. Find the area of the hexagon whose vertices are $A(4, 5)$, $B(7, 0)$, $C(4, -5)$, $D(-4, -5)$, $E(-7, 0)$ and $F(-4, 5)$.

7. Find the area of the quadrilateral whose vertices are $A(-4, -2)$, $B(0, 5)$, $C(9, 3)$, and $D(7, -4)$.

8. Find the area of the quadrilateral whose vertices are $M(-4, 2)$, $A(0, 5)$, $T(3, 3)$, and $H(1, -5)$.

9. Find the area of pentagon $ABCD$, whose vertices are $A(-2, -5)$, $B(-2, 2)$, $C(2, 4)$, $D(5, 2)$, and $E(5, -3)$.

10. Given the parallelogram whose vertices are $M(-3, 2)$, $A(4, 8)$, $T(15, 5)$, and $H(8, -1)$:
(a) find the area of $\square MATH$.
(b) find the area of $\triangle AMH$.

9.2 MIDPOINT AND DISTANCE FORMULAS

KEY IDEAS

If the coordinates of the endpoints of $\overline{AB}$ are $A(x_1, y_1)$ and $B(x_2, y_2)$, then:

- The **x**- and **y**-coordinates of the **midpoint** of $\overline{AB}$ may be found by using the formulas

$$\bar{x} = \frac{x_1 + x_2}{2}; \quad \bar{y} = \frac{y_1 + y_2}{2}.$$

- The **length** of $\overline{AB}$ (or the **distance** between points A and B) may be found by using the formula

$$AB = \sqrt{(x_2 - x_1)^2 + (y_2 - y_1)^2}.$$

USING THE MIDPOINT FORMULA. Figure 9.3 illustrates that the x- and y-coordinates of the midpoint of a line segment are equal to the *averages* of the corresponding coordinates of the endpoints of the segment.

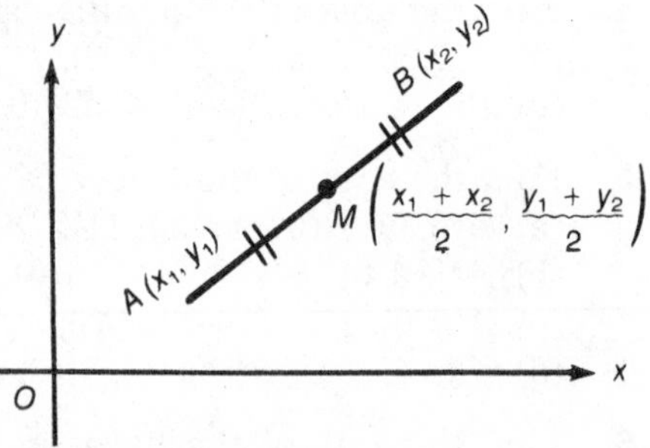

Figure 9.3 Using the Midpoint Formula

Examples

1. What are the coordinates of the center of a circle that has a diameter whose endpoints are (4, 9) and (−10, 1)?

Solution: The center of a circle is located at the midpoint of any diameter of the circle. Let $(x_1, y_1) = (4, 9)$ and $(x_2, y_2) = (-10, 1)$.

$$\bar{x} = \frac{x_1 + x_2}{2} = \frac{4 + (-10)}{2} = \frac{-6}{2} = -3$$

and

$$\bar{y} = \frac{y_1 + y_2}{2} = \frac{9 + 1}{2} = \frac{10}{2} = 5$$

The coordinates of the center of the circle are **(−3, 5)**.

2. If the midpoint of a line segment is (7, −1) and the coordinates of one endpoint are (5, 4), what are the coordinates of the other endpoint?

Solution: Let (x, y) represent the coordinates of the unknown endpoint. Then

$$7 = \frac{5+x}{2} \quad \text{and} \quad -1 = \frac{4+y}{2}.$$

Multiply each side of each equation by 2:

$$2 \cdot 7 = 2\left(\frac{5+x}{2}\right) \quad \text{and} \quad 2(-1) = 2\left(\frac{4+y}{2}\right)$$

$$14 = 5 + x \qquad -2 = 4 + y$$

$$9 = x \qquad -6 = y$$

The coordinates of the other endpoint are **(9, −6)**.

USING THE DISTANCE FORMULA. Figure 9.4 illustrates that points $A(x_1, y_1)$ and $B(x_2, y_2)$ determine a right triangle whose legs have lengths of $|x_2 - x_1|$ and $|y_2 - y_1|$. The length of the hypotenuse of this right triangle represents the distance between the two points and, using the Pythagorean theorem is given by the expression

$$\sqrt{(x_2 - x_1)^2 + (y_2 - y_1)^2}.$$

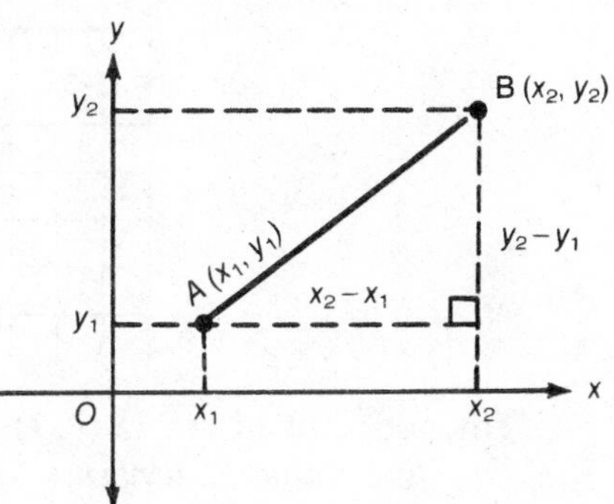

Figure 9.4 Using the Distance Formula

Since each difference is squared, the order in which the coordinates are subtracted does *not* matter.

Examples

3. Point (−2, 4) is on a circle whose center is at (1, 0). Find the length of the radius of the circle.

Solution: Use the distance formula to find the length of the segment joining points $(-2, 4)$ and $(1, 0)$. Let $(x_1, y_1) = (-2, 4)$ and $(x_2, y_2) = (1, 0)$.

$$\begin{aligned}\text{Radius length} &= \sqrt{(x_2 - x_1)^2 + (y_2 - y_1)^2} \\ &= \sqrt{[1 - (-2)]^2 + (0 - 4)^2} \\ &= \sqrt{(1 + 2)^2 + (-4)^2} \\ &= \sqrt{9 + 16} \\ &= \sqrt{25} = \mathbf{5}\end{aligned}$$

4. The coordinates of the vertices of quadrilateral *ABCD* are *A*(−3, 0), *B*(4, 7), *C*(9, 2), and *D*(2, −5). Prove that:

(a) *ABCD* is a parallelogram (b) *ABCD* is a rectangle

Solutions: (a) If the diagonals of a quadrilateral have the same midpoint, they bisect each other and the quadrilateral is a parallelogram.

To find the midpoint of $\overline{AC}$: Let $(x_1, y_1) = A(-3, 0)$, and $(x_2, y_2) = C(9, 2)$.

$$\bar{x} = \frac{x_1 + x_2}{2} = \frac{-3+9}{2} = \frac{6}{2} = 3$$

$$\bar{y} = \frac{y_1 + y_2}{2} = \frac{0+2}{2} = \frac{2}{2} = 1$$

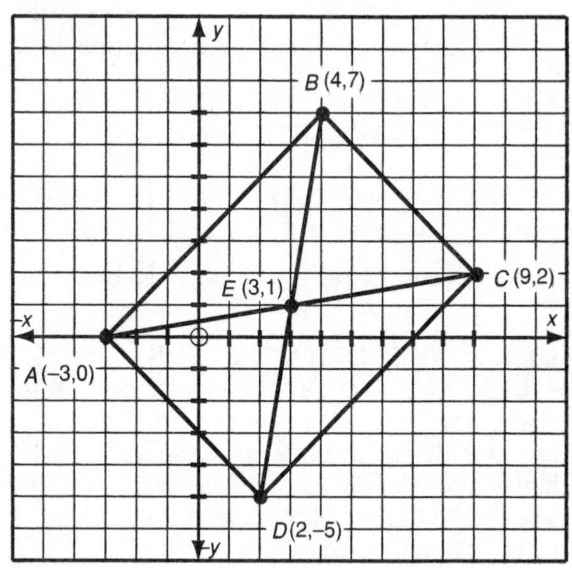

The midpoint of $\overline{AC}$ is (3, 1).

To find the midpoint of $\overline{BD}$: Let $(x_1, y_1) = B(4, 7)$, and $(x_2, y_2) = D(2, -5)$.

$$\bar{x} = \frac{x_1 + x_2}{2} = \frac{4+2}{2} = \frac{6}{2} = 3$$

$$\bar{y} = \frac{y_1 + y_2}{2} = \frac{7+(-5)}{2} = \frac{2}{2} = 1$$

The midpoint of $\overline{BD}$ is (3, 1).

Since the diagonals have the same midpoint, they bisect each other and **ABCD is a parallelogram.**

(b) If the diagonals of a parallelogram have the same length, then the parallelogram is a rectangle.

To find AC: Let $(x_1, y_1) = A(-3, 0)$, and $(x_2, y_2) = C(9, 2)$.

$$AC = \sqrt{(x_2 - x_1)^2 + (y_2 - y_1)^2} = \sqrt{[9-(-3)]^2 + (2-0)^2} = \sqrt{12^2 + 2^2} = \sqrt{144 + 4} = \sqrt{148}$$

To find BD: Let $(x_1, y_1) =$ B(4, 7), and $(x_2, y_2) = D(2, -5)$.

$$BD = \sqrt{(x_2 - x_1)^2 + (y_2 - y_1)^2} = \sqrt{(2-4)^2 + (-5-7)^2} = \sqrt{(-2)^2 + (-12)^2} = \sqrt{4 + 144} = \sqrt{148}$$

Since $AC = BD = \sqrt{148}$, the diagonals of ▱ABCD have the same length. Therefore, **▱ABCD is a rectangle.**

4. The coordinates of the vertices of a triangle are $A(-3, 7)$, $B(2, -2)$, and $C(11, 3)$.

(a) Show that $\triangle ABC$ is an isosceles right triangle.

(b) Find the area of $\triangle ABC$.

Solutions: (a) Find the length of each side of the triangle.

$$AC = \sqrt{[11-(-3)]^2+(3-7)^2} = \sqrt{14^2+(-4)^2} = \sqrt{212}$$
$$AB = \sqrt{[2-(-3)]^2+(-2-7)^2} = \sqrt{5^2+(-9)^2} = \sqrt{106}$$
$$BC = \sqrt{(11-2)^2+[3-(-2)]^2} = \sqrt{9^2+5^2} = \sqrt{106}$$

To prove that $\triangle ABC$ is a right triangle, show that the lengths of its sides satisfy the Pythagorean relationship.

$$(AC)^2 \overset{?}{=} (AB)^2 + (BC)^2$$
$$(\sqrt{212})^2 \mid (\sqrt{106})^2 + (\sqrt{106})^2$$
$$212 \mid 106 + 106$$
$$212 = 212 \checkmark$$

Since the square of the length of the longest side of $\triangle ABC$ is equal to the sum of the squares of the lengths of the remaining two sides, $\triangle ABC$ is a right triangle.

Since $AB = BC$, **$\triangle ABC$ is an isosceles right triangle.**

(b) The area of a right triangle is equal to one-half the product of the lengths of its legs.

$$\text{Area } \triangle ABC = \frac{1}{2} \times (\sqrt{106}) \times (\sqrt{106})$$
$$= \frac{1}{2} \times 106 = \mathbf{53 \text{ square units}}$$

5. The coordinates of the vertices of $\triangle RST$ are $R(9, 8)$, $S(-2, 5)$, and $T(4, -1)$. Find the length of the median drawn from R to $\overline{ST}$.

Solution: *Step 1.* Since the median to $\overline{ST}$ intersects $\overline{ST}$ at its midpoint, we must first find the coordinates of the midpoint $M(\bar{x}, \bar{y})$ of $\overline{ST}$. Let $(x_1, y_1) = (-2, 5)$, and $(x_2, y_2) = (4, -1)$.

$$\bar{x} = \frac{x_1+x_2}{2} \quad \text{and} \quad \bar{y} = \frac{y_1+y_2}{2}$$
$$= \frac{-2+4}{2} \quad \Big| \quad = \frac{5+(-1)}{2}$$
$$= \frac{2}{2} = 1 \quad \Big| \quad = \frac{4}{2} = 2$$

The coordinates of M are $(1, 2)$.

Step 2. Find the length of the segment joining $R(9, 8)$ and $M(1, 2)$.

$$RM = \sqrt{(9-1)^2+(8-2)^2}$$
$$= \sqrt{(8)^2 + (6)^2}$$
$$= \sqrt{64 + 36}$$
$$= \sqrt{100} = \mathbf{10}$$

LENGTHS OF HORIZONTAL AND VERTICAL SEGMENTS. Two points that have the same x- (or y-) coordinate determine a vertical (or horizontal) line segment. The distance between two points that determine a horizontal or vertical segment may be calculated either by using the distance formula or, more simply, by finding the difference of the *unequal* coordinates of the points, subtracting the smaller x- (or y-) coordinate from the larger x- (or y-) coordinate.

As illustrated in Figure 9.5, points $A(1, 3)$ and $B(5, 3)$ determine a horizontal segment since they have the same y-coordinate. Using the distance formula, we have

$$AB = \sqrt{(5-1)^2 + (3-3)^2}$$
$$= \sqrt{4^2 \qquad + 0}$$
$$= \sqrt{16} = \mathbf{4}$$

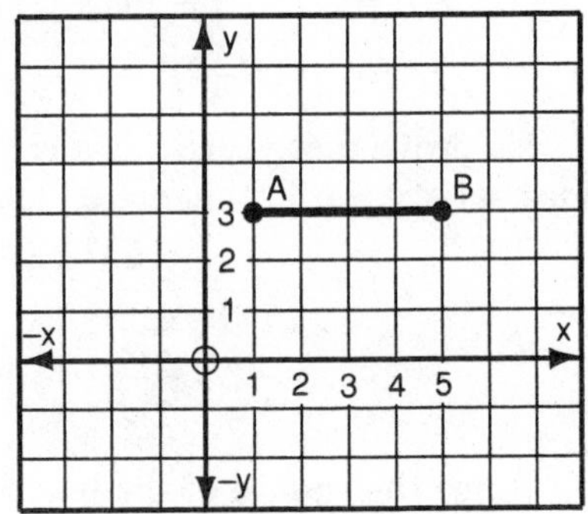

Figure 9.5 Determining the Length of a Horizontal Segment

The simpler method for determining the length of $\overline{AB}$ is to subtract the smaller x-coordinate from the larger x-coordinate: $AB = 5 - 1 = \mathbf{4}$.

Example

6. Find the distance between points (2, −1) and (2, 4).

Solution: Since the two points have the same x-coordinate, the two points determine a vertical line segment whose length is found by subtracting the smaller y-coordinate from the larger y-coordinate.

The length of the segment is $4 - (-1) = 4 + 1 = \mathbf{5}$.

EXERCISE SET 9.2

1. Fill in the missing items in the following table:

	Point A	Point B	Midpoint of $\overline{AB}$	Length of $\overline{AB}$
(a)	(3, 8)	(1, 2)	?	?
(b)	(−1, 3)	(−6, −9)	?	?
(c)	(2, 4)	?	(3, −2)	?
(d)	?	(−5, 1)	(−1, −4)	?
(e)	(0, 2a)	(2b, 0)	?	?
(f)	(a, b)	(c, d)	?	?

2. If $A(2, -1)$ and $B(6, 5)$ are the endpoints of a diameter of a circle, find the coordinates of the center of the circle.

3. What is the length of a radius of a circle whose center is at the origin and that passes through point $(-8, 15)$?

4. The coordinates of the endpoints of a diameter of a circle are $(-1, 7)$ and $(9, -17)$.
(a) Find the coordinates of the center of the circle.
(b) Find the length of the radius of the circle.

5. The vertices of $\triangle ABC$ are $A(2, 7)$, $B(8, 9)$, and $C(6, 3)$.
(a) Show that $\triangle ABC$ is an isosceles triangle.
(b) Find the length of the median drawn to the base of isosceles triangle ABC.
(c) Graph $\triangle ABC$ and find its area.

6. The coordinates of the vertices of $\square ABCD$ are $A(2, 1)$, $B(4, 3)$, $C(10, 3)$, and $D(x, 1)$. What is the value of x?

7. Show the quadrilateral $QRST$, with coordinates $Q(-5, 2)$, $R(7, 6)$, $S(8, 3)$, and $T(-4, -1)$ is a rectangle, and state a reason for your conclusion.

8. Show that quadrilateral $ABCD$, with vertices $A(-2, 8)$, $B(-4, 2)$, $C(8, 6)$, and $D(4, 10)$, is *not* a parallelogram.

9. The vertices of $\square ABCD$ are $A(-3, 1)$, $B(2, 6)$, $C(x, y)$, and $D(4, 0)$.
(a) Find the numerical coordinates of point C.
(b) Prove that $\square ABCD$ is a rhombus.

10. The vertices of $\triangle ABC$ are $A(5, 7)$, $B(11, -1)$, and $C(3, 3)$.
(a) Prove that $\triangle ABC$ is a right triangle.
(b) Show that the length of the median drawn to the hypotenuse is one-half the length of the hypotenuse.

11–13. In each case, do the following:
(a) Show that $\triangle$ABC *is a right triangle.*
(b) Find the area of $\triangle$ABC.

11. $A(2, 3)$, $B(6, 0)$, $C(12, 8)$
12. $A(2, 0)$, $B(11, 8)$, $C(6, 10)$
13. $A(-1, -2)$, $B(3, 1)$, $C(0, 5)$

9.3 SLOPE OF A LINE

KEY IDEAS

If you think of a line in the coordinate plane as a hill, then the *slope* of the line is a number that represents its steepness. The "steeper" a nonvertical line, the greater is the absolute value of its slope. The slope of a vertical line is undefined, and the slope of a horizontal line is 0.

Slope relationships can be used to demonstrate special properties of triangles and quadrilaterals.

DEFINING SLOPE. An **oblique line** is a line that is *not* parallel to a coordinate axis. Figure 9.6 illustrates that, when moving between two points on an oblique line, there is a change in vertical and horizontal distances. The change in the vertical distance between two points is measured by the difference in their y-coordinates, while the change in the horizontal distance is measured by the difference in their x-coordinates. In general, the slope of the line that contains points (x_1, y_1) and (x_2, y_2) is defined as a ratio:

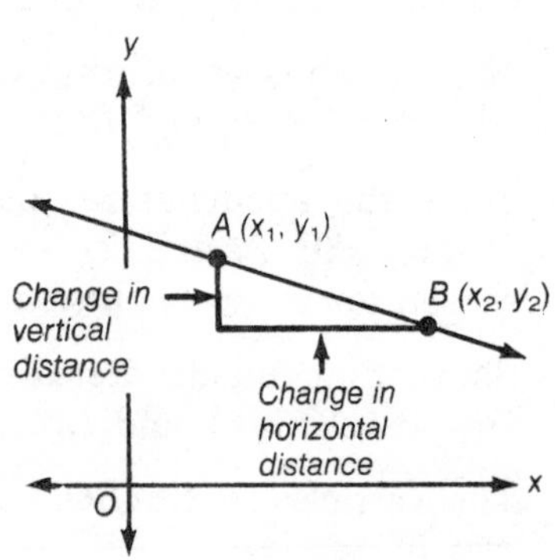

Figure 9.6 Slope of a Line

$$\textbf{Slope} = \frac{\textbf{change in } y\textbf{-coordinates}}{\textbf{change in } x\textbf{-coordinates}} = \frac{y_2 - y_1}{x_2 - x_1}.$$

Note: When calculating the differences in the numerator and the denominator of the slope fraction, the x- and y-coordinates of the *same* point must be in the *same* position, either both first or both second.

Example

1. What is the slope of the line that passes through points (1, −2) and (4, 7)?

Solution: Let $(x_1, y_1) = (1, -2)$, and $(x_2, y_2) = (4, 7)$.

$$\text{Slope} = \frac{y_2 - y_1}{x_2 - x_1}$$
$$= \frac{7 - (-2)}{4 - 1}$$
$$= \frac{7 + 2}{3} = \frac{9}{3} = \mathbf{3}$$

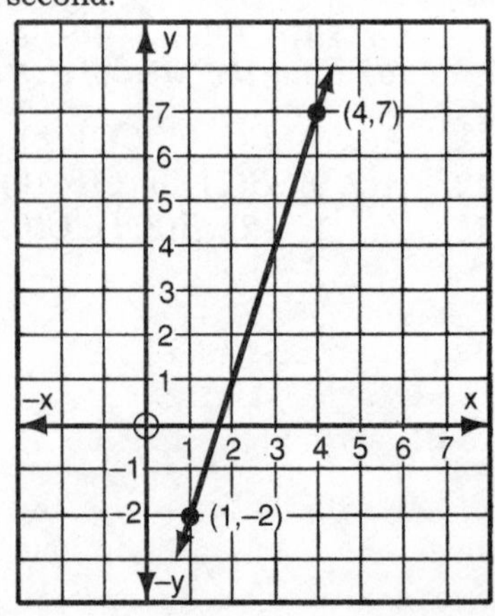

Note: When using the slope formula, either of the two given points may be considered the "second" point. For example, the same answer could be obtained by letting $(x_1, y_1) = (4, 7)$ and $(x_2, y_2) = (1, -2)$.

POSITIVE VERSUS NEGATIVE SLOPE. The slope of an oblique line may be either a positive or a negative number. Figure 9.7 illustrates that, as x increases:

- If the line rises, then its slope m is positive.
- If the line falls, then its slope m is negative.

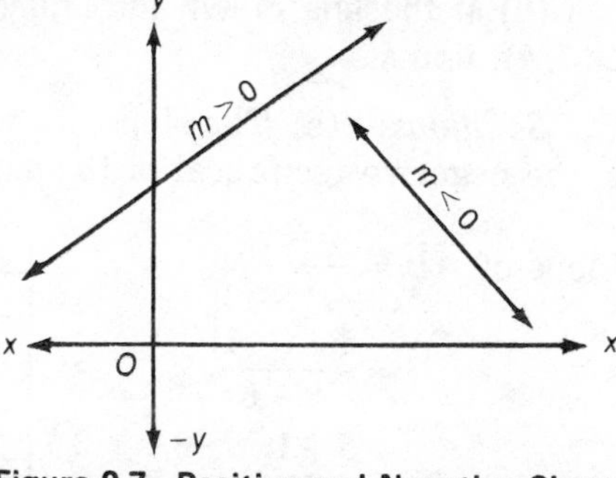

Figure 9.7 Positive and Negative Slopes

SLOPES OF HORIZONTAL AND VERTICAL LINES. On a horizontal line there is no change in y, and on a vertical line there is no change in x. Therefore, as shown in Figure 9.8:

- The slope m of a horizontal line is 0.
- The slope of a vertical line is *not* defined.

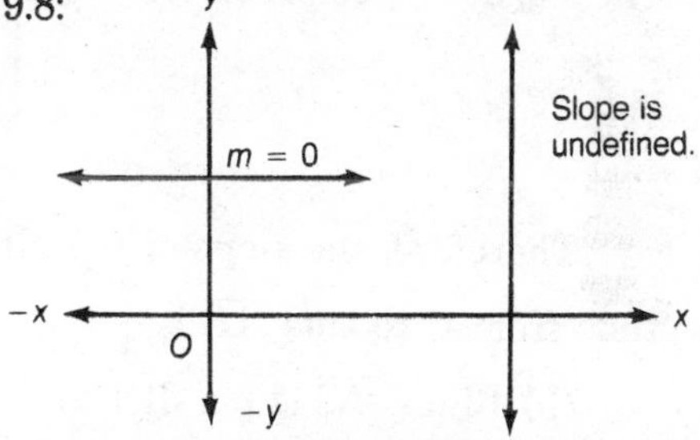

Figure 9.8 Slopes of Horizontal and Vertical Lines

SLOPES OF PARALLEL AND PERPENDICULAR LINES. Consider two lines that have slopes m_1 and m_2. As shown in Figure 9.9:

- If $m_1 = m_2$, then the two lines are parallel. Conversely, if two lines are parallel, then $m_1 = m_2$.

Parallel lines ↔ slopes are equal

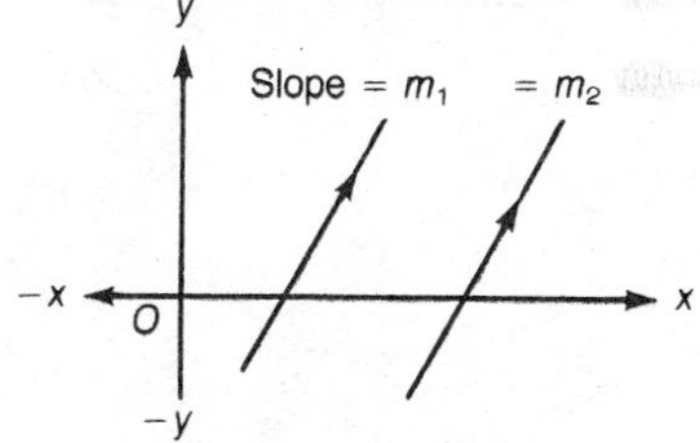

- If $m_1 = \dfrac{-1}{m_2}$ (or $m_1 \cdot m_2 = -1$), then the slopes are negative reciprocals and the lines are perpendicular. Conversely, if two lines are perpendicular, then their slopes are negative reciprocals.

Perpendicular lines ↔ slopes are negative reciprocals

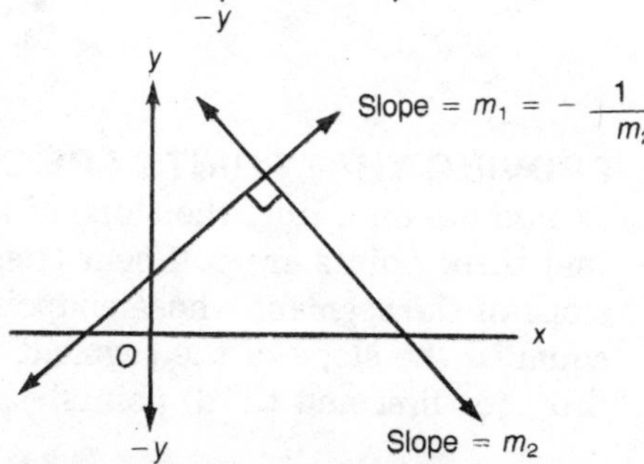

Figure 9.9 Slopes of Parallel and Perpendicular Lines

Example

2. The coordinates of the vertices of $\triangle ABC$ are $A(6, -1)$, $B(2, 5)$, and $C(1, 3)$.

(a) What is the slope of the altitude drawn from C to side $\overline{AB}$?

(b) If the line drawn through C and parallel to $\overline{AB}$ contains point $Q(7, k)$, find k.

Solutions: (a) The slope of the altitude drawn from C to side $\overline{AB}$ is the negative reciprocal of the slope $\overline{AB}$.

$$\text{Slope of } \overline{AB} = \frac{y_2 - y_1}{x_2 - x_1} = \frac{5-(-1)}{2-6} = \frac{5+1}{-4} = -\frac{6}{4} = -\frac{3}{2}$$

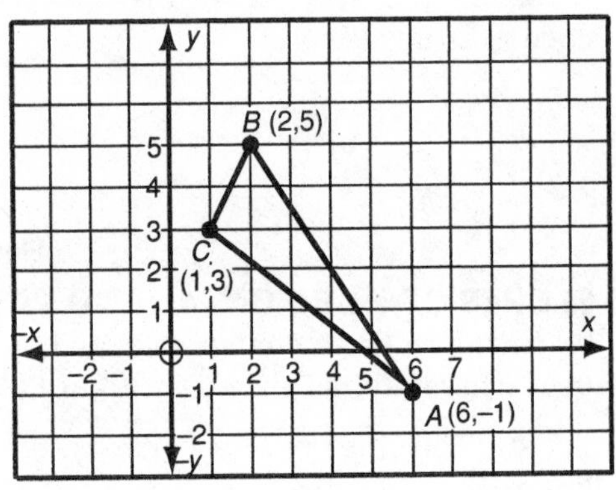

The negative reciprocal of $-\frac{3}{2}$ is $\frac{2}{3}$.

Therefore, the slope of the altitude from C to side $\overline{AB}$ is $\mathbf{\frac{2}{3}}$.

(b) Since $\overline{QC}$ is parallel to $\overline{AB}$, their slopes are equal.

$$\text{Slope of } \overline{QC} = \text{Slope of } \overline{AB}$$
$$\frac{y_2 - y_1}{x_2 - x_1} = \frac{k-3}{7-1} = -\frac{3}{2}$$
$$\frac{k-3}{6} = -\frac{3}{2}$$
$$2(k-3) = -3 \cdot 6$$
$$2k - 6 = -18$$
$$\frac{2k}{2} = \frac{-12}{2}$$
$$\mathbf{k = -6}$$

PROVING THAT POINTS ARE COLLINEAR. For any selection of two points on a line, the slope of the line is always the same. To show that three points are collinear (lie on the same line), we show that the slope of the segment whose endpoints are the first and second points is equal to the slope of the segment whose endpoints are the second and third (or first and third) points.

Example

3. In Example 2, determine whether point $E(-2, 11)$ lies on $\overleftrightarrow{AB}$.

Solution: Point E lies on $\overleftrightarrow{AB}$ if the slope of $\overline{BE}$ (or $\overline{AE}$) equals the slope of $\overline{AB}$. Let $(x_1, y_1) = B(2, 5)$, and $(x_2, y_2) = E(-2, 11)$.

$$\text{Slope of } \overline{BE} = \frac{y_2 - y_2}{x_2 - x_1} = \frac{11-5}{-2-2} = \frac{6}{-4} = -\frac{3}{2}$$

In Example 2 the slope of $\overline{AB}$ was calculated as $-\frac{3}{2}$. Since the slope of each line segment is equal to $-\frac{3}{2}$, the slope of $\overline{BE}$ equals the slope of $\overline{AB}$, so points A, B, and E are collinear, and **point $E(-2, 11)$ lies on $\overleftrightarrow{AB}$.**

SUMMARY OF METHODS FOR PROVING THAT A FIGURE IS A SPECIAL TRIANGLE OR QUADRILATERAL. Slope relationships provide additional methods for proving that a triangle or quadrilateral has a special property.

- *Prove that a triangle is a right triangle by using*:
  1. the distance formula to show that the lengths of the sides satisfy the converse of the Pythagorean theorem; *or*
  2. the slope formula to show that two sides are perpendicular.
- *Prove that a quadrilateral is a parallelogram by using*:
  1. the midpoint formula to show that the diagonals bisect each other; *or*
  2. the slope formula to show that both pairs of opposite sides are parallel, *or*
  3. the distance and slope formulas to show that the same pair of sides are congruent and parallel.
- *Prove that a parallelogram* is a rectangle by using*:
  1. the distance formula to show that the diagonals are congruent; *or*
  2. the slope formula to show that a pair of adjacent sides are perpendicular and, therefore, the parallelogram contains a right angle.
- *Prove that a parallelogram** is a rhombus by using:
  1. the distance formula to show that a pair of adjacent sides have the same length; *or*

2. the slope formula to show that the diagonals are perpendicular.

- *Prove that a parallelogram* is a square by using*:

1. the distance and slope formulas to show that the diagonals are congruent and perpendicular; *or*

2. the distance and slope formulas to show that a pair of adjacent sides are congruent and perpendicular.

- *Prove that a quadrilateral is a trapezoid by using*:

the slope formula to show that one pair of sides are parallel *and* the other pair of sides are *not* parallel.

- *Prove that a triangle or trapezoid is isosceles by using*:

1. the distance formula to show that two sides of the triangle are congruent.

2. the distance formula to show that the two nonparallel sides, or the diagonals, of the trapezoid are congruent.

****Note***: To prove that a *quadrilateral* is a rectangle, rhombus, or square, first show that it is a parallelogram.

Examples

4. The coordinates of the vertices of $\triangle PQR$ are $P(-1,-1)$, $Q(1,-2)$, and $R(3,2)$. Prove that $\triangle PQR$ is a right triangle.

Solution: A triangle contains a right angle if two of its sides are perpendicular. Find and then compare the slopes of sides $\overline{PQ}$, $\overline{PR}$, and $\overline{QR}$.

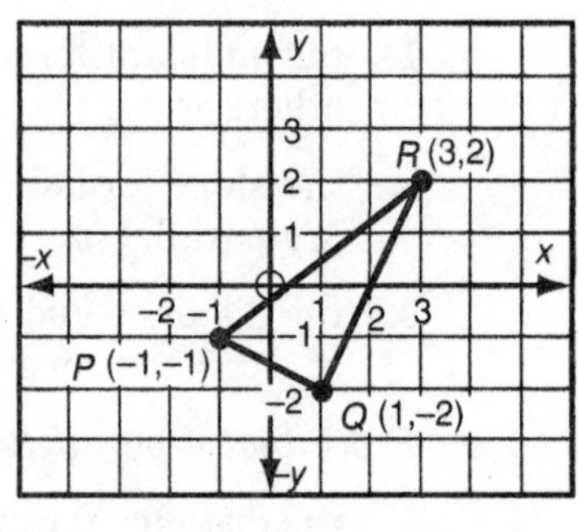

$$\text{Slope of } \overline{PQ} = \frac{-2-(-1)}{1-(-1)} = \frac{-2+1}{1+1} = \frac{-1}{2}$$

$$\text{Slope of } \overline{PR} = \frac{2-(-1)}{3-(-1)} = \frac{2+1}{3+1} = \frac{3}{4}$$

$$\text{Slope of } \overline{QR} = \frac{2-(-2)}{3-1} = \frac{2+2}{2} = \frac{4}{2} = \frac{2}{1}$$

Since the slopes of $\overline{PQ}$ and $\overline{QR}$ are negative reciprocals, $\overline{PQ} \perp \overline{QR}$. Therefore $\angle PQR$ is a right angle, and **$\triangle PQR$ is a right triangle.**

Note: An alternative method is to use the distance formula to find the length of each side of the triangle and then show that these lengths satisfy the Pythagorean relationship, $(PR)^2 = (PQ)^2 + (QR)^2$.

5. The coordinates of the vertices of a quadrilateral are $A(-2, 0)$, $B(10, 3)$, $C(5, 7)$, and $D(1, 6)$. Prove that quadrilateral $ABCD$ is a trapezoid.

Solution: A quadrilateral is a trapezoid if it has exactly one pair of parallel sides. Find and then compare the slopes of the four sides.

$$\text{Slope of } \overline{AB} = \frac{3-0}{10-(-2)} = \frac{3}{12} = \frac{1}{4}$$

$$\text{Slope of } \overline{BC} = \frac{7-3}{5-10} = \frac{4}{-5} = \frac{-4}{5}$$

$$\text{Slope of } \overline{CD} = \frac{6-7}{1-5} = \frac{-1}{-4} = \frac{1}{4}$$

$$\text{Slope of } \overline{AD} = \frac{6-0}{1-(-2)} = \frac{6}{3} = 2$$

Side $\overline{AB}$ is parallel to $\overline{CD}$ since their slopes are equal. Sides $\overline{BC}$ and AD are *not* parallel since their slopes are *not* equal.

Therefore, **$ABCD$ is a trapezoid** with parallel bases $\overline{AB}$ and $\overline{CD}$.

EXERCISE SET 9.3

1. Quadrilateral $ABCD$ is a parallelogram. Given that the slope of $\overline{AB}$ is $\frac{3}{4}$, find:
   (a) the slope of $\overline{DC}$
   (b) the slope of $\overline{BC}$ if $ABCD$ is a rectangle
   (c) the slope of $\overline{BD}$ if $ABCD$ is a rhombus and the slope of $\overline{AC}$ is $\frac{7}{6}$

2. In each of the following cases, determine whether $\overline{AB}$ is parallel to $\overline{CD}$, perpendicular to $\overline{CD}$, or neither.
   (a) $A(1, 5)$, $B(-1, 9)$; $C(2, 6)$, $D(1, 8)$
   (b) $A(-2, 7)$, $B(1, 4)$; $C(-8, 3)$, $D(-7, 4)$
   (c) $A(1, -5)$, $B(-4, 5)$; $C(0, -7)$, $D(4, 9)$
   (d) $A(-3, 6)$, $B(1, 1)$; $C(-7, 3)$, $D(1, -7)$
   (e) $A(3, 5)$, $B(7, 6)$; $C(-2, 1)$, $D(9, 5)$
   (f) $A(-1, -2)$, $B(-1, 5)$; $C(3, 4)$, $D(-2, 4)$

3. The slope of $\overleftrightarrow{AB}$ is $\frac{3}{5}$, and the slope of $\overleftrightarrow{CD}$ is $\frac{9}{k}$. Find the value of k if: (a) $\overleftrightarrow{AB} \parallel \overleftrightarrow{CD}$ (b) $\overleftrightarrow{AB} \perp \overline{CD}$

4. The coordinates of the vertices of $\square ABCD$ are $A(0, 0)$, $B(5, 0)$, $C(8, 1)$, and $D(x, 1)$. Find the value of x.

5. The coordinates of the vertices $\square ABCD$ are $A(1, y)$, $B(4, 10)$, $C(12, 10)$, and $D(9, 4)$. Find the value of y.

6. The line joining $A(-2, 0)$ and $B(10, 3)$ is parallel to the line joining $C(5, 7)$ and $D(1, k)$. Find the value of k.

7. Determine whether point C lies on line $\overleftrightarrow{AB}$.
   (a) $A(-4, -5)$; $B(0, -2)$; $C(8, 4)$
   (b) $A(-3, 2)$; $B(4, 2)$; $C(-5, 2)$
   (c) $A(1, 2)$; $B(5, 8)$; $C(-3, -4)$
   (d) $A(2, 1)$; $B(10, 7)$; $C(-4, -6)$

8. The coordinates of the vertices of quadrilateral $ABCD$ are $A(2, 0)$, $B(10, 2)$, $C(6, 7)$, and $D(2, 6)$. Prove that $ABCD$ is a trapezoid.

9. The coordinates of the vertices of trapezoid $ABCD$ are $A(1, 5)$, $B(7, k)$, $C(2, -4)$, and $D(-7, -1)$. If $\overline{AB}$ and $\overline{DC}$ are the bases of the trapezoid, find the value of k.

10. In trapezoid $ABCD$ with bases $\overline{AD}$ and $\overline{BC}$, the coordinates of the vertices are $A(3, 1)$, $B(1, 7)$, $C(4, 9)$, and $D(k, 5)$. Find the value of k.

11. The coordinates of the vertices of rectangle $ABCD$ are $A(-8, -1)$, $B(4, 3)$, $C(5, 0)$, and $D(-7, k)$. Find the value of k.

12. Parallelogram $ABCD$ has vertices $A(2, -1)$, $B(8, 1)$, and $D(4, k)$. The slope of $\overline{AD}$ is equal to 2.
    (a) Find k. (b) Find the coordinates of C.

13. The vertices of a triangle are $P(1, 2)$, $Q(-3, 6)$, and $R(4, 8)$.
    (a) Find the slope of $\overline{PR}$.
    (b) A line through Q is parallel to $\overline{PR}$. If this line contains point $(x, 14)$, find the value of x.

14. If $E(5, h)$ is a point on the line joining $A(0, 1)$ and $B(-2, -1)$, what is the value of h?

15. The coordinates of the vertices of quadrilateral $ABCD$ are $A(3, 0)$, $B(7, 0)$, $C(7, 11)$, and $D(3, 8)$.
    (a) Prove that $ABCD$ is a trapezoid.
    (b) Using graph paper, draw the trapezoid.
    (c) Find the area of the trapezoid.
    (d) Find the perimeter of the trapezoid.

16. Given points $A(1, -1)$, $B(5, 7)$, $C(0, 4)$, and $D(3, k)$.
    (a) Find the slope of $\overline{AB}$.
    (b) Find k if: (1) $\overline{AB} \parallel \overline{CD}$ (2) $\overline{AB} \perp \overline{CD}$

17. The coordinates of the vertices of $\triangle ABC$ are $A(1, 2)$, $B(5, 4)$, and $C(3, 8)$.
    (a) Find the coordinates of midpoint D of side $\overline{AC}$ and the coordinates of midpoint E of side $\overline{BC}$.
    (b) Show that $\overline{DE} \parallel \overline{AB}$.
    (c) Find the slope of the median to side $\overline{BC}$.
    (d) Find the slope of the altitude to side $\overline{AC}$.

18. Quadrilateral $ABCD$ has coordinates $A(-2, 3)$, $B(4, 6)$, $C(3, 2)$, and $D(-3, -1)$. Show that:
    (a) the opposite sides are parallel
    (b) the diagonals are *not* perpendicular

19. The vertices of $\triangle ABC$ are $A(1, 1)$, $B(10, 4)$, and $C(7, 7)$.
    (a) Find the slope of $\overleftrightarrow{AB}$.
    (b) If $D(7, k)$ is a point on $\overleftrightarrow{AB}$, find k.
    (c) Find the slope of the altitude from C to $\overline{AB}$.
    (d) Show by means of slope that $\triangle ABC$ is a right triangle.

20. The coordinates of the vertices of $\triangle ABC$ are $A(6, 2)$, $B(-4, 4)$, and $C(-2, -4)$.
    (a) Find the coordinates of midpoint D of side $\overline{AB}$ and the coordinates of midpoint E of side $\overline{BC}$.
    (b) Show that $\overline{DE} \parallel \overline{AC}$.
    (c) Show that $DE = \frac{1}{2}AC$.

21. The vertices of $\triangle ABC$ are $A(-1, 2)$, $B(7, 0)$, and $C(1, -6)$.
    (a) Show that point $D(4, -3)$ is on $\overline{BC}$.
    (b) Show that $\overline{AD}$ is the perpendicular bisector of $\overline{BC}$.
    (c) Show that $\triangle ABC$ is isosceles.

22. The vertices of quadrilateral $KLMN$ are $K(2, 3)$, $L(7, 3)$, $M(4, 7)$, and $N(-1, 7)$. Prove that quadrilateral $KLMN$ is a rhombus.

23. Quadrilateral $ABCD$ has vertices $A(-1, 0)$, $B(3, 3)$, $C(6, -1)$, and $D(2, -4)$. Prove that quadrilateral $ABCD$ is a square.

24. Quadrilateral $TRAP$ has vertices $T(0, 0)$, $R(0, 5)$, $A(9, 8)$, and $P(12, 4)$. Prove that quadrilateral $TRAP$ is an isosceles trapezoid.

9.4 EQUATIONS OF LINES

KEY IDEAS

The numerical relationship between the x- and y-coordinates of each point on a line is the same and can be expressed as an *equation* that defines the line. For example, if the sum of the x- and y-coordinates of every point on a certain line is 5, then an equation of this line is $x + y = 5$.

An equation of a line may be written in more than one way. For example, the equation $x + y = 5$ may also be written as $y = -x + 5$.

SLOPE-INTERCEPT FORM. Whenever an equation of a line is written in the form

$$y = mx + b,$$

the equation is said to be in **slope-intercept form** since the graph of this equation is a nonvertical line that has a slope of m and intersects the y-axis at b.

Examples

1. Line p is parallel to the line $y+4x=3$, and line q is perpendicular to the line $y+4x=3$.
   (a) What is the slope of line p?
   (b) What is the slope of line q?

Solutions: Write the equation of the given line in $y = mx + b$ form: $y = -4x + 3$. The slope of the line is $m = -4$.

(a) Since parallel lines have the same slope, the slope of line p is **−4.**

(b) Since perpendicular lines have slopes that are negative reciprocals, the slope of line q is $\mathbf{\frac{1}{4}}$.

2. The line whose equation is $y-3x=b$ passes through point $A(2,5)$.
   (a) Find the slope of the line.
   (b) Find the y-intercept of the line.

Solutions: (a) To find the slope of the line, put the equation of the line in $y = mx + b$ form. The equation $y - 3x = b$ may be written as $y = 3x + b$, which means that the slope of the line is **3**.

(b) In the equation $y = 3x + b$, b represents the y-intercept of the line. Since point A lies on the line, its coordinates must satisfy the equation. To find the value of b, replace x by 2 and y by 5 in the original equation.

$$\begin{aligned} y - 3x &= b \\ 5 - 3 \cdot 2 &= b \\ 5 - 6 &= b \\ -1 &= b \end{aligned}$$

The y-intercept is **−1**.

WRITING AN EQUATION OF A LINE: SLOPE-INTERCEPT FORM. If a line has a slope of 2 ($m = 2$) and a y-intercept of 5 ($b = 5$), then an equation of the line may be found by replacing m and b in $y = mx + b$ with their numerical values: $y = 2x + 5$.

Example

3. Write an equation of the line whose y-intercept is 1 and that is perpendicular to the line $y = \frac{1}{2}x - 4$.

Solution: The slope of the given line is $\frac{1}{2}$. Since the lines are perpendicular, the slope m of the desired line is the negative reciprocal of $\frac{1}{2}$, which is -2. The equation of the desired line is $\mathbf{y = -2x + 1}$.

WRITING AN EQUATION OF A LINE: POINT-SLOPE FORM. If the slope m of a line and the coordinates of a point $P(a, b)$ on the line are known, then an equation of the line may be written by using the point-slope form:

$$\mathbf{y - b = m(x - a).}$$

Examples

4. Write an equation of a line that is parallel to the line $y + 2x = 5$ and passes through point $(1, 4)$.

Solution: Since the lines are parallel, the slope m of the desired line is equal to the slope of the line $y + 2x = 5$. The line $y + 2x = 5$ may be written in the slope-intercept form as $y = -2x + 5$. Therefore, its slope is -2, so $m = -2$.

Method 1 (Slope-Intercept Form)	Method 2 (Point-Slope Form)
Find b by replacing m by -2, x by 1, and y by 4: $y = mx + b$ $4 = -2(1) + b$ $4 = -2 + b$ $6 = b$ Therefore, $\mathbf{y = -2x + 6}$	Since $(a, b) = (1, 4)$, replace m by -2, a by 1, and b by 4: $y - b = m(x - a)$ $\mathbf{y - 4 = -2(x - 1)}$

***Note*:** The slope-intercept and point-slope forms of an equation of a line are equivalent. For example, the equation $y - 4 = -2(x - 1)$ may be expressed in $y = mx + b$ form as follows:

$$\begin{aligned} y - 4 &= -2(x - 1) \\ &= -2x + 2 \\ y &= -2x + 2 + 4 \\ y &= -2x + 6 \end{aligned}$$

5. Write an equation of the line that passes through the origin and is perpendicular to the line whose equation is $y = 3x - 6$.

Solution: The slope of the line whose equation is $y = 3x - 6$ is 3, so the slope m of a perpendicular line is $-\frac{1}{3}$ (the negative reciprocal of 3). Since the desired line passes through the origin, $(a, b) = (0, 0)$. Therefore,

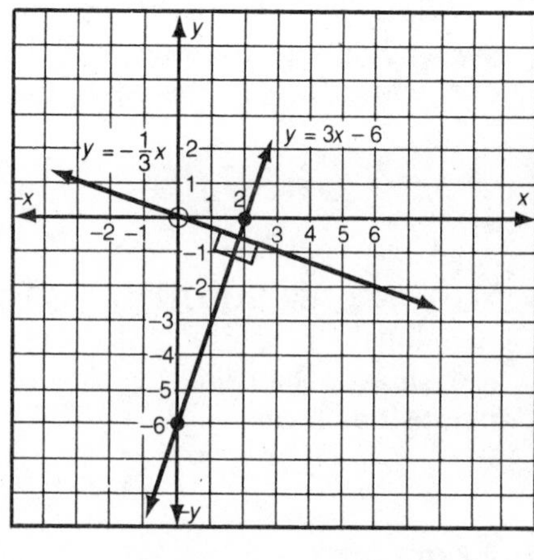

$$y - b = m(x - a)$$
$$y - 0 = -\frac{1}{3}(x - 0)$$

or

$$y = -\frac{1}{3}x$$

6. Write an equation of the line that contains points $A(6, 0)$ and $B(2, -6)$.

Solution: Use the point-slope form. First, find the slope m of $\overleftrightarrow{AB}$.

$$m = \frac{-6-0}{2-6} = \frac{-6}{-4} = \frac{3}{2}$$

Next, choose either point and then use its x- and y-coordinates. Consider point A. Then $(a, b) = (6, 0)$.

$$y - b = m(x - a)$$
$$y - 0 = \frac{3}{2}(x - 6)$$

$$y = \frac{3}{2}(x - 6) \quad \text{or} \quad y = \frac{3}{2}x - 9$$

EQUATIONS OF VERTICAL AND HORIZONTAL LINES. See Figure 9.10. The equation $x = a$ defines a vertical line that, extended if necessary, intersects the x-axis at a. Each point on a vertical line $x = a$ has an x-coordinate of a.

Similarly, the equation $y = b$ defines a horizontal line that, extended if necessary, intersects the y-axis at b. Each point on a horizontal line $y = b$ has a y-coordinate of b.

Example

7. Write an equation of a line that passes through point $(-3, 5)$ and is parallel to the:

(a) x-axis (b) y-axis

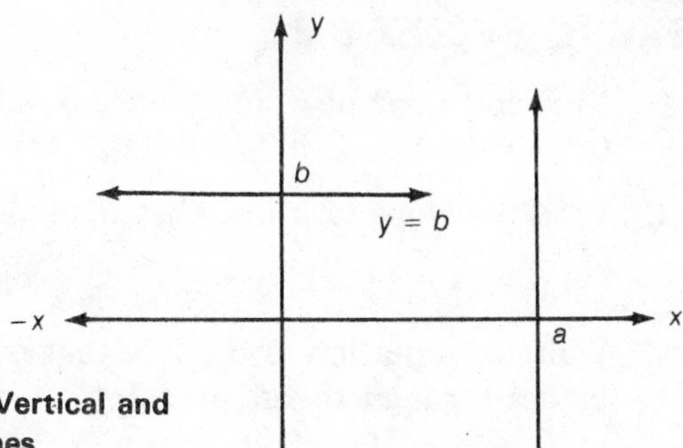

Figure 9.10 Equations of Vertical and Horizontal Lines

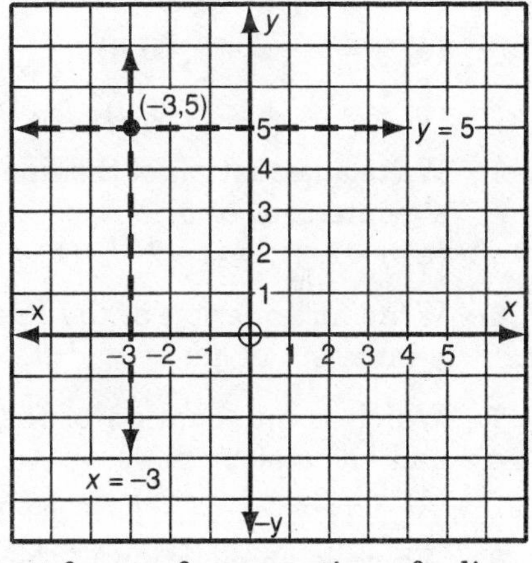

Solutions: (a) As shown in the accompanying diagram, a line parallel to the x-axis is a horizontal line, so its equation takes the general form $y = b$. Since the line passes through $(-3, 5)$, $b = 5$.

An equation of the line is $\mathbf{y = 5}$.

(b) Similarly, a line parallel to the y-axis is vertical, so its equation takes the form $x = a$. Since the line passes through $(-3, 5)$, $a = -3$.

An equation of the line is $\mathbf{x = -3}$.

Here is a summary of some different forms of an equation of a line.

General Form of Equation	Comments
$y = mx + b$	m = slope of line; b = y-intercept
$y - b = m(x - a)$	Form to use when the slope m of the line and the coordinates (a, b) of a point on the line are known. Also use when two points on the line are given, after first using their coordinates to calculate m.
$x = a$	Vertical line (parallel to y-axis) that intersects the x-axis at a.
$y = b$	Horizontal line (parallel to x-axis) that intersects the y-axis at b.

EXERCISE SET 9.4

1. Find the slope of a line that is parallel to the given line.
(a) $y = 2x + 5$ (b) $y - 3 = x$ (c) $2x + y = 8$ (d) $3x + 4y = 12$

2. Find the slope of a line that is perpendicular to the given line.
(a) $y = 3x - 1$ (b) $y = \frac{3}{5}x + 2$ (c) $x - 4y = 9$ (d) $3y + 4x = 12$

3. Write an equation of the line that is parallel to the given line and passes through the given point.
(a) $y = 2x - 1$; $(-4, 0)$ (c) $2y - 1 = 6x$; $(-1, 3)$
(b) $y - 4x = 2$; $(2, -5)$ (d) $3x - 2y = 6$; $(-2, 1)$

4. Write an equation of the line that is perpendicular to the given line and passes through the given point.
(a) $y = -\frac{1}{2}x + 3$; $(0, 4)$ (c) $y + x = 5$; $(-7, 3)$
(b) $y = 3x + 1$; $(-6, 2)$ (d) $3y + 2x = 12$; $(3, -4)$

5. Write an equation of the line that contains the given point and has a y-intercept of b.
(a) $(-2, 4)$; $b = -2$ (b) $(1, -3)$; $b = 5$ (c) $(3, 7)$; $b = -2$

6. Write an equation of the line that has an x-intercept of -3 and a y-intercept of 4.

7. Which is an equation of the line that passes through point $(0, 2)$ and has a slope of 4?
(1) $x = 2y - 4$ (2) $y = 2x + 4$ (3) $4x + y = 2$ (4) $y = 4x + 2$

8. Which is an equation of the line that is parallel to $y = 2x - 8$ and passes through point $(0, -3)$?
(1) $y = 2x + 3$ (2) $y = 2x - 3$ (3) $y = -\frac{1}{2}x + 3$ (4) $y = -\frac{1}{2}x - 3$

9. Write an equation of the line that passes through point $A(-4, 3)$ and is parallel to the line $2y - x = 3$.

10. Write an equation of the line that passes through point $B(3, 1)$ and is perpendicular to the line $3y + 2x = 15$.

11. Write an equation of the line that contains point $(-5, 2)$ and is parallel to:
(a) the x-axis (b) the y-axis (c) the line $y + x = 3$

12. Write an equation of the line that contains point $(4, -1)$ and is perpendicular to:
(a) the x-axis (b) the y-axis (c) the line $y - 2x = 3$

13. Write an equation of the line that passes through the two given points.
(a) $(-2, 7)$ and $(4, 7)$ (d) $(1, -3)$ and $(-1, 5)$
(b) $(-5, 8)$ and $(-5, -3)$ (e) $(2, -2)$ and $(-4, 4)$
(c) $(0, 3)$ and $(2, 1)$ (f) $(-4, 3)$ and $(1, -7)$

14. Which is an equation of the line that passes through points $(1, 3)$ and $(-1, 1)$?
(1) $x = 1$ (2) $y = 2x + 1$ (3) $y = x + 2$ (4) $y = 3$

15. The vertices of $\triangle ABC$ are $A(0, 6)$, $B(-8, 0)$ and $C(0, 0)$. Write an equation of the line that passes through one of the vertices of the triangle *and* is parallel to:
(a) $\overline{AC}$ (b) $\overline{BC}$ (c) $\overline{AB}$

16. Write an equation of the perpendicular bisector of the segment that joins points $(3, -7)$ and $(5, 1)$.

17. The coordinates of the vertices of $\triangle ABC$ are $A(-3, -4)$, $B(-1, 7)$, and $C(3, 5)$.
(a) Write an equation of the line drawn through B and parallel to $\overline{AC}$.
(b) Write an equation of the altitude drawn from C to side $\overline{AB}$.
(c) Write an equation of the median drawn from A to side $\overline{BC}$.
(d) Using graph paper, find the area of $\triangle ABC$.

18. The coordinates of the vertices of $\triangle ABC$ are $A(1, 2)$, $B(7, 0)$, and $C(3, -2)$.
(a) Show that $\triangle ABC$ is an isosceles triangle.
(b) Write an equation of the altitude drawn from vertex C to the base of the triangle.
(c) Write an equation of the median drawn from vertex C to the base of the triangle.
(d) Prove that $\triangle ABC$ is a right triangle.

9.5 GRAPHING LINEAR EQUATIONS

KEY IDEAS

Here are some general guidelines for graphing linear equations:

- Draw the coordinate axes on graph paper, using a straightedge.
- Label the horizontal axis x, and the vertical axis y.
- Label the origin, and number the boxes along each axis sequentially.
- Choose one of the two methods illustrated in this section.
- Graph the line, and label it with its equation.

THREE-POINT METHOD. To draw the graph of a linear equation, we must obtain the coordinates of at least two points that satisfy the equation. Graphing a third point and verifying that this point also lies on the line serves as a check. To graph the line whose equation is $y - 2x = 3$, using the three-point method, follow these steps:

Step	Example
1. Solve the original equation for y in terms of x.	$y - 2x = 3$ $y = 2x + 3$
2. Choose any three convenient values for x, and then calculate the corresponding values of y. The numbers -1, 0, and 1 are often good choices for x. It is helpful to organize your work in a table.	(see table below)
3. Graph the three points obtained in step 2. Use a straightedge to draw the line. Label the line with its equation, $y - 2x = 3$. If the three points do *not* lie on the same line, check your work since at least one of the points has been graphed incorrectly. Make certain that you have correctly numbered the axes and that each of the points has been accurately graphed. If the graphing of the points is correct, then return to step 2 and verify that the coordinates you are using are correct.	(see graph below)

x	$y = 2x + 3$	(x, y)
-1	$y = 2(-1) + 3$ $= -2 + 3 = 1$	$(-1, 1)$
0	$y = 2(0) + 3$ $= 0 + 3 = 3$	$(0, 3)$
1	$y = 2(1) + 3$ $= 2 + 3 = 5$	$(1, 5)$

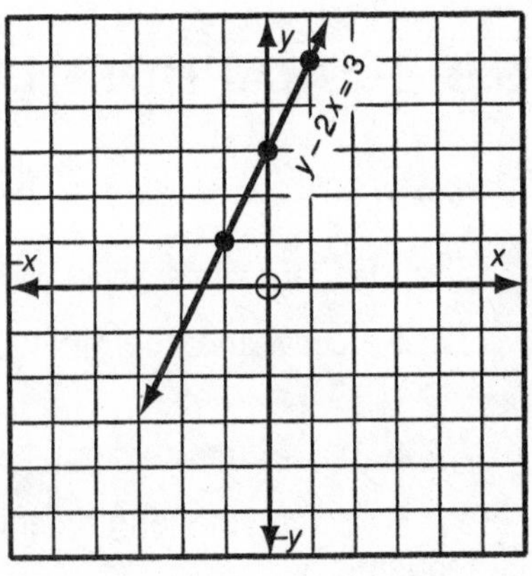

SLOPE-INTERCEPT METHOD. To graph a linear equation such as $y = \frac{2}{5}x + 1$, using the slope-intercept method, follow these steps:

Step	Example
1. Use the y-intercept to graph the point where the line crosses the y-axis. The coordinate of the y-intercept of $y = \frac{2}{5}x + 1$ is 1, so the line crosses the y-axis at $A(0, 1)$.	
2. Find another point of the given line, using the slope. The slope of $y = \frac{2}{5}x + 1$ is $\frac{2}{5}$. Start at $A(0, 1)$, and move up 2 units. Then move 5 units to the right. Label this point B. Note that the coordinates of B are (5, 3).	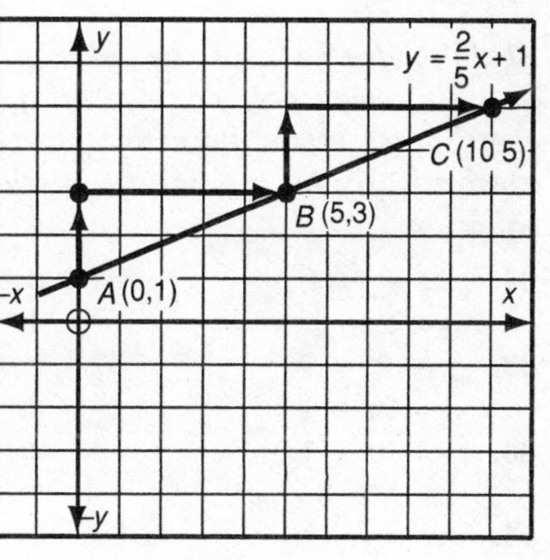
3. Find another point that will serve as a check point by moving up 2 units from point B, and then moving 5 units to the right. Label this point C. The coordinates of C are (10, 5).	
4. Draw $\overline{AB}$ and label this line with its equation, $y = \frac{2}{5}x + 1$.	

EXERCISE SET 9.5

1–18. Using graph paper, draw the graph of each equation using the three-point method.

1. $y = 3x$
2. $y = -3x$
3. $y = x + 2$
4. $y = -2x + 1$
5. $y + 3 = x$
6. $y - 3x + 5 = 0$
7. $y = \frac{1}{2}x$
8. $x + y = 7$
9. $2y - x = 8$
10. $y + 3x - 9 = 0$
11. $y = \frac{3}{5}x - 2$
12. $2y - 4x = 10$
13. $3y + 6x = 12$
14. $x - 2y = 8$
15. $2x - 5y = 15$
16. $3y - 2x = 6$
17. $6y + 18 = 4x$
18. $4x - 5y = 20$

19–30. Using graph paper, draw the graph of each equation using the slope-intercept method.

19. $y = 3x - 1$

20. $y = -2x + 4$

21. $y = \frac{3}{4}x + 8$

22. $2y - x = 10$

23. $x = 2y + 5$

24. $4x - 6y = 12$

25. $2x - 5y = 15$

26. $x - 2y = 3$

27. $\frac{y}{2} - x = 4$

28. $\frac{y}{3} - \frac{x}{2} = 1$

29. $\frac{y}{5} + \frac{x}{2} = 0$

30. $\frac{y - 1}{2} = 2x$

31–45. Solve each of the following systems of equations graphically by graphing each pair of lines on the same set of axes and determining the coordinates of the point of intersection of the lines. Check your answer algebraically by verifying that the coordinates of the point satisfy each equation.

31. $2x + y = 8$
$y = x + 2$

32. $x + 2y = 7$
$y = 2x + 1$

33. $y = 3x + 1$
$x = y - 3$

34. $2x - y = 10$
$x + 2y = 10$

35. $y = 2x - 4$
$x + 2y = -7$

36. $x + 2y = 6$
$4x + y = -4$

37. $x + y = 8$
$2x - y = 7$

38. $2x + y = 6$
$x - 2y = 8$

39. $2y = -5x$
$y - x = 7$

40. $x - 2y = 4$
$x = y + 2$

41. $2y = x + 6$
$y = 3x - 2$

42. $y - 2x = 5$
$x + 2y = 0$

43. $x - 3y = 9$
$2x + y = 4$

44. $3x - 2y = 4$
$3x + 2y = 8$

45. $y - x = -1$
$2y + x = 4$

46–60. Solve each system of equations in Exercises 31–45 algebraically.

9.6 TRANSFORMATIONS IN THE COORDINATE PLANE

KEY IDEAS

The process of moving each point of a figure according to some given rule is called a **transformation**. Each point of the new figure corresponds to exactly one point of the original figure and is called the **image** of that point. Sometimes the image of a point *P* is named as point *P'*, which is read as "*P* prime."

Reflections, translations, rotations, and *dilations* are special types of transformations; each uses a different rule for locating images of points of the figure that are undergoing the transformation.

REFLECTIONS IN A LINE. A **reflection** in a line may be thought of as the mirror image of a point or set of points with respect to a *line of symmetry* that serves as the "mirror." As illustrated in Figure 9.11, the image of a point P is determined in such a way that the line of symmetry (line l) is the perpendicular bisector of the segment ($\overline{PP'}$) determined by the original point (P) and its image (P').

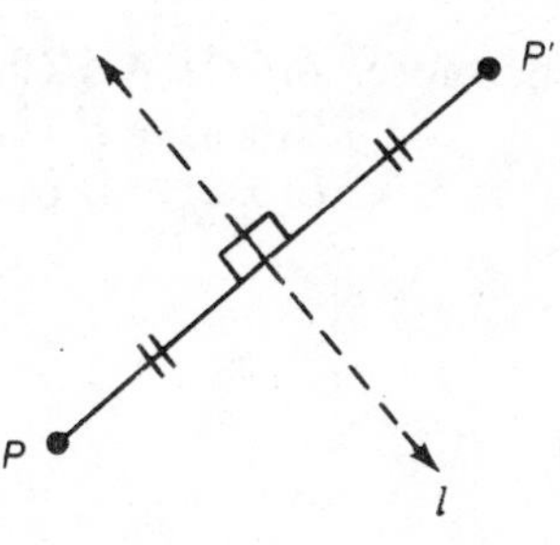

Figure 9.11 Reflection of a Point in a Line

RULES FOR REFLECTIONS IN LINES

- The reflection of point (a, b) in the x-axis is point $(a, -b)$.
- The reflection of point (a, b) in the y-axis is point $(-a, b)$.
- The reflection of point (a, b) in the line whose equation is $y = x$ is point (b, a).

Note: In each case (see the accompanying diagrams), the line of reflection is the perpendicular bisector of the line segment whose endpoints are the original point (a, b) and its image under the reflection.

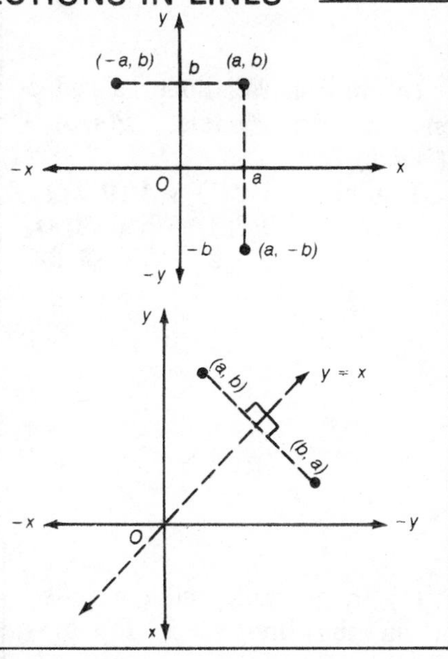

Example

1. The coordinates of the vertices of $\triangle ABC$ are $A(0, 1)$, $B(3, 4)$, and $C(5, 2)$. Determine the coordinates of the vertices of the image of $\triangle ABC$ under a reflection in the:

(a) x-axis (b) y-axis (c) line $y = x$

Solutions: (a) In general, under a reflection in the x-axis, $P(a, b) \rightarrow P'(a, -b)$.

Therefore, $A(0, 1) \rightarrow \mathbf{A'(0, -1)}$,
$B(3, 4) \rightarrow \mathbf{B'(3, -4)}$,
$C(5, 2) \rightarrow \mathbf{C'(5, -2)}$.

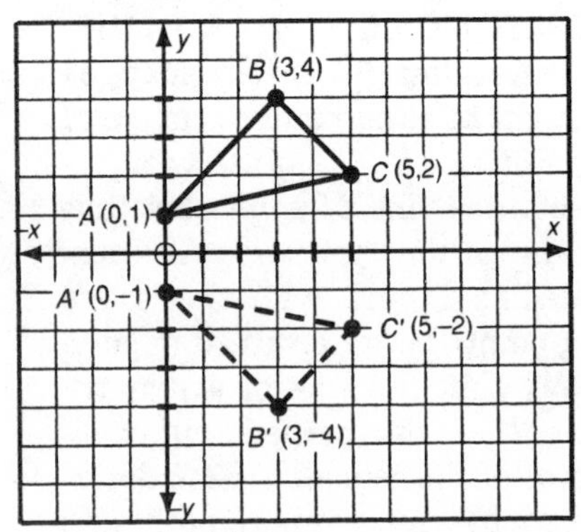

(b) In general, under a reflection in the y-axis, $P(a, b) \rightarrow P'(-a, b)$.

Therefore, $A(0, 1) \rightarrow \mathbf{A'(0, 1)}$,
$B(3, 4) \rightarrow \mathbf{B'(-3, 4)}$,
$C(5, 2) \rightarrow \mathbf{C'(-5, 2)}$.

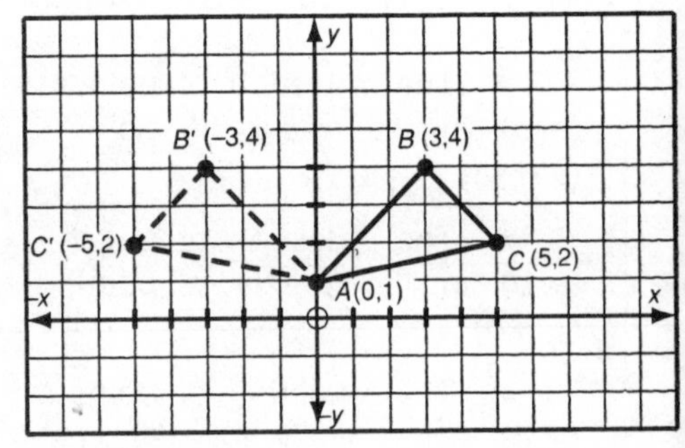

(c) In general, under a reflection in the line $y = x$, $P(a, b) \rightarrow P'(b, a)$.

Therefore, $A(0, 1) \rightarrow \mathbf{A'(1, 0)}$,
$B(3, 4) \rightarrow \mathbf{B'(4, 3)}$,
$C(5, 2) \rightarrow \mathbf{C'(2, 5)}$.

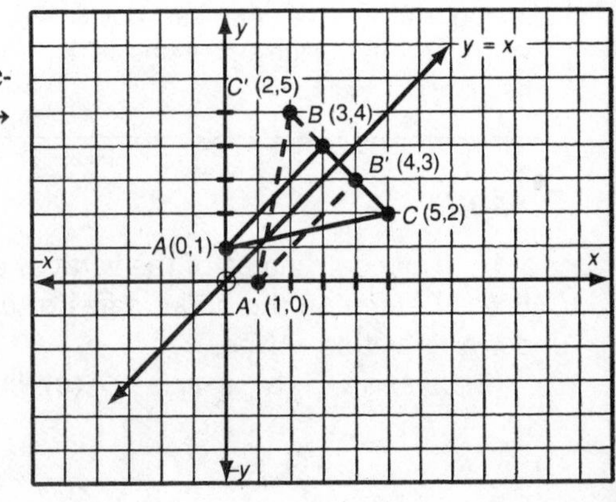

REFLECTIONS IN A POINT. If point P' is the image of point P under a reflection in point A, then A is the midpoint of $\overline{PP'}$. A point and its image under a reflection in the origin must have coordinates that have the same absolute value but are opposite in sign, so that the average of the x-coordinates and the average of the y-coordinates are each 0.

RULES FOR REFLECTIONS IN POINTS

- The reflection of point $P(a, b)$ in the origin $(0, 0)$ is the point $P'(-a, -b)$. (See the accompanying diagram.)

- In general, the reflection of point $P(x_1, y_1)$ in point $A(x, y)$ is point $P'(x_2, y_2)$ such that

$$x = \frac{x_1 + x_2}{2} \quad \text{and} \quad y = \frac{y_1 + y_2}{2}$$

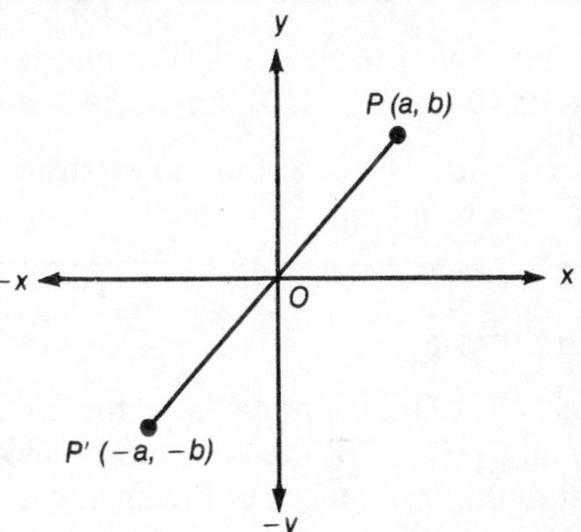

Example

2. Point $R(6, 1)$ is reflected in point $P(1, 2)$. What are the coordinates of the image of R under this reflection?

Solution: Point $P(1, 2)$ is the midpoint of the segment joining $R(6, 1)$ and its image $R'(x, y)$. Therefore,

$$1 = \frac{6 + x}{2} \quad \text{and} \quad 2 = \frac{1 + y}{2}$$

$$2 = 6 + x \qquad 4 = 1 + y$$

$$-4 = x \qquad 3 = y$$

The coordinates of the image of $R(6, 1)$ are **(−4, 3)**.

TRANSLATIONS. A **translation** may be thought of as a "slide" of a figure in a plane in which the image of each point of the figure is moved a fixed distance in the horizontal (x) direction and another fixed distance in the vertical (y) direction. The image of a figure that is translated is congruent to the original figure.

RULE FOR TRANSLATIONS

The image of a point $P(x, y)$ that is translated h units in the horizontal direction and k units in the vertical direction is point $P'(x+h, y+k)$. The accompanying diagram illustrates a translation in which both h and k are positive numbers.

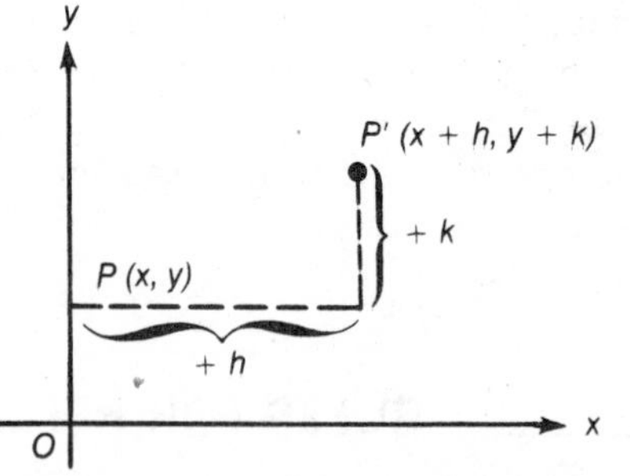

- If $h > 0$, then the translation shifts the figure horizontally to the right; if $h < 0$, the figure is shifted horizontally to the left.
- If $k > 0$, then the translation shifts the figure vertically up; if $k < 0$, the figure is shifted vertically down.

Example

3. The coordinates of the vertices of $\triangle ABC$ are $A(2, -3)$, $B(0, 4)$, and $C(-1, 5)$. If the image of point A under a translation is point $A'(0, 0)$, find the coordinates of the images of points B and C under this translation.

Solution: In general, after a translation of h units in the horizontal direction and k units in the vertical direction, the image of $P(x, y)$ is $P'(x+h, y+k)$. Since

$$A(2, -3) \rightarrow A'(2+h, -3+k) = A'(0, 0),$$

it follows that

$$\begin{aligned} 2+h=0 &\quad \text{and} \quad h=-2, \\ -3+k=0 &\quad \text{and} \quad k=3. \end{aligned}$$

Therefore,

$$B(0, 4) \rightarrow B'(0+[-2], 4+3) = B'(-2, 7),$$

$$C(-1, 5) \rightarrow C'(-1+[-2], 5+3) = C'(-3, 8).$$

The coordinates of point B' are $\mathbf{(-2, 7)}$, and the coordinates of point C' are $\mathbf{(-3, 8)}$.

ROTATIONS. A **rotation** is a transformation that may be thought of as a turn of a figure a given number of degrees about some fixed point in a plane. The point about which the figure is turned is called the *center of rotation* and is its own image. The image of a figure that is rotated is congruent to the original figure, although the direction it faces with respect to a fixed point of reference may be different.

Example

4. The coordinates of the vertices of rectangle $ABCD$ are $A(0, 0)$, $B(6, 0)$, $C(6, 3)$, and $D(0, 3)$. Find the coordinates of the vertices of the image of the rectangle after it is rotated 90° about point A in the counterclockwise direction.

Solution: Each vertex of the rectangle, except A, is rotated 90° with respect to A in such a way that the lengths of the sides of the rectangle are preserved. As the accompanying diagram illustrates, under a 90° counterclockwise rotation with respect to A,

$A(0, 0) \to \mathbf{A'(0, 0)}$, $C(6, 3) \to \mathbf{C'(-3, 6)}$,
$B(6, 0) \to \mathbf{B'(0, 6)}$, $D(0, 3) \to \mathbf{D'(-3, 0)}$.

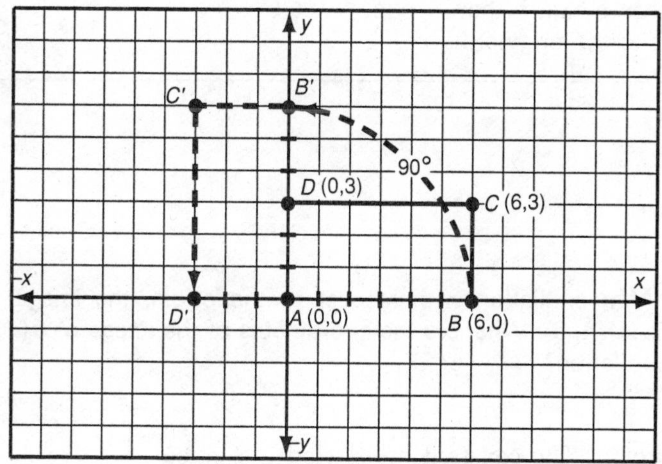

DILATIONS. Reflections, translations, and rotations produce figures that are *congruent* to the original figures, since under these transformations the lengths of the sides and the measures of the angles of the figures remain the same. A **dilation** is a transformation that produces a figure that is *similar* to the original figure. After a dilation the original figure may be enlarged or reduced with respect to a given point of reference, called the *center of dilation*. The numerical factor by which the original figure is enlarged or reduced is called the *constant of dilation*. For example, if two triangles are similar and their ratio of similitude is 2, then the larger triangle may be thought of as a dilation of the smaller triangle with a constant of dilation equal to 2.

RULE FOR DILATIONS

The image of $P(x, y)$ under a dilation with respect to the origin is $P'(cx, cy)$, where c is the constant of dilation. The accompanying figure illustrates a dilation in which $c > 1$, so that $OP' > OP$.

- If $c > 1$, the dilation produces a figure that is larger than the original figure.
- If $c < 1$, the dilation produces a figure that is smaller than the original figure.
- If $c = 1$, the dilation produces a figure that is congruent to the original figure.

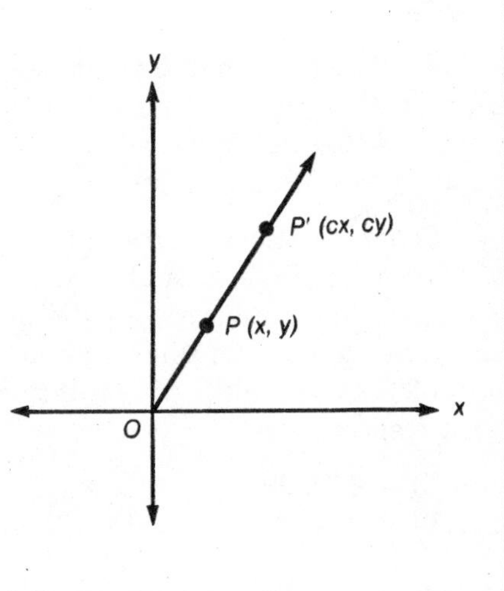

Example

5. After a dilation with respect to the origin, the image of point $A(2, 3)$ is $A'(4, 6)$. What are the coordinates of the image of $B(1, 5)$ after the same dilation?

Solution: The constant of dilation is 2 since

$$A(2, 3) \rightarrow A'(2 \cdot 2, 2 \cdot 3) = A'(4, 6).$$

Therefore, after the same dilation the x- and y-coordinates of point B are each multiplied by 2:

$$B(1, 5) \rightarrow B'(2 \cdot 1, 2 \cdot 5) = B'(2, 10).$$

The coordinates of point B' are **(2, 10)**.

EXERCISE SET 9.6

1. Insert the missing items in the following table:

	Point A	Image of Reflection of Point A in the:				
		x-axis	y-axis	line $y=x$	origin	point (1, 3)
(a)	$A(2, 5)$	?	?	?	?	?
(b)	$A(-3, -1)$	?	?	?	?	?
(c)	$A(?, ?)$	?	?	?	$(3, -2)$	?
(d)	$A(?, ?)$	?	?	$(4, 3)$	?	?
(e)	$A(?, ?)$	$(-1, -2)$	?	?	?	?
(f)	$A(?, ?)$	?	$(-2, 6)$	?	?	?
(g)	$A(?, ?)$	?	?	?	?	$(-1, 5)$

2. Point $(-4, -1)$ is reflected in point P. If the coordinates of its image are $(2, -3)$, what are the coordinates of point P?

3. The image of $(-2, 1)$ under a translation is $(1, 0)$. What are the coordinates of the image of $(2, 3)$ under the same translation?

4. Using the origin as the center of rotation, find the coordinates of the image of $(5, 0)$ after a counterclockwise rotation of:
(a) $90°$ (b) $180°$ (c) $270°$ (d) $360°$

5. Which of the following polygons coincides with its image after a counterclockwise rotation of $72°$ with respect to a given vertex of the polygon?
(1) Rhombus
(2) Square
(3) Regular pentagon
(4) Equilateral triangle

6. Which of the following transformations does *not* produce an image that is congruent to the original figure?
(1) Reflection (2) Dilation (3) Translation (4) Rotation

7. A point P is reflected in the origin, then in the x-axis, and then in the y-axis. This series of transformations is equivalent to a rotation of point P about the origin of:
(1) $90°$ (2) $180°$ (3) $270°$ (4) $360°$

8. Fill in the missing items in the following table:

	Point P	Translation Rule	Image of P
(a)	$(3, -5)$	$(x, y) \to (x-1, y+5)$	?
(b)	$(1, 2)$	$(x, y) \to ?$	$(4, 0)$
(c)	?	$(x, y) \to (x+2, y+2)$	$(-1, 3)$

9. Fill in the missing items in the following table:

	Point P	Dilation Rule	Image of P
(a)	$(2, -1)$	$(x, y) \rightarrow (3x, 3y)$	?
(b)	$(-8, 4)$	$(x, y) \rightarrow ?$	$(-4, 2)$
(c)	?	$(x, y) \rightarrow (2x, 2y)$	$(-3, 10)$

10. After a dilation, $(4, -2)$ is the image of $(2, -1)$. What are the coordinates of the image of $(-6, 0)$ after the same dilation?

11. Under a translation the image of the origin is point $(5, -3)$. What is the image of point $(3, 2)$ under the same translation?

12. The image of $A(2, 2)$ under a reflection in point P is $A'(-3, 5)$. What are the coordinates of the image of $B(4, -2)$ under a reflection in point P?

13. Verify that (b, a) is the image of (a, b) under a reflection in the line $y = x$ by showing that the line $y = x$ is the perpendicular bisector of the line segment determined by (a, b) and (b, a).

14. Given: hexagon $ABCDEF$ with coordinates $A(2, 2)$, $B(4, 0)$, $C(2, -2)$, $D(-2, -2)$, $E(-4, 0)$, $F(-2, 2)$, write the coordinates of the image of the given point after the transformation described.
    (a) The image of point A after a reflection in the line $y = -x$
    (b) The image of point F after a reflection in the y-axis
    (c) The image of point B after a reflection in the origin
    (d) The image of point C rotated 90° about the origin in the clockwise direction.
    (e) The image of point D after the transformation $(x, y) \rightarrow (x + 6, y + 2)$

15. (a) On graph paper, draw and label $\triangle ABC$ whose coordinates are $A(2, 1)$, $B(6, 4)$, $C(8, 1)$.
    (b) Graph and state the coordinates of $\triangle A'B'C'$, the reflection of $\triangle ABC$ in the x-axis.
    (c) Graph and state the coordinates of $\triangle A''B''C''$, the image of $\triangle A'B'C'$ after applying the translation rule $(x, y) \rightarrow (x - 6, y - 2)$.
    (d) Using the origin as the center of rotation, graph and state the coordinates of $\triangle A'''B'''C'''$, the result of rotating $\triangle A''B''C''$ 90° clockwise.

16. Given: $A(8, 5)$ and $B(6, 1)$ and transformations T, R, and S as described below:

$$T: (x, y) \rightarrow (x + 1, y - 5)$$
$$R: (x, y) \rightarrow (y, x)$$
$$S: (x, y) \rightarrow (-x, y)$$

(a) Graph $\overline{AB}$ and its image $\overline{A'B'}$ after transformation T.
(b) Graph $\overline{A''B''}$, the image of $\overline{AB}$ after transformation R.
(c) Graph $\overline{A'''B'''}$, the image of $\overline{AB}$ after transformation S.
(d) Compare the slopes of the pairs of segments listed below, and indicate whether these slopes are *equal*, *reciprocals*, *additive inverses*, or *negative reciprocals*.
(1) $\overline{AB}$ and $\overline{A'B'}$ (2) $\overline{AB}$ and $\overline{A''B''}$ (3) $\overline{AB}$ and $\overline{A'''B'''}$

17. Develop a general rule that gives the coordinates of the image of $P(x, y)$ after it has been reflected in the line:
(a) $y = -x$ (b) $x = 2$ (c) $y = 1$ (d) $x = a$ (e) $y = b$

9.7 ANALYTIC PROOFS USING GENERAL COORDINATES

KEY IDEAS

In an *analytic proof* a theorem about a figure is proved by placing a representative figure in the coordinate plane with variables and zeros used as general coordinates instead of specific nonzero numbers. The slope, midpoint, or distance formula is then applied to the general coordinates to demonstrate that the figure has the desired property.

POSITIONING FIGURES USING GENERAL COORDINATES. Since a figure may be reflected, rotated, or translated without changing its size or shape, there are infinitely many ways in which a given type of polygon can be placed in the coordinate plane. Here are two general guidelines for positioning a figure in the coordinate plane so that the calculations needed to prove a theorem about the figure are simplified.

1. Position the polygon so that the coordinates of its vertices can be represented by using a minimum number of different letters and a maximum number of zeros. Typically, this means that the polygon should be placed in the coordinate plane in such a way that the origin is one of the vertices of the polygon, and at least one of the sides of the polygon coincides with a coordinate axis.

2. Consider the defining characteristics of the given polygon when positioning it. For example, a right triangle is usually positioned so that its two legs coincide with the coordinate axes, thus forming a right angle at the origin. In Figure 9.12 two adjacent sides of a square are positioned so that they coincide with the coordinate axes, while opposite sides are drawn as horizontal and vertical segments with the coordinates of the vertices chosen so that

$AB = BC = CD = DA = a$ units.

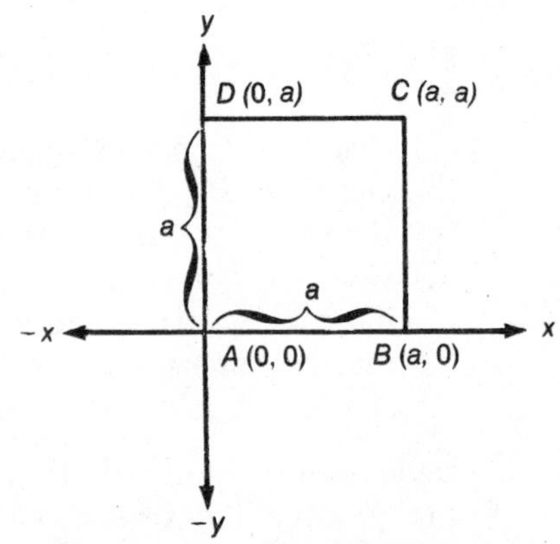

Figure 9.12 Positioning a Square

Examples

1. Quadrilateral $QRST$ has vertices $Q(a, b)$, $R(0, 0)$, $S(c, 0)$, and $T(a + c, b)$. Prove that quadrilateral $QRST$ is a parallelogram.

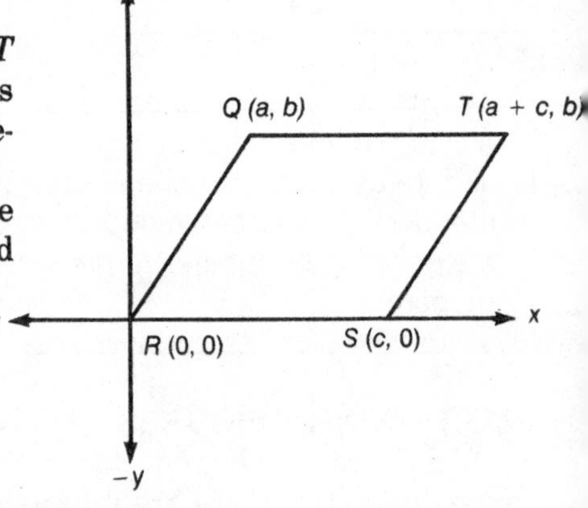

Solution: Quadrilateral $QRST$ is a parallelogram if its diagonals have the same midpoint and, therefore, bisect each other.

Step 1. Find the midpoint of the segment joining $R(0, 0)$ and $T(a + c, b)$.

$$\bar{x} = \frac{0 + (a + c)}{2} = \frac{a + c}{2},$$

$$\bar{y} = \frac{0 + b}{2} = \frac{b}{2}$$

The midpoint of diagonal $\overline{RT}$ is $\left(\frac{a + c}{2}, \frac{b}{2}\right)$.

Step 2. Find the midpoint of the segment joining $Q(a, b)$ and $S(c, 0)$.

$$\bar{x} = \frac{a + c}{2}, \qquad \bar{y} = \frac{b + 0}{2} = \frac{b}{2}.$$

The midpoint of diagonal $\overline{QS}$ is $\left(\frac{a + c}{2}, \frac{b}{2}\right)$.

Conclusion: Since diagonals $\overline{RT}$ and $\overline{QS}$ have the same midpoint, they bisect each other and quadrilateral $QRST$ is a parallelogram.

2. Using methods of coordinate geometry, prove that the diagonals of a rectangle are congruent.

Solution: In the accompanying diagram, the rectangle is positioned so that one of the vertices coincides with the origin and a pair of adjacent sides coincide with the coordinate axes. To show that $\overline{AC} \cong \overline{BD}$, find the lengths of $\overline{AC}$ and $\overline{BD}$ by applying the distance formula, $\sqrt{(x_2 - x_1)^2 + (y_2 - y_1)^2}$:

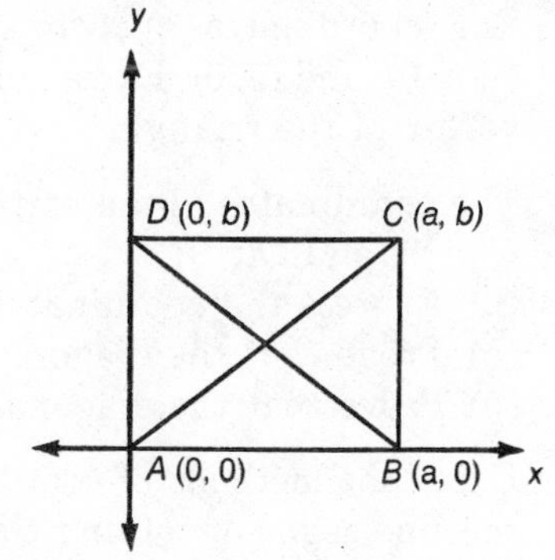

$$AC = \sqrt{(a-0)^2 + (b-0)^2} = \sqrt{a^2 + b^2}$$

$$BD = \sqrt{(0-a)^2 + (b-0)^2} = \sqrt{a^2 + b^2}$$

Conclusion: Since the lengths of $\overline{AC}$ and $\overline{BD}$ are both equal to $\sqrt{a^2 + b^2}$, $\overline{AC} \cong \overline{BD}$.

3. Prove that the line segment joining the midpoints of two sides of a triangle is parallel to the third side of the triangle.

Solution: In the accompanying diagram, the choice of coordinates of $2a$, $2b$, and $2c$ simplifies the calculations since it eliminates fractions for the coordinates of the vertices of the midpoints of sides $\overline{AB}$ and $\overline{BC}$:

$$L(x, y) = L\left(\frac{0 + 2a}{2}, \frac{0 + 2b}{2}\right) = L\left(\frac{2a}{2}, \frac{2b}{2}\right) = L(a, b)$$

$$M(x, y) = M\left(\frac{2a + 2c}{2}, \frac{2b + 0}{2}\right) = M\left(\frac{2(a + c)}{2}, \frac{2b}{2}\right) = M(a + c, b)$$

$$\text{Slope of } \overline{LM} = \frac{y_2 - y_1}{x_2 - x_1} = \frac{b - b}{a + c - c} = \frac{0}{a} = 0$$

$$\text{Slope of } \overline{AC} = \frac{y_2 - y_1}{x_2 - x_1} = \frac{0 - 0}{2c - 0} = \frac{0}{2c} = 0$$

Conclusion: Since $\overline{LM}$ and $\overline{AC}$ have the same slope, they are parallel.

EXERCISE SET 9.7

1. The coordinates of two vertices of an equilateral triangle are $(-a, 0)$, and $(a, 0)$. Express in terms of a the coordinates of the third vertex of the triangle.

2. The coordinates of the vertices of square $ABCD$ are $A(0, 0)$, $B(0, t)$, $C(x, y)$, and $D(t, 0)$.
   (a) Express the coordinates of C in terms of t.
   (b) Prove that the diagonals of square $ABCD$ are congruent.
   (c) Prove that the diagonals of square $ABCD$ are perpendicular.

3. Using the methods of coordinate geometry, prove that the length of the line segment joining the midpoints of two sides of a triangle is one-half the length of the third side of the triangle.

4. The coordinates of the vertices of right triangle ABC are $A(0, 2a)$, $B(2b, 0)$, and $C(0, 0)$. Using the methods of coordinate geometry, prove that:
   (a) the length of the median drawn to the hypotenuse of the triangle is one-half the length of the hypotenuse
   (b) the midpoint of the hypotenuse is equidistant from the three vertices of the right triangle

5. The coordinates of the vertices of quadrilateral $ABCD$ are $A(0, 0)$, $B(a, b)$, $C(c, b)$, and $D(a + c, 0)$. Prove that quadrilateral $ABCD$ is an isosceles trapezoid.

6. The coordinates of the vertices of quadrilateral $MATH$ are $M(0, 0)$, $A(r, t)$, $T(s, t)$, and $H(s - r, 0)$. Prove that:
   (a) quadrilateral $ABCD$ is a parallelogram
   (b) the diagonals of quadrilateral $ABCD$ are not necessarily perpendicular
   (c) the diagonals of quadrilateral $ABCD$ are not necessarily congruent

7. The coordinates of the vertices of $\square ABCD$ are $A(0, 0)$, $B(b, y)$, $C(a + b, y)$, and $D(a, 0)$.
   (a) If $ABCD$ is a rhombus, express y in terms of a and b.
   (b) Prove that the diagonals of rhombus $ABCD$ are perpendicular.

CHAPTER 9 REVIEW EXERCISES

1. The coordinates of the vertices of right triangle ABC are $A(6, 0)$, $B(0, 8)$, and $C(0, 0)$. Find the coordinates of the midpoint of the hypotenuse.

2. Find the slope of the line that contains points $(2, 1)$ and $(4, 1)$.

3. The vertices $\triangle ABC$ are $A(-4, 0)$, $B(2, 4)$, and $C(4, 0)$. What is the area of the triangle?
(1) 8 (2) 16 (3) 32 (4) 64

4. In $\square ABCD$, the coordinates of A are (2, 3) and the coordinates of B are (4, 8). The slope of $\overline{CD}$ is:
(1) $\frac{2}{5}$ (2) $-\frac{2}{5}$ (3) $\frac{5}{2}$ (4) $-\frac{5}{2}$

5. The slope of the line $y + 3x = 6$ is:
(1) -6 (2) -3 (3) 3 (4) 6

6. An equation of the line that passes through point (0, 2) and has a slope of 4 is:
(1) $y = 4x + 2$ (2) $x = 4y + 2$ (3) $x = 2y + 4$ (4) $y = 2x + 4$

7. Line segment $\overline{AB}$ has midpoint M. If the coordinates of A are $(-3, 2)$ and the coordinates of M are $(-1, 5)$, what are the coordinates of B?
(1) (1, 10) (2) (1, 8) (3) (0, 7) (4) $(-5, 8)$

8. The length of the line segment connecting the points whose coordinates are $(3, -1)$ and (6, 5) is:
(1) $\sqrt{45}$ (2) 5 (3) 3 (4) $\sqrt{97}$

9. The coordinates of the endpoints of the base of an isosceles triangle are (2, 1) and (8, 1). The coordinates of the vertex of this triangle may be:
(1) (1, 5) (2) (2, 5) (3) $(2, -6)$ (4) $(5, -6)$

10. The point whose coordinates are $(4, -2)$ lies on a line whose slope is $\frac{3}{2}$. The coordinates of another point on this line may be:
(1) (1, 0) (2) (2, 1) (3) (6, 1) (4) (7, 0)

11. Which pair of points will determine a line parallel to the y-axis?
(1) (1, 1) and (2, 3)
(2) (1, 1) and (3, 3)
(3) (2, 3) and (2, 5)
(4) (2, 5) and (4, 5)

12. Which line is parallel to the line $y = 2x + 4$?
(1) $y = 2x + 6$ (2) $y = 4 - 2x$ (3) $y = 4x - 2$ (4) $2y = x - 2$

13. The coordinates of the vertices of rectangle $ABCD$ are $A(2, 2)$, $B(2, 6)$, $C(8, 6)$ and $D(8, 2)$. The area of rectangle $ABCD$ is:
(1) 16 (2) 24 (3) 36 (4) 48

14. Which pair of points will determine a line parallel to the x-axis?
(1) (1, 3) and $(-2, 3)$
(2) $(1, -1)$ and $(-1, 1)$
(3) (1, 3) and $(1, -1)$
(4) (1, 1) and $(-3, -3)$

15. Triangle PQR has vertices $P(-1, -1)$, $Q(1, -2)$, and $R(3, 2)$.
(a) Using graph paper, draw $\triangle PQR$.
(b) Show that $\triangle PQR$ is a right triangle, and state a reason for the conclusion.
(c) Find the area of $\triangle PQR$.

16. Given: $\triangle ABC$ with vertices $A(2, 1)$, $B(10, 7)$, and $C(4, 10)$.
(a) Find the area of $\triangle ABC$.
(b) Find the length of side $\overline{AB}$.
(c) Using the answers from parts (a) and (b), find the length of the altitude drawn from C to $\overline{AB}$.

17. In $\triangle ABC$, the coordinates of the vertices are $A(-6, -8)$, $B(6, 4)$, and $C(-6, 10)$.
(a) Write an equation of the altitude of $\triangle ABC$ from C to $\overline{AB}$.
(b) Write an equation of the altitude of $\triangle ABC$ from B to $\overline{AC}$.
(c) Find the x-coordinate of the point of intersection of the two altitudes in parts (a) and (b).

18. The vertices of $\triangle ABC$ are $A(4, 4)$, $B(12, 10)$, and $C(6, 13)$.
(a) Show that $\triangle ABC$ is *not* equilateral.
(b) Find the area of $\triangle ABC$.

19. The coordinates of the vertices of quadrilateral $ABCD$ are $A(-4, 0)$, $B(6, 0)$, $C(8, 5)$, and $D(-2, 5)$.
(a) Show by means of coordinate geometry that quadrilateral $ABCD$ is a parallelogram and state a reason for your conclusion.
(b) Find the length of the altitude from D to $\overline{AB}$.
(c) Find the area of $ABCD$.

20. The coordinates of the vertices of rhombus $ABCD$ are $A(-3, 1)$, $B(2, 6)$, $C(x, y)$, and $D(4, 0)$.
(a) Find the numerical coordinates of point C.
(b) Verify, by means of coordinate geometry, that $AB = AD$.
(c) Verify, by means of coordinate geometry, that diagonals $\overline{AC}$ and $\overline{BD}$ are perpendicular to each other.

21. The vertices of $\square STWU$ are $S(1, 1)$, $T(-2, 3)$, $W(0, b)$, and $U(3, -5)$.
(a) Find the slope of $\overline{ST}$.
(b) Express the slope of $\overline{UW}$ in terms of b.
(c) Find the value of b.
(d) Write an equation of the line passing through point S and perpendicular to $\overline{ST}$.

22. Given: points $A(1, -1)$, $B(5, 7)$, $C(0, 4)$, and $D(3, k)$.
(a) Find the slope of $\overleftrightarrow{AB}$.
(b) Express the slope of $\overleftrightarrow{CD}$ in terms of k.
(c) If $\overleftrightarrow{AB} \parallel \overleftrightarrow{CD}$, find k.
(d) Write an equation of $\overleftrightarrow{CD}$.

23. The vertices of $\triangle PQR$ are $P(1, 2)$, $Q(-3, 6)$, and $R(4, 8)$.
(a) Find the coordinates of S, the midpoint of $\overline{PQ}$.

(b) Express in radical form the length of median $\overline{RS}$.
(c) Find the slope of $\overline{PR}$.
(d) A line through point Q is parallel to $\overline{PR}$. If this line passes through point $(x, 14)$, find the value of x.

24. The vertices of an isosceles trapezoid with bases $\overline{BC}$ and $\overline{AD}$ are $A(0, 0)$, $B(b, c)$, $C(h, k)$, and $D(a, 0)$.
(a) Express h and k in terms of a, b, and/or c.
(b) Use the methods of coordinate geometry to prove that the diagonals of an isosceles trapezoid are congruent.

25. Given: points $A(3, 0)$ and $B(-4, 6)$, write the coordinates of the images of points A and B after each transformation described.
(a) The images of points A and B after a reflection in the line $y = x$
(b) The image of point A after a 90° counterclockwise rotation about the origin
(c) The images of points A and B after a reflection in the origin
(d) The images of points A and B after a dilation with respect to the origin such that $(x, y) \rightarrow (\frac{1}{2}x, \frac{1}{2}y)$

26. For each transformation listed in (a) through (e), select the description, *chosen from the list below*, that best fits the transformation.

Descriptions

(1) reflects all the points in the origin
(2) reflects all the points in the x-axis
(3) reflects all the points in the y-axis
(4) reflects all the points in the line $y = x$
(5) reflects all the points in the line $x = 1$
(6) translates all the points 3 units to the right
(7) translates all the points 3 units to the left
(8) rotates all the points 90° counterclockwise about the origin

Transformations

(a) $(x, y) \rightarrow (x + 3, y)$
(b) $(x, y) \rightarrow (x, -y)$
(c) $(x, y) \rightarrow (2 - x, y)$
(d) $(x, y) \rightarrow (-x, -y)$
(e) $(x, y) \rightarrow (y, x)$

27. Given $A(3, 3)$, answer the following questions:
(a) In a reflection in point P, the image of A is $(2, 1)$. What are the coordinates of the image of $(4, -3)$ in a reflection in point P?
(b) In a dilation with respect to the origin, the image of A is $(2, 2)$. What are the coordinates of the image of $(6, 9)$ under the same dilation?
(c) Under a translation, the image of A is $(0, 2)$. What are the coordinates of the image of $(5, -1)$ under the same translation?
(d) What are the coordinates of the image of A after a 90° counterclockwise rotation about the origin?

CHAPTER 10

Locus and Constructions

10.1 SIMPLE LOCUS

KEY IDEAS

A **locus** (plural: loci) may be thought of as a path consisting of the set of all points, and only those points, that satisfy a given set of conditions.

FINDING A SIMPLE LOCUS. Finding a *simple locus* refers to the process of identifying and then describing the set of all points that satisfy a single condition.

Example: Find the locus of points that are 3 inches from point *K*.

Step *1.* Identify the condition that the points must satisfy. *Condition*: All points must be 3 inches from point *K*.

Step 2. Draw a diagram. Draw point *K* and enough representative points that satisfy the stated condition so that you are able to discover a pattern. Connect these points with a broken line or a broken curve.

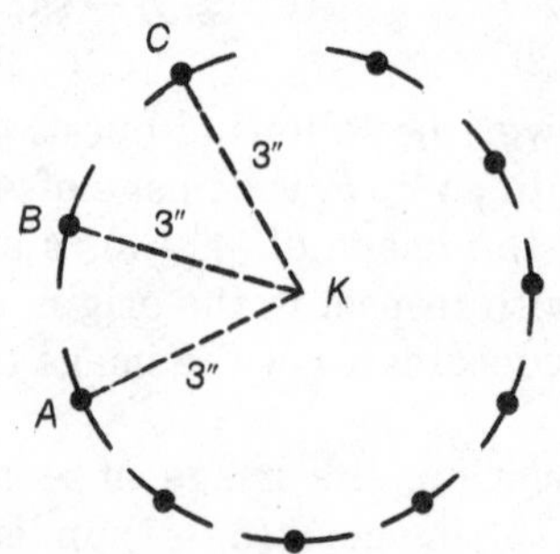

Step 3. Write a sentence that describes the locus. *Sentence*: The locus of points that are 3 inches from point K is a circle having point K as its center and a radius of 3 inches.

COMMONLY ENCOUNTERED SIMPLE LOCI. Tables 10.1, 10.2, and 10.3 describe the loci for various given conditions. In each case, the accompanying diagram shows the locus as a broken line or curve.

TABLE 10.1 Simple Loci Involving Circles

Locus	Diagram
1. **Condition:** All points d units from a fixed point P. **Locus:** A circle having P as its center and a radius of d units.	 d P
2. **Condition:** All points d units from a circle having a radius of r units. **Locus:** Two concentric circles having the same center as the original circle, the smaller circle having a radius of $r - d$ units and the larger circle having a radius of $r + d$ units.	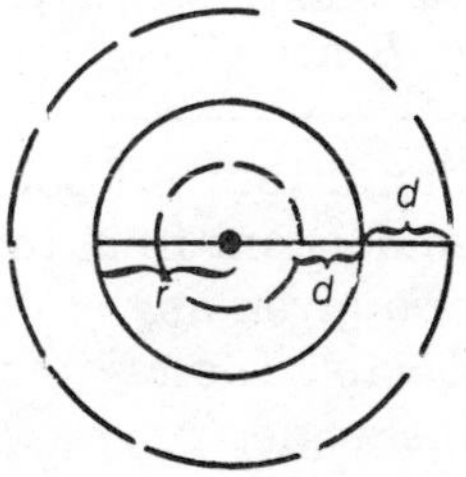 d r d
3. **Condition:** All points equidistant from two concentric circles having radii of p and q units. **Locus:** A circle having the same center as the given circles and a radius of $\frac{p+q}{2}$ units.	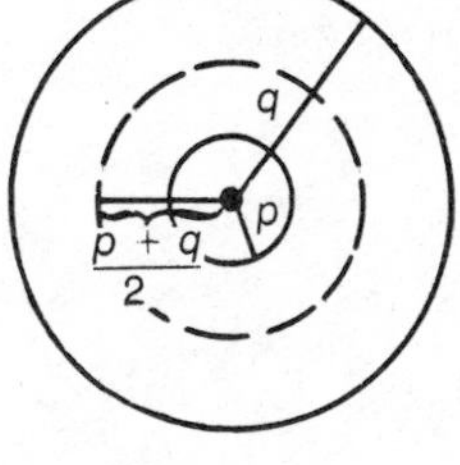 q p $\frac{p+q}{2}$

TABLE 10.2 Simple Loci Involving Given Points and Angles

Locus	Diagram
1. **Condition:** All points equidistant from two given points, A and B. **Locus:** The perpendicular bisector of the line segment determined by points A and B.	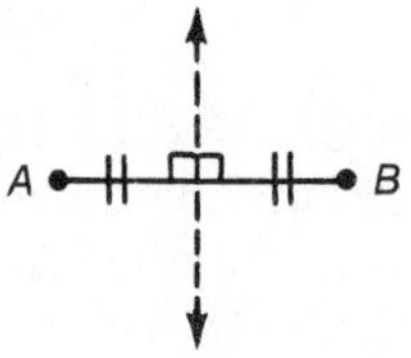 A B
2. **Condition:** All points equidistant from three given noncollinear points, A, B, and C. **Locus:** The points at which the perpendicular bisectors of $\overline{AB}$ and $\overline{BC}$ intersect. (*Note*: This point represents the center of the circle that can be circumscribed about the triangle whose vertices are A, B, and C.)	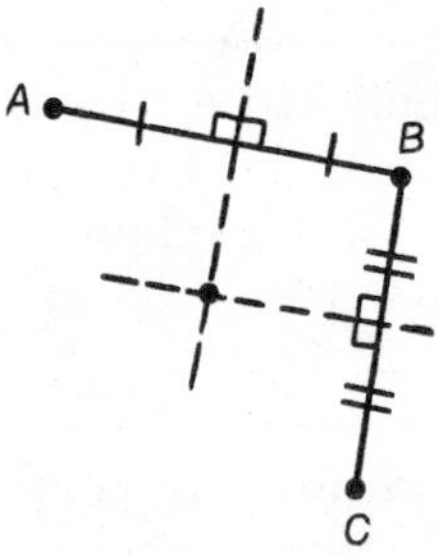 A B C
3. **Condition:** All points equidistant from the sides of a given angle. **Locus:** The ray that bisects the angle.	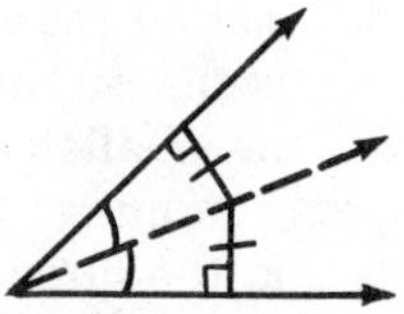

TABLE 10.3 Simple Loci Involving Given Lines

Locus	Diagram
1. **Condition:** All points d units from a given line l. **Locus:** Two lines each parallel to line l, one on each side of line l and at a distance of d units from line l.	
2. **Condition:** All points equidistant from two parallel lines. **Locus:** A line that is parallel to the given pair of lines and midway between them.	
3. **Condition:** All points equidistant from two intersecting lines. **Locus:** Two lines, each of which bisects a pair of vertical angles formed by the two intersecting lines.	

THE PARABOLA AS A LOCUS. A **parabola** is a smooth curve that arises from finding the locus of points equidistant from a fixed point, called the **focus**, and a given line, called the **directrix**. In Figure 10.1, point F is the focus and line l is the directrix. The curve that connects the set of all points equidistant from point F and line l is a parabola. Equations whose graphs are parabolas are discussed in Section 11.2.

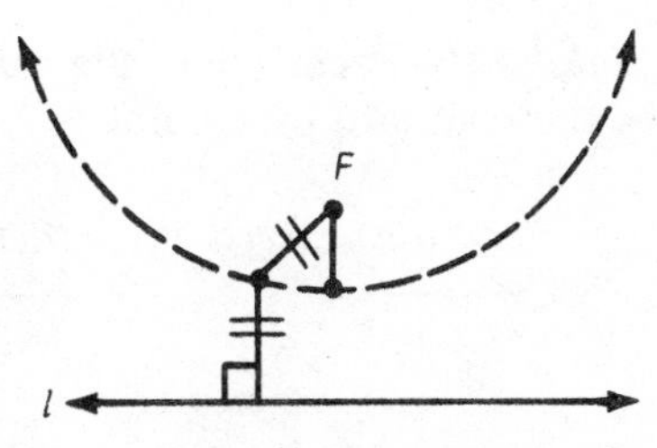

Figure 10.1 Parabola, with Focus (F) and Directrix (l)

Example

1. Find the locus of points:

(a) equidistant from two concentric circles having radii of lengths 3 and 7 centimeters, respectively;

(b) equidistant from two parallel lines that are 8 inches apart.

Solutions: (a) The locus of points equidistant from two concentric circles is another circle having the same center as the original circles and a radius length equal to 5 centimeters, which is the average of the lengths of the radii of the two given circles $\left(\frac{3+7}{2}=5\right)$.

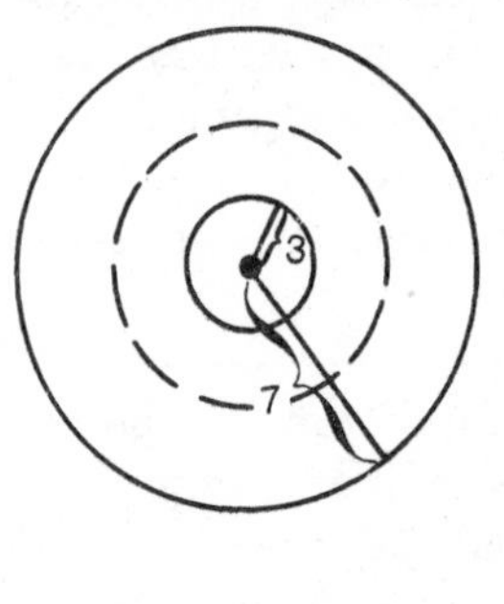

(b) The locus of points equidistant from two parallel lines that are 8 inches apart is a line that is parallel to the given pair of lines and is at a distance of 4 inches from each of them.

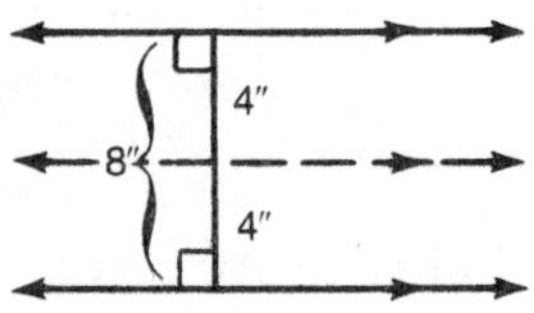

LOCUS AND COORDINATES. As Examples 2–5 illustrate, coordinates may also be used to specify a locus.

Examples

2. Find an equation that describes the locus of points whose ordinates exceed twice their abscissas by 1.

Solution: Recall that the abscissa of a point is its x-coordinate, while the ordinate of a point is its y-coordinate. Translate the words

ordinates exceed twice their abscissas by 1

as

$$\underbrace{\text{ordinates}}_{y} \ \ \underbrace{\text{exceed}}_{=} \ \ \underbrace{\text{twice their abscissas}}_{2x} \ \ \underbrace{\text{by 1}}_{+1}$$

The required locus is the set of all points on the line whose equation is $\boldsymbol{y=2x+1}$.

3. Find the locus of points that are 2 units from the line whose equation is $x=3$.

Solution: The required locus consists of two lines on either side of $x = 3$ that are parallel to $x = 3$, and each at a distance of 2 units from $x = 3$. As the accompanying diagram shows, the line $x = 1$ is 2 units to the left of $x = 3$ and the line $x = 5$ is 2 units to the right of $x = 3$.

The locus of points that are 2 units from the line whose equation is $x = 3$ is the lines whose equations are $\mathbf{x = 1}$ and $\mathbf{x = 5}$.

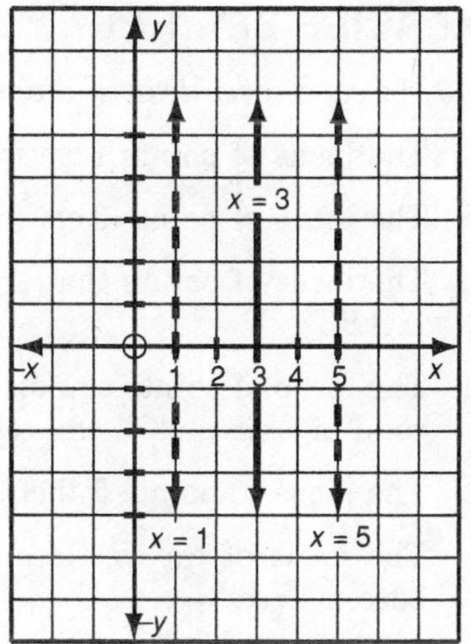

4. Find an equation of a line that describes the locus of points equidistant from the lines whose equations are $y = 3x - 1$ and $y = 3x + 5$.

Solution: The lines given by the equations $y = 3x - 1$ and $y = 3x + 5$ are parallel since the slope of each line is 3. The line parallel to these lines and midway between them must have a slope of 3 and a y-intercept of 2 since 2 is midway between (the average of) the y-intercepts of the original pair of lines. Hence, the locus of points equidistant from the lines whose equations are $y = 3x - 1$ and $y = 3x + 5$ is the line whose equation is $\mathbf{y = 3x + 2}$.

5. Find the locus of points equidistant from points $A(4, 5)$ and $B(4, -1)$.

Solution: The required locus is the perpendicular bisector of $\overline{AB}$. Since $\overline{AB}$ is a vertical line segment, the perpendicular bisector of $\overline{AB}$ is a horizontal line that contains the midpoint of $\overline{AB}$, which is $M(4, 2)$. The equation of a horizontal line that contains $M(4, 2)$ is $y = 2$. Therefore, the locus of points equidistant from points A and B is the line whose equation is $\mathbf{y = 2}$.

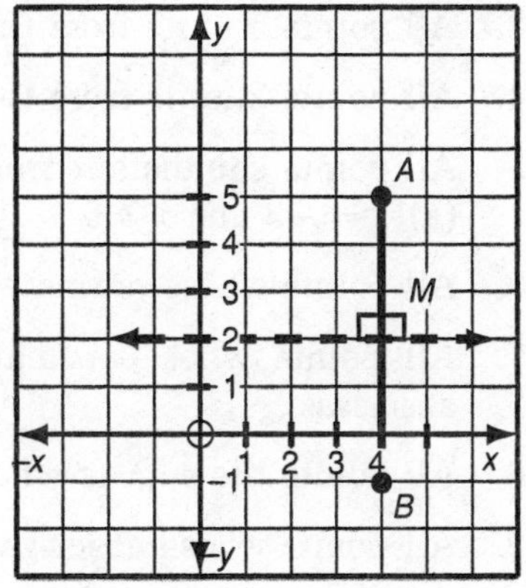

EXERCISE SET 10.1

1–10. In each case, draw a diagram and describe the locus.

1. The locus of points 1 cm from a circle whose radius is 8 cm.
2. The locus of points 3 cm from a given line.
3. The locus of points that are equidistant from sides $\overline{AB}$ and $\overline{AC}$ of $\triangle ABC$.
4. The locus of points equidistant from two concentric circles having radii of 7 cm and 11 cm, respectively.
5. The locus of points 5 inches from vertex B of $\triangle ABC$.
6. The locus of the vertices of isosceles triangles having the same base.
7. The locus of the centers of circles tangent to each of two parallel lines that are 10 inches apart.
8. The locus of the center of a circle that rolls along a flat surface in a plane.
9. The locus of points in the coordinate plane that are equidistant from the x- and y-axes.
10. The locus of points in the coordinate plane that are 7 units from the origin.

11–21. Find an equation of the line or lines that satisfies each of the following conditions:

11. All points 4 units from the x-axis
12. All points 3 units from the y-axis
13. All points 5 units from the line $x = 3$
14. All points 2 units from the line $y = -1$
15. All points equidistant from the lines:
    (a) $x = -3$ and $x = 5$ (b) $y = -7$ and $y = -1$
16. All points whose ordinates are one-half as great as their abscissas
17. All points whose ordinates are 2 more than three times their abscissas
18. All points the sum of whose ordinates and abscissas is -4
19. All points whose abscissas exceed twice their ordinates by 3
20. All points equidistant from the lines whose equations are:
    (a) $y = x$ and $y = x - 4$ (b) $y = 2x + 7$ and $y + 1 = 2x$

21. All points equidistant from points:
(a) $P(1, -4)$ and $Q(5, -4)$ (b) $R(2, 7)$ and $S(2, -1)$

22–26. Complete each sentence by using one of the following choices:
(1) one point *(3) two points* *(5) a circle*
(2) one line *(4) two lines* *(6) a parabola*

22. The locus of points equidistant from any two points is ____.

23. The locus of points equidistant from two parallel lines is ____.

24. The locus of points equidistant from two intersecting lines is ____.

25. The locus of points equidistant from a fixed point and a line is ____.

26. The locus of points at a fixed distance from a fixed point is ____.

27. Prove that any point on the perpendicular bisector of a line segment is equidistant from the endpoints of the segment.

28. Prove that any point on the bisector of an angle is equidistant from the sides of the angle.

10.2 COMPOUND LOCI

KEY IDEAS

To find the number of points that satisfy two or more conditions at the same time, use the same diagram to describe the locus for each condition and then locate the points, if any, at which the loci intersect.

COMPOUND LOCI THAT SATISFY TWO CONDITIONS. *Compound loci* are represented by the set of points that satisfy two or more conditions at the same time. To determine compound loci having two conditions, proceed as follows:

Step 1. Identify the first locus condition, and describe this locus using a diagram.

Step 2. Identify the second locus condition, and describe this locus using the same diagram.

Step 3. Determine the point or points, if any, at which the loci intersect.

Examples

1. Point P on the line l. Find the locus of points that are 2 inches from line l and also 2 inches from point P.

Solution:

Step *1. Locus condition 1*: All points 2 inches from line *l*. The desired locus is a pair of parallel lines on either side of line *l*, each line at a distance of 2 inches from line *l*.

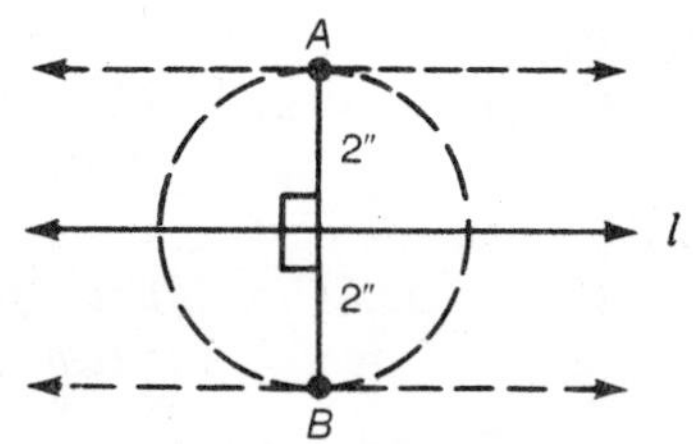

Step 2. Locus condition 2: All points 2 inches from point *P*. The desired locus is a circle that has *P* as its center and has a radius of 2 inches. See the accompanying diagram.

Step 3. The required locus is the **2 points, *A* and *B*,** that satisfy both conditions.

2. Two parallel lines are 8 inches apart. Point *P* is located on one of the lines. Find the number of points that are equidistant from the parallel lines and are also at a distance from point *P* of:

(a) 5 inches (b) 4 inches (c) 3 inches

Solutions: (a) *Step 1. Locus condition 1*: All points equidistant from the two parallel lines. The desired locus is a line parallel to the original pair of lines and midway between them.

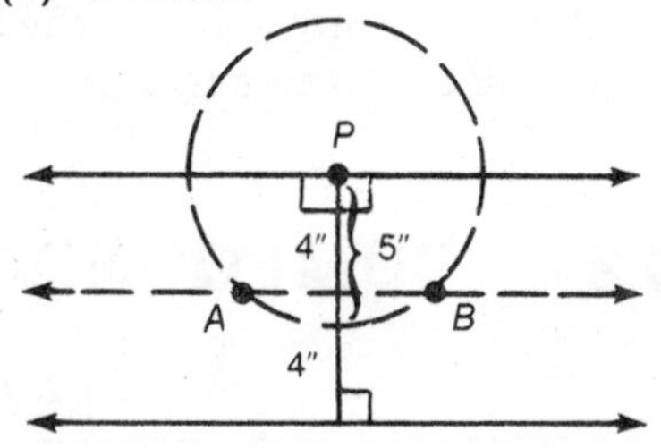

Step 2. Locus condition 2: All points 5 inches from point *P*. The desired locus is a circle that has *P* as its center and has a radius of 5 inches. See the accompanying diagram.

Step 3. Since the loci intersect at points *A* and *B*, there are **2** points that satisfy both conditions.

(b) Since the circle has a radius of 4 inches, it is tangent to the line that is midway between the original pair of parallel lines. The loci intersect at point *A*. There is **1** point that satisfies both conditions.

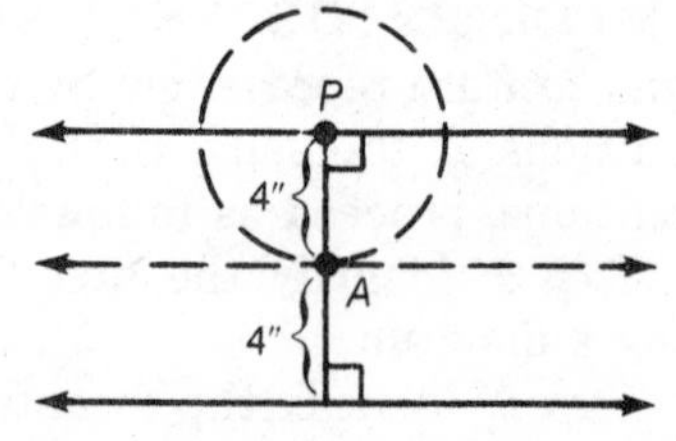

(c) Since the circle has a radius of 3 inches, it does *not* intersect the line that is midway between the original pair of parallel lines. Since the loci do *not* intersect, there is **0** point that satisfies both conditions.

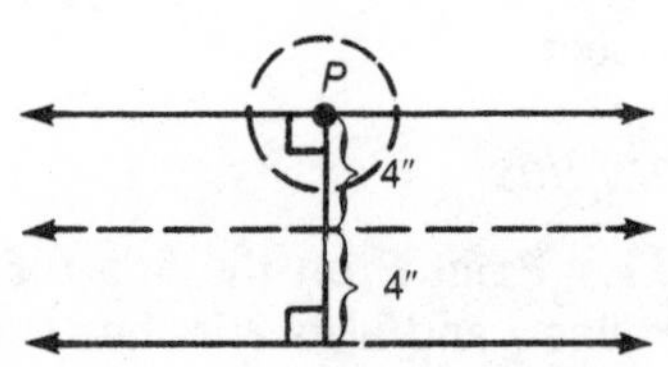

3. How many points are 3 units from the origin and 2 units from the *y*-axis?

Solution:

Step 1. Locus condition 1: All points 3 units from the origin. The desired locus is a circle with the origin as its center and a radius of 3 units. Note that the circle intersects each coordinate axis at 3 and −3.

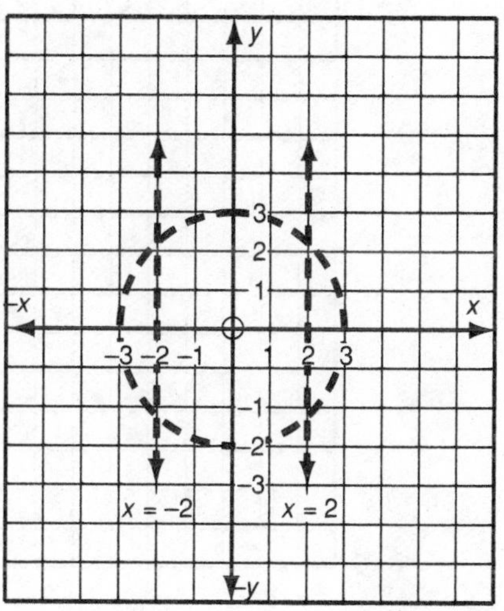

Step 2. Locus condition 2: All points 2 units from the *y*-axis. The desired locus is a pair of parallel lines; one line is 2 units to the right of the *y*-axis ($x = 2$), and the other line is 2 units to the left of the *y*-axis ($x = -2$). See the accompanying diagram.

Step 3. Since the loci intersect at points *A*, *B*, *C*, and *D*, there are **4** points that satisfy both conditions.

4. Given: points $A(2, 7)$ and $B(6, 7)$.

(a) Write an equation of $\overleftrightarrow{AB}$.

(b) Describe the locus of points equidistant from:

(1) points *A* and *B* *(2)* the *x*- and *y*-axes

(c) How many points satisfy both conditions obtained in part (b)?

Solutions: (a) Since the *y*-coordinates of points *A* and *B* are the same, points *A* and *B* determine a horizontal line, $\overleftrightarrow{AB}$ (see the accompanying diagram on page 328), whose equation is $\mathbf{y = 7}$.

(b) *(1)* The locus of points equidistant from two points is the perpendicular bisector of the line segment determined by the given points. The midpoint of $\overline{AB}$ is $M(4, 7)$. Since $\overleftrightarrow{AB}$ is a horizontal line, a vertical line that contains (4, 7) will be the perpendicular bisector of $\overline{AB}$ (see the accompanying diagram on page 328). An equation of this line is $x = 4$.

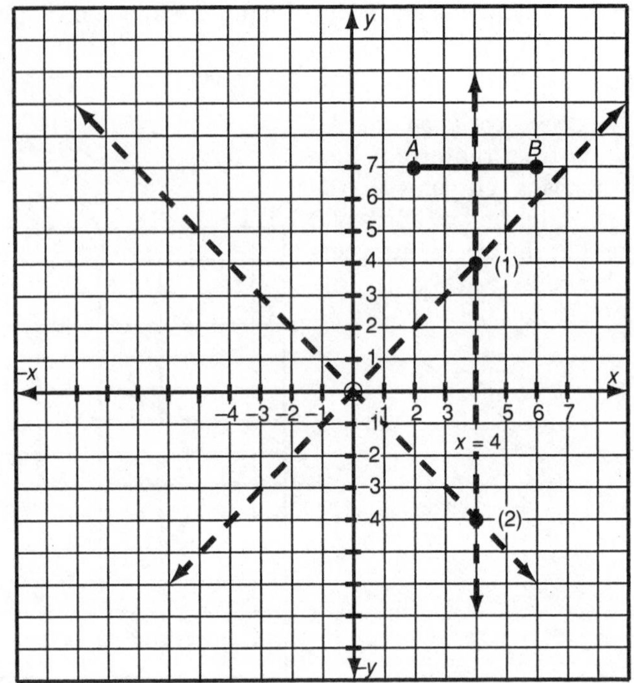

(2) The locus of points equidistant from the x- and y-axes is the pair of lines that bisect the pairs of vertical angles formed by the intersecting coordinate axes (see the accompanying diagram).

(c) There are 2 points that satisfy the two conditions in part (b).

EXERCISE SET 10.2

1. How many points are equidistant from points A and B and also 4 inches from $\overline{AB}$?

2. How many points are equidistant from two intersecting lines and also 3 inches from their points of intersection?

3. Point A is 4 inches from line k. Find the number of points that are 1 inch from line k and also at a distance from point A of:
(a) 1 inch (c) 4 inches (e) 6 inches
(b) 7 inches (d) 5 inches (f) 3 inches

4. Points J, K, and L are noncollinear. If points K and L are 8 units apart, find the number of points that are equidistant from points K and L and also:
(a) 3 units from L (c) 2 units from $\overleftrightarrow{KL}$
(b) 4 units from K (d) equidistant from the sides of $\angle JKL$

5. Point P is x inches from line l. If there are exactly 3 points that are 2 inches from line l and also 6 inches from point P, find the value of x.

6. Point X is the midpoint of $\overline{PQ}$. Find the number of points that are equidistant from points P and Q and also 5 cm from point X if:
(a) $PQ = 8$ (b) $PQ = 10$ (c) $PQ = 12$

7. Lines p and q are parallel and 10 cm apart. Point A is between lines p and q and is 2 cm from line q. Find the number of points equidistant from lines p and q and also d cm from point A if:
(a) $d = 7$ cm (b) $d = 1$ cm (c) $d = 3$ cm

8. Complete Exercise 7 assuming that point A is *not* between lines p and q.

9. Parallel lines $\overleftrightarrow{AB}$ and $\overleftrightarrow{CD}$ are 6 inches apart. Find the number of points that are equidistant from the two lines and are also:
(a) 3 inches from point A
(b) 4 inches from point C
(c) 2 inches from point D
(d) equidistant from points A and B

10. How many points are equidistant from the sides of $\angle ABC$ and are also 2 inches from point B?

11. How many points are equidistant from the three vertices of a triangle?

12. Find the number of points that are:
(a) 2 units from the x-axis and 3 units from the y-axis
(b) equidistant from points $A(0, 3)$ and $B(4, 3)$ and also 2 units from the origin
(c) 4 units from the origin and also 4 units from the x-axis
(d) 3 units from the line $y = -1$ and also 3 units from the origin
(e) equidistant from the lines $x = -3$ and $x = 1$ and also 2 units from the origin
(f) equidistant from the lines $y = 5$ and $y = -3$ and also 2 units from the origin
(g) 2 units from the line $x = 4$ and also 3 units from the origin
(h) equidistant from points $P(2, 1)$ and $Q(2, 5)$ and also 3 units from the origin

13. Point P is the center of two concentric circles having radii of 4 inches and 10 inches, respectively. If point A lies on the larger circle, find the number of points equidistant from the two circles and also:
(a) equidistant from points P and A (b) 3 units from point A

14. (a) On graph paper, draw the locus of points 2 units from the line whose equation is $y = 1$.
(b) Write the equation(s) of the locus described in part (a).
(c) Describe fully the locus of points at a distance p from point $(2, 5)$.
(d) How many points satisfy the conditions in parts (a) and (c) simultaneously if:
(1) $p = 2$ *(2)* $p = 3$ *(3)* $p = 4$

10.3 BASIC CONSTRUCTIONS

KEY IDEAS

Geometric constructions, unlike *drawings*, are performed only with a straightedge (for example, an unmarked ruler) and compass. The point at which the pivot point of the compass is placed is sometimes referred to as the **center**, while the fixed compass setting that is used is called the **radius length**; the part of the circle made by the compass is termed an **arc**.

COPYING SEGMENTS AND ANGLES. Given a line segment or angle, it is possible to construct another line segment or angle that is congruent to the original segment or angle without using a ruler or protractor.

Construction 1

Given line segment $\overline{AB}$, construct a congruent segment.

Step	Diagram
1. Draw any line, and choose any convenient point on it. Label the line as l and the point as C. 2. Using a compass, measure $\overline{AB}$ by placing the compass point on A and the pencil point on B. 3. Using the same compass setting, place the compass point on C and draw an arc that intersects line l. Label the point of intersection as D.	A B l C D

Conclusion: $\overline{AB} \cong \overline{CD}$.

Construction 2

Given $\angle ABC$, construct a congruent angle.

Step	Diagram
1. Draw any line and choose any point on it. Label the line as l and the point as S. 2. Using any convenient compass setting, place the compass point on B and draw an arc, intersecting $\overrightarrow{BC}$ at X and $\overrightarrow{BA}$ at Y. 3. Using the same compass setting, place the compass point at S and draw arc WT, intersecting line l at T. 4. Adjust the compass setting to measure the line segment determined by points X and Y by placing the compass point at X and the pencil at Y. 5. Using the same compass setting, place the compass point at T and construct an arc intersecting arc WT at point R. 6. Using a straightedge, draw $\overrightarrow{SR}$.	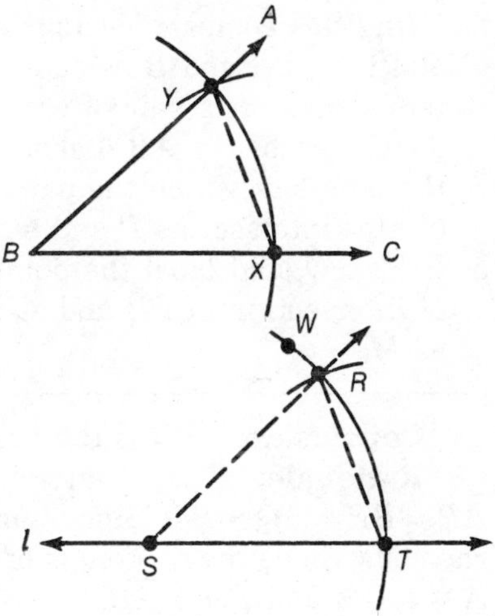 A, Y, B, X, C, W, R, l, S, T

Conclusion: $\angle ABC \cong \angle RST$.

Rationale: The arcs were constructed so that $\overline{BX} \cong \overline{ST}$, $\overline{BY} \cong \overline{SR}$, and $\overline{XY} \cong \overline{TR}$. Therefore, $\triangle XYB \cong \triangle TRS$ by the SSS postulate. By CPCTC, $\angle ABC \cong \angle RST$.

CONSTRUCTING BISECTORS OF SEGMENTS AND ANGLES. To find by construction the locus of points equidistant from two points, construct the line that is the perpendicular bisector of the segment determined by the two points. To find by construction the locus of points equidistant from the sides of an angle, construct the ray that bisects the angle.

Construction 3

Given two points, construct the perpendicular bisector of the segment determined by the two points.

Step	Diagram
1. Label points A and B, and draw $\overline{AB}$. Choose any compass setting (radius length) that is more than one-half the length of $\overline{AB}$. 2. Using this compass setting, and points A and B as centers, construct a pair of arcs above and below $\overline{AB}$. Label the points at which the pairs of arcs intersect as P and Q. 3. Draw $\overleftrightarrow{PQ}$, and label the point of intersection of $\overleftrightarrow{PQ}$ and $\overline{AB}$ as M.	P A M B Q

Conclusion: $\overleftrightarrow{PQ}$ is the perpendicular bisector of $\overline{AB}$.

Rationale: The arcs were constructed so that $AP = BP = AQ = BQ$. Since quadrilateral $APBQ$ is equilateral, it is a rhombus. Since the diagonals of a rhombus are perpendicular bisectors, $\overline{AM} \cong \overline{BM}$ and $\overline{PQ} \perp \overline{AB}$.

Construction 4

Given an angle, construct the bisector of the angle.

Step	Diagram
1. Name the angle, $\angle ABC$. Using B as a center, construct an arc, using any convenient radius length, that intersects $\overrightarrow{BA}$ at point P and $\overrightarrow{BC}$ at point Q. 2. Using points P and Q as centers and the same radius length, draw a pair of arcs that intersect. Label the point at which the arcs intersect as D. 3. Draw $\overrightarrow{BD}$.	A P D B Q C

Conclusion: $\overrightarrow{BD}$ is the bisector of $\angle ABC$.

Rationale: See Exercise 1(a).

CONSTRUCTING PERPENDICULAR LINES. A line can be constructed perpendicular to a given line at a given point on the line or through a given point not on the line.

Construction 5

Given a line l and a point P *not* on the line, construct a line through P and perpendicular to line l.

Step	Diagram
1. Using P as a center and any convenient radius length, construct an arc that intersects line l at two points. Label these points as A and B. 2. Choose a radius length greater than one-half the length of $\overline{AB}$. Using points A and B as centers, construct a pair of arcs that intersect at point Q. 3. Draw $\overleftrightarrow{PQ}$, intersecting line l at point M.	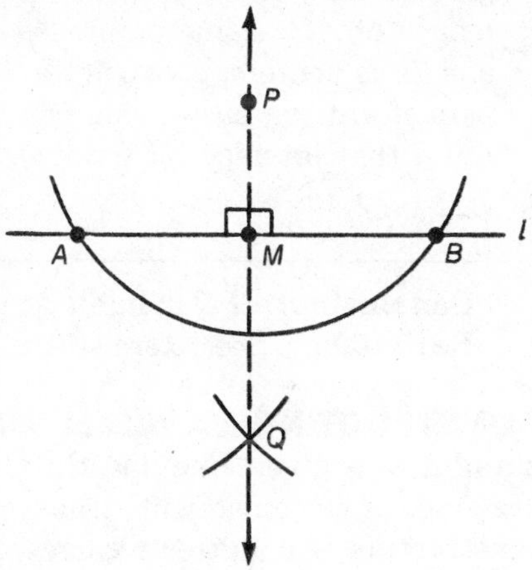

Conclusion: $\overleftrightarrow{PQ} \perp$ line l at point M.
Rationale: See Exercise 1(b).

Construction 6

Given a line l and a point P on line l, construct a line through P and perpendicular to line l.

Step	Diagram
1. Using P as a center and any convenient radius length, construct an arc that intersects line l at two points. Label these points as A and B. 2. Choose a radius length greater than one-half the length of $\overline{AB}$. Using points A and B as centers, construct a pair of arcs on either side of line l that intersect at point Q. 3. Draw $\overleftrightarrow{PQ}$.	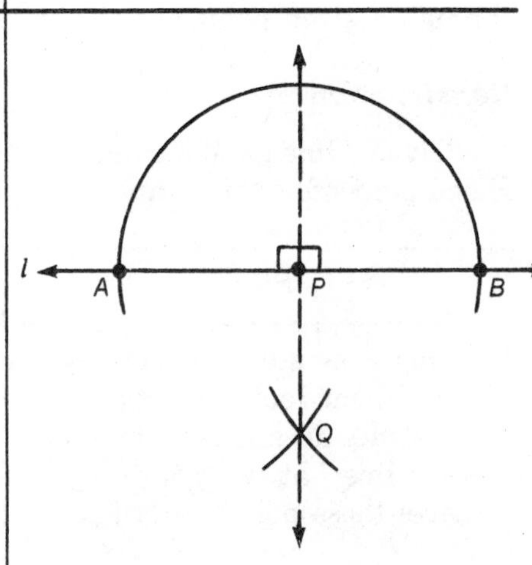

Conclusion: $\overleftrightarrow{PQ} \perp$ line l at point P.
Rationale: See Exercise 1(c).

CONSTRUCTING PARALLEL LINES. A line can be constructed parallel to a given line and through a given point not on the line by drawing any convenient transversal through the point and then constructing a congruent corresponding angle, using this point as a vertex.

Construction 7

Given a line $\overleftrightarrow{AB}$ and a point P *not* on the line, construct a line through P and parallel to $\overleftrightarrow{AB}$.

Step	Diagram
1. Through P draw any convenient line, extending it so that it intersects $\overleftrightarrow{AB}$. Label the point of intersection as Q. 2. Using P and Q as centers, draw arcs having the same radius length. Label the point at which the arc intersects the ray opposite $\overrightarrow{PQ}$ as R. 3. Construct an angle at P, one of whose sides is $\overrightarrow{PR}$, congruent to $\angle PQB$. 4. Draw $\overleftrightarrow{PS}$.	

Conclusion: Since $\angle PQB$ and $\angle RPS$ are congruent corresponding angles, $\overleftrightarrow{PS}$ is parallel to $\overleftrightarrow{AB}$.

Example

1. (a) Construct the altitude to side $\overline{RT}$ of acute triangle *RST*.
(b) Construct the median to side $\overline{ST}$ of acute triangle *RST*.
(c) Construct a line through vertex *T* of acute triangle *RST* and parallel to side $\overline{RS}$.

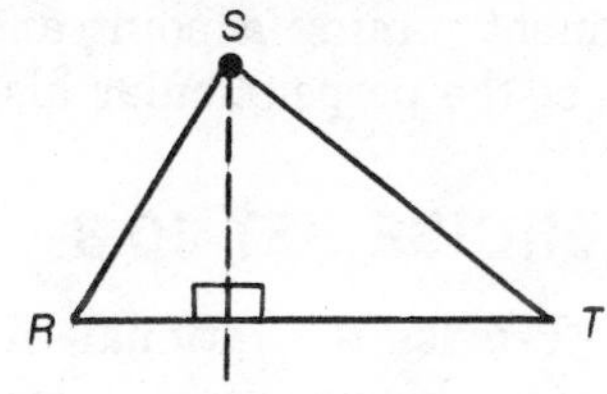

Solutions: (a) Construct a line through *S* and perpendicular to $\overline{RT}$ (refer to Construction 5). The actual construction is left for you.

(b) Locate the midpoint *M* of $\overline{ST}$ by constructing the perpendicular bisector of $\overline{ST}$ (refer to Construction 3). Draw the line segment (median) whose endpoints are *R* and *M*. The actual construction is left for you.

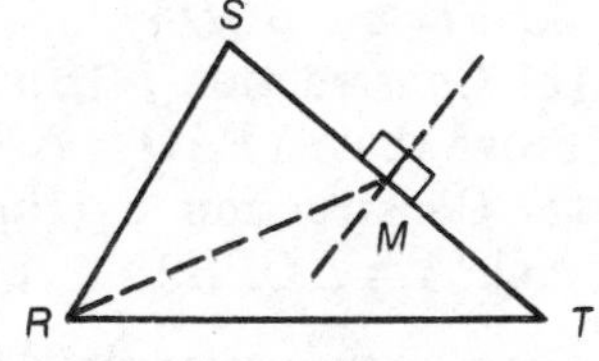

(c) At vertex *T*, using $\overrightarrow{RT}$ as a transversal, construct $\angle 1$ congruent to $\angle R$ as shown in the accompanying diagram. Since $\angle 1$ and $\angle R$ are corresponding angles, $\overrightarrow{TA}$ is parallel to $\overline{RS}$.

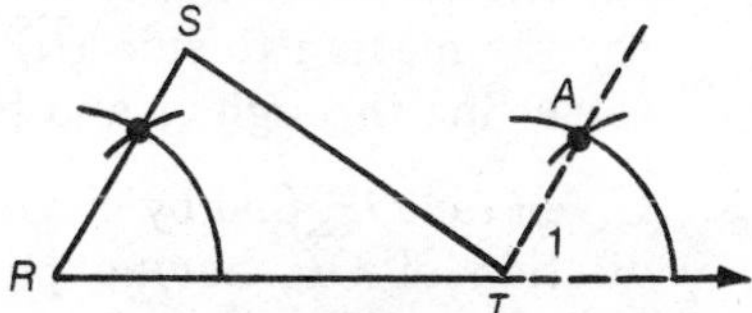

SOME CONSTRUCTIONS RELATED TO TRANSFORMATIONS. If a point *A* is not on a line *l*, then its image *B* under a reflection in line *l* is the endpoint of $\overline{AB}$ such that line *l* is the perpendicular bisector of $\overline{AB}$. Therefore, to locate by construction the image of point *A* under a reflection in line *l*, proceed as follows:

Step	Diagram
1. Construct a line *k* through *A* and perpendicular to line *l*. (Refer to Construction 5.) 2. Label the point at which line *k* intersects line *l* as *M*. 3. Locate image point *B* on line *k* by constructing $\overline{MB} \cong \overline{MA}$. (Refer to Construction 1.)	A, M, B, k, l

Example

2. Given two different points, A and B, if point B is the image of point A under a reflection in line l, find line l by construction.

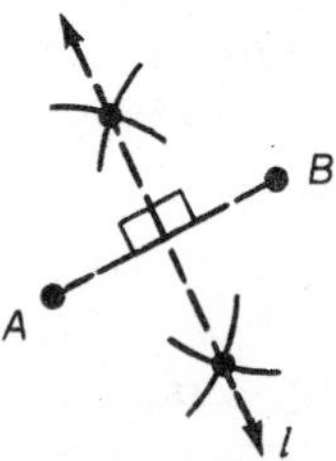

Solution: The line of reflection is the perpendicular bisector of the segment joining a point and its image. Therefore, draw $\overline{AB}$ and construct the perpendicular bisector of $\overline{AB}$. (Refer to Construction 3.)

EXERCISE SET 10.3

1. Provide the rationale for:
   (a) Construction 4 (*Hint*: Draw $\overline{PD}$ and $\overline{QD}$, and prove that $\triangle BPD \cong \triangle BQD$.)
   (b) Construction 5 (*Hint*: Draw segments $\overline{PA}$, $\overline{PB}$, $\overline{AQ}$ and $\overline{BQ}$. Prove that $\triangle PAQ \cong \triangle PBQ$. Next, prove that $\triangle PMA \cong \triangle PMB$.)
   (c) Construction 6 (*Hint*: Draw $\overline{AQ}$ and $\overline{BQ}$, and prove that $\triangle QPA \cong \triangle QPB$.)

2. Draw obtuse triangle ABC, where $\angle B$ is obtuse. Construct:
   (a) the altitude to side $\overline{BC}$ (extended, if necessary)
   (b) the median to side $\overline{BC}$
   (c) a line through A and parallel to $\overline{BC}$

3. Given ray OP, find by construction the segment that represents the dilation of $\overrightarrow{OP}$, having point O as the center of dilation and a constant of dilation of:

   (a) 2 (b) $\frac{1}{2}$ (c) 1.5 (d) -1

4. Given two segments having lengths of a units and b units $(b > a)$, respectively, construct a segment whose length is:
   (a) $(a + b)$ units (b) $(b - a)$ units

5. Given two angles having degree measures of $a°$ and $b°$ $(b > a)$, respectively, construct an angle whose degree measure is:
   (a) $(a + b)°$ (b) $(b - a)°$

6. Draw three noncollinear points.
   (a) Construct the locus of points equidistant from the three points.
   (b) Construct the circle that contains the three points.

7. Given a circle, find *by construction* the center of the circle. (*Hint*: Construct the locus of points equidistant from *any* three points on the circle.)

8. Given a triangle, construct the circumscribed circle. (*Hint*: See Exercises 6 and 7.)

9. Given ▱$ABCD$:
   (a) Construct an altitude from point B to $\overline{AD}$.
   (b) If point H is between points A and D, construct an altitude at point H.

10. Construct an angle having a degree measure of 45.

11. A *tangent* to a circle is a line that intersects the circle at exactly one point, called *the point of tangency*. A radius drawn to the point of tangency forms a right angle with the tangent. Given a circle whose center is point O and any point P on the circle, construct a line that is tangent to the circle at point P. (*Hint*: Draw $\overline{OP}$, and construct a line perpendicular to $\overline{OP}$ at point P.)

12. Given trapezoid $RSTW$, in which $\overline{RW} \parallel \overline{ST}$, construct the median of the trapezoid.

13. Construct a parallelogram that is:
   (a) not a rectangle
   (b) a rectangle
   (c) a rhombus but not a square
   (d) a square

10.4 CONSTRUCTING TRIANGLES AND PROPORTIONAL AND CONGRUENT SEGMENTS

KEY IDEAS

Triangles and special segments can be constructed by using the seven basic constructions presented in Section 10.3.

CONSTRUCTING TRIANGLES. Given a triangle, you can construct a congruent triangle by copying any of the following sets of three parts: (1) three sides; (2) two sides and the included angle; (3) two angles and the included side; (4) two angles and the side opposite one of them.

Example

Given $\triangle ABC$, construct a triangle that is congruent to $\triangle ABC$.

Solution: Draw line l, and label any convenient point on l as P. Construct $\triangle PQR$ in such a way that its three sides have the same lengths as the corresponding sides of $\triangle ABC$.

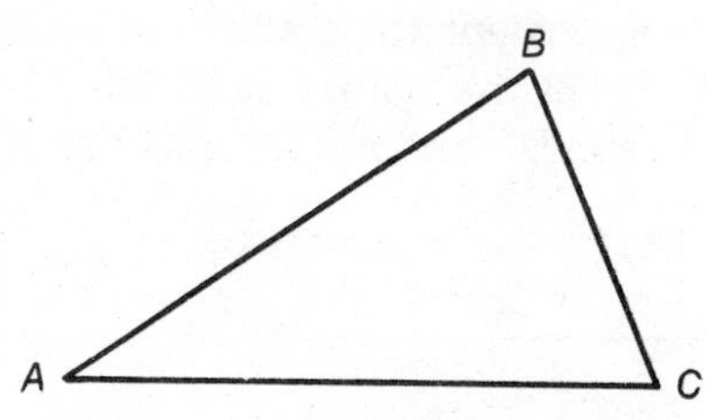

Step 1. Using P as a center and a radius length of AC, mark off segment $\overline{PR}$ on line l so that $\overline{PR} \cong \overline{AC}$.

Step 2. Using R as a center, construct an arc having a radius length of BC.

Step 3. Using P as a center, construct an arc having a radius length of AB.

Step 4. Label the point at which the arcs constructed in steps 3 and 4 intersect as Q.

Step 5. Draw $\overrightarrow{PQ}$ and $\overrightarrow{RQ}$.

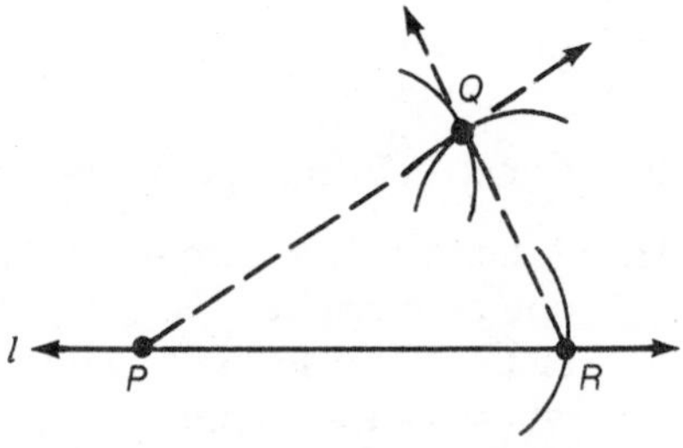

Conclusion: Since $\overline{AC} \cong \overline{PR}$, $\overline{AB} \cong \overline{PQ}$, and $\overline{BC} \cong \overline{QR}$, $\triangle ABC \cong \triangle PQR$ by the SSS $\cong$ SSS postulate.

Note: The methods illustrated in this example can be used to construct: (1) an equilateral triangle whose sides have a given length; (2) an isosceles triangle whose leg and base each have a given length.

CONSTRUCTING SIMILAR TRIANGLES. To construct a triangle similar to a given triangle, construct a triangle having two angles congruent to two angles of the original triangle. For example, a triangle similar to $\triangle ABC$ can be constructed as follows:

Step	Diagram
1. Draw line l. Since the ratio of similtude is not specified, label any convenient points on this line as D and E. 2. At point D, construct ray DP such that $\angle EDP$ is congruent to $\angle BAC$. 3. At point E, construct ray EQ such that $\angle DEQ$ is congruent to $\angle ABC$. 4. Label the points at which rays DP and EQ intersect as F.	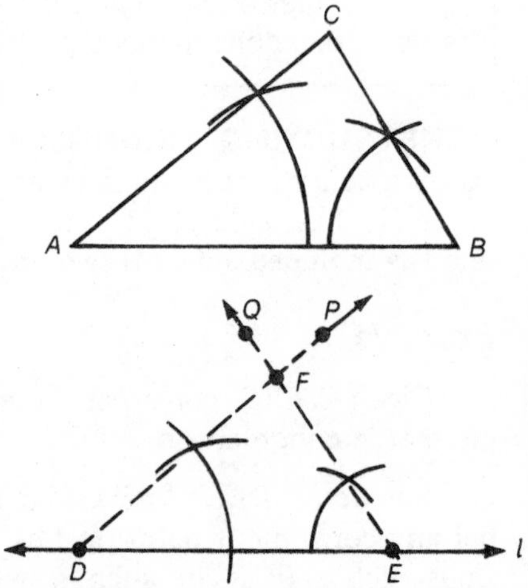

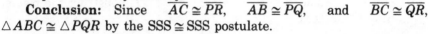

Conclusion: $\triangle DEF \sim \triangle ABC$ by the AA theorem of similarity.

DIVIDING A LINE SEGMENT INTO PROPORTIONAL SEGMENTS. Given two line segments having lengths p and q,

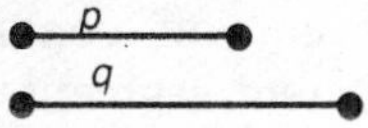

we can divide line segment $\overline{AB}$ into two segments that have the ratio $p:q$ by proceeding as follows:

Step	Diagram
1. Draw ray $\overrightarrow{AP}$, making any convenient angle, $\angle BAP$. 2. On $\overrightarrow{AP}$ mark off successive arcs such that $AC = p$ and $CD = q$. 3. Draw $\overline{DB}$. 4. At C, construct $\overline{CE} \parallel \overline{DB}$. Label the point at which $\overrightarrow{CE}$ intersects $\overline{AB}$ as X.	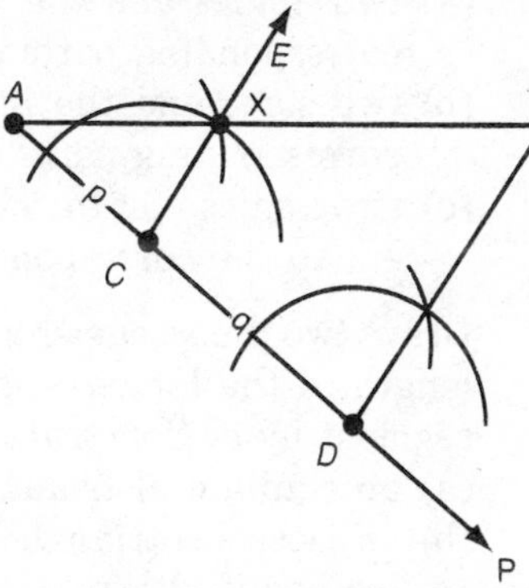

Conclusion: $AX : XB = p : q$ since a line parallel ($\overline{CX}$) to one side of a triangle (side $\overline{BD}$ of $\triangle DAB$) and intersecting the other two sides ($\overline{AB}$ and $\overline{AD}$) divides these sides proportionally.

DIVIDING A LINE SEGMENT INTO CONGRUENT SEGMENTS. To divide $\overline{AB}$ into n congruent parts, follow these steps:

Step	Diagram
1. Draw ray AP, making any convenient angle, $\angle BAP$. 2. On $\overrightarrow{AP}$ mark off n consecutive arcs. For example, to divide $\overline{AB}$ into three congruent segments, mark off three equal arcs on $\overrightarrow{AP}$ so that $\overline{AC} \cong \overline{CD} \cong \overline{DE}$. 3. Draw $\overline{EB}$. 4. Through C, construct $\overrightarrow{CQ} \parallel \overline{EB}$, where point Q is on $\overline{AB}$. 5. Through D, construct $\overrightarrow{DR} \parallel \overline{EB}$, where point R is on $\overline{AB}$.	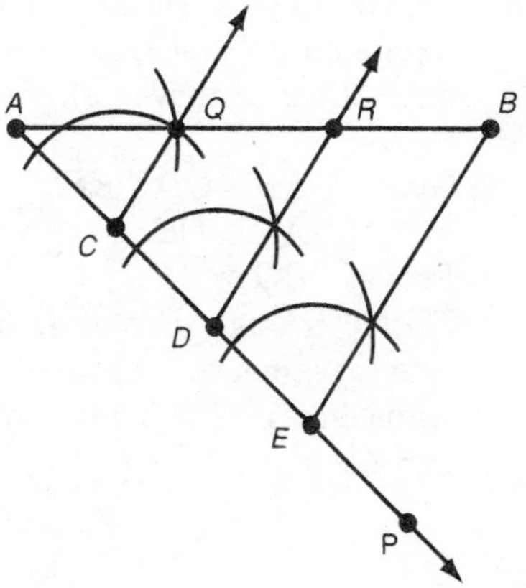

Conclusion: $\overline{AQ} \cong \overline{QR} \cong \overline{RB}$ because, if three (or more) parallel lines cut off congruent segments on one transversal, they cut off congruent segments on any other transversal. (See Exercise 6.)

EXERCISE SET 10.4

1. Draw $\triangle ABC$. If point R is any point on line l, construct $\triangle RST$ congruent to $\triangle ABC$ by constructing:
   (a) two angles and the included side of $\triangle RST$ congruent to the corresponding parts of $\triangle ABC$
   (b) two sides and the included angle of $\triangle RST$ congruent to the corresponding parts of $\triangle ABC$
   (c) two angles and the side opposite one of them in $\triangle RST$ congruent to the corresponding parts of $\triangle ABC$.

2. Draw two line segments having different lengths. Represent the length of the longer segment by p, and the length of the shorter segment by q. Construct:
   (a) an equilateral triangle in which each side has length p
   (b) an isosceles triangle in which the base has length q, and each leg has length p.

3. Draw a line l. If A is any point on l construct an angle having a degree measure of 60. (*Hint*: First construct an equilateral triangle having A as one of its vertices.)

4. Given $\triangle ABC$, if point R is any point on line l, construct $\triangle RST$ similar to $\triangle ABC$ so that $AC:RT = 2:1$.

5. Given $\overline{LM}$, divide $\overline{LM}$ into two segments whose lengths have the ratio 1:2. (*Hint*: Draw ray LP, and mark off on $\overrightarrow{LP}$ arc LA using any convenient radius length; then mark off arc AB such that $AB = 2AL$.)

6. Prove that, if three (or more) parallel lines cut off congruent segments on one transversal, they cut off congruent segments on any other transversal. Use the following diagram, Given, and Prove.
   Given: $\overleftrightarrow{AB} \parallel \overleftrightarrow{CD} \parallel \overleftrightarrow{EF}$,
   $\overline{AC} \cong \overline{CE}$.
   Prove: $\overline{BD} \cong \overline{DF}$.
   Hint: Through B and D draw line segments parallel to transversal t. Then prove $\triangle BXD \cong \triangle DYF$.

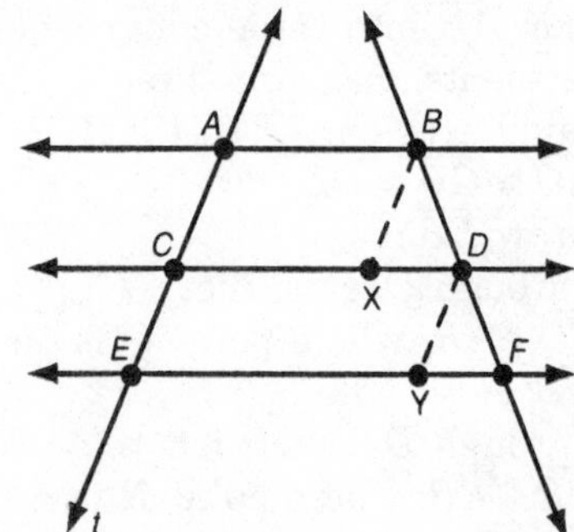

CHAPTER 10 REVIEW EXERCISES

1. How many points are equidistant from two parallel lines and also equidistant from two points on one of the lines?

2. Find the number of points that are 3 inches from point A and 5 inches from point B if:
(a) $AB = 8$ (b) $AB = 6$ (c) $AB = 10$

3. How many points are 3 inches from line l and also 3 inches from a point on line l?

4. Two parallel lines are 12 inches apart. Point P is located on one of the lines. What is the number of points that are equidistant from the parallel lines and are also at a distance of 5 inches from point P?

5. Lines p and q are parallel and d inches apart. Point A is between lines p and q and 4 inches from line q. How many points are equidistant from lines p and q and also 1 inch from point A if:
(a) $d = 8$? (b) $d = 10$? (c) $d = 12$?

6. What is an equation of the locus of points whose ordinates are 2 less than three times their abscissas?

7. What is an equation of the locus of points equidistant from points $A(3, 1)$ and $B(7, 1)$?

8. Give an equation or equations that describe the locus of points:
(a) equidistant from the lines $x = -1$ and $x = 5$
(b) equidistant from the lines $y = -3$ and $y = -7$
(c) equidistant from the lines $y = 3x + 1$ and $y = 3x + 9$

9. Find the number of points that are 3 units from the origin and also 2 units from the x-axis.

10. Find the number of points that are equidistant from points $(-1, 0)$ and $(3, 0)$ and are also 2 units from the origin.

For Exercises 11–15 first draw $\triangle$RST *as given.*

11. Given acute triangle RST, construct a line through R and parallel to $\overline{ST}$.

12. Given obtuse triangle RST with obtuse angle S.
(a) Construct the bisector of $\angle S$.
(b) Construct the altitude from R to $\overline{ST}$.

13. Given right triangle RST, construct the median from right angle S to $\overline{RT}$.

14. Given scalene triangle RST, construct the circle that circumscribes (contains the vertices of) $\triangle RST$.

15. Given $\triangle RST$, construct a triangle that is:
(a) similar to $\triangle RST$ (b) congruent to $\triangle RST$

16. The coordinates of point P are (3, 5).
(a) Describe fully the locus of points at a distance of:
(1) d units from P *(2)* 1 unit from the y-axis
(b) How many points satisfy the conditions in part (a) simultaneously for the following values of d?
(1) $d = 2$ *(2)* $d = 4$ *(3)* $d = 5$

CHAPTER 11

Quadratic Equations in Two Variables

11.1 GENERAL EQUATION OF A CIRCLE

KEY IDEAS

The graph of the equation $x^2 + y^2 = r^2$ is a circle whose center is at the origin and whose radius is r. The general equation of a circle whose center is *not* necessarily located at the origin can be derived by using the distance formula to find the length of the segment drawn from the center $O(h, k)$ of the circle to any point $P(x, y)$ on the circle:

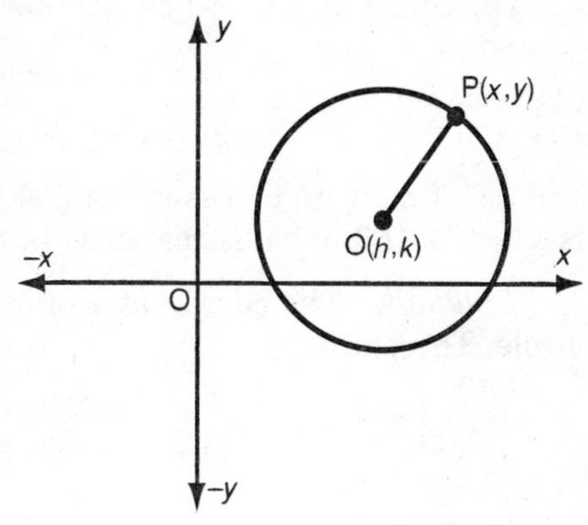

$$\sqrt{(x-h)^2 + (y-k)^2} = OP$$

By substituting the radius length r for OP and then squaring both sides of the equation, the general equation of the circle

$$(x-h)^2 + (y-k)^2 = r^2$$

is obtained.

The dilation of a circle produces another circle that has the same center but a different radius length. The translation of a circle produces another circle that has the same radius length but a different center.

EQUATIONS OF CIRCLES. The graph of the equation $(x-h)^2+(y-k)^2=r^2$ is a circle having a radius of r units and a center located at (h, k). If $h=0$ and $k=0$, then the center of the circle is at the origin, and the equation of the circle simplifies to $x^2+y^2=r^2$.

Examples

1. Write an equation that describes the locus of points 5 units from point (2, −1).

Solution: The locus of points 5 units from the given point (2, −1) is a circle having (2, −1) as its center and a radius length of 5. An equation of the circle is

$$(x-2)^2+(y-[-1])^2=5^2$$
$$\mathbf{(x-2)^2+(y+1)^2=25}$$

2. Determine the center and the radius length of a circle whose equation is: (a) $(x-1)^2+y^2=13$ (b) $(x+3)^2+(y-4)^2=36$

Solutions: (a) Since $(x-1)^2+y^2=13$ may be rewritten as

$$(x-1)^2+(y-0)^2=(\sqrt{13})^2,$$

the center of this circle is **(1, 0)** and the length of its radius is $\mathbf{\sqrt{13}}$.

(b) Since $(x+3)^2+(y-4)^2=36$ may be rewritten as

$$(x-(-3))^2+(y-4)^2=6^2,$$

the center of this circle is **(−3, 4)** and the length of its radius is **6**.

3. If point $(t, 5)$ lies in the first quadrant of a circle whose equation is $x^2+y^2=169$, what is the value of t?

Solution: The coordinates of $(t, 5)$ must satisfy the equation of the circle. Therefore,

$$x^2+y^2=169$$
$$t^2+5^2=169$$
$$t^2=169-25=144$$
$$t=\pm\sqrt{144}=\pm 12$$

Since the point is in the first quadrant, $t=\mathbf{12}$.

4. What is an equation of a circle that is tangent to the y-axis and whose center is (2, 3)?

Solution: As the accompanying diagram illustrates, the radius drawn to the point of tangency is a horizontal segment whose length is 2 units.

Hence, an equation of the circle is $(x-2)^2+(y-3)^2=4$.

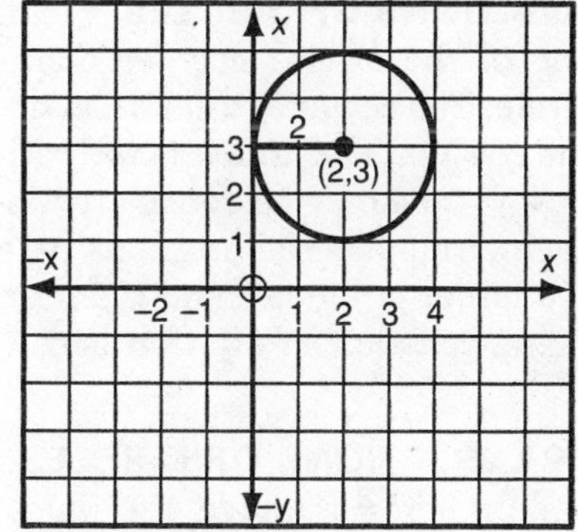

5. The coordinates of the endpoints of a diameter of a circle are $A(1, 2)$ and $B(-7, -4)$.

(a) Find an equation of this circle.

(b) Determine whether the circle passes through point $P(-6, -5)$.

Solutions: (a) The center $O(h, k)$ of the circle is found by finding the midpoint of diameter $\overline{AB}$:

$$h=\frac{1+(-7)}{2}=\frac{-6}{2}=-3, \qquad k=\frac{2+(-4)}{2}=\frac{-2}{2}=-1$$

The coordinates of the center of the circle are $O(-3, -1)$. To find the length of the radius, find the distance between the center and *any* point on the circle, say point A.

Let $(x_1, y_1)=A(1, 2)$ and $(x_2, y_2)=O(-3, -1)$. Then

$$\begin{aligned} OA &= \sqrt{(x_2-x_1)^2+(y_2-y_1)^2} \\ &= \sqrt{(-3-1)^2+(-1-2)^2} \\ &= \sqrt{(-4)^2+(-3)^2} \\ &= \sqrt{16+9}=\sqrt{25}=5 \end{aligned}$$

An equation of a circle whose center is at $(-3, -1)$ and whose radius is 5 is

$$(x-[-3])^2+(y-[-1])^2=5^2$$
$$\mathbf{(x+3)^2+(y+1)^2=25}$$

(b) Point $P(-6, -5)$ lies on the circle if its coordinates satisfy the equation of the circle:

$$\begin{array}{rl} (x+3)^2+(y+1)^2 & =25 \\ (-6+3)^2+(-5+1)^2 & \stackrel{?}{=} 25 \\ (-3)^2+(-4)^2 & \quad 25 \\ 9+16 & \quad 25 \\ 25 & \downarrow\!= 25 \end{array}$$

Therefore, **$P(-6, -5)$ lies on the circle.**

DILATIONS OF CIRCLES. If two circles have the same center, then one of the circles may be considered to be the dilation of the other circle. The *center of dilation* is the common center of the two circles, and the *constant of dilation* is the ratio of their radii. For example, the circle $x^2 + y^2 = 64$ may be considered a dilation of the circle $x^2 + y^2 = 16$ since each circle has the origin as its center. In this case, since the radius of the larger circle is 8 and the length of the radius of the smaller circle is 4, the constant of dilation is $8 \div 4$ or 2.

TRANSLATIONS OF CIRCLES. The circles $(x - h)^2 + (y - k)^2 = r^2$ and $x^2 + y^2 = r^2$ have the same size and shape, but differ in their locations in the coordinate plane. Compared to the center of the circle $x^2 + y^2 = r^2$, the center of the circle $(x - h)^2 + (y - k)^2 = r^2$ is shifted h units in the horizontal direction and k units in the vertical direction.

- If $h > 0$, the circle is shifted horizontally to the right; if $h < 0$, the circle is shifted horizontally to the left.

- If $k > 0$, the circle is shifted vertically up; if $k < 0$, the circle is shifted vertically down.

For example, since the center of the circle $(x - 1)^2 + (y + 3)^2 = 100$ is $(1, -3)$, the circle $(x - 1)^2 + (y + 3)^2 = 100$ is a translation of the circle $x^2 + y^2 = 100$, shifted 1 unit horizontally to the right and 3 units vertically down.

Examples

6. Write an equation of the circle that is the dilation of the circle $x^2 + y^2 = 9$, using a constant of dilation of 4.

Solution: The length of the radius of the original circle is 3. The dilation of this circle, using a dilation constant of 4, is a circle whose radius has a length of 3×4 or 12.

Hence, the dilation of the circle $x^2 + y^2 = 9$, using a constant of dilation of 4, is the circle $\mathbf{x^2 + y^2 = 144}$.

7. Under a certain translation the image of the circle $x^2 + y^2 = 25$ is the circle $(x - 1)^2 + (y - 4)^2 = 25$. What is an equation of the circle that is the image of the circle $(x + 2)^2 + (y - 3)^2 = 49$ under the same translation?

Solution: Since the center of the circle $(x - 1)^2 + (y - 4)^2 = 25$ is $(1, 4)$, the translation shifts the original circle 1 unit horizontally to the right and 4 units vertically up. Therefore, to find the image of the circle $(x + 2)^2 + (y - 3)^2 = 49$ under the same translation, we add 1 unit to the x-coordinate of its center, and add 4 units to the y-coordinate of its center. The center of the circle $(x + 2)^2 + (y - 3)^2 = 49$ is $(-2, 3)$, so the center of its image is $(-2 + 1, 3 + 4) = (-1, 7)$.

An equation of the image of this circle is $(x - [-1])^2 + (y - 7)^2 = 49$, or $\mathbf{(x + 1)^2 + (y - 7)^2 = 49}$.

EXERCISE SET 11.1

1. Fill in the missing items in the following table:

	Equation	Center	Radius
(a)	$x^2 + y^2 = 81$	?	?
(b)	$(x - 5)^2 + y^2 = 121$	?	?
(c)	$(x + 3)^2 + (y + 1)^2 = 50$	?	?
(d)	?	(0, 2)	5
(e)	?	(−1, −4)	6
(f)	?	(3, −2)	$4\sqrt{2}$

2. Write an equation that describes the locus of points 7 units from point (−2, 3).

3. Write an equation of the circle that has a diameter whose end-points are:
   (a) (0, 3) and (0, −3) (b) (−2, 5) and (−8, 5)

4. For each of the following, state whether point P lies on the given circle:
   (a) $(x - 4)^2 + (y + 5)^2 = 49$; $P(4, -12)$
   (b) $(x + 2)^2 + (y - 3)^2 = 37$; $P(-8, 2)$
   (c) $(x - 1)^2 + (y + 6)^2 = 25$; $P(3, -9)$

5. Point $P(a, -12)$ lies on the circle $x^2 + y^2 = 169$. If point P lies in the third quadrant, find the value of a.

6. Find an equation of the dilation of the circle $x^2 + y^2 = 100$ if the constant of dilation is:
   (a) 2 (b) $\frac{1}{2}$ (c) 1

7. The center of a circle that has a radius of 4 is the origin. Find an equation of the translation of this circle if the original circle is shifted:
   (a) 2 units to the right and 3 units up
   (b) 1 unit to the left and 4 units up
   (c) 3 units to the left and 1 unit down
   (d) 2 units down

8. Given the circle $(x + 1)^2 + (y - 2)^2 = 36$, write an equation of the circle that is:
   (a) a translation of the original circle 1 unit horizontally to the left, and 3 units vertically down
   (b) a dilation of the original circle with a dilation constant of 3

9. Equations of a circle and of its dilation are $(x-1)^2+(y+1)^2=4$ and $(x-1)^2+(y+1)^2=36$, respectively. If the circle $x^2+y^2=1$ undergoes the same dilation, what is an equation of its image?

10. What is an equation of a circle that is tangent to the line $y=-3$ and whose center is at the origin?

11. What is an equation of a circle that is tangent to the x-axis and whose center is at $(-4, 3)$?

12. How many points, if any, do the graphs of the following pairs of equations have in common?
(a) $x^2+y^2=4$ and $x=-1$
(b) $x^2+y^2=9$ and $y=3$
(c) $x^2+y^2=25$ and $y=6$
(d) $x^2+y^2=16$ and $y=x$

13–16. In each case, write an equation of the circle that has center O *and passes through point* P.

13. $O(1, 1)$; $P(7, -7)$
14. $O(2, -3)$; $P(-2, 0)$
15. $O(-3, 5)$; $P(1, 9)$
16. $O(-1, -4)$; $P(-6, 8)$

17. In how many points does each of the following circles intersect the x-axis?
(a) $x^2+y^2=16$
(b) $x^2+(y-3)^2=4$
(c) $(x-3)^2+y^2=4$
(d) $(x-2)^2+(y-1)^2=1$

18. Fill in the missing items in the following table, where A and B are endpoints of a diameter of a circle whose center is at point P.

	Equation of Circle	A	B	P
(a)	?	(6, 2)	(−4, 2)	?
(b)	?	(−3, 1)	(5, −5)	?
(c)	?	(2, −7)	(−2, −3)	?
(d)	?	(−5, 2)	?	(3, −4)

19. Find the x- and y-intercepts, if any, of each of the following circles:
(a) $x^2+y^2=49$
(b) $x^2+(y-1)^2=16$
(c) $(x-3)^2+(y-1)^2=4$
(d) $(x-3)^2+(y-4)^2=25$
(e) $(x+2)^2+(y-5)^2=85$
(f) $(x+3)^2+(y-7)^2=45$

11.2 GRAPHING $y = ax^2 + bx + c$ $(a \neq 0)$

KEY IDEAS

The graph of the quadratic equation $y = ax^2 + bx + c$ $(a \neq 0)$ is a smooth curve called a **parabola** that is symmetric with respect to a vertical line, the *axis of symmetry*. The axis of symmetry intersects the parabola at a point called the *turning point* or *vertex* of the parabola.

Each point on the parabola has a corresponding point on the parabola that has the same y-coordinate but lies on the opposite side of the axis of symmetry and at the same distance from it. Either of these two points may be considered the image of the other point in its reflection across the axis of symmetry.

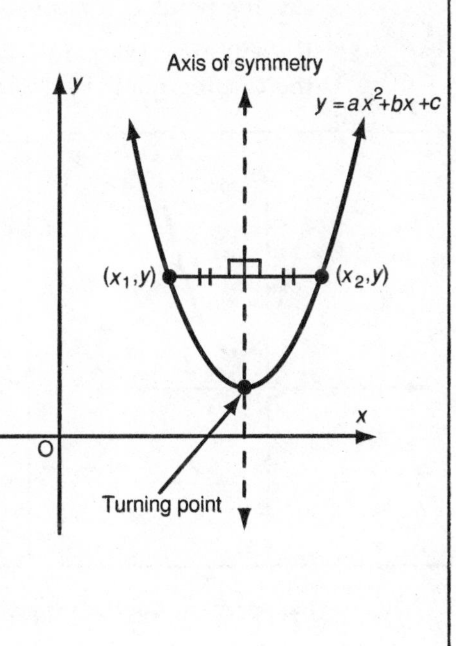

AXIS OF SYMMETRY AND TURNING POINT. Every parabola has a line called the **axis of symmetry** that divides the parabola into two parts that are mirror images of each other. The point at which the axis of symmetry intersects the parabola is called the **turning point** or **vertex**. For the parabola $y = ax^2 + bx + c$ $(a \neq 0)$:

- The axis of symmetry (see Figure 11.1) is the vertical line an equation of which is

$$x = -\frac{b}{2a}.$$

The x-coordinate of the turning point is also $-\frac{b}{2a}$.

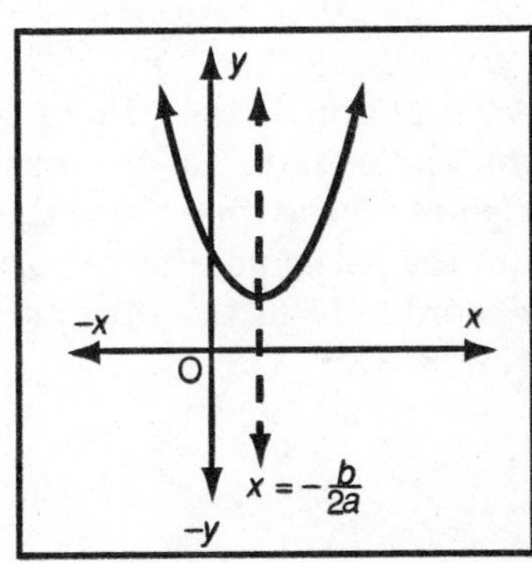

Figure 11.1 Axis of Symmetry of a Parabola

● The sign of a, the coefficient of x^2, determines whether the turning point of a parabola is a minimum point or maximum point on the graph. Note in Figure 11.2 that:

1. If $a > 0$, then the parabola opens up ("holds water") and the turning point is a *minimum* point of the parabola.
2. If $a < 0$, then the parabola opens down ("spills water") and the turning point is a *maximum* point on the parabola.

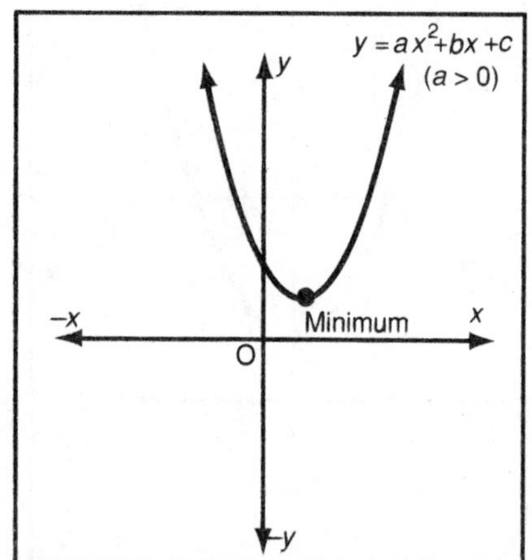

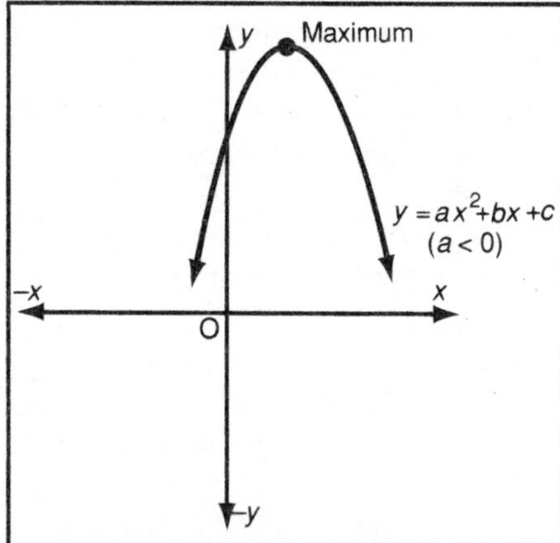

Figure 11.2 Effect of Coefficient *a* on the Turning Point of a Parabola

Examples

1. For the parabola $y = 3x^2 - 6x + 2$, find:
(a) an equation of the axis of symmetry
(b) the coordinates of the turning point of the parabola

Solutions: (a) Let $a = 3$ and $b = -6$. Then

$$x = -\frac{b}{2a} = -\frac{(-6)}{2 \cdot 3} = \frac{6}{6} = 1.$$

An equation of the axis of symmetry is $\boldsymbol{x = 1}$.

(b) The equation of the axis of symmetry gives the x-coordinate of every point on the line. Since the axis of symmetry contains the turning point of the parabola, the x-coordinate of the turning point is 1. To find the y-coordinate of the turning point, replace x by 1 in the quadratic equation $y = 3x^2 - 6x + 2$:

$$\begin{aligned} y &= 3x^2 - 6x + 2 \\ y &= 3 \cdot 1^2 - 6 \cdot 1 + 2 \\ &= 3 - 6 + 2 \\ &= -3 + 2 = -1 \end{aligned}$$

The coordinates of the turning point are **(1, −1)**.

2. If a parabola intersects the x-axis at $x = 1$ and $x = 5$, what is an equation of its axis of symmetry?

Solution: The axis of symmetry bisects each horizontal segment whose endpoints are two points on the parabola. As shown in the accompanying diagram, since $x = 3$ is midway between $x = 1$ and $x = 5$, the axis of symmetry intersects the x-axis at $x = 3$.

An equation of the axis of symmetry is $\boldsymbol{x = 3}$.

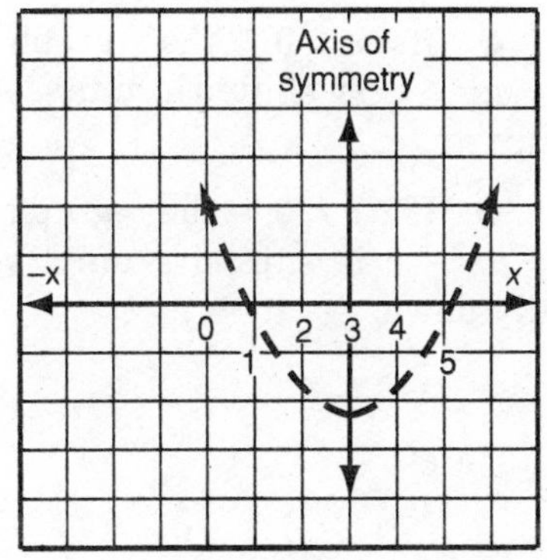

Figure 11.3 Parabola $y = ax^2$ ($a = 1$)

THE GRAPH OF $y = ax^2$ ($a \neq 0$). The generalizations that follow are illustrated in Figures 11.3 and 11.4.

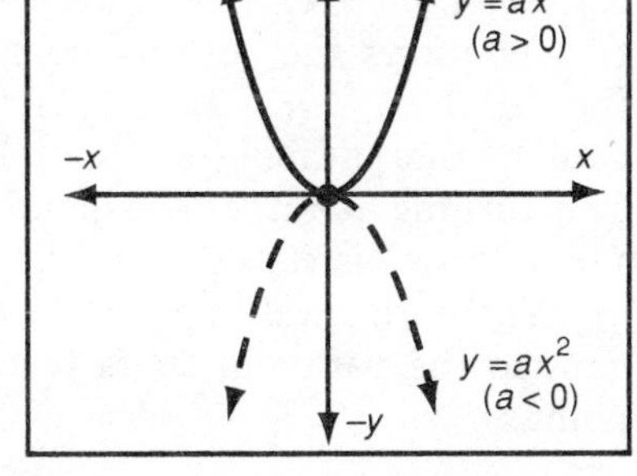

- The turning point of the parabola $y = ax^2$ is $(0, 0)$, and its axis of symmetry is the y-axis.
- The parabola $y = -ax^2$ is a reflection of the parabola $y = ax^2$ in the x-axis.
- Compared with the parabola $y = x^2$ where coefficient $a = 1$:
  1. As $|a|$ gets larger, the parabola $y = ax^2$ becomes narrower horizontally.
  2. As $|a|$ gets smaller, the parabola $y = ax^2$ becomes broader horizontally.

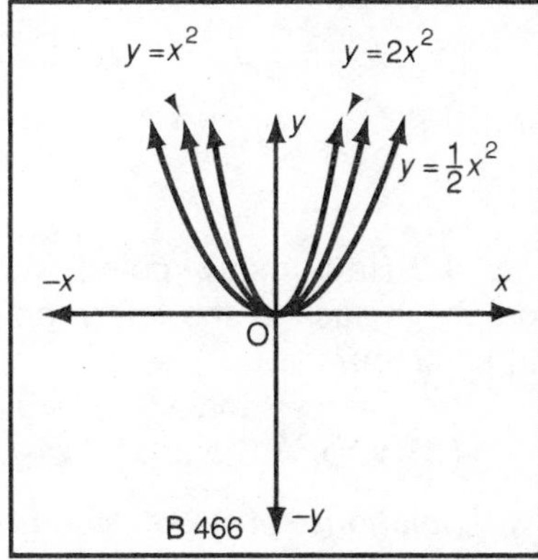

Figure 11.4 Comparing the Width of $y = ax^2$ for Different Values of a

THE GRAPH OF $y = ax^2 + c$ ($a \neq 0$). The graph of $y = ax^2 + c$ is a translation of the graph of $y = ax^2$, shifted c units in the vertical direction. The effect of c is illustrated in Figure 11.5 for the cases in which $c = 2$ and $c = -2$. In general, compared with the graph of $y = ax^2$:

- If $c > 0$, the graph of $y = ax^2 + c$ is shifted c units vertically *up*.

- If $c < 0$, the graph of $y = ax^2 + c$ is shifted c units vertically down.

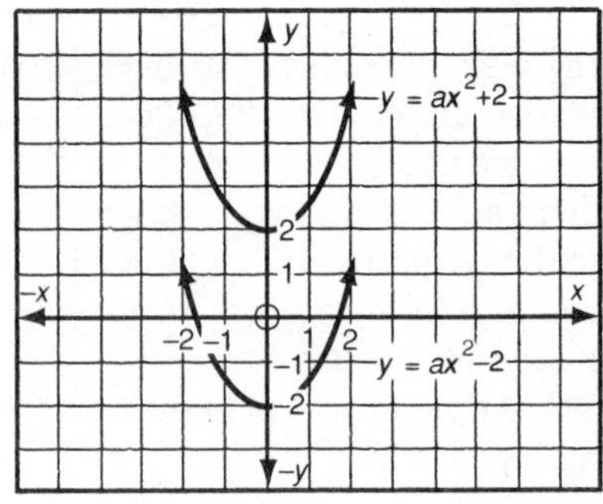

Examples

3. What is the equation of the horizontal line that is tangent to the graph of $y = -2x^2 + 1$?

Solution: A horizontal line that contains the turning point of the parabola $y = -2x^2 + 1$ is tangent to the parabola at this point. The turning point of the parabola $y = -2x^2 + 1$ is (0, 1).

The horizontal line $\mathbf{y = 1}$ is tangent to the parabola at its turning point.

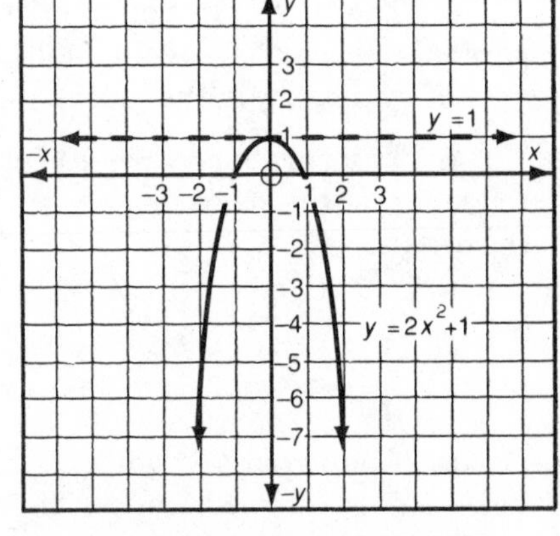

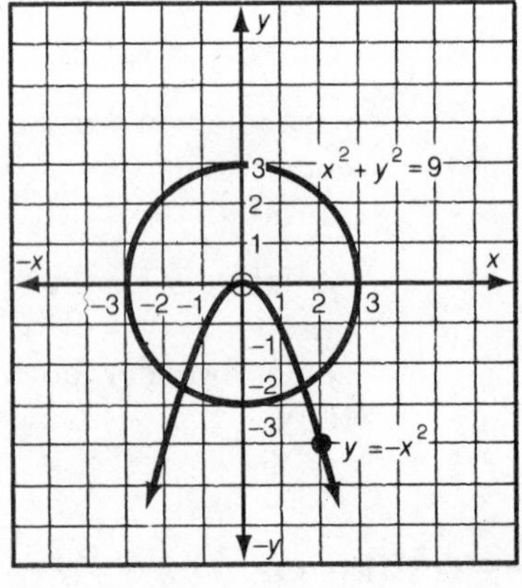

4. How many points do the graphs in each of the following have in common?

(a) $y = -x^2$ and $x^2 + y^2 = 9$
(b) $y = x^2 + 3$ and $x^2 + y^2 = 4$

Solutions: (a) As the accompanying diagram illustrates the circle $x^2 + y^2 = 9$ and the parabola $y = -x^2$ intersect at **2** points.

(b) The length of the radius of the circle $x^2 + y^2 = 4$ is 2, and the turning point of the parabola is (0, 3), which is a minimum point on the graph. Therefore, as shown in the accompanying figure, these graphs have **0** points in common.

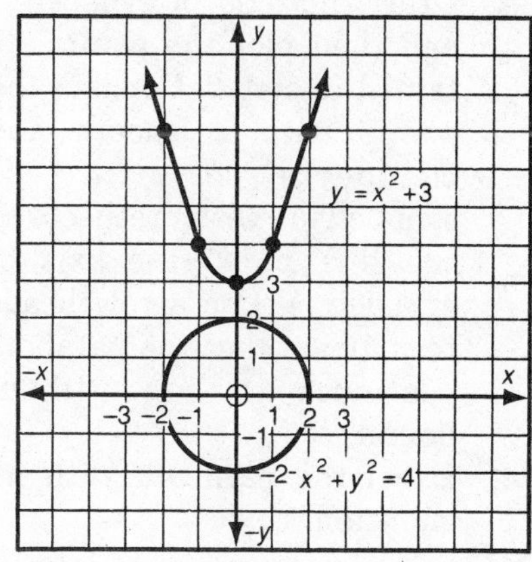

GRAPHING PARABOLAS. A parabola can be drawn by locating its turning point (vertex) and then graphing several pairs of corresponding points on either side of its axis of symmetry. For example, to graph the parabola $y = x^2 - 4x + 1$, proceed as follows:

Step	Example
1. Find the x-coordinate of the turning point.	1. Since $y = x^2 - 4x + 1$, let $a = 1$ and $b = 4$. $x = -\frac{b}{2a} = -\frac{(-4)}{2(1)} = \frac{4}{2} = 2$
2. Make a table of x and y values. Include *at least* seven values of x: the x-coordinate of the turning point and the next three consecutive integer values of x on either side of it. *Note*: Corresponding points on either side of the turning point have the same y-coordinate.	2. (see table below)

x	$x^2 - 4x \quad +1$	$= y$
−1	$(-1)^2 - 4(-1) + 1$	6
0	$0^2 - 4 \cdot 0 \quad +1$	1
1	$1^2 - 4 \cdot 1 \quad +1$	−2
2	$2^2 - 4 \cdot 2 \quad +1$	−3
3	$3^2 - 4 \cdot 3 \quad +1$	−2
4	$4^2 - 4 \cdot 4 \quad +1$	1
5	$5^2 - 4 \cdot 5 \quad +1$	6

3. Draw the axis of symmetry, and then plot the points obtained in step 2. Connect the points with a smooth and continuous curve that is symmetric with respect to the line $x = 2$ (the axis of symmetry). Use arrow heads to indicate that the parabola continues to rise without bound.
4. Label the parabola with its equation.

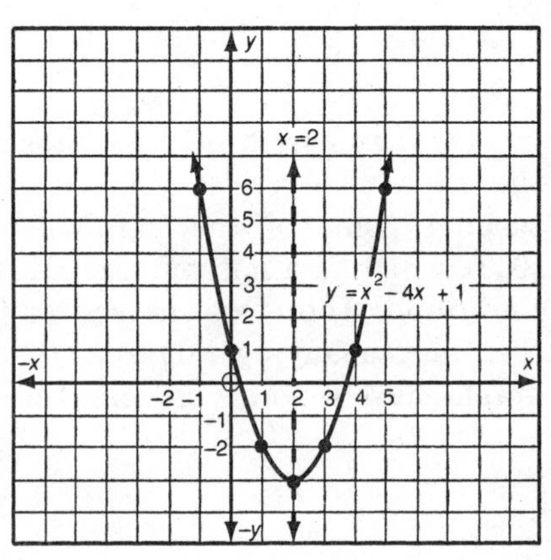

Figure 11.5 Parabola $y = ax^2 + c$ $(a \neq 0)$

Example

5. Draw the graph of $y = -x^2 + 3x + 5$, including all values of x such that $-1 \leq x \leq 4$.

Solution: Begin by finding the equation of the axis of symmetry and, therefore, the x-coordinate of the turning point of the parabola. Since $y = -x^2 + 3x + 5$, let $a = -1$ and $b = 3$. Then

$$x = -\frac{b}{2a} = -\frac{3}{2(-1)} = \frac{-3}{-2} = \frac{3}{2}.$$

In order to include all values of x such that $-1 \leq x \leq 4$, construct a table of values that contain all integer values in this interval, as well the x-coordinate of the turning point of the parabola: $-1, 0, 1, \frac{3}{2}, 2, 3, 4$. Notice that there are three pairs of values of x such that the numbers in each pair are on either side of $\frac{3}{2}$ and the same distance from it.

x	$-x^2$	$+3x$	$+5$	$= y$
-1	$-(-1)^2$	$+3(-1)$	$+5$	1
0	0	$+3\cdot 0$	$+5$	5
1	$-(1)^2$	$+3\cdot 1$	$+5$	7
$\frac{3}{2}$	$-\left(\frac{3}{2}\right)^2$	$+3\left(\frac{3}{2}\right)$	$+5$	$\frac{29}{4}$
2	$-(2)^2$	$+3\cdot 2$	$+5$	7
3	$-(3)^2$	$+3\cdot 3$	$+5$	5
4	$-(4)^2$	$+3\cdot 4$	$+5$	1

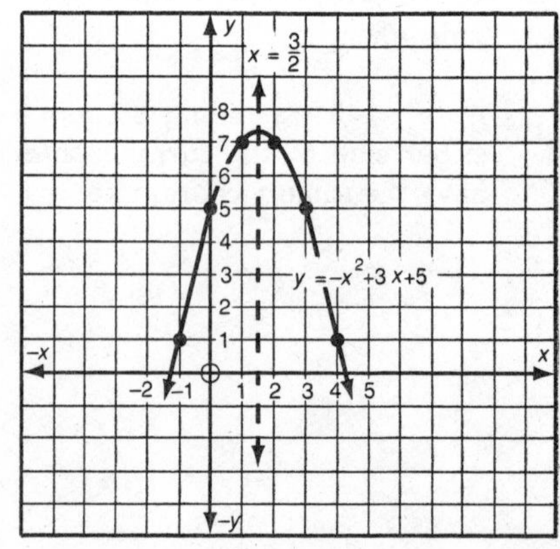

EXERCISE SET 11.2

1–20. In each case, do the following:

(a) Determine the equation of the axis of symmetry of the parabola whose equation is given.

(b) Find the coordinates of the turning point (vertex).

(c) Draw the graph of the given equation in such a way that all values of x in the given interval are included.

(d) From the graph determine the coordinates of the point(s), if any, at which the parabola crosses the x-axis. If necessary, estimate these values to the nearest tenth.

1. $y = x^2 - x - 6;\ -2 \le x \le 3$

2. $y = 2x - x^2;\ -2 \le x \le 4$

3. $y = -x^2 + 2x - 3;\ -2 \le x \le 4$

4. $y = -x^2 + 4x - 4;\ -1 \le x \le 5$

5. $y = x^2 + 4x - 1;\ -5 \le x \le 1$

6. $y = -2x^2 + 8x + 3;\ -1 \le x \le 5$

7. $y = x^2 - 6x + 9;\ 0 \le x \le 6$

8. $y = x^2 + 4x + 3;\ -5 \le x \le 1$

9. $y = x^2 - 4x - 5;\ -1 \le x \le 5$

10. $y = -x^2 - 4x + 5;\ -5 \le x \le 1$

11. $y = -x^2 + 4x - 3;\ -1 \le x \le 5$

12. $y = -x^2 + 3x + 5;\ -1 \le x \le 4$

13. $y = x^2 - x;\ -2 \le x \le 3$

14. $y = -x^2 + x;\ -2 \le x \le 3$

15. $y = x^2 + 3x + 2;\ -4 \le x \le 1$

16. $y = -x^2 - 2x + 8;\ -4 \le x \le 2$

17. $y = -x^2 + 3x + 10;\ -1 \le x \le 4$

18. $y = (x - 1)^2;\ -2 \le x \le 4$

19. $y = (x - 1)^2 + 2;\ -2 \le x \le 4$

20. $y = (x - 1)^2 - 3;\ -2 \le x \le 4$

21. For the graph of $y = -x^2 + 2x + 5$:

(a) Find the coordinates of the turning point.

(b) Write an equation of the line that contains the turning point and is:

(1) parallel to the x-axis

(2) parallel to the line whose equation is $y = 3x + 4$.

22. Write an equation (or equations) of the locus of points 3 units from the axis of symmetry of the graph of $y = x^2 - 4x + 5$.

23. Write an equation of the locus of points 4 units from the turning point of the graph of $y = x^2 + 6x - 5$.

24. Which is an equation of the parabola shown in the accompanying graph?

(1) $y = \frac{1}{2}x^2$

(2) $y = -\frac{1}{2}x^2$

(3) $y = 2x^2$

(4) $y = -2x^2$

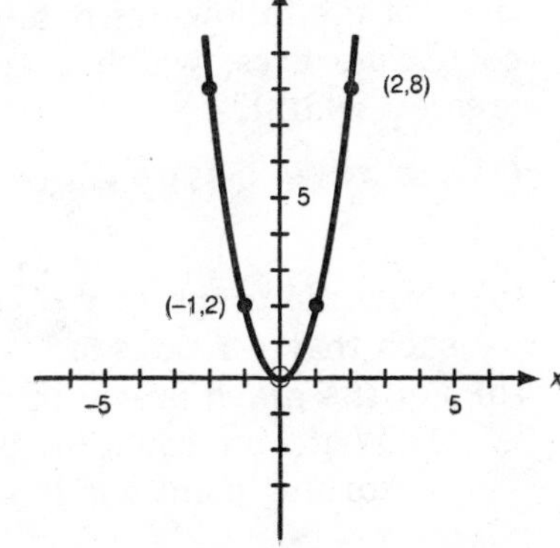

25. Which is an equation of the parabola graphed in the accompanying diagram?
(1) $y = x^2 + 4$
(2) $y = x^2 - 4$
(3) $y = -x^2 + 4$
(4) $y = -x^2 - 4$

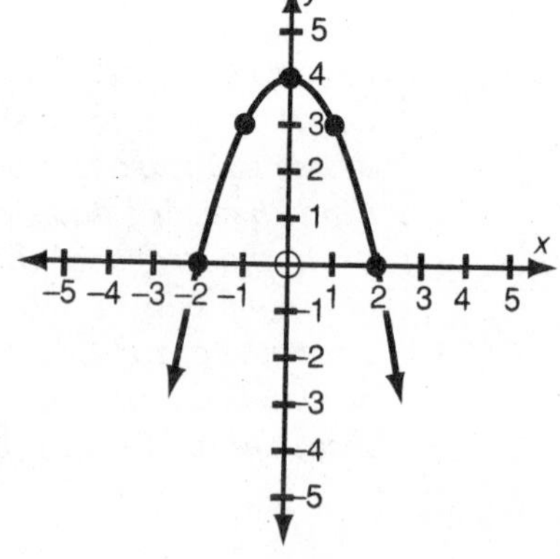

26. Which is *not* true for the parabola in the accompanying diagram?
(1) An equation of the axis of symmetry is $x = 1$.
(2) The x-intercepts are at 0 and at 2.
(3) An equation of the parabola is $y = -x^2 + 2x$.
(4) The coordinates of the turning point are $(1, -1)$.

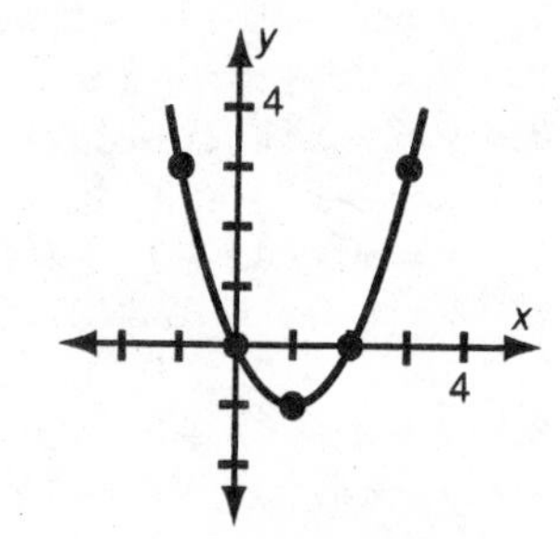

27. If a parabola intersects the x-axis at $x = -1$ and $x = 3$, what is an equation of the axis of symmetry?

28. Write an equation of the horizontal line that is tangent to the graph of:
(a) $y = x^2 - 2$ (b) $y = -x^2 + 1$ (c) $y = -x^2 - 3$ (d) $y = 3x^2 + 2$

29. Write an equation of the horizontal line that is tangent to the graph of $y = 2x^2 + 8x - 5$.

30. Write an equation of the line that contains the turning point of the graph of $y = 2x^2 - 8x + 3$ and is perpendicular to its axis of symmetry.

31. If all of the following equations were graphed on the same set of coordinate axes, which graph would be the broadest (have the greatest width)?

(1) $y = -3x^2$ (2) $y = \frac{1}{2}x^2$ (3) $y = x^2$ (4) $y = -x^2$

32. (a) Draw the graph of $y = \frac{1}{2}x^2 - 2x + 3$, including all values of x such that $-1 \leq x \leq 5$.
(b) For the graph drawn in part (a):
(1) Write an equation of the line that passes through the turning point and is perpendicular to the axis of symmetry.

(2) Write an equation of the locus of points 5 units from the turning point.

(3) Write an equation (or equations) of the locus of points 3 units from the axis of symmetry.

33. For each of the following pairs of equations, determine the number of points the graphs of the equations have in common, if any.

(a) $y = 3$ and $y = -x^2 + 3$
(b) $y = 2$ and $y = x^2 + 1$
(c) $y = x^2$ and $x^2 + y^2 = 25$
(d) $y = -2$ and $y = 2x^2 - 1$
(e) $y = x^2$ and $y = -2x^2$
(f) $y = x^2 + 1$ and $y = x^2 - 1$
(g) $y = x^2 + 4$ and $x^2 + y^2 = 16$
(h) $y = x^2 - 3$ and $x^2 + y^2 = 9$

34. (a) Find the equations of the axis of symmetry of the graphs of the equations $y = x^2$, $y = x^2 + 2x$, and $y = x^2 - 2x$. Sketch each parabola, using the same set of axes.

(b) On the basis of the graphs drawn in part (a), draw a conclusion about the effect of changes in the value of coefficient b on the position of the graph of $y = ax^2 + bx + c$ $(a \neq 0)$.

11.3 SOLVING QUADRATIC EQUATIONS GRAPHICALLY

KEY IDEAS

The point or points, if any, at which the parabola $y = ax^2 + bx + c$ $(a \neq 0)$ crosses the x-axis have y-coordinates of 0, so that their x-coordinates represent the roots of the corresponding quadratic equation, $0 = ax^2 + bx + c$.

If the parabola does not cross the x-axis, then the roots of $ax^2 + bx + c = 0$ are *not* real.

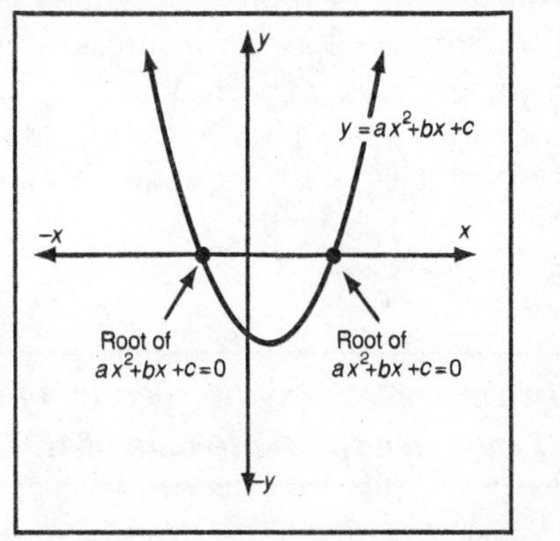

FINDING THE y-INTERCEPT OF A PARABOLA. The y-coordinate of a point at which a graph crosses the y-axis is called a **y-intercept** of the graph. To find the y-intercept of a parabola algebraically, replace x in the equation of the parabola with 0 and then solve for y.

	Example	General Equation
Write the equation of the parabola:	$y = x^2 - 6x + 5$	$y = ax^2 + bx + c$
Let $x = 0$:	$y = 0^2 - 6 \cdot 0 + 5$	$y = a \cdot 0^2 + b \cdot 0 + c$
Simplify:	$y = 0 - 0 + 5$	$y = 0 + c$
	$y = 5$	$y = c$

In general, every parabola* $\mathbf{y = ax^2 + bx + c}$ $\mathbf{(a \neq 0)}$ *has a* y*-intercept numerically equal to* c *(the constant term). For example, the y-intercept of the parabola $y = 2x^2 + x - 3$ is -3. The y-intercept of the parabola $y = -x^2 + 4x$ is 0 since the equation of the parabola can be rewritten as $y = -x^2 + 4x + 0$.

FINDING THE x-INTERCEPTS OF A PARABOLA. The x-coordinate of a point at which a graph crosses the x-axis is called an **x-intercept** of the graph. Figure 11.6 shows that the parabola $y = x^2 + x - 6$ has two x-intercepts, $x = 2$ and $x = -3$. The x-intercepts of a parabola may also be determined algebraically by replacing y in the equation of the parabola with 0 and then solving the resulting quadratic equation for x.

$$0 = x^2 + x - 6$$
$$(x - 2)(x + 3) = 0$$
$$(x - 2 = 0) \vee (x + 3 = 0)$$
$$x = 2 \vee x = -3$$

Since the points at which the parabola $y = x^2 + x - 6$ crosses the x-axis have a y-coordinate of 0, the x-coordinates of these points (x-intercepts) satisfy the quadratic equation $x^2 + x - 6 = 0$.

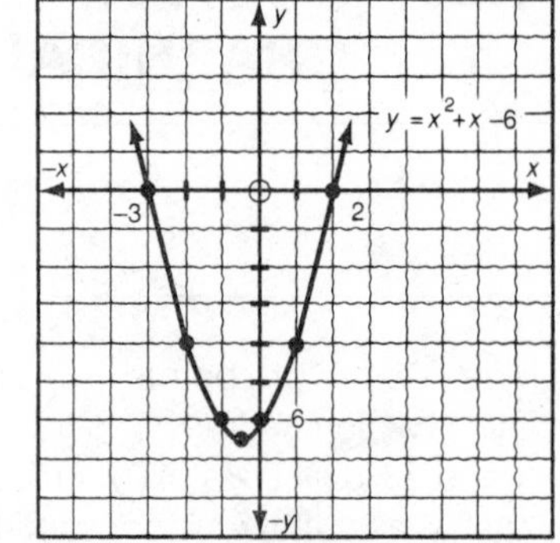

Figure 11.6 Graph of $y = x^2 + x - 6$

***In general, the* x*-intercept(s) of the parabola* $\mathbf{y = ax^2 + bx + c}$ $\mathbf{(a \neq 0)}$*, if any, are the real roots of the quadratic equation* $\mathbf{ax^2 + bx + c = 0}$.**

THE RELATIONSHIP BETWEEN THE PARABOLA $y = ax^2 + bx + c$ AND THE NATURE OF THE ROOTS OF $ax^2 + bx + c = 0$ $(a \neq 0)$. A parabola may intersect the x-axis at two different points, at exactly one point (be *tangent* to the x-axis), or at no points. These situations are summarized in Table 11.1.

TABLE 11.1 Using the Parabola $y = ax^2 + bx + c$ to Determine the Nature of the Roots of $ax^2 + bx + c = 0$

Number of x Intercepts	A Possible Graph of $y = ax^2 + bx + c$ $(a > 0)$	$ax^2 + bx + c = 0$ $(a \neq 0)$	
		Nature of Roots	Value of Discriminant
2	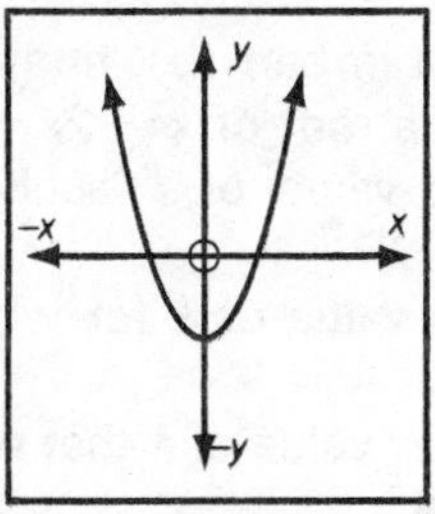	Real and unequal	$b^2 - 4ac > 0$
1	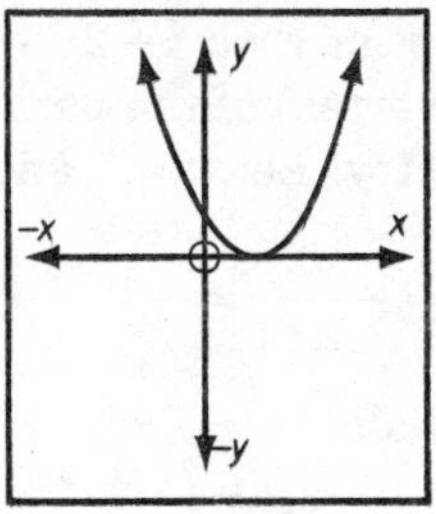	Real and equal	$b^2 - 4ac = 0$
0	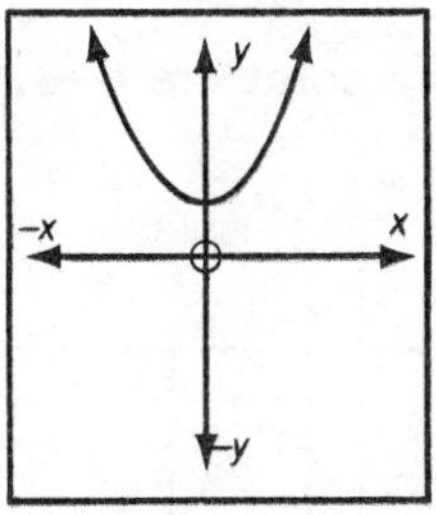	*Not* real	$b^2 - 4ac < 0$

Examples

1. Which is true of the graph of the equation $y = x^2 - 6x + 9$?
(1) It is tangent to the x-axis.
(2) It intersects the x-axis at only two distinct points.
(3) It intersects the x-axis at more than two distinct points.
(4) It has no points that lie on the x-axis.

Solution: Calculate the discriminant of $x^2 - 6x + 9 = 0$, where $a = 1$, $b = -6$, and $c = 9$.

$$
\begin{aligned}
b^2 - 4ac &= (-6)^2 - 4(1)(9) \\
&= 36 - 36 \\
&= 0
\end{aligned}
$$

The discriminant is equal to 0, meaning that the roots of $x^2-6x+9=0$ are real and *equal.* Therefore, the parabola $y=x^2-6x+9$ has one x-intercept, so the parabola is tangent to the x-axis. The correct answer is **choice (1)**.

2. (a) What are the y-intercept and the coordinates of the turning point of the parabola $y=x^2+2x-5$?

(b) Draw the graph of $y=x^2+2x-5$, including all values of x such that $-4 \le x \le 2$.

(c) From the graph drawn in part (b), find the two consecutive integers between which the positive root of $x^2+2x-5=0$ lies.

(d) What is the smallest value of k such that the line $y=k$ intersects the graph of $y=x^2+2x-5$?

(e) What is the minimum value of k for which the roots of the equation $x^2+2x-5=k$ are real?

(f) Give the greatest integer value of k that will make the roots of the equation $x^2+2x-5=k$ not real.

Solutions: (a) The y-intercept of $y=x^2+2x-5$ is **−5**. To find the coordinates of the turning point (vertex) of the parabola, first determine the equation of the axis of symmetry. Let $a=1$ and $b=2$. Then

$$x=-\frac{b}{2a}=-\frac{2}{2\cdot 1}=-1.$$

Rewrite the equation of the parabola: $y = x^2 + 2x - 5$
Replace x by -1: $y=(-1)^2+2(-1)-5$
Simplify: $y = 1 + -2 - 5 = -6$

The coordinates of the turning point are **(−1, −6)**.

(b)

x	x^2 $+2x$ -5 =	y
−4	$(-4)^2+2(-4)-5$	3
−3	$(-3)^2+2(-3)-5$	−2
−2	$(-2)^2+2(-2)-5$	−5
−1	*Turning point*	**−6**
0	$0^2 \quad +2\cdot 0 \quad -5$	−5
1	$1^2 \quad +2\cdot 1 \quad -5$	−2
2	$2^2 \quad +2\cdot 2 \quad -5$	3

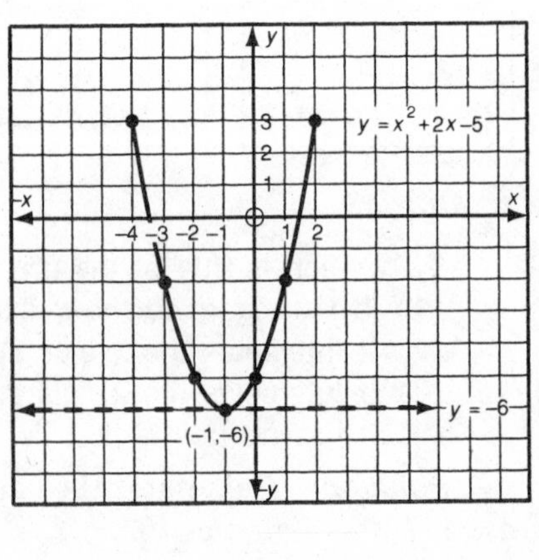

(c) Since the parabola intersects the positive x-axis between $x = 1$ and $x = 2$, the positive root of $x^2 + 2x - 5 = 0$ lies between **1** and **2**.

(d) The lowest point on the parabola is the turning point whose coordinates are $(-1, -6)$. Therefore, the smallest value of k such that the horizontal line $y = k$ intersects the graph of $y = x^2 + 2x - 5$ is $\mathbf{-6}$ (the y-coordinate of the turning point).

(e) In general, the point or points at which the parabola $y = ax^2 + bx + c$ and the horizontal line $y = k$ intersect, if any, represent the real roots of the quadratic equation $ax^2 + bx + c = k$, where k is some constant number.

The *minimum* value of k for which the roots of the equation $x^2 + 2x - 5 = k$ are real corresponds to the *smallest* value of k for which the graphs $y = k$ and $y = x^2 + 2x - 5$ intersect, which is $\mathbf{-6}$.

(f) In general, if the graphs of $y = k$ and $y = ax^2 + bx + c$ do *not* intersect, then the roots of $ax^2 + bx + c = k$ are *not* real (are imaginary). For all values of k less than -6, the graphs of $y = k$ and $y = x^2 + 2x - 5$ do not intersect.

Therefore, $\mathbf{-7}$ is the *largest integer* value of k such that the roots of $x^2 + 2x - 5 = k$ are *not* real.

3. (a) Determine the nature of the roots of $x^2 + 2x - 5 = 0$.
(b) From the graph drawn in Example 2, give a rational approximation of the roots of $x^2 + 2x - 5 = 0$.

Solutions: (a) Calculate the discriminant of the quadratic equation, where $a = 1$, $b = 2$, and $c = -5$.

$$\begin{aligned} b^2 - 4ac &= 2^2 - 4(1)(-5) \\ &= 4 + 20 \\ &= 24 \end{aligned}$$

Since the discriminant is greater than 0 and is not a perfect square, the roots are **real, unequal**, and **irrational**.

(b) A rational approximation of the roots of $x^2 + 2x - 5 = 0$ can be obtained by referring to the parabola drawn in Example 2 and approximating the values of the x-intercepts from the graph. The parabola crosses the x-axis at approximately -3.4 and 1.4.

The roots of $x^2 + 2x - 5 = 0$, estimated to the nearest tenth, are $\mathbf{-3.4}$ and **1.4**.

Note: The estimates obtained in this manner may vary slightly from student to student since they will depend on how carefully and accurately the parabola has been drawn.

EXERCISE SET 11.3

1–8. In each case, do the following:

(a) Determine the y-intercept and the coordinates of the turning point of the graph of the given equation.

(b) *Draw the graph of each equation, including all values of* x *in the given interval.*

(c) *Use the graph of* $y = ax^2 + bx + c$ *drawn in part (b) to find the roots of the corresponding quadratic equation,* $ax^2 + bx + c = 0$.

(d) *Determine the value of* k *that will make the roots of* $ax^2 + bx + c = k$ *real and equal.*

(e) *Give an integer value of* k *that will make the roots of* $ax^2 + bx + c = k$ not *real.*

1. $y = x^2 + 6x + 8;\ -6 \le x \le 0$

2. $y = x^2 - 4x;\ -1 \le x \le 5$

3. $y = x^2 + 4x;\ -5 \le x \le 1$

4. $y = x^2 - 6x - 7;\ 0 \le x \le 6$

5. $y = -x^2 - 2x + 3;\ -4 \le x \le 2$

6. $y = -x^2 + 2x - 1;\ -2 \le x \le 4$

7. $y = x^2 - x - 6;\ -2 \le x \le 3$

8. $y = \frac{1}{2}x^2 - x - 4;\ -2 \le x \le 4$

9–18. In each case, do the following:

(a) *Determine the y-intercept and the coordinates of the turning point of the graph of the given equation.*

(b) *Draw the graph of each equation, including all values of* x *in the given interval.*

(c) *If the graph of* $y = ax^2 + bx + c$ *drawn in part (b) intersects the x-axis, find the consecutive integers between which the real roots of* $ax^2 + bx + c = 0$ *lie; if the graph does not intersect the x-axis, state "The roots of the equation are not real."*

(d) *Determine the value of* k *that will make the roots of* $ax^2 + bx + c = k$ *real and equal.*

(e) *Give an integer value of* k *that will make the roots of* $ax^2 + bx + c = k$ not *real.*

9. $y = x^2 + 2x - 1;\ -4 \le x \le 2$

10. $y = x^2 - 2x - 4;\ -2 \le x \le 4$

11. $y = x^2 - 4x + 1;\ -1 \le x \le 5$

12. $y = -x^2 + 4x - 2;\ -1 \le x \le 5$

13. $y = x^2 + 2x + 3;\ -4 \le x \le 2$

14. $y = -x^2 + 4x + 3;\ -1 \le x \le 5$

15. $y = -x^2 + 2x + 5;\ -2 \le x \le 4$

16. $y = x^2 + x - 3;\ -3 \le x \le 2$

17. $y = -x^2 + 4x + 7;\ -1 \le x \le 5$

18. $y = \frac{1}{2}x^2 - 3x + 5;\ 0 \le x \le 6$

19–36. For each equation $y = ax^2 + bx + c$ *given in Exercises 1–18, find the roots of the corresponding quadratic equation,* $ax^2 + bx + c = 0$, *algebraically either by factoring or by using the quadratic formula. Express irrational roots in simplest radical form. If the discriminant is less than 0, state, "The roots of the equation are not real."*

37. (a) Draw the graph of $y = x^2 - 4x + 2$, including all values of x such that $-1 \le x \le 5$.

(b) Using the graph drawn in part (a):

(1) Estimate the roots of $x^2 - 4x + 2 = 0$ to the nearest tenth.

(2) Find the smallest integral value of k for which the roots of $x^2 - 4x + 2 = 0$ are real.

38. (a) Draw the graph of $y = -2x^2 + 4x - 9$, including all values of x such that $-2 \le x \le 4$.
(b) Using the graph drawn in part (a):
(1) Estimate the roots of $-2x^2 + 4x - 9 = 0$ to the nearest tenth.
(2) Find the largest integral value of k for which the roots of $x^2 - 4x + 2 = 0$ are real.

39. In each case, find the value of k such that the given parabola will be tangent to the x-axis.
(a) $y = x^2 + 10x + k$ (b) $y = x^2 + kx + 16$ (c) $y = kx^2 - 12x + 9$

40. In each case, find the smallest integer value of k such that the given parabola will *not* intersect the x-axis.
(a) $y = x^2 - 4x + k$ (b) $y = kx^2 + 6x + 3$

41. In each case, find the largest integer value of k such that the given parabola will *not* intersect the x-axis.
(a) $y = -x^2 + kx - 4$ (b) $y = -x^2 + 3x + k$

11.4 SOLVING LINEAR-QUADRATIC SYSTEMS OF EQUATIONS

KEY IDEAS

Solving a system of equations means finding the set of all ordered pairs of numbers that satisfy two or more equations at the same time. A system of equations consisting of one linear equation and one quadratic equation in two variables may be solved in either of two ways:

- *Graphically*, by graphing each equation on the same set of coordinate axes. The coordinates of the point(s) of intersection of the graphs, if any, represent the solution set of the system of equations.

- *Algebraically*, by using the linear equation and the substitution principle to eliminate one of the two variables in the quadratic equation. The resulting quadratic equation can then be solved by factoring or by using the quadratic formula.

SOLVING A LINEAR-QUADRATIC SYSTEM OF EQUATIONS GRAPHICALLY. As Figure 11.7 illustrates, a line may intersect a parabola (or circle) in no, one, or two points. Therefore, the solution set of a linear-quadratic system of equations may consist of two ordered pairs of numbers, one ordered pair of numbers, or no ordered pairs.

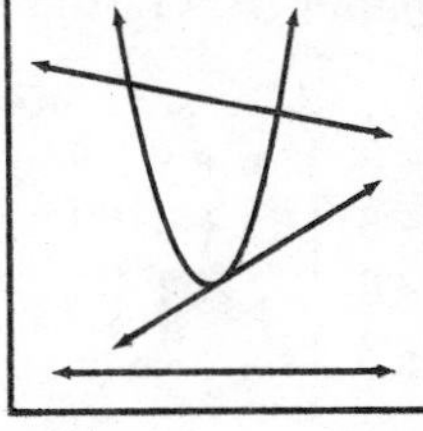

Figure 11.7 The Possible Number of Points in Which a Line and Parabola May Intersect

Example

1. (a) Draw the graph of $y=-x^2+4x-3$ for all values of x such that $-1 \le x \le 5$.

(b) On the same set of axes, draw the graph of $x+y=1$.

(c) Determine the solution of the system:

$$y=-x^2+4x-3$$
$$x+y=1.$$

(d) Check the answer obtained in part (c) *algebraically*.

Solutions:

(a) Make a table of values, using all integral values of x such that $-1 \le x \le 5$, and then draw the parabola.

Note: (2, 1) is the turning point of the parabola.

x	$-x^2 \quad +4x \quad -3$	$= y$
-1	$-(-1)^2+4(-1)-3$	-8
0	$-0^2 \quad +4\cdot 0 \quad -3$	-3
1	$-(1)^2 \quad +4\cdot 1 \quad -3$	0
2	$-(2)^2 \quad +4\cdot 2 \quad -3$	**1**
3	$-(3)^2 \quad +4\cdot 3 \quad -3$	0
4	$-(4)^2 \quad +4\cdot 4 \quad -3$	-3
5	$-(5)^2 \quad +4\cdot 5 \quad -3$	-8

(b) Solve for y: $y=-x+1$. Make a table of values, using any three convenient values for x.

x	$-x+1$	$= y$
0	$0+1$	1
1	$-1+1$	0
2	$-2+1$	-1

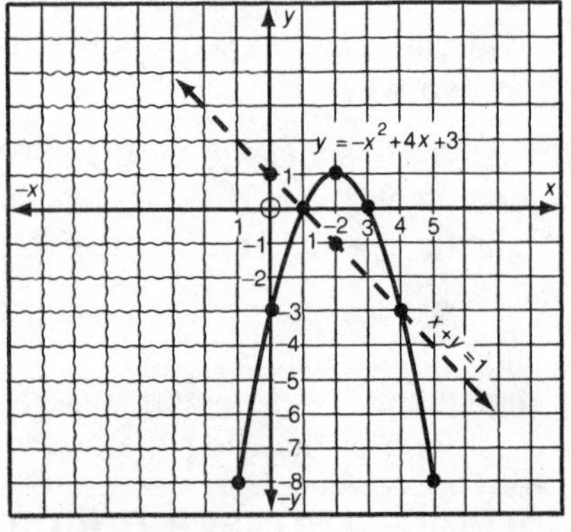

(c) Since the graphs intersect at (1, 0) and (4, −3), the solution set of the system of equations $y=-x^2+4x-3$ and $x+y=1$ is $\{(1,0),(4,-3)\}$.

(d) In an algebraic check, we must demonstrate that each ordered pair of the solution set satisfies each of the *original* equations.

To check (1, 0), let $x=1$ and $y=0$:

$$\begin{array}{l|l} y \overset{?}{=} -x^2 + 4x - 3 & x+y \overset{?}{=} 1 \\ 0 \quad -(1)^2+4\cdot 1-3 & 1+0 \overset{?}{=} 1 \\ 0 \quad -1 \quad +4 \quad -3 & 1 \overset{\checkmark}{=} 1 \\ 0 \quad 3 \quad -3 & \\ 0 \overset{\checkmark}{=} 0 & \end{array}$$

To check (4, −3), let $x=4$ and $y=-3$:

$$\begin{array}{l|l} y \overset{?}{=} -x^2 + 4x - 3 & x+y \overset{?}{=} 1 \\ -3 \quad -(4)^2+4\cdot 4-3 & 4+(-3) \overset{?}{=} 1 \\ -3 \quad -16 \quad +16 \quad -3 & 1 \overset{\checkmark}{=} 1 \\ -3 \quad 0 \quad -3 & \\ -3 \overset{\checkmark}{=} -3 & \end{array}$$

SOLVING A LINEAR-QUADRATIC SYSTEM OF EQUATIONS ALGEBRAICALLY. A linear-quadratic system of equations such as $y = -x^2 + 4x - 3$ and $x + y = 1$ may be solved algebraically as follows:

Step	Example
1. If necessary, rewrite the linear equation so that one of the variables is expressed in terms of the other. When the quadratic equation has the form $y = ax^2 + bx + c$, it is usually easier to solve the system by solving the linear equation for y in terms of x.	1. $x + y = 1$ $y = -x + 1$
2. Substitute into the quadratic equation the expression obtained by solving for the variable in the linear equation.	2. Since $y = -x + 1$, replace y by $-x + 1$: $y = -x^2 + 4x - 3$ $-x + 1 = -x^2 + 4x - 3$
3. Express the quadratic equation in standard form. Solve the resulting quadratic equation, usually by factoring.	3. $0 = -x^2 + 4x + x - 3 - 1$ $0 = -x^2 + 5x - 4$ Multiply each term by -1: $0 = x^2 - 5x + 4$ $0 = (x - 1)(x - 4)$ $(x - 1 = 0) \quad \vee (x - 4 = 0)$ $x = 1 \quad \vee x = 4$
4. Find the value of the remaining variable of each ordered pair by substituting each root of the quadratic equation into the linear equation.	4. Let $x = 1$. Then $y = -x + 1 = -1 + 1 = 0$ Let $x = 4$. Then $y = -x + 1 = -4 + 1 = -3$ The solution set is $\{(\mathbf{1, 0}), (\mathbf{4, -3})\}$.
5. Check algebraically by verifying that each ordered pair satisfies each of the *original* equations.	5. The check is left for you. (See previous solution by graphing on p. 364.)

Example

2. Solve the following system of equations algebraically and check:

$$x^2 + y^2 = 50$$
$$y + 10 = 3x.$$

Solution:

Solve for y in the linear equation: $y = 3x - 10$

Substitute for variable y in the quadratic equation:

$$x^2 + y^2 = 50$$
$$x^2 + (3x - 10)^2 = 50$$

Simplify and express the quadratic equation in standard form:

$$x^2 + 9x^2 - 60x + 100 = 50$$
$$10x^2 - 60x + 100 = 50$$
$$10x^2 - 60x + 100 - 50 = 0$$
$$10x^2 - 60x + 50 = 0$$
$$\frac{10x^2}{10} - \frac{60x}{10} + \frac{50}{10} = 0$$
$$x^2 - 6x + 5 = 0$$

Solve by factoring:

$$(x - 1)(x - 5) = 0$$
$$(x - 1 = 0) \vee (x - 5 = 0)$$
$$x = 1 \quad \vee \quad x = 5$$

Find the corresponding value of y by substituting each value of x into the linear equation:

$y = 3x - 10$	$y = 3x - 10$
$= 3(1) - 10$	$= 3(5) - 10$
$= 3 - 10$	$= 15 - 10$
$= -7$	$= 5$

Write the solution as a set of ordered pairs: The solution set is $\{(\mathbf{1}, \mathbf{-7}), (\mathbf{5}, \mathbf{5})\}$.

The check is left for you.

EXERCISE SET 11.4

1–15. Solve each of the following systems of equations algebraically and check. If the quadratic equation in one variable cannot be factored, find the discriminant. If the discriminant is less than 0, write, "The system does not have real roots."

1. $y = x^2 - 2$
$y = 2x + 1$

2. $y = x^2 - x$
$y = 3x$

3. $y = -x^2 + 3x$
$y + 3 = x$

4. $y = 3x - 2$
$y = x^2 + 2x$

5. $y = x^2 - 2x + 3$
$y = 2x - 1$

6. $y = x + 9$
$y = x^2 - 4x + 3$

7. $y = x^2 - 6x - 1$
$y - x = 7$

8. $y + 7x = 3$
$y = 2x^2 - 4x + 1$

9. $y = -x^2 - 2x + 9$
$y = 2x - 3$

10. $y = x^2 + 3x - 5$
$y - 2x = 1$

11. $y = -x^2 + 4x - 2$
$y + 5 = 2x$

12. $y = x^2 - 6x + 9$
$y + 2x = 5$

13. $y = \frac{1}{2}x^2 - 2x + 1$
$y + 3 = x$

14. $y = -x^2 - 2x + 3$
$y + x = 6$

15. $y = x^2 + 3x + 2$
$y + x + 2 = 0$

16–30. For each system of equations given in Exercises 1–15, do the following:

(a) Draw the graph of the quadratic equation, using the turning point and three consecutive integer values of x on either side of it.

(b) On the same set of axes used in part (a), draw the graph of the linear equation.

(c) Using the graphs drawn in parts (a) and (b), determine the solution set of the system of equations.

31–40. Solve each of the following systems of equations algebraically, and check.

31. $y = 2x^2 - 5x + 5$
$y - x = 5$

32. $x^2 + y^2 = 50$
$x - y = 8$

33. $y + 11 = 3x$
$x^2 - 4x - y - 5 = 0$

34. $x^2 + y^2 = 20$
$y + 2x = 10$

35. $x^2 + y^2 = 17$
$x - y = 5$

36. $x^2 + y^2 = 25$
$x + 2y = 10$

37. $x^2 + y^2 - 8 = 0$
$x + 3y = 4$

38. $x - y = 1$
$x^2 + y^2 - 4y = 5$

39. $x^2 - 3y^2 = 6$
$x + 2y = -1$

40. $(x - 2)^2 + (y - 3)^2 = 25$
$x - y + 2 = 0$

41. (a) Using a compass, draw the graph of $x^2 + y^2 = 25$.
(b) On the same set of axes used in part (a), draw the graph of $x + y = -1$.
(c) On the basis of the graphs drawn in parts (a) and (b), determine the solution set of the system of equations $x^2 + y^2 = 25$ and $x + y = -1$.

42. (a) Using a compass, draw the graph of $x^2 + y^2 = 100$.
(b) On the same set of axes used in part (a), draw the graph of $x + y = 2$.
(c) On the basis of the graph drawn in parts (a) and (b), determine the solution set of the system of equations $x^2 + y^2 = 100$ and $x + y = 2$.

CHAPTER 11 REVIEW EXERCISES

1. Which is an equation of the circle whose center is at (4, −2) and whose radius is 3?
(1) $(x + 4)^2 + (y - 2)^2 = 9$
(2) $(x - 4)^2 + (y + 2)^2 = 9$
(3) $(x + 4)^2 + (y - 2)^2 = 3$
(4) $(x - 4)^2 + (y + 2)^2 = 3$

2. An equation of the axis of symmetry of the parabola $y = -2x^2 + 12x - 19$ is
(1) $x = 6$ (2) $x = -6$ (3) $x = 3$ (4) $x = -3$

3. Which is an equation of the circle that is a translation of the circle $x^2 + y^2 = 25$ shifted 3 units horizontally to the right and 2 units vertically down?
(1) $(x+3)^2 + (y-2)^2 = 25$
(2) $(x-3)^2 + (y+2)^2 = 25$
(3) $(x-3)^2 + (y-2)^2 = 25$
(4) $(x+3)^2 + (y+2)^2 = 25$

4. Which is an equation of the parabola that has the y-axis as its axis of symmetry and its maximum point at $(0, -4)$?
(1) $y = x^2 - 4$ (2) $y = -x^2 + 4$ (3) $y = x^2 + 4$ (4) $y = -x^2 - 4$

5. Which is an equation of the parabola that has the y-axis as its axis of symmetry and its minimum point at $(0, 3)$?
(1) $y = x^2 - 3$
(2) $y = -x^2 + 3$
(3) $y = x^2 + 3$
(4) $y = -x^2 - 3$

6. Which is a possible equation of the parabola that crosses the x-axis at $(-2, 0)$ and $(2, 0)$?
(1) $y = x^2 + 2x$ (2) $y = x^2 - 2x$ (3) $y = x^2 - 4$ (4) $y = x^2 + 4$

7. Which is an equation of a parabola that is tangent to the x-axis?
(1) $y = x^2 - 2x + 1$
(2) $y = x^2 + 2x - 1$
(3) $y = -x^2 + 4x + 4$
(4) $y = x^2 + 4x$

8. Write an equation of the locus of points 4 units from the y-intercept of the parabola whose equation is $y = 2x^2 - 8x + 3$.

9. Write an equation of the locus of points 5 units from the turning point of the graph of $y = -3x^2 - 6x + 1$.

10. How many points are 3 units from the turning point of the parabola $y = x^2$ and also 2 units from its axis of symmetry?

11. Which is an equation of a parabola that does *not* cross the x-axis?
(1) $y = x^2 + 3x - 4$
(2) $y = -x^2 + 3x$
(3) $y = -x^2 - 6x - 5$
(4) $y = x^2 + 3x + 4$

12. Which is an equation of the axis of symmetry of the accompanying graph?
(1) $x = 1$
(2) $x = 0$
(3) $y = 1$
(4) $y = 8$

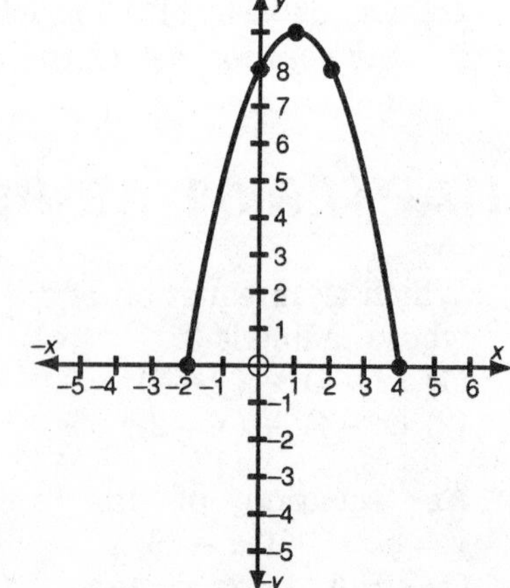

13. (a) Draw the graph of the equation $y = 2x^2 + 4x - 3$, using all integral values of x from $x = -4$ to $x = 2$.
(b) On the same set of axes used in part (a), draw the graph of the equation $y = 3$.
(c) Write the coordinates of the points of intersection of the graphs for parts (a) and (b).

14. (a) Draw the graph of the equation $y = -x^2 + 2x + 4$, using all integral values of x from $x = -2$ to $x = 4$ inclusive.
(b) Write an equation of the axis of symmetry.
(c) Write an equation of the circle whose center is the origin and that passes through the y-intercept of the graph for part (a).
(d) Between which two consecutive positive integers does a root of $-x^2 + 2x + 4 = 0$ lie?

15. Solve each of the following systems of equations algebraically, and check:

(a) $y = x^2 - 2x - 2$
$y = 2x + 3$

(b) $y = x^2 - 7x + 6$
$y + 3x = 2$

(c) $x^2 + y^2 = 130$
$y + 2x = 25$

(d) $x^2 + y^2 = 65$
$y - 5x = 13$

UNIT V: PROBABILITY AND COMBINATIONS

CHAPTER 12

Probability and Combinations

12.1 BASIC PROBABILITY CONCEPTS

KEY IDEAS

Consider finding the probability of obtaining an even number if a single die is rolled. Here is some common terminology, based on this example.

Event, E = rolling a single die and obtaining an even number
Sample space, S = set of all possible outcomes = $\{1, 2, 3, 4, 5, 6\}$
$n(S)$ = number of elements in sample space $S = 6$
$n(E)$ = number of ways in which E can occur ("successful outcomes")
= 3 (there are three even numbers in the sample space)
$P(E)$ = probability of event E occurring

$$= \frac{n(E)}{n(S)} = \frac{\text{Number of ways E can occur}}{\text{Number of possible outcomes}} = \frac{3}{6} = \frac{1}{2}$$

SOME PROBABILITY FACTS. Since the probability of an event is defined as a fraction, certain conclusions can be drawn regarding the possible range of values of the probability of a specific event occurring.

- **P(an impossibility occurs) = 0.** Event E is impossible if there is no way in which E can happen, so that the numerator of the probability fraction is 0. For example, the probability of rolling a 7 on a single roll of a die is $\frac{0}{6} = 0$.

- **P(a certainty occurs) = 1.** Event E is a certainty whenever the number of ways in which event E can occur equals the number of elements in the sample space. For example, the probability of rolling a single die and getting a number less than 7 is $\frac{6}{6} = 1$.

- **$0 \leq P(E) \leq 1$.** The probability of an event occurring cannot be less than 0 or greater than 1.

- **$P(\text{not } E) = 1 - P(E)$.** Subtracting the probability that an event will occur from 1 gives the probability that the event will *not* occur. For example, suppose there is a 30% chance that it will rain tomorrow. Then the probability that it will *not* rain is 70%, or 0.7, since

$$P(\text{not rain}) = 1 - P(\text{rain}) = 1 - 30\% = 1 - 0.3 = \mathbf{0.7}$$

THE COUNTING PRINCIPLE. The counting principle states that, if one activity can be performed in p ways and another activity in q ways, then there are $p \times q$ possible ways in which both activities can be performed.

Example

1. A man has one white, one blue, one gray, and one pink shirt. If he has one blue, one black, and one red necktie, find:

(a) the number of ways that he can choose a shirt and a necktie

(b) the probability that, if he chooses a shirt and a necktie at random, they are the same color

Solutions: (a) There are four shirts and three ties. Using the counting principle, we find that there are $4 \times 3 = \mathbf{12}$ ways in which the man can choose a shirt and a necktie.

(b) There is only one selection for which the shirt and the necktie have the same color (a blue shirt and a blue necktie), so

$$P(\text{same color}) = \mathbf{\frac{1}{12}}.$$

PROBABILITY INVOLVING *OR*. In probability problems involving two (or more) events, the events may or may not have outcomes that are successes for both events.

● If events A and B have no successful outcomes in common, then

$$P(A \text{ or } B) = P(A) + P(B).$$

Example: What is the probability of obtaining a 1 or an even number on a single roll of a die?

Let A = event of rolling a 1;
B = event of rolling an even number.

Since there are six possible outcomes and three even numbers, $P(A) = \frac{1}{6}$ and $P(B) = \frac{3}{6}$.

$$P(A \text{ or } B) = P(A) + P(B)$$
$$= \frac{1}{6} + \frac{3}{6} = \frac{4}{6} = \frac{2}{3}$$

● If events A and B have outcomes in common that are successes for both events, then

$$P(A \text{ or } B) = P(A) + P(B) - P(A \text{ and } B)$$

Example: What is the probability of obtaining a 2 or an even number on a single roll of a die?

Let A = event of rolling a 2;
B = event of rolling an even number.

● Since there are six possible outcomes, $P(A) = \frac{1}{6}$. There are three even numbers, so $P(B) = \frac{3}{6}$. Since 2 is an even number, it is reflected in both $P(A)$ and in $P(B)$; that is, events A and B have one successful outcome in common, so $P(A \text{ and } B) = \frac{1}{6}$.

$$P(A \text{ or } B) = P(A) + P(B) - P(A \text{ and } B)$$
$$= \frac{1}{6} + \frac{3}{6} - \frac{1}{6} = \frac{3}{6} = \frac{1}{2}$$

Example

2. Find the probability of drawing from a standard deck of 52 playing cards a card that is red or is a 5.

Solution: In a standard deck of playing cards there are 26 red cards (13 diamonds and 13 hearts) and four 5's. However, included in the 26 red cards are the 5 of diamonds and the 5 of hearts. Therefore,

$$P(\text{red card or } 5) = P(\text{red card}) + P(5) - P(\text{red card and } 5)$$
$$= \frac{26}{52} + \frac{4}{52} - \frac{2}{52}$$
$$= \frac{28}{52} = \frac{7}{13}$$

EXERCISE SET 12.1

1. The numbers from 1 to 20, inclusive, are written on individual slips of paper and placed in a hat. What is the probability of selecting each of the following?
(a) An even number
(b) A number divisible by 5
(c) A number divisible by 10
(d) A number divisible by 5 and by 10
(e) A number that is at least 16
(f) A prime number

2. A letter is selected at random from the alphabet. What is the probability that it is *not* a vowel?

3. One of the angles of a right triangle is selected at random. What is the probability of each of the following?
(a) The angle is obtuse. (b) The angle is acute.

4. A large parking lot has five entrances and seven exits. In how many different ways can a driver enter the parking lot and exit from it?

5. An ice cream parlor makes a sundae using one of six different flavors of ice cream, one of three different flavors of syrup, and one of four different toppings. What is the total number of different sundaes that this ice cream parlor sells?

6. If $P(A) = 0.7$, $P(B) = 0.5$, and $P(A \text{ and } B) = 0.35$, then what is $P(A \text{ or } B)$?

7. Find the probability that, when a single die is rolled, the number rolled is:
(a) odd and greater than 3
(b) odd or greater than 3
(c) prime and greater than 3
(d) prime or greater than 3
(e) prime and less than 3
(f) prime or even

8. One number is selected from the set of integers from 1 to 10, inclusive. Find the probability that the number selected is:
(a) even and prime
(b) even or prime
(c) prime and greater than 5
(d) divisible by 3 or by 5
(e) divisible by 3 and by 5
(f) even or greater than 5
(g) divisible by 2 and by 3
(h) divisible by 2 or by 3

9. A card is selected at random from a standard deck of 52 playing cards. Find the probability that the card selected is:
(a) red and an ace
(b) red or an ace
(c) a club or a diamond
(d) a 7 of hearts or a 9 of clubs
(e) a 7 or a 9
(f) a picture card or a black card
(g) a picture card and a black card
(h) not a club and not a jack

10. Given four geometric figures: an equiangular triangle, a square, a trapezoid, and a rhombus. If one of the figures is selected at random, what is the probability that the figure will be equilateral?

11. On a test the probability of getting the correct answer to a certain question is represented by $\frac{x}{10}$. Which *cannot* be a value of x?

(1) -1 (2) 0 (3) 1 (4) 10

12. What is the probability that the average of two consecutive even integers is also an even integer?

(1) 0 (2) $\frac{1}{2}$ (3) 1 (4) Impossible to determine

13. Express, in terms of x, the probability that an event will *not* happen if the probability that the event will happen is represented by:

(a) x (b) $\frac{x}{4}$ (c) $x + 4$ (d) $4x - 1$

14. A jar contains x red marbles, $2x - 1$ blue marbles, and $2x + 1$ white marbles. One marble is drawn at random.

(a) Express in terms of x the total number of marbles in the jar.
(b) Express in terms of x the probability of drawing a blue marble.
(c) If the probability of drawing a blue marble is $\frac{1}{3}$, find the value of x.
(d) What is the probability of *not* drawing a red marble?

12.2 FINDING PROBABILITIES OF EVENTS THAT OCCUR JOINTLY

KEY IDEAS

Two events are said to occur *jointly* when they both happen. Multiplying the probability that event *A* will occur by the probability that event *B* will occur gives the probability that events *A* and *B* will occur jointly.

JOINT PROBABILITIES. As the counting principle suggests, to find the probability of two or more events occurring jointly, we multiply the probabilities of the individual events. If the probability of one event occurring is p and the probability of another event occurring is q, then the probability of these events occurring together is $p \times q$.

Examples

1. A coin and a single die are tossed. What is the probability of getting a head *and* a number less than 3?

Solution: $P(H) = \frac{1}{2}$. There are two numbers less than 3 (1 and 2), so $P(N < 3) = \frac{2}{6} = \frac{1}{3}$.

$$\begin{aligned} P(H \text{ and } N > 3) &= P(H) \times P(N < 3) \\ &= \frac{1}{2} \times \frac{1}{3} \\ &= \mathbf{\frac{1}{6}}. \end{aligned}$$

2. A card is drawn at random from a standard deck of 52 playing cards, looked at, and then replaced. Another card is drawn. What is the probability that two aces are drawn?

Solution: There are four aces in a deck of 52 playing cards, so

$$\begin{aligned} P(2 \text{ aces}) &= P(1 \text{ ace}) \times P(1 \text{ ace}) \\ &= \frac{4}{52} \times \frac{4}{52} \\ &= \frac{1}{13} \times \frac{1}{13} \\ &= \mathbf{\frac{1}{169}}. \end{aligned}$$

PROBLEMS WITH AND WITHOUT REPLACEMENT. An urn is a vase whose contents are not readily visible. If an object is taken out of an urn, examined, and then put back, we classify this as a problem "with replacement." The sample space for each withdrawal of an object with replacement remains the same.

In probability problems "without replacement," the object is *not* replaced after being drawn, so that the sample space changes for each succeeding withdrawal of an object.

Examples

3. An urn contains three red marbles, two white marbles, and four blue marbles. A marble is chosen at random from the urn and then replaced. Another marble is chosen.

(a) Find the probability that both marbles are white.

(b) If the first marble is not replaced, find the probability that both marbles are the same color.

Solutions: (a) There are nine marbles in the run, so that the sample space for each selection with replacement is 9. On each selection from the urn, two white marbles are available to be picked. The probability that a white marble is selected on the first try is $\frac{2}{9}$, and with replacement the probability that a white marble is picked on the second try is also $\frac{2}{9}$. The probability that these events will occur jointly is, therefore,

$$P(\text{both white}) = P(\text{first white}) \times P(\text{second white})$$
$$= \frac{2}{9} \times \frac{2}{9} = \frac{4}{81}.$$

(b) $P(\text{same color}) = P(\text{both red}) + P(\text{both white}) + P(\text{both blue})$. Since there is no replacement, there are nine outcomes in the sample space for the first pick, and eight possible outcomes for the second pick. The outcome that is removed is assumed to be the color that the second pick is trying to match.

$$P(\text{both red}) = \frac{3}{9} \times \frac{2}{8} = \frac{6}{72}$$
$$P(\text{both white}) = \frac{2}{9} \times \frac{1}{8} = \frac{2}{72}$$
$$P(\text{both blue}) = \frac{4}{9} \times \frac{3}{8} = \frac{12}{72}$$
$$P(\text{same color}) = \frac{20}{72} = \frac{5}{18}.$$

4. From a standard deck of 52 playing cards, two cards are drawn at random. Find the probability that both cards are jacks.

Solution: Drawing two cards from the deck implies that the second card is drawn without the first card being replaced. Since there are four jacks and a total of 52 playing cards, $P(\text{first jack}) = \frac{4}{52}$. If it is assumed that the first card picked is a jack and is not replaced. $P(\text{second jack}) = \frac{3}{51}$. To find the probability that these events occur jointly, multiply their individual probabilities:

$$P(\text{both jacks}) = P(\text{first jack}) \times P(\text{second jack})$$
$$= \frac{4}{52} \times \frac{3}{51}$$
$$= \frac{1}{13} \times \frac{1}{17} = \frac{1}{221}.$$

EXERCISE SET 12.2

1. An urn contains three yellow marbles, five white marbles, and two black marbles. Two marbles are randomly selected, and their colors are noted. Find the probability of selecting without replacement:
   (a) one white and one black marble
   (b) two white marbles
   (c) two marbles having the same color
   (d) two marbles having different colors

2. Answer each part of Exercise 1 assuming replacement.

3. The integers from 1 to 10, inclusive, are written on slips of paper and placed in an urn. Two slips of paper are drawn without replacement. Find the probability that:
(a) both numbers are even
(b) one number is even and one number is odd
(c) both numbers are prime
(d) both numbers are divisible by 3
(e) both numbers are at least 5

4. Answer each part of Exercise 3 assuming replacement.

5. The letters of the word PARALLELOGRAM are written on individual slips of paper and placed in an urn. Find the probability of drawing two letters at random and obtaining:
(a) an L on both selections, assuming replacement after the first pick
(b) an L on both selections without replacement
(c) two letters that are *not* vowels, assuming replacement after the first pick
(d) two letters that are *not* vowels without replacement
(e) two letters that are the same, assuming replacement after the first pick
(f) two letters that are the same without replacement

6. Two cards are drawn at random without replacement from a standard deck of 52 playing cards. Find the probability that the two cards:
(a) are both spades
(b) are in different suits
(c) are picture cards
(d) have the same face value

7. Answer each part of Exercise 6 assuming replacement.

8. A softball team plays two games each weekend, one on Saturday and the other on Sunday. The probability of winning on Saturday is $\frac{3}{5}$, and the probability of winning on Sunday is $\frac{4}{7}$. Find the probability of:
(a) losing a Saturday game and winning a Sunday game
(b) winning a Sunday game after already winning a Saturday game
(c) winning both games
(d) losing both games

9. John has 10 navy blue socks and 14 black socks in a drawer. If John selects two socks at random, what is the probability they will be the same color?

10. A coach has to purchase uniforms for a team. A uniform consists of one pair of pants and one shirt. The colors available for the pants are black and white. The colors available for the shirt are green, orange, and yellow. Find the probability that in the uniform the coach chooses:

(a) the pants are black and the shirt is orange
(b) the shirt is green
(c) the pants and the shirt are of different colors

12.3 COUNTING ARRANGEMENTS OF OBJECTS: PERMUTATIONS

KEY IDEAS

A **permutation** is an arrangement of objects in which order matters. A special notation is useful when discussing permutations. The product of the integers from n to 1, inclusive, is called **n factorial** and is written as $n!$ For example.

$$5! = 5 \cdot 4 \cdot 3 \cdot 2 \cdot 1 = 120.$$

Note that n is defined only if n is a positive integer. 0! is defined to be equal to 1. Alternatively, $n!$ may be written as ${}_nP_n$. For example,

$${}_4P_4 = 4! = 4 \cdot 3 \cdot 2 \cdot 1 = 24.$$

ARRANGING OBJECTS. When finding the number of possible different arrangements of a set of objects in which the order of the objects is considered, two important situations arise:

1. *All of the objects are used in each arrangement.* If every object is used in each arrangement, then the number of different arrangements possible is $n!$, where n represents the number of objects being arranged. This situation is sometimes symbolized by the notation ${}_n\boldsymbol{P}_n$, which is read as "the permutation of n objects taken n at a time." The notations $n!$ and ${}_nP_n$ are mathematically equivalent.

Example: (a) In how many different ways can an algebra book, a geometry book, a trigonometry book, and a calculus book be arranged on a bookshelf?

(b) In how many different ways can these books be arranged if the geometry book must appear first?

(a) Since there are four different books, they can be arranged on a shelf in

$${}_4P_4 = 4 \cdot 3 \cdot 2 \cdot 1 = \mathbf{24} \text{ ways.}$$

(b) If the geometry book must come first, then there is only one choice for the first position on the shelf; the next three positions may be filled by any of the *three* remaining books. Therefore, the books can be arranged in

$$1 \cdot {}_3P_3 = 1 \cdot 3! = 1 \cdot 3 \cdot 2 \cdot 1 = \mathbf{6} \text{ ways.}$$

2. *Not all of the objects are used in each arrangement.* If n objects are being used to fill r positions and $r < n$, then every object is not used in each arrangement. The process of arranging n objects in r positions ($r \leq n$) is symbolized by the notation ${}_nP_r$, which is read as "the permutation of n objects taken r at a time" and is defined as the product of the r greatest factors of $n!$:

$$\boldsymbol{{}_nP_r = n(n-1)(n-2)(n-3)\ldots(n-r+1).}$$

Example: In how many ways can five students be seated in a row that has three chairs?

The students can be seated in

$$\underset{\textit{Chair} \rightarrow}{{}_5P_3 =} \underset{\textit{1st}}{\underline{5}} \cdot \underset{\textit{2nd}}{\underline{4}} \cdot \underset{\textit{3rd}}{\underline{3}} = \mathbf{60} \text{ ways}$$

Examples

1. In how many different ways can the digits 1, 3, 5, and 7 be arranged to form a four-digit number if repetition of digits:

(a) is *not* allowed (b) is allowed

Solutions: (a) Since each digit can be used only once, the number of ways in which the four digits can be used to form a four-digit number is

$${}_4P_4 = 4 \cdot 3 \cdot 2 \cdot 1 = \mathbf{24}.$$

(b) Each position of the number may be filled by any of the four original digits, so a four-digit number can be formed from the original four digits in

$$4 \cdot 4 \cdot 4 \cdot 4 = \mathbf{256} \text{ ways.}$$

2. How many even four-digit numbers can be formed using the digits 1, 2, 3, and 9 if repetition of digits is *not* allowed?

Solution: An integer is even if it ends in an even number. Therefore, in this case the last digit of the number must be 2. The first three positions of the number can be filled in ${}_3P_3$ ways, so the number of different even numbers that can be formed using these digits is

$${}_3P_3 \cdot 1 = 3 \cdot 2 \cdot 1 \cdot 1 = \mathbf{6}.$$

3. How many three-digit numbers greater than 500 can be formed from the digits, 1, 2, 3, 4, 5, and 6:

(a) without repetition of digits? (b) with repetition of digits?

Solution: (a) The first digit of the number must be greater than or equal to 5. Since two digits in the set of digits are greater than or equal to 5, the first position of the number can be filled in ${}_2P_1$ ways. After the digit for the first position is chosen, the middle and last positions of the three-digit number can be filled from the remaining five digits in ${}_5P_2$ ways. Using the counting principle, we find that the number of three-digit numbers greater than 500 can be formed without repetition of digits is

$${}_2P_1 \cdot {}_5P_2 = 2 \cdot (5 \cdot 4) = \mathbf{40}.$$

(b) Again, the first position of the number can be filled in ${}_2P_1$ ways. After this digit is chosen, any one of the original six digits can be used to fill both the middle and the last position since repetition of digits is allowed. Therefore, the number of different three-digit numbers greater than 500 that can be formed when repetition of digits is allowed is

$${}_2P_1 \cdot 6 \cdot 6 = 2 \cdot (6 \cdot 6) = \mathbf{72}.$$

ARRANGEMENTS OF OBJECTS WITH SOME IDENTICAL. If in a set of n objects some are exactly alike, then the n objects can be arranged in *fewer* different ways than n objects that are all different.

- If in a set of n objects, a objects are identical, then the n objects taken all at a time can be arranged in

$$\frac{n!}{a!} \text{ different ways.}$$

Example: The word BETWEEN has seven letters, and three of these letters (E) are identical. The number of possible different arrangements of these seven letters is

$$\frac{7!}{3!} = \frac{7 \cdot 6 \cdot 5 \cdot 4 \cdot \cancel{3} \cdot \cancel{2} \cdot \cancel{1}}{\cancel{3} \cdot \cancel{2} \cdot \cancel{1}} = \mathbf{840}$$

- If, in a set of n objects, a objects are identical, b objects are identical, c objects are identical, and so forth, then the number of different ways in which the n objects taken all at a time can be arranged is

$$\frac{n!}{a!b!c!\ldots}$$

Examples

4. In how many different ways can four red flags, three blue flags, and one green flag be arranged on a vertical flagpole?

Solution: The eights flags include four alike and three alike, so they can be arranged in

$$\frac{8!}{4! \cdot 3!} = \frac{8 \cdot 7 \cdot \cancel{6} \cdot 5 \cdot \cancel{4} \cdot \cancel{3} \cdot \cancel{2} \cdot \cancel{1}}{(\cancel{4} \cdot \cancel{3} \cdot \cancel{2} \cdot \cancel{1})(\cancel{3} \cdot \cancel{2} \cdot 1)} = \mathbf{280} \text{ ways.}$$

5. The letters of the word CIRCLE are randomly rearranged.
(a) In how many different ways can the six letters be arranged?
(b) In how many different ways can the six letters be arranged if:
(1) the letter *E* is first? *(2)* the first and last letters are *C*?

Solution: (a) Six letters, two of which are identical, can be arranged in

$$\frac{6!}{2!} = \mathbf{360} \text{ ways.}$$

(b) *(1)* If the letter *E* is first, then the remaining five letters include two *C*'s. Five letters, two of which are identical, can be arranged in

$$\frac{5!}{2!} = \mathbf{60} \text{ ways.}$$

(2) If one letter *C* is first and the other letter *C* is last, then the remaining four letters can be arranged in $4! = \mathbf{24}$ ways.

PROBABILITY AND PERMUTATIONS. It may be helpful to think in terms of permutations when solving some types of probability problems, including those in which a random selection of objects (or digits) is made without replacement from a given set of objects (or digits).

Examples

6. An urn contains three red marbles, two white marbles, and four blue marbles. Two marbles, one at a time, are to be chosen from the urn without replacement.

(a) What is the total number of ways in which the two marbles can be selected?

(b) What is the probability that the two marbles selected will be the same color?

Solutions: (a) Two marbles can be selected without replacement from the nine marbles in the urn in ${}_9P_2 = \mathbf{72}$ ways.

(b) Two red marbles can be selected from three red marbles, without replacement, in ${}_3P_2$ ways.

Two white marbles can be selected from two white marbles, without replacement, in ${}_2P_2$ ways.

Two blue marbles can be selected from four blue marbles, without replacement, in ${}_4P_2$ ways. Therefore,

$$\begin{aligned} P(\text{same color}) &= P(\text{both red}) + P(\text{both white}) + P(\text{both blue}) \\ &= \frac{{}_3P_2}{{}_9P_2} + \frac{{}_2P_2}{{}_9P_2} + \frac{{}_4P_2}{{}_9P_2} = \frac{{}_3P_2 + {}_2P_2 + {}_4P_2}{{}_9P_2} \\ &= \frac{3 \cdot 2 + 2 \cdot 1 + 4 \cdot 3}{9 \cdot 8} \\ &= \frac{20}{72} = \mathbf{\frac{5}{18}} \end{aligned}$$

Note: You should compare this solution to the solution of Example 3(b) of Section 12.2 on page 376, which solves the same problem without using permutations.

7. From the set $\{1, 2, 3, 4, 5\}$, a three-digit number is formed by randomly selecting three digits, one at a time, without repetition of digits.
   (a) How many different three-digit numbers can be formed?
   (b) What is the probability that an odd number is formed?

Solutions: (a) A three-digit number can be formed from five digits, without the same digit being used more than once, in ${}_5P_3 = \mathbf{60}$ ways.

(b) An odd number must end in an odd digit. Since there are three odd digits in the given set of numbers, there are ${}_3P_1$ ways in which the last digit of the number can be selected.

The first two digits of the number can be selected from the remaining four digits in ${}_4P_2$ ways.

Using the counting principle, the number of ways in which an odd number can be formed is ${}_4P_2 \cdot {}_3P_1$. Therefore,

$$P(\text{forming an odd number}) = \frac{{}_4P_2 \cdot {}_3P_1}{{}_5P_3} = \frac{(\cancel{4 \cdot 3}) \cdot 3}{5 \cdot \cancel{4 \cdot 3}} = \mathbf{\frac{3}{5}}$$

8. Two cards are randomly drawn, one at a time without replacement, from a standard deck of 52 playing cards. Find the probability of each of the following events:
   (a) Two hearts are drawn.
   (b) The ace of spades and a red card are drawn in that order.

Solution: (a) Method 1: Using Permutations. Two cards can be selected from 52 cards, without replacement, in ${}_{52}P_2$ ways. Two hearts, one followed by another, can be selected from a total of 13 hearts in ${}_{13}P_2$ ways. Therefore,

$$P(\text{two hearts are drawn}) = \frac{{}_{13}P_2}{{}_{52}P_2} = \frac{\overset{1}{\cancel{13}} \cdot 12}{\underset{4}{\cancel{52}} \cdot 51} = \frac{3}{51} = \mathbf{\frac{1}{17}}.$$

Method 2: Without Permutations. To find the joint probability of drawing a heart on the first pick *and* drawing a heart on the second pick, multiply the probabilities of drawing a heart on the two card selections.

$$\begin{aligned} P(\text{two hearts}) &= P(\text{one heart}) \times P(\text{second heart}) \\ &= \frac{13(\text{hearts})}{52(\text{cards})} \times \frac{12(\text{hearts remaining})}{51(\text{cards remaining})} \\ &= \frac{1}{\cancel{4}} \times \frac{\overset{1}{\cancel{4}}}{17} = \mathbf{\frac{1}{17}}. \end{aligned}$$

(b) Method 1: Using Permutations. The ace of spades can be selected in only one way, and one red card can be selected from a total of 26 red cards (13 hearts + 13 diamonds) in ${}_{26}P_1$ ways. Using the counting principle, we find that the ace of spades and one red card can be drawn in $1 \cdot {}_{26}P_1 =$ ways. Therefore,

$$P(\text{ace of spades, a red card}) = \frac{{}_{26}P_1}{{}_{52}P_2} = \frac{\overset{1}{\cancel{26}}}{\underset{2}{\cancel{52}} \cdot 51} = \frac{1}{102}.$$

Method 2: Without Permutations. To find the joint probability of drawing the ace of spades on the first pick *and* drawing a red card on the second pick, multiply the probability of selecting the ace of spades on the first pick by the probability of selecting a red card on the second pick.

$$\begin{aligned} P(\text{ace of spades, a red card}) &= P(\text{ace of spades}) \times P(\text{red card}) \\ &= \frac{1}{52} \times \frac{26}{51} \\ &= \frac{1}{\underset{2}{\cancel{52}}} \times \frac{\overset{1}{\cancel{26}}}{51} = \frac{1}{102}. \end{aligned}$$

EXERCISE SET 12.3

1. Evaluate each of the following:
   (a) $6!$
   (b) $(9-1)!$
   (c) $\dfrac{9!}{4!2!}$
   (d) ${}_4P_4$
   (e) $\dfrac{7!}{6!}$
   (f) $\dfrac{(4+1)!}{(5-2)!}$
   (g) ${}_5P_2$
   (h) ${}_7P_3$
   (i) ${}_5P_4$
   (j) $2({}_3P_3)$
   (k) $\dfrac{10!}{3!2!}$
   (l) $\dfrac{14!}{3!2!}$

2. Show that ${}_8P_3$ and $\dfrac{8!}{(8-3)!}$ are equivalent.

3–12. Find the number of ways in which each of the following activities can be performed:

3. Arranging a chemistry book, a calculus book, a history book, a poetry book, and a dictionary on a shelf so that:
   (a) the dictionary appears last
   (b) the chemistry or history book appears first

4. Forming a three-digit number (without using the same digit more than once) using the digits:
   (a) 1, 3, 5 (b) 2, 4, 6, 8 (c) 0, 1, 2, 3, 4, 5

5. Seating seven students in a row of:
   (a) seven chairs (b) five chairs

6. Using the letters of the word SQUARE (without using the same letter more than once) to form:
   (a) six-letter arrangements
   (b) four-letter arrangements beginning with a vowel

7. Arranging in random order the letters of the word:
   (a) FREEZE
   (b) ARRAY
   (c) PARALLEL
   (d) ELLIPSE
   (e) COMMITTEE

8. Using the digits 2, 4, 6, and 8 to form a three-digit number less than 500:
   (a) without repetition of digits
   (b) with repetition of digits

9. Arranging the letters of the word TRIANGLE so that a vowel comes first.

10. Arranging three black flags, two red flags, and one green flag on a vertical flagpole.

11. Arranging three black flags, two red flags, and one green flag on a vertical flagpole so that:
    (a) the green flag is first
    (b) a black flag is last

12. Arranging the letters of the word PARABOLA so that:
    (a) a vowel is not the first letter
    (b) the three A's are consecutive

13. How many three-digit numbers can be formed from the digits 1, 2, 3, 4, and 5 if:
    (a) each number must be less than 400 and no digit may be repeated?
    (b) each number must be less than 400 and digits may be repeated?
    (c) the middle digit of each number must be an odd number and no digit may be repeated?

14. How many four-digit numbers greater than 1000 can be formed from the digits 0, 1, 2, 3, 4, and 5 if:
    (a) repetition of digits is not allowed?
    (b) repetition of digits is allowed?
    (c) each number must be even and must begin with an odd number, and digits may not be repeated?

15. Find the probability that an odd number will be formed from the digits 2, 4, 5, 6, and 8 if:
    (a) all the digits are used without repetition
    (b) all the digits are used with repetition of digits allowed
    (c) three of the digits are used without repetition
    (d) a three-digit number is formed with repetition of digits allowed

16. What is the probability that, when Allan, Barbara, José, Steve, George, and Maria line up:
    (a) Barbara is first?
    (b) the boys are before the girls?
    (c) a girl is first and last?

17. What is the probability that, when one red, one white, one blue, one green, and one orange marble are placed in a line, the red marble is first and the blue marble is last?

18. Two cards are randomly drawn from a standard deck of 52 playing cards. Find the probability of each of the following events. (Answers may be expressed using permutation notation.)
(a) Two kings are drawn.
(b) A heart and then a club are drawn.
(c) A spade followed by a card of a different suit is drawn.

19. An urn contains three orange marbles, four brown marbles, and two red marbles. Three marbles are drawn in sequence, without replacement. Find the probability of each of the following events. (Answers may be expressed using permutation notation.)
(a) Two orange marbles and a red marble are drawn in that order.
(b) Three marbles having the same color are drawn.

20. If three black flags, two red flags, and one green flag are randomly arranged on a vertical flagpole, find the probability that:
(a) the first and last flags are red
(b) the three black flags are next to each other

21. Three letters of the English alphabet are randomly selected in such a way that the same letter cannot be chosen more than once. Find the probability of each of the following events. (Answers may be expressed using permutation notation.)
(a) Three vowels are selected.
(b) Three vowels are *not* selected.
(c) The letter L and then two vowels are selected.
(d) The letters X, Y, and Z are selected, in that order.

12.4 COMBINATIONS

KEY IDEAS

If from a group of five people a committee consisting of Joe, Susan, and Elizabeth is selected, then this *combination* of three people is the *same* as a committee consisting of Elizabeth, Joe, and Susan. A **combination** is a selection of people or objects in which the identity, rather than the order, of the people or objects is important.

A *permutation*, unlike a combination, is an arrangement of objects in which order is considered. For example, the arrangement in a line of Joe followed by Susan followed by Elizabeth is *different* from the arrangement in a line of Elizabeth followed by Joe followed by Susan.

COMBINATION NOTATION. The notation $_nC_r$ is read as "the combination of n objects taken r at a time," and represents the number of different ways r objects can be selected from a group of n objects ($r \leq n$) where the order of the r objects does not matter. For example, the number of different ways in which three people can be selected from a group of five people to serve on a committee can be represented by $_5C_3$. On the other hand, $_5P_3$ may be interpreted as the number of different ways in which three people selected from a group of five people can be arranged in a line.

EVALUATING COMBINATIONS. Here are four rules for evaluating expressions that use combination notation.

- To evaluate $_nC_r$, use the formula

$$_nC_r = \frac{_nP_r}{r!}.$$

Sometimes the notation $\binom{n}{r}$ or $C(n, r)$ is used instead of $_nC_r$.

Example: To evaluate $_5C_3$, let $n = 5$ and $r = 3$; then

$$_5C_3 = \frac{_5P_3}{3!} = \frac{5 \cdot \overset{2}{\cancel{4}} \cdot \cancel{3}}{\cancel{3} \cdot \cancel{2} \cdot 1} = \mathbf{10}.$$

- $_nC_n = 1$.

Example: There is exactly one way in which a committee of five people can be chosen from a group of five people. Mathematically this situation is represented by

$$_5C_5 = \frac{_5P_5}{5!} = \frac{5 \cdot 4 \cdot 3 \cdot 2 \cdot 1}{5 \cdot 4 \cdot 3 \cdot 2 \cdot 1} = \mathbf{1}.$$

- $_nC_r = {_nC_{n-r}}$.

Example: The number of different committees of 13 people that can be chosen from a group of 15 people is given by $_{15}C_{13}$, where

$$_{15}C_{13} = \frac{_{15}P_{13}}{13!} = \frac{15 \cdot \overset{7}{\cancel{14}} \cdot \cancel{13 \cdot 12 \cdot 11 \cdot 10 \cdot 9 \cdot 8 \cdot 7 \cdot 6 \cdot 5 \cdot 4 \cdot 3}}{\cancel{13 \cdot 12 \cdot 11 \cdot 10 \cdot 9 \cdot 8 \cdot 7 \cdot 6 \cdot 5 \cdot 4 \cdot 3} \cdot \cancel{2} \cdot 1} = \mathbf{105}.$$

This situation is equivalent to selecting two people *not* to serve on the committee, which can be accomplished in

$$_{15}C_2 = \frac{_{15}P_2}{2!} = \frac{15 \cdot \overset{7}{\cancel{14}}}{\cancel{2} \cdot 1} = \mathbf{105} \text{ ways.}$$

Notice that $_{15}C_{13} = {_{15}C_2}$ and that $_{15}C_2$ is easier to calculate than $_{15}C_{13}$. Therefore, when n and r are both large and close in value, calculate $_nC_{n-r}$ rather than $_nC_r$.

- $_nC_0 = 1$ since $_nC_0 = {_nC_{n-0}} = {_nC_n} = 1$.

Examples

1. An urn contains two red marbles, three white marbles, and five blue marbles. In how many ways can a set of seven marbles be selected?

Solution: A group of seven marbles ($r = 7$) can be selected from a group of 10 marbles ($n = 2 + 3 + 5 = 10$) in

$${}_{10}C_7 = {}_{10}C_{10-7} = \frac{{}_{10}P_3}{3!} = \frac{10 \cdot \overset{3}{\cancel{9}} \cdot \overset{4}{\cancel{8}}}{\cancel{3} \cdot \cancel{2} \cdot 1} = \mathbf{120} \text{ ways.}$$

2. The coach of a team is going to select five players at random from a group of 11 students trying out for the team. If Lois is one of the 11 students trying out, how many five-player combinations will:
(a) include Lois? (b) not include Lois?

Solutions: (a) Since Lois must be on the team, the other four team players can be selected from the remaining 10 players in ${}_{10}C_4$ ways. Therefore, the number of five-player combinations that include Lois is:

$${}_{10}C_4 = \frac{{}_{10}P_4}{4!} = \frac{10 \cdot \overset{3}{\cancel{9}} \cdot \overset{2}{\cancel{8}} \cdot 7}{\underset{1}{\cancel{4}} \cdot \cancel{3} \cdot \cancel{2} \cdot 1} = \mathbf{210}\,.$$

(b) Since Lois cannot be on the team, the five team players must be selected from the remaining 10 students. Therefore, the number of five-player combinations that do *not* include Lois is:

$${}_{10}C_5 = \frac{{}_{10}P_5}{5!} = \frac{\overset{2}{\cancel{10}} \cdot 9 \cdot \overset{2}{\cancel{8}} \cdot 7 \cdot \cancel{6}}{\cancel{5} \cdot \cancel{4} \cdot \cancel{3} \cdot \cancel{2} \cdot 1} = \mathbf{252}\,.$$

3. Solve for x: ${}_xC_2 = 21$.

Solution:

$$\begin{aligned} {}_xC_2 = \frac{x(x-1)}{2!} &= 21 \\ \frac{x^2 - x}{2} &= 21 \\ x^2 - x &= 2 \cdot 21 \\ x^2 - x - 42 &= 0 \\ (x-7) \quad (x+6) &= 0 \\ (x - 7 = 0) \vee (x + 6 &= 0) \\ x = \mathbf{7} \;\big|\; x &= -6 \text{ (Reject since } x \text{ must be positive.)} \end{aligned}$$

COMBINATIONS AND THE COUNTING PRINCIPLE. To find the number of ways two or more selection processes can occur together, apply the counting principle and find the product of the number of ways each can occur.

Example

4. From a group of six boys and three girls, how many six-member committees can be formed if:

(a) each committee must have four boys and two girls?
(b) each committee must have the same number of boys and girls?
(c) each committee must have *at least* one girl?

Solutions: (a) Four boys can be selected from six boys in ${}_6C_4$ ways. Two girls can be selected from three girls in ${}_3C_2$ ways.

Use the counting principle: the number of ways in which four boys *and* two girls can be selected for a six-member committee is

$${}_6C_4 \cdot {}_3C_2 = \frac{6 \cdot 5 \cdot 4 \cdot 3}{4 \cdot 3 \cdot 2 \cdot 1} \cdot \frac{3 \cdot 2}{2 \cdot 1} = 15 \cdot 3 = \mathbf{45}.$$

(b) Each committee must have three boys and three girls.
Three boys can be selected from six boys in ${}_6C_3$ ways.
Three girls can be selected from three girls in ${}_3C_3$ ways.
Use the counting principle: the number of ways in which equal numbers of boys *and* girls can be selected for each committee is

$${}_6C_3 \cdot {}_3C_3 = 20 \cdot 1 = \mathbf{20}.$$

(c) Here are the possible compositions of the committee:

Committee	Number of Different Committees
1 girl and 5 boys	${}_3C_1 \cdot {}_6C_5 = 3 \cdot 6 = 18$
2 girls and 4 boys	${}_3C_2 \cdot {}_6C_4 = 3 \cdot 15 = 45$
3 girls and 3 boys	${}_3C_3 \cdot {}_6C_3 = 1 \cdot 20 = 20$

The number of possible committees with *at least* 1 girl $= 18 + 45 + 20 = \mathbf{83}$.

PROBABILITY AND COMBINATIONS. Combinations can also be used to solve probability problems involving selections in which order does not matter.

Examples

5. From a group of six boys and three girls, a six-member committee is formed.

(a) How many six-member committees can be formed?
(b) If Maria is one of the girls, what is the *probability* that Maria is *not* selected for the committee?

(c) What is the *probability* that a six-member committee will have more boys than girls?

(d) What is the *probability* that the six-member committee will *not* include a girl?

(e) What is the *probability* that the six-member committee will include a boy?

Solutions: (a) The number of six-member committees that can be selected from nine people (six boys + three girls) is ${}_9C_6$ or **84**.

(b) If Maria is excluded, then the number of ways in which a six-member committee can be selected from the remaining eight students is ${}_8C_6$.

$$P(\text{Maria is } \textit{not} \text{ selected}) = \frac{{}_8C_6}{{}_9C_6} = \frac{28}{84} = \mathbf{\frac{1}{3}}$$

(c) A six-member committee having more boys than girls must include four or more boys.

Committee	Number of Different Committees
4 boys and 2 girls	${}_6C_4 \cdot {}_3C_2 = 15 \cdot 3 = 45$
5 boys and 1 girl	${}_6C_5 \cdot {}_3C_1 = 6 \cdot 3 = 18$
6 boys and 0 girls	${}_6C_6 \cdot {}_3C_0 = 1 \cdot 1 = 1$

$$P(\text{more boys than girls}) = \frac{45 + 18 + 1}{{}_9C_6} = \frac{64}{84} = \mathbf{\frac{16}{21}}$$

(d) $P(\text{no girl}) = P(\text{6 boys}) = \mathbf{\frac{1}{84}}$.

(e) Since there are only three girls, a six-member committee must include a boy, so $P(\text{committee includes a boy}) = \mathbf{1}$.

6. A golf bag contains seven white balls, three orange balls, and two yellow balls. At random, three golf balls are selected.

(a) How many different selections of three golf balls can be made?

(b) What is the probability that the three golf balls selected will all be the same color?

Solutions: (a) Three golf balls can be selected from the 12 golf balls in the bag in ${}_{12}C_3 = \mathbf{220}$ ways.

(b) Since there are only two yellow golf balls, the probability that three yellow balls are selected is 0.

Three white balls can be selected from seven white balls in ${}_7C_3$ ways.

Three orange balls can be selected from three orange balls in ${}_3C_3$ ways.

$$P(\text{3 balls are same color}) = P(\text{3 white}) + P(\text{3 orange}) + P(\text{3 yellow})$$
$$= \frac{{}_7C_3}{{}_{12}C_3} + \frac{{}_3C_3}{{}_{12}C_3} + 0$$
$$= \frac{35 + 1}{220} = \frac{36}{220} = \frac{9}{55}$$

7. Two cards are drawn at random (without replacement) from a standard deck of 52 playing cards. Find the probability of each of the following events. (Leave the answer in notation form.)
(a) Both cards are aces.
(b) One card is 7, and the other card is a picture card.
(c) Both cards are in the same suit.
(d) A heart is *not* drawn.

Solutions: In each of parts (a)–(d), two cards are drawn out of 52 with restrictions, so that ${}_{52}C_2$ represents the number of elements in the sample space.

(a) Two aces can be selected out of four aces in ${}_4C_2$ ways.

$$P(\text{two aces are drawn}) = \frac{{}_4C_2}{{}_{52}C_2}$$

(b) A deck includes four 7's and 12 picture cards.
A 7 can be selected in ${}_4C_1$ ways.
A picture card can be selected in ${}_{12}C_1$ ways.

$$P(\text{7 and picture card}) = \frac{{}_4C_1 \cdot {}_{12}C_1}{{}_{52}C_2}$$

(c) Each of the four suits contains 13 cards, so two cards can be selected from each suit in ${}_{13}C_2$ ways.

$$P(\text{same suit}) = \frac{\overset{\textit{Hearts}}{{}_{13}C_2} + \overset{\textit{Diamonds}}{{}_{13}C_2} + \overset{\textit{Spades}}{{}_{13}C_2} + \overset{\textit{Clubs}}{{}_{13}C_2}}{{}_{52}C_2}$$

(d) A deck has 13 hearts. If a heart is not drawn, there are 39 cards left from which to draw two cards.

$$P(\textit{not}\text{ a heart}) = \frac{{}_{39}C_2}{{}_{52}C_2}$$

EXERCISE SET 12.4

1. Evaluate each of the following:

(a) ${}_8C_3$	(d) ${}_{17}C_{17}$	(g) $4! \cdot {}_6C_4$
(b) ${}_{11}C_8$	(e) ${}_7C_5 \cdot {}_{12}C_{10}$	(h) $7! \cdot {}_{15}C_8$
(c) ${}_9C_0$	(f) $\frac{{}_{10}C_3}{{}_5C_3}$	(i) $\binom{10}{4}$

2. In each case, find x:
(a) ${}_6C_x = 1$ (c) ${}_xC_7 = {}_xC_2$ (e) ${}_xC_2 = x$
(b) ${}_xC_2 = 15$ (d) ${}_xC_2 = 45$ (f) $6({}_xC_5) = {}_{(x+2)}C_5$

3. How many committees of three can be chosen from a class of 10 students?

4. How many triangles can be formed using as vertices nine points, no three of which are collinear?

5. How many different diagonals can be drawn in a polygon having nine sides?

6. If a student wants to take four pens and three books from a desk on which there are six pens and seven books, how many choices does the student have?

7. In how many ways can a committee of four be selected from five women and three men if:
(a) the same person is always included in the committee?
(b) the same person is always excluded from the committee?
(c) the committee includes more women than men?
(d) the committee includes all men?
(e) the committee has no more than two men?

8. A three-person committee is selected from three men and four women. What is the probability that:
(a) the members of the committee are all men or all women?
(b) Juan, one of the men, is *always* selected for the committee?
(c) Christine, one of the women, is *never* included on the committee?
(d) at least two women are selected?

9. Three letters of the English alphabet are selected at random, so that the same letter cannot be picked more than once. Find the probability of each of the following events. (Leave answer in notation form.)
(a) No vowels are selected.
(b) Exactly one vowel is selected.
(c) At least one vowel is selected.
(d) The letter *A* is selected.

10. A drawer holds five navy socks, seven brown socks, and six black socks. If two socks are selected without looking, what is the probability that they match?

11–12. Three marbles are randomly selected without replacement from an urn that contains seven red, five blue, and three white marbles. (Leave answers in notation form.)

11. In how many different ways can:
    (a) the three marbles be selected?
    (b) a blue marble *not* be selected?
    (c) three marbles of the same color be selected?
    (d) three marbles of different colors be selected?

12. Find the probability of each of the following events:
    (a) Three red marbles are selected.
    (b) One red and two white marbles are selected.
    (c) At least one of the marbles is blue.
    (d) No more than two marbles are red.

13. A five-card poker hand is dealt from a deck of 52 shuffled playing cards. Find the probability of each of the following events. (Leave answer in notation form.)
    (a) The five-card hand consists of two aces and three kings.
    (b) The five-card hand is a flush (all cards are in the same suit).
    (c) The five-card hand includes four aces.
    (d) The five-card hand includes any four of a kind.

14. The math department of a certain high school has five classes of Course *A*, four classes of Course *B*, and three classes of Course *C*.
    (a) If a teacher's program consists of five classes, how many combinations of five classes are possible?
    (b) How many five class teacher programs will have three Course *A*'s and two Course *B*'s?
    (c) If a teacher's program consists of five classes, what is the probability that a teacher's program will have three Course *A*'s and two Course *B*'s?

15. Coach Euclid will select six players at random from a group of 10 students trying out for a team.
    (a) How many different six-player combinations are possible?
    (b) If Jill is one of the 10 students trying out for the team, how many of the six-player combinations will include Jill?
    (c) What is the probability that Jill will be selected as one of the six players?
    (d) After selecting the team, Coach Euclid asked the six members to stand in a straight line. How many different lineups are possible?
    (e) If Jill is on the team, what is the probability that she will be standing first in the lineup?

CHAPTER 12 REVIEW EXERCISES

1. How many three-digit numbers can be formed using the digits 1, 2, and 3 if repetition is *not* allowed?

2. How many different committees of five students can be chosen from eight students?

3. How many different five-letter permutations are there of the letters of the word PROOF?

4. There are seven points on a circle. How many straight lines can be drawn joining pairs of these seven points?

5. A set contains five quadrilaterals: a rectangle, a rhombus, a parallelogram, a square, and an isosceles trapezoid. If one quadrilateral is selected from the set at random, what is the probability that its diagonals are congruent?

6. Three students are chosen to form a committee from the membership of a club of four seniors and six juniors. How many different committees consisting of one senior and two juniors can be formed?

7. A three-digit number is formed by selecting from the digits 1, 2, 5, and 6, with no repetition. What is the probability that the number formed is greater than 500?

8. If a committee of two boys and two girls is seated in a line, in how many different ways can it be arranged so that two members of the same sex do *not* sit next to each other?

9. The expression ${}_{25}C_{20}$ is equivalent to:
(1) ${}_{25}P_{20}$ (2) ${}_{25}C_5$ (3) ${}_{25}P_5$ (4) 5!

10. In a 52-card deck, the probability of drawing either a jack or a 3 in one draw is:
(1) $\frac{4}{52}$ (2) $\frac{8}{52}$ (3) $\frac{13}{52}$ (4) $\frac{26}{52}$

11. An urn contains six red balls and four white balls, all of equal size. If three balls are selected at random (no replacement), what is the probability that all three balls selected are the same color?
(1) $\frac{1}{5}$ (2) $\frac{1}{6}$ (3) $\frac{2}{3}$ (4) $\frac{1}{30}$

12. A track team consists of six athletes. For a relay race, four runners have been chosen.
(a) In how many orders can the four athletes run the race?
(b) What is the probability that a particular athlete (from the original six athletes) will lead off the race?

13. A committee of four is to be chosen from two men and four women.
(a) What is the probability that the committee will contain exactly one man?
(b) What is the probability that a woman will be on the committee?

14. A committee of five people chosen for a class function has three male and two female members.
(a) How many three-person subcommittees having *at least* two males can be formed?
(b) What is the probability of a three-person subcommittee having exactly one male member?

15. From a standard deck of 52 cards, two cards are drawn at random with no replacement. State the probability of selecting:
(a) the ace of spades and the king of spades (in that order)
(b) a pair of aces
(c) any ace and any king (in that order)
(d) any ace and any king (in either order)
(e) two spades

16. (a) From the letters in the word OCTAGON, how many distinct arrangements using all seven letters are possible?
(b) From the letters in the word TRAPEZOID, how many distinct combinations using seven letters at a time are possible?
(c) From the letters in the word SQUARE, what is the probability that in a six-letter arrangement the first two letters are vowels?
(d) From the letters in the word RHOMBUS, what is the probability that in a seven-letter arrangement the first three letters are vowels?

17. The student government at Central High School consists of four seniors, three juniors, three sophomores, and two freshmen.
(a) How many committees of four students can be formed?
(b) How many committees of four students will have exactly one student from each grade?
(c) What is the probability that a committee of four students will have exactly one student from each grade?
(d) Nine students will be chosen from the student government to go to a convention. What is the probability that no senior will be chosen to go?

18. On an examination a student is to select any four out of nine problems. All of the problems are of equal difficulty. The examination contains one geometry, three logic, one locus, and four probability problems.
(a) How many four-problem selections can be made?
(b) How many of those selections will contain one logic, one locus, and two probability problems?
(c) What is the probability that a four-problem selection contains one logic, one locus, and two probability problems?
(d) What is the probability that a four-problem selection will contain all logic problems?

REGENTS TUNE-UP: CHAPTERS 9–12

Here is an opportunity for you to review Chapters 9–12 and, at the same time, prepare for the Course II Regents Examination. Problems included in this section are similar in form and difficulty to those found on the New York State Regents Examination for Course II of the Three-Year Sequence for High School Mathematics. Problems preceded by an asterisk have actually appeared on a previous Course II Regents Examination.

***1.** What is the slope of the line that passes through points (4, 9) and (−1, 12)?

***2.** How many different five-letter permutations can be formed from the letters in the word ERROR?

***3.** Find the distance between the points whose coordinates are (2, 7) and (8, −1).

***4.** The coordinates of the vertices of $\square ABCD$ are $A(2, 1)$, $B(4, 3)$, $C(10, 3)$, $D(x, 1)$. What is the value of x?

***5.** Write an equation of the line that passes through point (2, 7) and has a slope of 4.

***6.** A set contains five quadrilaterals: a rectangle, a rhombus, a parallelogram, a square, and an isosceles trapezoid. If one quadrilateral is selected from the set at random, what is the probability that its diagonals bisect each other?

***7.** The endpoints of $\overline{AB}$ are $A(x, 3)$ and $B(4, 7)$. If the coordinates of the midpoint M of $\overline{AB}$ are (−1, 5), find x.

***8.** From a menu of five sandwiches and five beverages, how many different lunches consisting of two different sandwiches and one beverage can be selected?

***9.** Write an equation of the line that passes through points (2, 1) and (6, 3).

***10.** The line that passes through points (−2, 3) and (5, y) has a slope of $\frac{4}{7}$. Find y.

***11.** For the graph of which equation is $x = 2$ an equation of the axis of symmetry?

(1) $3x^2 + 6x - 8 = y$ (3) $x^2 - 4x - 6 = y$

(2) $x^2 + 2x - 3 = y$ (4) $4x^2 - 2x + 10 = y$

***12.** Two points, L and P, are 8 units apart. How many points are equidistant from L and P and also 3 units from L?
(1) 1 (2) 2 (3) 0 (4) 4

***13.** Which are the coordinates of the turning point of the parabola whose equation is $y = x^2 - 8x + 5$?

(1) (0, 5) (2) (5, 0) (3) (8, 5) (4) (4, −11)

***14.** A bag of marbles contains two green, one blue, and three red marbles. If two marbles are chosen at random without replacement, what is the probability that both will be red?

(1) $\frac{1}{5}$ (2) $\frac{1}{6}$ (3) $\frac{1}{10}$ (4) $\frac{1}{12}$

***15.** How many points do the graphs of $x^2 + y^2 = 9$ and $y = 4$ have in common?
(1) 1 (2) 2 (3) 0 (4) 4

***16.** An equation of the line that passes through point (3, 2) and has a slope of 4 is:
(1) $y - 2 = 4(x - 3)$ (3) $y - 2 = -4(x - 3)$

(2) $y + 2 = 4(x - 3)$ (4) $y - 2 = \frac{1}{4}(x - 3)$

***17.** Which pair of points will determine a line parallel to the x-axis?
(1) (1, 3) and (−2, 3) (3) (1, 3) and (1, −1)
(2) (1, −1) and (−1, 1) (4) (1, 1) and (−3, −3)

***18.** Point C is 3 centimeters from line $\overleftrightarrow{AB}$. The number of points on $\overleftrightarrow{AB}$ that are also 5 centimeters from C is:
(1) 1 (2) 2 (3) 3 (4) 0

***19.** Which is a solution for the system of equations $y = 2x - 15$ and $y = x^2 - 6x$?
(1) (3, −9) (2) (0, 0) (3) (5, 5) (4) (6, 0)

***20.** Point P lies between two parallel lines a and b, which are 3 centimeters apart. What is the total number of points equidistant from a and b, and also 2 centimeters from P?
(1) 1 (2) 2 (3) 3 (4) 4

***21.** The graph of which equation is perpendicular to the graph of

$y = \frac{1}{2}x + 3$?

(1) $y = -\frac{1}{2}x + 5$ (2) $2y = x + 3$ (3) $y = 2x + 5$ (4) $y = -2x + 3$

***22.** What is the total number of points that are 2 units from the y-axis and also 3 units from the origin?
(1) 0 (2) 2 (3) 3 (4) 4

***23.** Which is a point of intersection of the graphs of the line $y = x$ and the parabola $y = x^2 - 2$?
(1) (1, 1) (2) (2, 2) (3) (0, 0) (4) (2, −1)

***24.** The equation of the locus of points equidistant from the graphs of $x = 3$ and $x = -5$ is:
(1) $y = 1$ (2) $y = -1$ (3) $x = 1$ (4) $x = -1$

***25.** Which is an equation of the line that is parallel to $y = 2x - 8$ and passes through point (0, −3)?
(1) $y = 2x + 3$ (2) $y = 2x - 3$ (3) $y = -\frac{1}{2}x + 3$ (4) $y = -\frac{1}{2}x - 3$

***26.** (a) Draw the graph of the equation $y = \frac{1}{2}x^2 - 4x + 4$, including all values of x such that $0 \le x \le 8$.
(b) Write the coordinates of the turning point.
(c) Between which pair of consecutive integers does one root of the equation $\frac{1}{2}x^2 - 4x + 4 = 0$ lie?
(1) 0 and 1 (2) 1 and 2 (3) 2 and 3 (4) −1 and 0

***27.** The vertices of $\triangle ABC$ are $A(-4, -2)$, $B(2, 6)$, $C(2, -2)$.
(a) Write an equation of the locus of points equidistant from vertex B and vertex C.
(b) Write an equation of the line parallel to $\overline{BC}$ and passing through vertex A.
(c) Find the coordinates of the point of intersection of the locus in part (a) and the line determined in part (b).
(d) Write an equation of the locus of points that are 4 units from vertex C.
(e) What is the total number of points that satisfy the loci described in parts (a) and (d)?

28. Given points $A(0, 0)$, $B(8, 6)$, and $C(8, 0)$.
(a) Graph $\triangle ABC$.
(b) Graph and state the coordinates of $\triangle A'B'C'$, the image of $\triangle ABC$ under the transformation $(x, y) \rightarrow \left(\frac{3}{2}x, \frac{3}{2}y\right)$.
(c) Graph and state the coordinates of $\triangle A''B''C''$, the image of $\triangle ABC$ under the transformation $(x, y) \rightarrow (x + 3, y - 7)$.

29. The coordinates of the endpoints of line segment $\overline{AB}$ are $A(4, 1)$ and $B(5, 4)$.
(a) Graph $\overline{AB}$.
(b) Graph $\overline{A'B'}$, the image of $\overline{AB}$, after a reflection over the line $y = x$.
(c) Graph $\overline{A''B''}$, the image of $\overline{A'B'}$, after the transformation $(x, y) \rightarrow (x - 5, y - 5)$.
(d) Graph $\overline{A'''B'''}$, the image of $\overline{A''B''}$, after a reflection through the origin.
(e) Write a translation rule that will make $\overline{B'''A'''}$ the image of $\overline{A'B'}$.

***30.** Solve the following system of equations algebraically and check:

$$y = 3x^2 - 8x + 5$$
$$x + y = 3$$

***31.** Quadrilateral $ABCD$ has vertices $A(-4, -2)$, $B(0, 5)$, $C(9, 3)$, $D(7, -4)$.
(a) On graph paper, plot the vertices and draw quadrilateral $ABCD$.
(b) Find the area of quadrilateral $ABCD$.

***32.** (a) Describe completely the locus of points n units from point $P(3, 2)$.
(b) Describe completely the locus of points 2 units from the line whose equation is $x = 3$.
(c) What is the total number of points that satisfy the conditions in parts (a) and (b) simultaneously for the following values of n?
(1) $n < 2$ *(2)* $n = 2$.

***33.** (a) Draw a graph of the equation $y = -x^2 + 4x - 3$ for all values of x such that $-1 \leq x \leq 5$.
(b) On the same set of axes, draw the graph of $y + 1 = x$.
(c) Using the graphs drawn in parts (a) and (b), determine the solution of the system:

$$y = -x^2 + 4x - 3$$
$$y + 1 = x$$

***34.** From the members of a band consisting of five clarinet players, four trumpet players, and three tuba players, a three-member group is to be formed.

(a) How many three-member groups can be formed?

(b) What is the probability that the three-member group formed consists of one clarinet player, one trumpet player, and one tuba player?

(c) What is the probability that the three-member group formed consists of three clarinet players, three trumpet players, or three tuba players?

***35.** (a) Find, in radical form, the roots of $x^2 - 6x + 3 = 0$.

(b) Draw the graph of the equation $y = x^2 - 6x + 3$, including all values of x such that $0 \leq x \leq 6$.

***36.** Quadrilateral $JAKE$ has coordinates $J(0, 3a)$, $A(3a, 3a)$, $K(4a, 0)$, $E(-a, 0)$. Prove by coordinate geometry that quadrilateral $JAKE$ is an isosceles trapezoid.

***37.** There are five boys and three girls in a math class.

(a) In how many different ways may the eight students be arranged in a line?

(b) In how many different ways may the eight students be arranged in a line if all three girls are to precede the five boys?

(c) How many six-member committees can be formed from the eight students?

(d) How many of these committees consist of exactly four boys and two girls?

(e) What is the probability that one of the six-member committees consists of exactly four boys and two girls?

Table of Values of Trigonometric Ratios

Angle	Sin	Cos	Tan	Angle	Sin	Cos	Tan
0°	0.0000	1.0000	0.0000	45°	0.7071	0.7071	1.0000
1°	0.0175	0.9998	0.0175	46°	0.7193	0.6947	1.0355
2°	0.0349	0.9994	0.0349	47°	0.7314	0.6820	1.0724
3°	0.0523	0.9986	0.0524	48°	0.7431	0.6691	1.1106
4°	0.0698	0.9976	0.0699	49°	0.7547	0.6561	1.1504
5°	0.0872	0.9962	0.0875	50°	0.7660	0.6428	1.1918
6°	0.1045	0.9945	0.1051	51°	0.7771	0.6293	1.2349
7°	0.1219	0.9925	0.1228	52°	0.7880	0.6157	1.2799
8°	0.1392	0.9903	0.1405	53°	0.7986	0.6018	1.3270
9°	0.1564	0.9877	0.1584	54°	0.8090	0.5878	1.3764
10°	0.1736	0.9848	0.1763	55°	0.8192	0.5736	1.4281
11°	0.1908	0.9816	0.1944	56°	0.8290	0.5592	1.4826
12°	0.2079	0.9781	0.2126	57°	0.8387	0.5446	1.5399
13°	0.2250	0.9744	0.2309	58°	0.8480	0.5299	1.6003
14°	0.2419	0.9703	0.2493	59°	0.8572	0.5150	1.6643
15°	0.2588	0.9659	0.2679	60°	0.8660	0.5000	1.7321
16°	0.2756	0.9613	0.2867	61°	0.8746	0.4848	1.8040
17°	0.2924	0.9563	0.3057	62°	0.8829	0.4695	1.8807
18°	0.3090	0.9511	0.3249	63°	0.8910	0.4540	1.9626
19°	0.3256	0.9455	0.3443	64°	0.8988	0.4384	2.0503
20°	0.3420	0.9397	0.3640	65°	0.9063	0.4226	2.1445
21°	0.3584	0.9336	0.3839	66°	0.9135	0.4067	2.2460
22°	0.3746	0.9272	0.4040	67°	0.9205	0.3907	2.3559
23°	0.3907	0.9205	0.4245	68°	0.9272	0.3746	2.4751
24°	0.4067	0.9135	0.4452	69°	0.9336	0.3584	2.6051
25°	0.4226	0.9063	0.4663	70°	0.9397	0.3420	2.7475
26°	0.4384	0.8988	0.4877	71°	0.9455	0.3256	2.9042
27°	0.4540	0.8910	0.5095	72°	0.9511	0.3090	3.0777
28°	0.4695	0.8829	0.5317	73°	0.9563	0.2924	3.2709
29°	0.4848	0.8746	0.5543	74°	0.9613	0.2756	3.4874
30°	0.5000	0.8660	0.5774	75°	0.9659	0.2588	3.7321
31°	0.5150	0.8572	0.6009	76°	0.9703	0.2419	4.0108
32°	0.5299	0.8480	0.6249	77°	0.9744	0.2250	4.3315
33°	0.5446	0.8387	0.6494	78°	0.9781	0.2079	4.7046
34°	0.5592	0.8290	0.6745	79°	0.9816	0.1908	5.1446
35°	0.5736	0.8192	0.7002	80°	0.9848	0.1736	5.6713
36°	0.5878	0.8090	0.7265	81°	0.9877	0.1564	6.3138
37°	0.6018	0.7986	0.7536	82°	0.9903	0.1392	7.1154
38°	0.6157	0.7880	0.7813	83°	0.9925	0.1219	8.1443
39°	0.6293	0.7771	0.8098	84°	0.9945	0.1045	9.5144
40°	0.6428	0.7660	0.8391	85°	0.9962	0.0872	11.4301
41°	0.6561	0.7547	0.8693	86°	0.9976	0.0698	14.3007
42°	0.6691	0.7431	0.9004	87°	0.9986	0.0523	19.0811
43°	0.6820	0.7314	0.9325	88°	0.9994	0.0349	28.6363
44°	0.6947	0.7193	0.9657	89°	0.9998	0.0175	57.2900
45°	0.7071	0.7071	1.0000	90°	1.0000	0.0000	—

Selected Answers to Chapter Review and Regents Tune-up Exercises

CHAPTER 1

1. (2)
2. (4)
3. (3)
4. (4)
5. (2)
6. (2)
7. (1)
8. (2)
9. (1)
10. (2)
11. (4)
12. (1)
13. (2)
14. (a) (1)
(b) (4)
(c) 2
(d) (5)
(e) (3)
(f) (3)

15. (a) It will not snow this weekend.
(b) Flowers will not grow in May.
(c) Mary owes the library a quarter.
(d) If I play tennis, then I will not go swimming.
(e) I will stay home.
(f) $\sim r \rightarrow \sim t$ (g) y
(h) Jim will get an A. (i) $x = 8$
(j) Crunchies are good to eat.

16. (b)

Statement	Reason
1. $\sim S$	1. Given.
2. $P \vee S$	2. Given.
3. P	3. Law of Disjunctive Inference.
4. $A \rightarrow \sim P$	4. Given.
5. $\sim A$	5. Modus Tollens (3, 4).
6. $R \rightarrow A$	6. Given.
7. $\sim R$	7. Modus Tollens (5, 6).

CHAPTER 2

1. 1 **2.** d **3.** (1)
4. (a) A
(b) N
(c) A
(d) N
(e) I
5. (b) 0
(c) 3
(d) 1
(e) (1) 2 (2) 0 (3) 3
(f) 4

CHAPTER 3

1. $2a - 3b - 3c$
2. $2x^2 + x - 1$
3. $6x^3 - 15x^2$
4. $4x^2 + 4x + 1$
5. $8x + 4$
6. $8x^2 + 14x + 3$
7. $k = 5$
8. $x^2 - 2x - 15$
9. $2x^4 - x^3 - 6x^2 + 13x - 5$
10. (a) $(y - 7)(y + 4)$ (b) $2(x - 7)(x + 6)$ (c) $-(t - 5)(t + 2)$
11. (a) $(2p + 3)(p - 2)$ (b) $(3q - 2)(q + 4)$ (c) $(2w - 1)^2$
12. (a) $3(b - 2)(b + 2)$ (b) $y(y - 4)(y + 4)$ (c) $-r(r + 4)(r - 1)$
13. $3x - 5$
14. $2a - 5$
15. $-x^2 + 5x - 1$
16. $2mp(6m^2 + 3mp - 2)$
17. $3x^2 + 7x - 4$
18. $x^2 + 3$
19. $-3x^2 - 5x + 7$
20. $7x^2 - 8x + 2$
21. (2)
22. (4)
23. (3)
24. (1)
25. (3)
26. (3)
27. (3)
28. (2)
29. (4)
30. (1)
31. (4)
32. (3)
33. (4)

CHAPTER 4

1. $x^2 - x - 12 = 0$
2. $x = 2$
3. 48
4. 32
5. (a) 3
   (b) $-2, 1$
   (c) 0
   (d) $\emptyset$
6. 25
7. $\dfrac{6 \pm \sqrt{24}}{2}$
8. (a) $x^2 + 4x + 4$
   (b) $4x^2 - 6x - 4$
   (c) 4
9. 13
10. (3)
11. (2)
12. (3)
13. (4)
14. (1)
15. (4)
16. (a) $x^2 + (8 - x)^2 = 36$
    (b) $\dfrac{8 \pm \sqrt{120}}{2}$
17. Length = 8
    Width = 6
18. $\dfrac{9 \pm \sqrt{57}}{2}$

CHAPTER 5

1. $x + 2$
2. $\dfrac{b^2 - 1}{8b}$
3. 1/2
4. $\dfrac{3(x - 1)}{4(x - 2)}$
5. $\dfrac{-x + 18}{12}$
6. $\dfrac{5a - b}{6}$
7. $\dfrac{-4}{x - 4}$
8. $\dfrac{3a - 10}{a - 5}$
9. $\dfrac{x - 4}{x + 2}$
10. $\dfrac{3(x - 2)}{2x(x - 1)}$
11. $5(x - 1)$
12. $\dfrac{3x}{x + 3}$
13. $\dfrac{2}{a - 1}$
14. $-3(x + y)$
15. $\dfrac{3a(x - 8)}{b^2(x - 9)}$
16. $x - 3$
17. $\dfrac{-4xy}{x^2 - y^2}$
18. $\dfrac{x(x + y)}{y}$
19. $\dfrac{5}{1 - a}$
20. $\dfrac{3p}{(p - 3)(p + 2)}$
21. 48
22. 3
23. $\dfrac{1}{4}$
24. -1
25. 1/8
26. 8
27. $7, -1$
28. $4, -3$

REGENTS TUNE-UP: CHAPTERS 1–5

1. 1
2. 8
3. $2/a$
4. $\dfrac{y}{1-y}$
5. $\dfrac{x}{2x-3}$
6. $\dfrac{x+2}{x(x-1)}$
7. (1)
8. (1)
9. (2)
10. (2)
11. (1)
12. (2)
13. (4)
14. (4)
15. (4)
16. (4)
17. (2)
18. (4)
20. (a) 3
    (b) 2
    (c) 3
    (d) 1 lacks an inverse.

CHAPTER 6

1. 25
2. 13
3. 90
4. 18
5. (3)
6. (1)
7. (2)
8. 15
9. Plan: Show $\triangle AFB \cong \triangle DEC$ by SAS since $\overline{AF} \cong \overline{DE}$, $\angle GFE \cong \angle GEF$, and $\overline{BF} \cong \overline{CE}$.
11. Plan: Show $\triangle BFD \cong \triangle BFE$ by SAS so that $\overline{DB} \cong \overline{EB}$ and $\angle FDB \cong \angle FEB$ by CPCTC. $\triangle ADF \cong \triangle CEF$ by ASA since $\angle AFD \cong \angle CFE$, $\overline{DF} \cong \overline{EF}$, and $\angle FDA \cong \angle FEC$. By CPCTC, $\overline{AD} \cong \overline{CE}$. $\overline{AB} \cong \overline{CB}$ by addition.
13. Plan: Show $\triangle KRO \cong \triangle JLO$ by AAS. By CPCTC, $\overline{OR} \cong \overline{OL}$ so that right $\triangle ORG \cong$ right $\triangle OLG$ by Hy-Leg. By CPCTC, $\angle OGR \cong \angle OGL$.
15. Plan: Show $\triangle CAE = \triangle CAF$ by SAS. By CPCTC, $\angle CEA \cong \angle CFB$ so that $\angle DEA \cong \angle BFA$. $\triangle AED \cong \triangle AFB$ by ASA. $\overline{AD} \cong \overline{AB}$ by CPCTC.

CHAPTER 7

1. 20
2. 900
3. 4
4. 8
5. 36
6. *AB*
7. 6
8. 15
9. 12
10. $\angle MQN$
11. 78
12. *ST*
13. 13
14. 16
15. 7
16. 12
17. 4
18. 120
19. (2)
20. (3)
21. (3)
22. (1)
23. (3)
24. (4)
25. (2)
26. (2)
27. (2)
28. (3)
29. (3)
30. (1)
31. (3)
32. (1)
43. 124
44. 27
45. 9
46. 66

CHAPTER 8

1. 24
2. 22°
3. $\sqrt{65}$
4. 35°
5. (a) 24
   (b) 10
6. $4\sqrt{3}$
7. (a) 4
   (b) 36
8. 36
9. 10
10. 12
11. 20
12. 30
13. 7
14. 56°
15. 670
16. (a) 6
    (b) 37°
17. (a) 224
    (b) 54°
18. 28
19. 39
20. 16
21. $6\sqrt{3}$
22. 8
23. 40
24. 4
25. 35°
26. 8
27. $5\sqrt{2}$
28. (a) 4
    (b) $\sqrt{52}$
    (c) 34°
29. (a) 24
    (b) 120
    (c) 23°

REGENTS TUNE-UP: CHAPTERS 6–8

1. 10
2. 50
3. $5\sqrt{2}$
4. 6
5. 3
6. 4
7. BC
8. 12
9. 55
10. 6
11. 6
12. $5\sqrt{3}$
13. 15
14. 24
15. 50
16. 6
17. 9
18. 8
19. (1)
20. (4)
21. (3)
22. (1)
23. (3)
24. (2)
25. (2)
26. (1)
27. (4)
28. (1)
29. (2)
30. $\dfrac{-5+\sqrt{61}}{2}$
31. 4
32. 4
38. (a) 8 (b) 27
39. (a) 56°
    (b) 7

CHAPTER 9

1. (3, 4)
2. 0
3. 16
4. (3)
5. (2)
6. (1)
7. (2)
8. (1)
9. (4)
10. (3)
11. (3)
12. (1)
13. (2)
14. (1)
15. (c) 5
16. (a) 30
    (b) 10
    (c) 6
17. (a) $y = -x + 4$
    (b) $y = 4$
    (c) $x = 0$
18. (b) 30
19. (b) 5
    (c) 50
20. (a) $C(9, 5)$
21. (a) $-\dfrac{2}{3}$
    (b) $-\dfrac{(b+5)}{3}$
    (c) -3
    (d) $y - 1 = \dfrac{3}{2}(x - 1)$
22. (a) 2
    (b) $(k-4)/3$
    (c) 10
    (d) $y = 2x + 4$
23. (a) $(-1, 4)$
    (b) $\sqrt{41}$
    (c) 2
    (d) 1
24. (a) $k = c$; $h = a - b$
25. (a) $A'(0, 3)$; $B'(6, -$
    (b) $A'(0, 3)$
    (c) $A'(-3, 0)$; $B'(4, -6)$
    (d) $A'\left(\dfrac{3}{2}, 0\right)$; $B'(-2, 3)$
26. (a) 6
    (b) 2
    (c) 5
    (d) (1)
    (e) (3)
27. (a) (5/2, −2)
    (b) (4, 6)
    (c) (2, −2)
    (d) (−3, −3)

CHAPTER 10

1. 1
2. (a) 1 (b) 2 (c) 0
3. 2
4. 0
5. (a) 2 (b) 1 (c) 0
6. $y = 3x - 2$
7. $x = 5$
8. (a) $x = 2$ (b) $y = -5$ (c) $y = 3x + 5$
9. 4
10. 2

16. (b) (1) 1 (2) 3 (3) 4

CHAPTER 11

1. (2)
2. (3)
3. (3)
4. (4)
5. (3)
6. (3)
7. (1)
8. $x^2 + (y-3)^2 = 16$
9. $(x+1)^2 + (y-4)^2 = 25$
10. 4
11. (4)
12. (1)
13. (c) $(-3, 3)$
    $(1, 3)$
14. (b) $x = 1$
    (c) $x^2 + y^2 = 16$
    (d) 3 and 4
15. (a) $\{(5, 13), (-1, 1)\}$
    (b) $\{(2, -4)\}$
    (c) $\{(11, 3), (9, 7)\}$
    (d) $\{(-4, -7), (-1, 8)\}$

CHAPTER 12

1. 6
2. 56
3. 60
4. 21
5. 3/5
6. 45
7. 1/2
8. 8
9. (2)
10. (2)
11. (1)
12. (a) 24 (b) 1/6
13. (a) $\dfrac{{}_2C_1 \cdot {}_4C_3}{{}_6C_4}$
    (b) $\dfrac{{}_4C_4 + {}_4C_3 \cdot {}_2C_1 + {}_4C_2 \cdot {}_2C_2}{{}_6C_4}$
14. (a) ${}_3C_2 \cdot {}_2C_1 + {}_3C_3$
    (b) $\dfrac{{}_2C_2 \cdot {}_3C_1}{{}_5C_3}$
15. (a) $\dfrac{1}{52} \cdot \dfrac{1}{51}$
    (b) $\dfrac{4}{52} \cdot \dfrac{3}{51}$
    (c) $\dfrac{4}{52} \cdot \dfrac{4}{51}$
    (d) $2 \cdot \dfrac{4}{52} \cdot \dfrac{4}{51}$
    (e) $\dfrac{13}{52} \cdot \dfrac{12}{51}$
16. (a) $\dfrac{7!}{2!}$
    (b) ${}_9C_7$
    (c) $\dfrac{{}_3P_2 \cdot {}_3P_3}{{}_6P_6}$
    (d) 0
17. (a) ${}_{12}C_4$
    (b) ${}_4C_1 \cdot {}_3C_1 \cdot {}_3C_1 \cdot {}_2C_1$
    (c) $\dfrac{{}_4C_1 \cdot {}_3C_1 \cdot {}_3C_1 \cdot {}_2C_1}{{}_{12}C_4}$
    (d) 0
18. (a) ${}_9C_4$
    (b) ${}_3C_1 \cdot {}_1C_1 \cdot {}_4C_2$
    (c) $\dfrac{{}_3C_1 \cdot {}_1C_1 \cdot {}_4C_2}{{}_9C_4}$
    (d) 0

REGENTS TUNE-UP: CHAPTERS 9–12

1. $-\frac{3}{5}$
2. 20
3. 10
4. 8
5. $y-7=4(x-2)$
6. $\frac{4}{5}$
7. -6
8. ${}_5C_2 \cdot {}_5C_1(=50)$
9. $y-1=\frac{1}{2}(x-2)$
10. 7
11. (3)
12. (3)
13. (4)
14. (1)
15. (3)
16. (1)
17. (1)
18. (2)
19. (1)
20. (2)
21. (4)
22. (4)
23. (2)
24. (4)
25. (2)
26. (b) $(4, -4)$ (c) (2)
27. (a) $y=2$
    (b) $x=-4$
    (c) $(-4, 2)$
    (d) $(x-2)^2+(y+2)^2=16$
    (e) 1
28. (b) $A'(0,0)$; $B'(12,9)$; $C'(12,0)$
    (c) $A''(3,-7)$; $B''(11,-1)$; $C''(11,-7)$
29. (b) $A'(1,4)$; $B'(4,5)$
    (c) $A''(-4,-1)$; $B''(-1,0)$
    (d) $A'''(4,1)$; $B'''(1,0)$
    (e) $(x,y)\rightarrow(5-x,5-y)$
30. $\left\{\left(\frac{1}{3},\frac{8}{3}\right),(2,1)\right\}$
31. (b) 76
32. (a) $(x-3)^2+(y-2)^2=n^2$
    (b) parallel lines $x=5$ and $x=1$
    (c) (1) 0 (2) 2
33. $\{(1,0),(2,1)\}$
34. (a) 220 (b) 3/11 (c) $\frac{15}{220}$
35. $\frac{6\pm\sqrt{24}}{2}$
37. (a) 40320 (b) 720 (c) 28 (d) 15 (e) $\frac{15}{28}$

REGENTS EXAMINATIONS

mination January 1987

Three-Year Sequence for High School Mathematics—Course II

Part I

Answer 30 questions from this part. Each correct answer will receive 2 credits. No partial credit will be ed. Write your answers on a separate answer sheet. Where applicable, answers may be left in radical form. [60]

Jsing the accompanying table, solve for x if $a * b = c * x$.

*	a	b	c	d
a	a	b	c	d
b	b	c	d	a
c	c	d	a	b
d	d	a	b	c

In triangle ABC, D is a point on $\overline{AB}$ and E is a point on $\overline{AC}$ such that $\overline{DE}$ is parallel to $\overline{BC}$. If $AB = 12$, $AC = 15$, and $AD = 8$, find the length of $\overline{AE}$.

n the accompanying diagram, parallel lines $\overleftrightarrow{AB}$ nd $\overleftrightarrow{CD}$ are intersected by transversal $\overleftrightarrow{EF}$ at G nd H, respectively. If $m\angle AGH = 75$, find $m\angle FHD$.

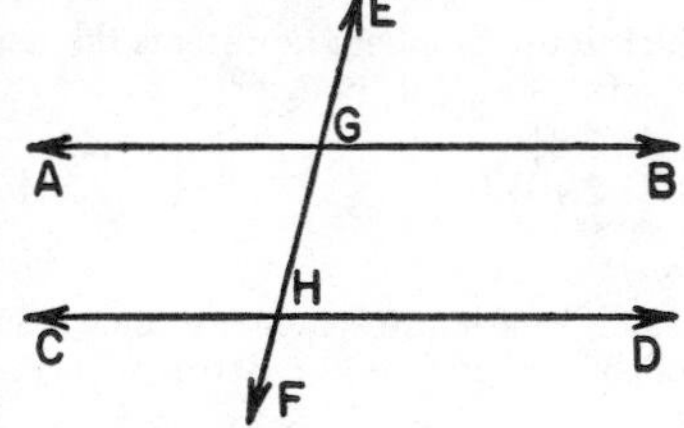

The opposite angles of a parallelogram have measures of $3x - 20$ and $x + 15$. Find x.

n $\triangle ABC$, $m\angle B$ is three times as large as $m\angle A$. An exterior angle at C measures 140. Find $m\angle A$.

Vhat is the area of rectangle $ABCD$ whose vertices are $A(3,2)$, $B(3,-2)$, $C(-3,-2)$, and $D(-3,2)$?

egment AB is the diameter of a circle whose enter is the point (2,5). If the coordinates of A re (1,3), find the coordinates of B.

8 The corresponding sides of two similar triangles are 8 and 12. If the perimeter of the smaller triangle is 28, find the perimeter of the larger triangle.

9 Find a positive value for x such that $\frac{x}{2} = \frac{3}{x-5}$.

10 In $\triangle ABC$, $\angle A \cong \angle C$, $AB = 10x - 7$, $BC = 2x + 33$, and $AC = 4x - 6$. Find x.

11 How many different 6-letter permutations can be formed from the letters in the word "KOODOO"?

12 Evaluate: $\quad {}_7C_3$

13 What is the slope of the line which passes through the points (3,–8) and (–1,0)?

14 If the coordinates of A are (–2,3) and the coordinates of B are (7,–1), find, in radical form, the length of $\overline{AB}$.

15 In the accompanying diagram, the altitude of trapezoid $ABCD$ is 5, $CD = 2$, $m\angle A = 45$, and $m\angle B = 90$. Find AB.

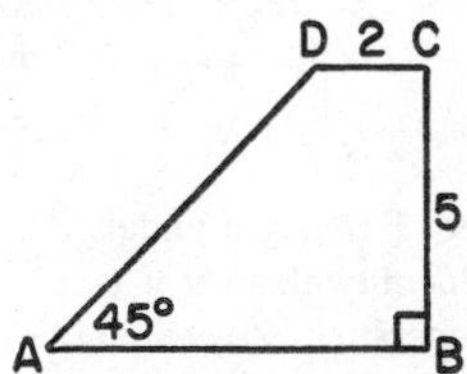

16 The diagonals of a rhombus measure 6 meters and 8 meters. Find the number of meters in the perimeter of the rhombus.

17 In right triangle ABC, altitude $\overline{CD}$ is drawn to the hypotenuse $\overline{AB}$. If $CD = 6$ and $AD = 3$, find the length of $\overline{DB}$.

Directions (18–34): For *each* question chosen, write on the separate answer sheet the *numeral* preceding the word or expression that best completes the statement or answers the question.

18 If * is a binary operation defined by $x * y = x^2 + y$, what is the value of $2 * 3$?

(1) 6 (3) 8
(2) 7 (4) 9

19 In the accompanying diagram, $\overrightarrow{BD} \perp \overleftrightarrow{ABC}$ at B and $\overrightarrow{BE} \perp \overrightarrow{BF}$ at B. If $m\angle FBC = 20$, what is $m\angle EBD$?

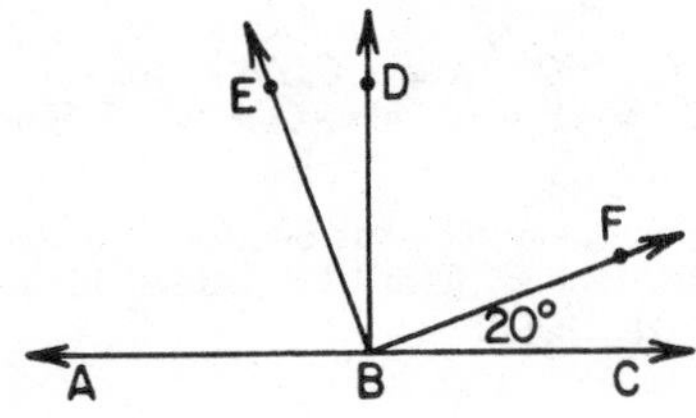

(1) 20 (3) 90
(2) 70 (4) 110

20 The accompanying table defines the operation ♡ for the set $P = \{H,E,A,R,T\}$. According to this table, which statement is true?

♡	H	E	A	R	T
H	E	T	R	H	A
E	R	A	T	E	H
A	T	R	O	A	E
R	H	E	A	R	T
T	A	H	E	T	R

(1) The set P is closed under ♡.
(2) The identity element is R.
(3) The operation ♡ is commutative.
(4) The inverse of T is R.

21 Which statement is the negation of $m \vee \sim n$?

(1) $\sim m \wedge n$ (3) $\sim m \vee \sim n$
(2) $\sim m \vee n$ (4) $\sim m \wedge \sim n$

22 Which sentence is *not* true?

(1) $\exists_x\, x^2 > 5$ (3) $\forall_x\, x + 1 =$
(2) $\forall_x\, x^2 \geq 0$ (4) $\exists_x\, x - 2 =$

23 The graph of which equation does *not* through the origin?

(1) $y = x$ (3) $y = 0$
(2) $y = -x$ (4) $y = 1$

24 The coordinates of a point on the graph o equation $y = x^2 - 4$ are

(1) (0,0) (3) (−4,0)
(2) (−2,0) (4) (2,2)

25 An equation of the circle whose center (2,−3) and whose radius measures 4 is

(1) $(x - 2)^2 + (y + 3)^2 = 16$
(2) $(x + 2)^2 + (y - 3)^2 = 16$
(3) $(x - 2)^2 + (y - 3)^2 = 4$
(4) $(x + 2)^2 + (y - 3)^2 = 4$

26 What are the roots of the quadratic equati $2x^2 + 7x + 4 = 0$?

(1) $\frac{7 \pm \sqrt{17}}{4}$ (3) $\frac{-7 \pm \sqrt{17}}{4}$

(2) $\frac{7 \pm \sqrt{17}}{2}$ (4) $\frac{-7 \pm \sqrt{17}}{2}$

27 Which set of numbers represents the leng the sides of a right triangle?

(1) {2,3,6} (3) {4,5,6}
(2) {5,5,10} (4) {5,12,13}

28 If r and $\sim s$ are true statements, then what probabilty that $s \rightarrow \sim r$ is true?

(1) 1 (3) $\frac{3}{4}$

(2) $\frac{1}{4}$ (4) 0

29 If $p \rightarrow q$ is a true statement and $\sim q$ is a statement, then it follows that

(1) p must be a true statement
(2) p must be a false statement
(3) p may be either a true or a false staten
(4) q must be a true statement

Which is an equation of a line whose slope is 0?
(1) $3x = y$
(2) $x + y = 3$
(3) $y = 3$
(4) $x = 3$

What is the total number of points of intersection of the graphs of the equations $x^2 + y^2 = 9$ and $y = x$?
(1) 1
(2) 2
(3) 3
(4) 4

If the midpoints of two consecutive sides of a rhombus are joined, the triangle formed must be
(1) isosceles
(2) acute
(3) equilateral
(4) right

What is the turning point of the parabola whose equation is $y = 2x^2 + 4x - 3$?
1) (1,3)
2) (-2,-3)
(3) (2,13)
(4) (-1,-5)

34 Under which operation is the set of positive rational numbers *not* closed?
(1) addition
(2) subtraction
(3) multiplication
(4) division

Directions (35): Leave all construction lines on the answer sheet.

35 *On the answer sheet,* construct the bisector of angle C in parallelogram *ABCD.*

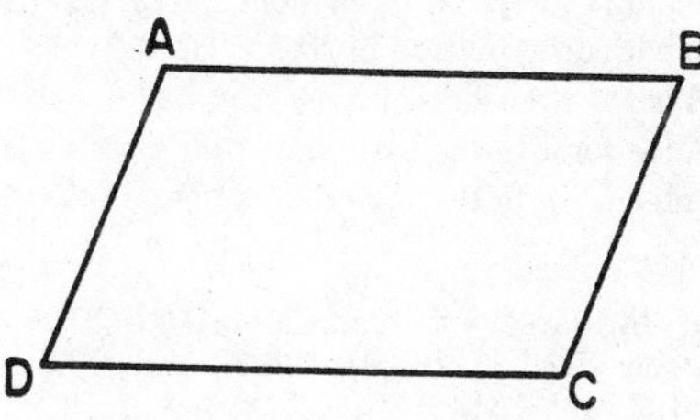

Part II

Answer three questions from this part. Show all work unless otherwise directed. [30]

36 The accompanying table defines the operation @ on the set $\{N,I,T,A\}$.

@	N	I	T	A
N	I	T	A	N
I	T	A	N	I
T	A	N	I	T
A	N	I	T	A

a What is the identity element for @? [2]
b What is the inverse of T? [2]
c What is the value of N @ A @ T? [2]
d Solve for x in the system: N @ $x = I$ [2]
e Solve for y in the system: y @ $y = A$ [1,1]

37 Find the area of quadrilateral $ABCD$ with vertices $A(-1,1)$, $B(3,4)$, $C(8,5)$, and $D(5,-3)$. [10]

38 *a* Draw the graph of the equation $y = x^2 - 4x$ for all values of x such that $-1 \leq x \leq 5$. [6]
b On the same set of axes used in part *a*, draw the graph of the equation $y = -x + 4$. [2]
c Using the graphs drawn in parts *a* and *b*, what is the positive value of x for which $x^2 - 4x = -x + 4$. [2]

39 There are 6 boys and 4 girls of equal ability t ing out for a 4-player school bowling team.
a What is the probability this team will con of 2 boys and 2 girls? [6]
b How many all-boy teams can be formed? [2]
c How many all-girl teams can be formed? [2]

40 In a trapezoid, the length of one base is 5 tin the length of the other base. The height of trapezoid is 1 less than the length of the shor base. If the area of the trapezoid is 90, find length of the *shorter* base. [*Only an algebraic lution will be accepted.*] [4,6]

Part III

Answer one question from this part. Show all work unless otherwise directed. [10]

41 Given: Either Evan or Rona went out on Saturday night.
If Evan went out on Saturday night, then he studied on Tuesday.
If Evan studied on Tuesday, then Anita did not do her schoolwork on Sunday.
Anita did her schoolwork on Sunday.

Let E represent: "Evan went out on Saturday night."
Let R represent: "Rona went out on Saturday night."
Let T represent: "Evan studied on Tuesday."
Let A represent: "Anita did her schoolwork on Sunday."

Prove: Rona went out on Saturday night. [10]

42 Quadrilateral $QRST$ has vertices $Q(a,b)$, $R(0,$ $S(c,0)$, and $T(a+c,b)$. Prove that $QRST$ is parallelogram. [10]

amination June 1987

Three-Year Sequence for High School Mathematics—Course II

Part I

Answer 30 questions from this part. Each correct answer will receive 2 credits. No partial credit will be ed. Write your answers on a separate answer sheet. Where applicable, answers may be left in radical form. [60]

he slope of line $\overleftrightarrow{AB}$ is $\frac{2}{3}$ and the slope of ine $\overleftrightarrow{CD}$ is $\frac{k}{12}$. If $\overleftrightarrow{AB}$ is parallel to $\overleftrightarrow{CD}$, what is he value of k?

n the accompanying figure, $\overleftrightarrow{WX} \parallel \overleftrightarrow{YZ}$; $\overleftrightarrow{AB}$ and $\overleftrightarrow{CD}$ intersect $\overleftrightarrow{WX}$ at E and $\overleftrightarrow{YZ}$ at F and G, re- pectively. If $m\angle CEW = m\angle BEX = 50$, find $n\angle EGF$.

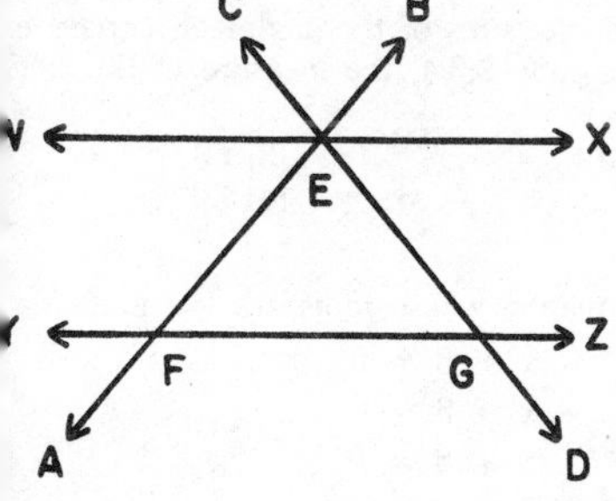

n parallelogram $ABCD$, $m\angle A = 5x - 10$ and $\angle C = 3x + 4$. Find the value of x.

Vhat is the inverse of a in the system defined elow?

&	a	b	c	d
a	b	d	a	c
b	d	c	b	a
c	a	b	c	d
d	c	a	d	b

orresponding sides of two similar triangles are and 15. If the perimeter of the smaller tri- ngle is 22, what is the perimeter of the larger iangle?

6 Determine the value of $a \# (d \# b)$ in the system defined below.

#	a	b	c	d
a	d	c	a	b
b	c	d	b	a
c	a	b	c	d
d	b	c	d	a

7 In the accompanying diagram, $ABCD$ is a parallelogram with altitude $\overline{DE}$ drawn to side $\overline{AB}$. If $m\angle B = 140$, find $m\angle ADE$.

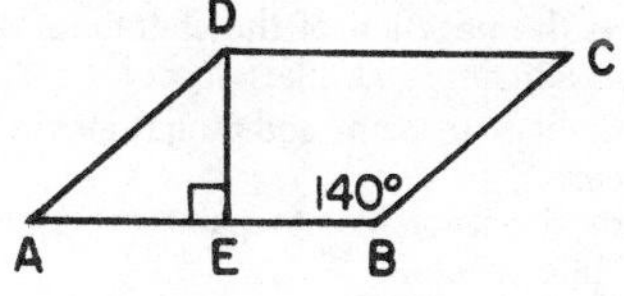

8 If @ is a binary operation defined as $p @ q = p^2 - 2pq + q^2$, evaluate 2 @ 3.

9 In $\triangle ABC$, $\overline{BC}$ is extended through C to D. If $m\angle A = 65$ and $m\angle B = 70$, what is the measure of $\angle DCA$?

10 The perimeter of $\triangle ABC$ is 36. If $\angle A \cong \angle B$, $AC = x$, and $AB = x + 3$, find the value of x.

11 How many different 6-letter permutations can be formed from the letters in the word "DELETE"?

12 How many different sets of 3 pictures can be chosen from a set of 8 pictures?

13 What is the area of a parallelogram if the coordinates of its vertices are $(0, -2)$, $(3,2)$, $(8,2)$, and $(5, -2)$?

14 In the accompanying diagram of $\triangle ABD$, $\overline{AB} \perp \overline{AD}$ and $\overline{EC} \perp \overline{AD}$. If $AB = 6$, $EC = 4$, and $ED = 8$, find AE.

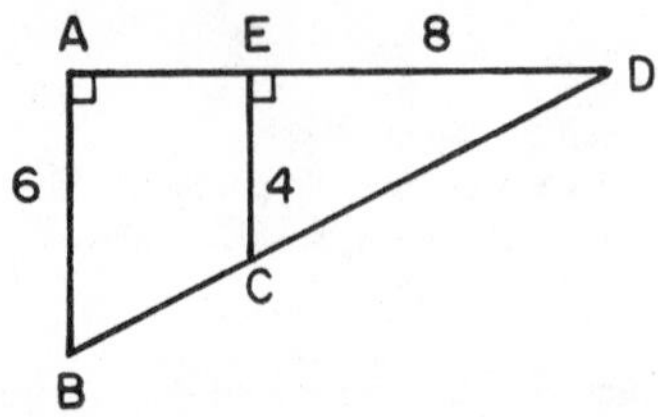

15 Write an equation of the locus of points equidistant from the graphs of the equations $y = 3$ and $y = -5$.

Directions (16–34): For *each* question chosen, write on the separate answer sheet the *numeral* preceding the word or expression that best completes the statement or answers the question.

16 Which is the negation of the statement, "Chris likes to paint or Rona likes soccer"?

(1) Chris likes to paint and Rona does not like soccer.
(2) Chris does not like to paint or Rona does not like soccer.
(3) Chris does not like to paint and Rona does not like soccer.
(4) Chris does not like to paint or Rona likes soccer.

17 If two cards are drawn from a standard deck of 52 cards without replacement, what is the probability that both cards are fives?

(1) $\frac{2}{52}$ (3) $\frac{1}{4} \cdot \frac{1}{3}$
(2) $\frac{4}{52} \cdot \frac{3}{51}$ (4) $\frac{5}{52} \cdot \frac{4}{51}$

18 Which number property is illustrated by the equation $\frac{35}{4} + 0 = \frac{35}{4}$?

(1) associative property for addition
(2) commutative property for addition
(3) identity property for addition
(4) inverse property for addition

19 If the lengths of the legs of a right triangle 2 and 3, then the length of its hypotenuse

(1) $\sqrt{13}$ (3) 5
(2) $\sqrt{5}$ (4) 4

20 If $A \rightarrow C$ and $R \rightarrow \sim C$, which statement valid conclusion?

(1) $A \rightarrow R$ (3) $C \rightarrow R$
(2) $\sim A \rightarrow \sim R$ (4) $A \rightarrow \sim R$

21 Given the true statement, "If a person drive car, he or she must be at least 16 years o Which statement is logically correct?

(1) David is 16 years old; therefore he drive car.
(2) Maria does not drive a car; therefore sh not yet 16 years old.
(3) Noel is not yet 16 years old; therefore does not drive a car.
(4) Irene drives a car; therefore she is not 16 years old.

22 If the measures of the angles of a triangle in the ratio 2:3:4, the measure of the *smal* angle is

(1) 20 (3) 60
(2) 40 (4) 80

23 An equation whose roots are -3 and 2 is

(1) $x^2 - x + 6 = 0$
(2) $x^2 + x - 6 = 0$
(3) $x^2 + x + 6 = 0$
(4) $x^2 + 5x - 6 = 0$

24 The distance between the points $A(a,0)$ $B(0,b)$ is

(1) $\sqrt{a^2 + b^2}$ (3) $a + b$
(2) $a^2 + b^2$ (4) $\sqrt{a} + \sqrt{b}$

25 Which point is on the graph of the paral whose equation is $y = 2x^2 - 3x + 4$?

(1) (0,3) (3) (4,0)
(2) (3,0) (4) (0,4)

26 What is the negation of $\forall_x\ x < 7$?

(1) $\exists_x\ x < 7$ (3) $\exists_x\ x \geq 7$
(2) $\forall_x\ x \geq 7$ (4) $\forall_x\ x > 7$

In $\triangle ABC$, if $AB = 10$ and $BC = 6$, AC can *not* be equal to

(1) 10 (3) 6
(2) 8 (4) 4

In the accompanying diagram, $\triangle RST$ is a right triangle and $\overline{SP}$ is the altitude to hypotenuse $\overline{RT}$. If $SP = 6$ and the lengths of $\overline{RP}$ and $\overline{PT}$ are in the ratio 1:4, what is the length of $\overline{RP}$?

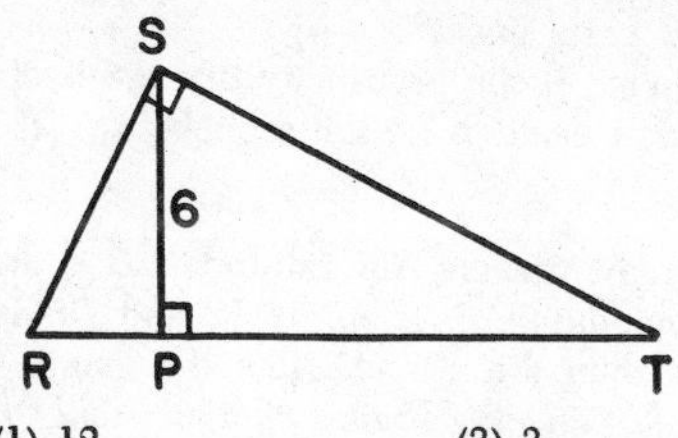

(1) 12 (3) 3
(2) 15 (4) 9

Circle O has center (7,–3) and diameter $\overline{AB}$. The coordinates of A are (–2,4). What are the coordinates of B?

(1) (16,–10) (3) (16,–2)
(2) (12,–2) (4) (12,–10)

Which is an equation of the circle whose center is (3,–2) and whose radius is 5?

(1) $(x + 3)^2 + (y - 2)^2 = 5$
(2) $(x + 3)^2 + (y - 2)^2 = 25$
(3) $(x - 3)^2 + (y + 2)^2 = 5$
(4) $(x - 3)^2 + (y + 2)^2 = 25$

Which statement is true?

(1) Every square is a rhombus.
(2) Every rhombus is a square.
(3) Every trapezoid is a parallelogram.
(4) Every parallelogram is a rectangle.

32 Which is an equation of a line perpendicular to the line whose equation is $y = \frac{1}{3}x - 5$?

(1) $y = \frac{1}{3}x + 5$ (3) $y = -3x - 5$
(2) $y = -\frac{1}{3}x - 5$ (4) $y = 3x + 5$

33 Which is an equation of the axis of symmetry of the parabola whose equation is $y = 2x^2 - 3x + 4$?

(1) $x = -\frac{3}{4}$ (3) $y = -\frac{3}{4}$
(2) $x = \frac{3}{4}$ (4) $y = \frac{3}{4}$

34 What are the roots of the equation $x^2 - 3x - 1 = 0$?

(1) $x = \frac{3 \pm \sqrt{5}}{2}$ (3) $x = \frac{3 \pm \sqrt{13}}{2}$
(2) $x = \frac{-3 \pm \sqrt{5}}{2}$ (4) $x = \frac{-3 \pm \sqrt{13}}{2}$

Directions (35): Leave all construction lines on the answer sheet.

35 *On the answer sheet*, construct the locus of points equidistant from the sides of $\angle A$.

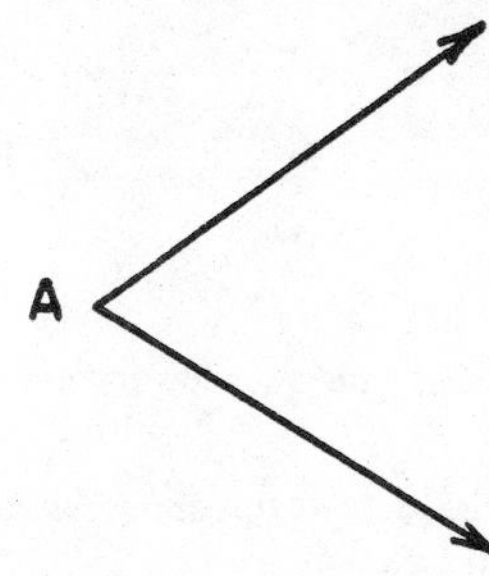

Part II

Answer three questions from this part. Show all work unless otherwise directed. [30]

36 *a* Draw the graph of the equation $y = x^2 - 4x - 5$, including all values of x such that $-2 \le x \le 6$. [6]

b On the same set of axes, draw the graph of the equation $y = x - 5$. [2]

c What are the coordinates of the points of intersection of the graphs drawn in parts *a* and *b*? [2]

37 The diagonals of rhombus $ABCD$ intersect at E. The vertices of $ABCD$ are $A(1,2)$, $B(9,8)$, $C(15,0)$, and $D(7,-6)$.

a Find the length of a side of rhombus $ABCD$. [2]

b Write an equation of the circle whose center is at E and whose radius is equal in length to a side of rhombus $ABCD$. [4]

c What is the slope of diagonal $\overline{BD}$? [2]

d Write an equation of diagonal $\overline{BD}$. [2]

38 Solve the following system of equations algebraically and check.

$$y = x^2 + 3x + 4$$
$$y - x = 7$$

[8,2]

39 Danny has 3 English, 5 Spanish, and 4 books.

a How many different selections of 5 b can be made? [2]

b What is the probability that a 5-book s tion contains 1 English, 2 Spanish, 2 Latin books? [4]

c What is the probability that a 5-book s tion contains no Latin books? [4]

40 In right triangle ABC, altitude $\overline{CD}$ is draw hypotenuse $\overline{AB}$. If AD is 12 and DB is less than the altitude, find the length of [*Only an algebraic solution will be cepted.*] [5,5]

Part III

Answer one question from this part. Show all work unless otherwise directed. [10]

41 Given: Either I go to summer school or I go on vacation.
If I go on vacation, then I will swim every day.
If I swim every day, then I will try out for the team.
I did not try out for the team.

Let S represent: "I go to summer school."
Let G represent: "I go on vacation."
Let E represent: "I swim every day."
Let T represent: "I try out for the team."

Prove: I go to summer school. [10]

42 Quadrilateral $JAKE$ has coordinates J(0 $A(3a,3a)$, $K(4a,0)$, and $E(-a,0)$. Prove by dinate geometry that quadrilateral $JAKE$ i isosceles trapezoid. [10]

xamination January 1988

Three-Year Sequence for High School Mathematics—Course II

Part I

Answer 30 questions from this part. Each correct answer will receive 2 credits. No partial credit will be owed. Write your answers on a separate answer sheet. Where applicable, answers may be left in radical form. [60]

1 If the measures of the angles of a triangle are represented by x, $3x + 6$, and $2x - 6$, find the value of x.

2 In the system defined in the accompanying table, what is the inverse element of 2?

□	1	2	3	4
1	1	2	3	4
2	2	4	1	3
3	3	1	4	2
4	4	3	2	1

3 If * is a binary operation defined by $a * b = \frac{5a + b}{2}$, find the value of $3 * 4$.

4 The sides of a triangle have lengths 2, 3, and 4. The perimeter of a similar triangle is 36. Find the length of the *longest* side of the larger triangle.

5 In parallelogram $ABCD$, $m\angle A = 2x$ and $m\angle B = 2x - 20$. Find x.

6 What is the positive root of the equation $x^2 + 5x - 14 = 0$?

7 Solve the equation $d * (x * a) = c$ for x in the system defined below.

*	a	b	c	d	e
a	c	d	b	e	a
b	d	a	e	c	b
c	b	e	d	a	c
d	e	c	a	b	d
e	a	b	c	d	e

8 In $\triangle ABC$, an exterior angle at C measures 85. What is the longest side of $\triangle ABC$?

9 In the accompanying diagram, $\overleftrightarrow{AB}$ intersects $\overleftrightarrow{PQ}$ and $\overleftrightarrow{RS}$ at C and D, respectively. If $\overleftrightarrow{PQ} \parallel \overleftrightarrow{RS}$, $m\angle RDB = 2x - 10$, and $m\angle QCA = 3x - 65$, find x.

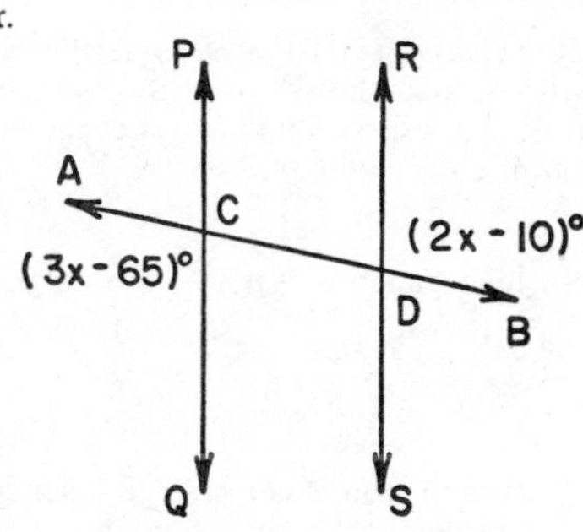

10 What is the slope of a line parallel to the line whose equation is $y = \frac{2}{5}x - 3$?

11 How many different 5-letter permutations can be formed from the letters in the word "JELLO"?

12 Find the coordinates of the midpoint of the line segment whose endpoints are (2,–6) and (10,4).

13 The point (2,–3) is on a circle whose center is at (6,0). Find the length of the radius of the circle.

14 In the accompanying figure, $\overline{AB} \cong \overline{BC}$, $m\angle A = 40$, and $\overline{CD}$ bisects $\angle ACB$. Find $m\angle CDB$.

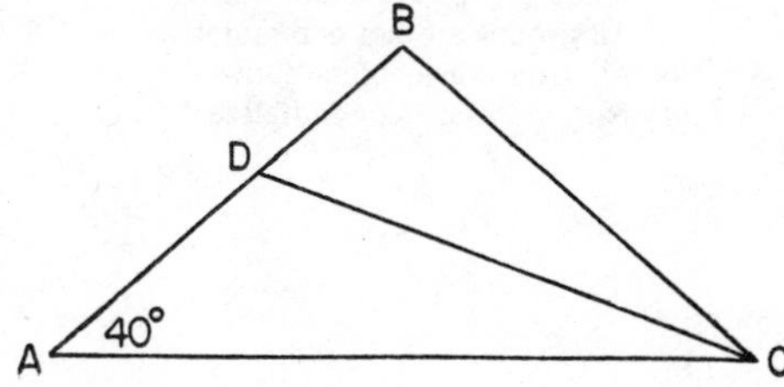

15 The equation of the axis of symmetry of a parabola is $x = 2$. If the parabola intersects the x-axis at the points whose coordinates are $(-1,0)$ and $(k,0)$, find k.

16 The bases of a trapezoid have lengths 10 and 18. If the height of the trapezoid is 6, what is the area of the trapezoid?

17 If the sides of a rectangle are 8 and 5, find the length of a diagonal of the rectangle.

18 Evaluate: $_{15}C_{13}$

Directions (19–35): For *each* question chosen, write on the separate answer sheet the *numeral* preceding the word or expression that best completes the statement or answers the question.

19 The negation of $\sim p \vee \sim q$ is

(1) $p \wedge q$
(2) $\sim p \vee q$
(3) $\sim p \wedge q$
(4) $\sim(p \wedge q)$

20 Given the true statements: "If Janet does not pass the Chemistry examination, she will take Chemistry again next year," and "Janet will not take Chemistry next year." Which statement must be true?

(1) Janet does not pass the Chemistry examination.
(2) Janet passes the Chemistry examination.
(3) Janet takes Physics next year.
(4) No conclusion is possible.

21 What is the total number of points equidistant from two intersecting lines and 1 centimeter from their point of intersection?

(1) 1
(2) 2
(3) 3
(4) 4

22 What is the negation of the statement, "Some groups are not commutative"?

(1) Some groups are commutative.
(2) All groups are not commutative.
(3) All groups are commutative.
(4) No groups are commutative.

23 In $\triangle ABC$, if an exterior angle drawn at verte measures 45, then angle B

(1) must be an acute angle
(2) must be a right angle
(3) must be an obtuse angle
(4) may be either an obtuse angle or an ac angle

24 If the diagonals of a parallelogram are perpen ular and not congruent, then the parallelogra

(1) a rectangle
(2) a rhombus
(3) a square
(4) an isosceles trapezoid

25 Which is an equation of the line that is paralle the x-axis and that passes through the point (5

(1) $x = 5$
(2) $y = 5$
(3) $x = 3$
(4) $y = 3$

26 In right triangle ABC, $m\angle C = 90$, D is a poin $\overline{AB}$, and $\overline{CD} \perp \overline{AB}$. If $AB = 20$ and $AD = 5$, length of $\overline{AC}$ is

(1) 10
(2) 25
(3) $\sqrt{300}$
(4) 4

27 Which line has a slope of $-\frac{1}{2}$?

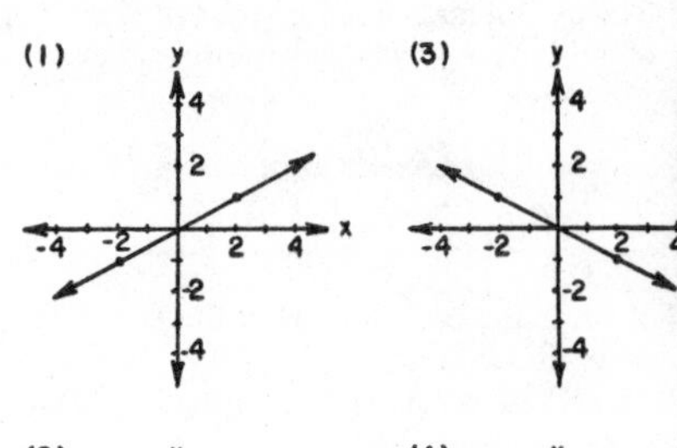

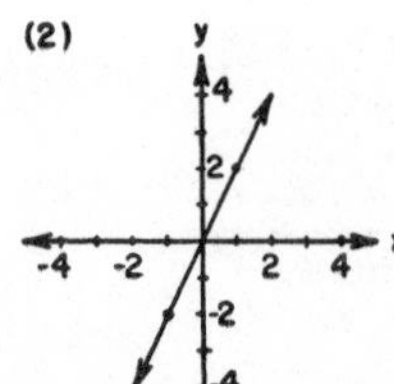

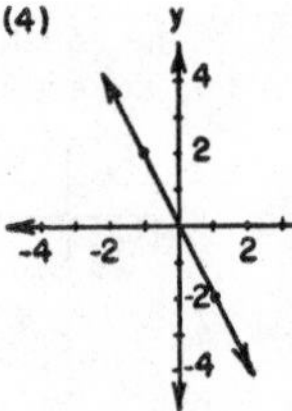

Vhich set of numbers could represent the engths of the sides of a triangle?

1) {6,8,9} (3) {2,2,5}
2) {1,2,3} (4) {2,6,9}

f $\sim a \rightarrow \sim b$ and $\sim c \rightarrow b$, which statement is a alid conclusion?

1) $a \rightarrow c$ (3) $\sim b \rightarrow a$
2) $b \rightarrow \sim c$ (4) $\sim a \rightarrow c$

he solution to the quadratic equation $x^2 + 5x - 1 = 0$ is

1) $\frac{5 \pm \sqrt{17}}{4}$ (3) $\frac{5 \pm \sqrt{33}}{4}$

2) $\frac{-5 \pm \sqrt{17}}{4}$ (4) $\frac{-5 \pm \sqrt{33}}{4}$

Vhich is true of the graph of the parabola whose quation is $y = x^2 - 2x - 8$?

1) The x-intercepts are at $x = 2$ and $x = -4$.
2) The only x-intercept is at $x = 4$.
3) The x-intercepts are at $x = 4$ and $x = -2$.
4) There are no x-intercepts.

32 Under which operation is the set of all odd integers closed?

(1) division (3) addition
(2) multiplication (4) subtraction

33 A student wants to take 3 pens and 4 notebooks from a desk on which there are 6 pens and 8 notebooks. How many choices does the student have?

(1) 1400 (3) 48
(2) 90 (4) 12

34 What are the coordinates of the turning point of the parabola whose equation is $y = x^2 - 2x - 4$?

(1) (–2,4) (3) (–1,–1)
(2) (1,–5) (4) (2,–4)

35 The center of the circle that can be circumscribed about a scalene triangle is located by constructing the

(1) medians of the triangle
(2) altitudes of the triangle
(3) perpendicular bisectors of the sides of the triangle
(4) angle bisectors of the triangle

Part II

Answer three questions from this part. Show all work unless otherwise directed. [30]

36 Quadrilateral *MATH* has vertices $M(-4,2)$, $A(0,5)$, $T(3,3)$, and $H(1,-5)$. Find the area of quadrilateral *MATH*. [10]

37 *a* Draw the graph of the equation $y = -x^2 + 4x + 3$ for all values of x from $x = -1$ to $x = 5$. [6]

b Write an equation of the axis of symmetry of the graph drawn in part *a*. [2]

c Write an equation of a circle whose radius is 5 and whose center is at the y-intercept of the graph drawn in part *a*. [2]

38 A committee of 5 is to be chosen from 4 men and 3 women.

a What is the probability that the committee will consist of 2 men and 3 women? [6]

b What is the probability that the committee will include all 4 men? [2]

c What is the probability that the committee will consist of men only? [2]

39 The coordinates of the vertices of $\triangle ABC$ $A(2,3)$, $B(9,-2)$, and $C(5,10)$. Median $\overline{AE}$ is dra

a Find the length of $\overline{AE}$. [Answer may be le radical form.] [4]

b Find the slope of $\overline{BD}$, the altitude from vertex B. [4]

c Write an equation of $\overleftrightarrow{BD}$. [2]

40 In the accompanying diagram of right triar *RST*, altitude $\overline{TP}$ is drawn to hypotenuse $\overline{RS}$ $TP = 6$ and RP is 5 less than PS, find the lengt hypotenuse $\overline{RS}$. [*Only an algebraic solution be accepted.*] [4,6]

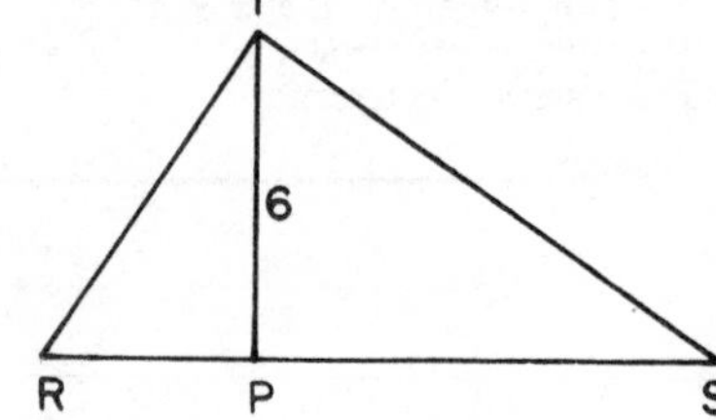

Part III

Answer one question from this part. Show all work unless otherwise directed. [10]

en: If the sun is shining, I will go to the ball game.
If it rains, I will not go to the ball game.
Either the sun is shining or I will need an umbrella.
I will not need an umbrella.

S represent: "The sun is shining."
B represent: "I will go to the ball game."
R represent: "It rains."
U represent: "I will need an umbrella."

e: It will not rain. [10]

42 Given: $\overline{AEC}$, $\overline{BFC}$, $\overline{EGB}$, $\overline{FGA}$, $\overline{FG} \cong \overline{EG}$, and $\angle EGC \cong \angle FGC$.

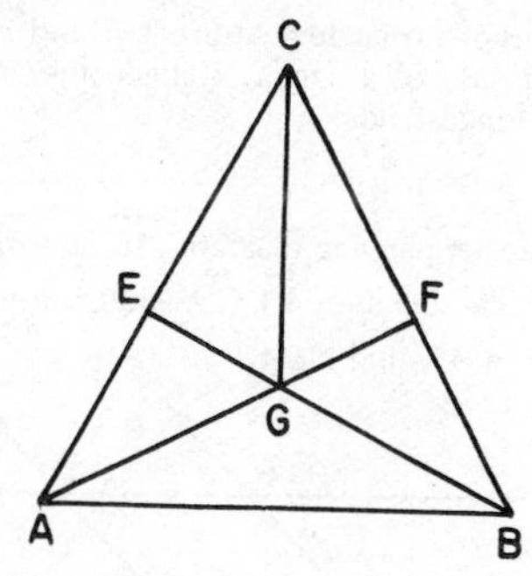

Prove: $\overline{AC} \cong \overline{BC}$ [10]

Examination June 1988

Three-Year Sequence for High School Mathematics—Course II

Part I

Answer 30 questions from this part. Each correct answer will receive 2 credits. No partial credit will allowed. Write your answers on a separate answer sheet. Where applicable, answers may be left in radical form.

1 The sides of a triangle measure 3, 5, and 7. If the smallest side of a similar triangle measures 9, find its longest side.

2 In the accompanying diagram, $\overleftrightarrow{AC}$ and $\overleftrightarrow{BDE}$ are parallel. Parallel lines $\overleftrightarrow{AB}$, $\overleftrightarrow{CD}$, and $\overleftrightarrow{EF}$ are drawn. If $m\angle 1 = 45$, find $m\angle 4$.

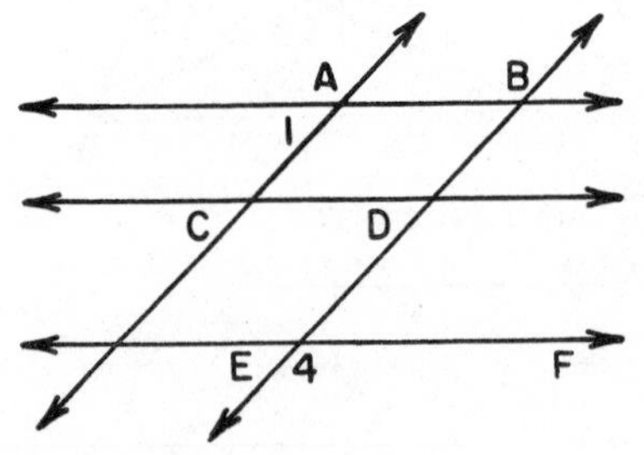

3 If $\triangle MAR$ is an equilateral triangle, find the measure of an exterior angle at R.

4 An urn contains 3 red marbles and 1 green marble. If 2 marbles are selected at random, without replacement, what is the probability that they are *both* green?

5 In $\triangle ABC$, $m\angle A = 80$ and $m\angle B = 50$. If $AB = 4x - 4$ and $AC = 2x + 16$, what is the value of x?

6 In the accompanying diagram of $\triangle ABC$, $\overline{AD} \cong \overline{DC}$, $\overline{DE} \parallel \overline{AB}$, and $DE = 4$. Find AB.

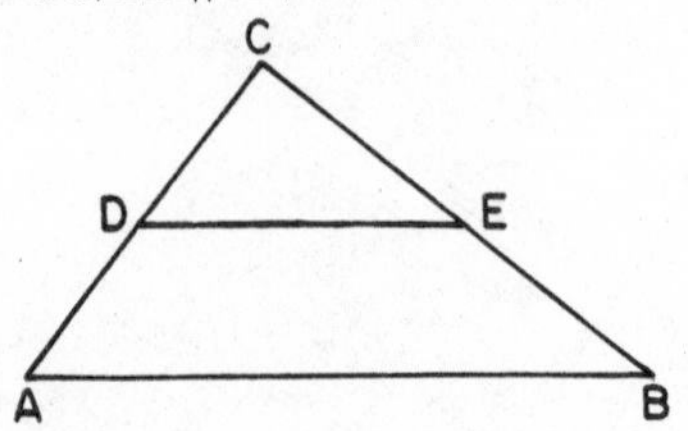

7 What is the area of the triangle whose verti are (0,0), (3,0), and (0,4)?

8 If 3 is a root of the equation $x^2 - 4x + k =$ find k.

9 How many different four-digit numerals can formed from the digits in 1988?

10 The length of the diagonal of a square is 6. F the length of a side of the square.

11 In right triangle ABC, $\overline{CD}$ is the altitude to hyp enuse $\overline{AB}$. If $AD = 2$ and $DB = 18$, find CI

12 A set contains five quadrilaterals: a parall gram, a rectangle, a rhombus, a square, an trapezoid. If one quadrilateral is selected random from the set, what is the probability the figure selected will have congruent oppo angles?

13 Solve the following system of equations for positive value of y:

$$x + y^2 = 13$$
$$x = -3$$

14 Line m and line n are parallel and 6 units ap Point Q lies on line m. How many points 3 units from point Q and equidistant from line and n?

Directions (15–34): For *each* question chosen, w on the separate answer sheet the *numeral* prece the word or expression that best completes the st ment or answers the question.

15 A statement that is logically equivalent to $(a \rightarrow b) \wedge (b \rightarrow c)$ is

(1) $a \rightarrow c$
(2) $c \rightarrow a$
(3) $a \rightarrow \sim b$
(4) $c \rightarrow \sim b$

If ■ is a binary operation defined by $a \blacksquare b = \dfrac{a^2 - b^2}{a + b}$, what is the value of $5 \blacksquare 2$?

(1) 1 (3) 3

(2) $\frac{6}{7}$ (4) $\frac{3}{7}$

The statement $\sim(\sim p \wedge q)$ is logically equivalent to

(1) $p \wedge \sim q$ (3) $p \vee \sim q$

(2) $\sim p \vee \sim q$ (4) $\sim p \vee q$

Which is an equation of a line that is parallel to the line whose equation is $y = 3x + 7$?

(1) $y = -\frac{1}{3}x + 6$ (3) $y = \frac{1}{3}x - 5$

(2) $y = -3x + 6$ (4) $y = 3x - 5$

The coordinates of three vertices of parallelogram $ABCD$ are $A(-1,0)$, $B(4,0)$, and $C(5,4)$. What are the coordinates of vertex D?

(1) (0,4) (3) (0,3)

(2) (1,4) (4) (1,3)

Using the accompanying table, which is the solution set for the equation $2 \odot x = 2$?

⊙	2	4	6	8
2	8	6	4	2
4	2	8	6	4
6	2	2	8	6
8	4	2	2	8

(1) {6} (3) {6,8}

(2) {8} (4) {4,6}

Which is an equation of the locus of points 3 units below the line whose equation is $y = 2$?

(1) $x = -1$ (3) $x = 5$

(2) $y = -1$ (4) $y = 5$

Under which operation is the set {1,2,4,8,16...} closed?

(1) addition (3) multiplication

(2) subtraction (4) division

23 What are the roots of the equation $ax^2 + bx + c = 0$?

(1) $x = \dfrac{-b \pm \sqrt{b^2 - 2ac}}{4a}$

(2) $x = \dfrac{b \pm \sqrt{b^2 - 4ac}}{2a}$

(3) $x = \dfrac{-b + \sqrt{b^2 \pm 4ac}}{2a}$

(4) $x = \dfrac{-b \pm \sqrt{b^2 - 4ac}}{2a}$

24 If the coordinates of A are (3,4) and the coordinates of B are (−3,−4), then the length of $\overline{AB}$ is

(1) 5 (3) 20

(2) 10 (4) 100

25 Given the true statements:

$p \vee q$

$p \rightarrow r$

$\sim q$

Which statement must also be true?

(1) q (3) $\sim p$

(2) r (4) $p \wedge q$

26 If $M(-2,5)$ is the midpoint of $\overline{AB}$ and the coordinates of A are (4,7), what are the coordinates of B?

(1) (1,6) (3) (−8,6)

(2) (2,12) (4) (−8,3)

27 What is the y-intercept of the graph of the equation $y = x^2 - 2x + 3$?

(1) 1 (3) 3

(2) 2 (4) −2

28 If ${}_nC_2 = 45$, what is the value of n?

(1) 5 (3) 9

(2) 8 (4) 10

29 Which statement is *always* true?

(1) A square is a rhombus.

(2) A parallelogram is a square.

(3) A rhombus is a rectangle.

(4) A rectangle is a rhombus.

30 What are the coordinates of the center of the circle whose equation is $(x - 3)^2 + (y + 5)^2 = 16$?

(1) (0,4) (3) (–3,5)
(2) (4,0) (4) (3,–5)

31 In $\triangle ABC$, $\angle C$ is a right angle. If the slope of $\overline{AC}$ is $\frac{2}{3}$, then the slope of $\overline{BC}$ equals

(1) $\frac{2}{3}$ (3) $-\frac{3}{2}$
(2) $-\frac{2}{3}$ (4) $\frac{3}{2}$

32 If the measures of the angles of a triangle are represented by $x + 30$, $4x + 30$, and $10x - 30$, the triangle must be

(1) isosceles (3) right
(2) obtuse (4) scalene

33 If the lengths of two sides of a triangle are 6 and 8, the length of the third side may be

(1) 7 (3) 14
(2) 2 (4) 15

34 Which is an equation of the axis of symmetry the graph whose equation is $y = x^2 + 8x -$

(1) $y = -4$ (3) $x = -4$
(2) $y = 4$ (4) $x = 4$

Directions (35): Leave all construction lines on answer sheet.

35 *On the answer sheet*, in parallelogram ABC locate by construction the point on side $\overline{DC}$ is equidistant from points A and B.

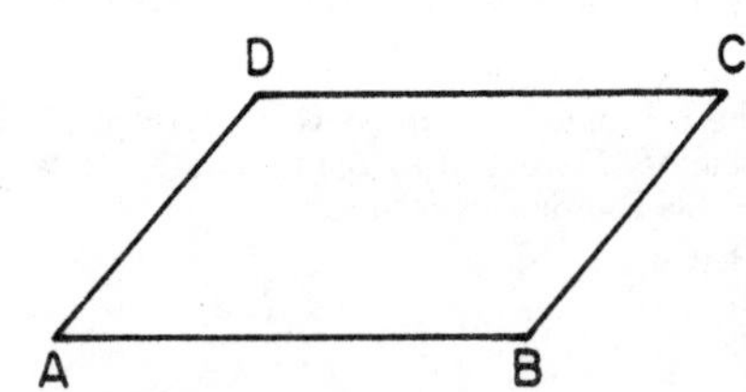

Part II

Answer three questions from this part. Show all work unless otherwise directed. [30]

iven: $\{r,s,t,a\}$ and the operations $*$ and # de-ned by the accompanying tables.

*	*r*	*s*	*t*	*a*
r	*a*	*r*	*s*	*t*
s	*r*	*s*	*t*	*a*
t	*s*	*t*	*a*	*r*
a	*t*	*a*	*r*	*s*

#	*r*	*s*	*t*	*a*
r	*t*	*a*	*r*	*s*
s	*a*	*r*	*s*	*t*
t	*r*	*s*	*t*	*a*
a	*s*	*t*	*a*	*r*

What is the identity element for $*$? [2]

What is the inverse of a under the operation #? [2]

Find the value of $(r * t) \# (a * s)$. [3]

Solve for x: $(t * s) \# x = r$. [3]

n a box, there are five balls. Three are red and wo are white.

Two balls are drawn without replacement. Find the probability that
(1) both are red [2]
(2) both are white [2]
(3) one is red and one is white [3]

Three balls are drawn without replacement. Find the probability of drawing a white ball followed by two red balls. [3]

n right triangle ABC, altitude $\overline{CD}$ is drawn to ypotenuse $\overline{AB}$. If AB is 4 times as large as AD nd AC is 3 more than AD, find the length of $\overline{AD}$. *Only an algebraic solution will be ccepted.*] [4,6]

39 Given: Bill vacations in Canada or the United States.
If the metric system is not used, then gasoline is not sold in liters.
If Bill vacations in Canada, then gasoline is sold in liters.
Bill does not vacation in the United States.

Let C represent: "Bill vacations in Canada."
Let S represent: "Bill vacations in the United States."
Let M represent: "The metric system is used."
Let L represent: "Gasoline is sold in liters."

Prove: The metric system is used. [10]

40 *a* Draw a graph of the equation $y = x^2 - 4x + 3$ for all values of x such that $-1 \le x \le 5$. [6]

b On the same set of axes, sketch the graph of the equation $x^2 + (y - 3)^2 = 4$. [3]

c Using the graphs drawn in parts *a* and *b*, determine the number of points these graphs have in common. [1]

Part III

Answer one question from this part. Show all work unless otherwise directed. [10]

41 Quadrilateral *KLMN* has coordinates *K*(–2,3), *L*(4,6), *M*(3,2), and *N*(–3,–1). Using coordinate geometry, prove that

a the diagonals bisect each other [4]

b the opposite sides are congruent [6]

42 Given: quadrilateral $\overline{ABCD}$, $\overline{AEC}$, $\overline{BED}$, $\overline{AB} \cong \overline{AD}$, and $\overline{BC} \cong \overline{DC}$.

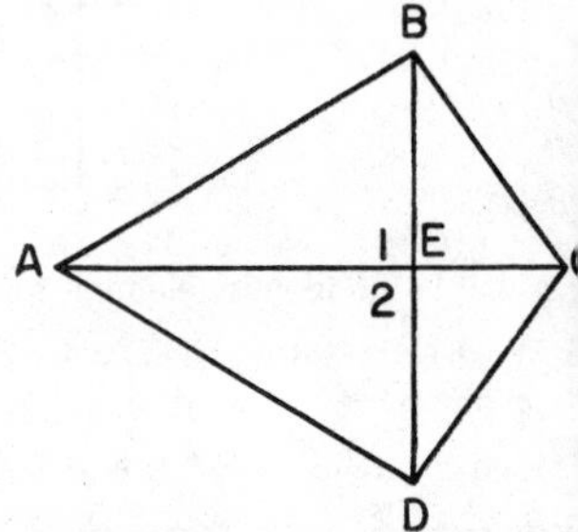

Prove: $\angle 1 \cong \angle 2$ [10]

INDEX

E

F

G

H

I

L

MAXIMIZE YOUR MATH SKILLS!

BARRON'S EASY WAY SERIES

Specially structured to maximize learning with a minimum of time and effort, these books promote fast skill building through lively cartoons and other fun features.

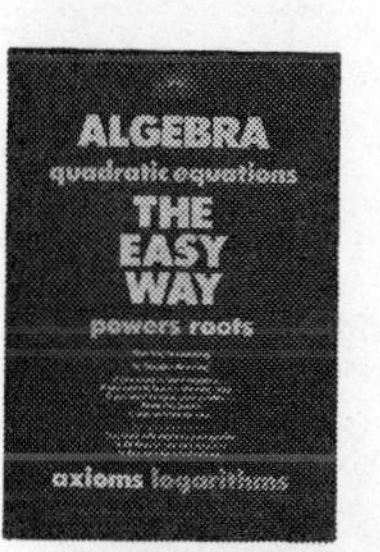

ALGEBRA THE EASY WAY
Douglas Downing
In this one-of-a-kind algebra text, all the fundamentals are discovered through the fascinating adventures that took place in a faraway land! Written as a fantasy novel, this book includes plenty of exercises with solutions, and over 100 illustrations. **352 pp. (2716-7) $8.95, Can. $11.95**

CALCULUS THE EASY WAY, ***Revised***
Douglas Downing
Here, a journey through a fantasy land leads to calculus mastery. All principles are taught in an easy-to-follow adventure tale. Included are numerous exercises, diagrams, and cartoons which aid comprehension. **228 pp. (4078-3) $9.95, Can. $12.95.**

GEOMETRY THE EASY WAY
Lawrence Leff
While other geometry books simply summarize basic principles, this book focuses on the "why" of geometry: why you should approach a problem a certain way, and why the method works. Each chapter concludes with review exercises. **288 pp. (2718-3) $8.95, Can. $11.95**

TRIGONOMETRY THE EASY WAY
Douglas Downing
In this adventure story, the inhabitants of a faraway kingdom use trigonometry to solve their problems. Covered is all material studied in high school or first-year college classes. Practice exercises, explained answers, and illustrations enhance understanding. 288 pp. (2717-5) **$8.95, Can. $11.95**

FM: FUNDAMENTALS OF MATHEMATICS
Cecilia Cullen and Eileen Petruzillo, editors
Volume 1 (2501-6)—Formulas; Introduction to Algebra; Metric Measurement; Geometry; Managing Money; Probability and Statistics.
Volume 2 (2508-3)—Solving Simple Equations; Tables, Graphs and Coordinate Geometry; Banking; Areas; Indirect Measurement and Scaling; Solid Geometry.
The ideal text/workbooks for pre-algebra students and those preparing for state minimum competency exams. They conform with the New York State curriculum, and follow units recommended by the New York City Curriculum Guide. Each book 384 pp., **$12.95, Can. $17.95**

SURVIVAL MATHEMATICS
Edward Williams
Presented here are refreshing, practical new math concepts for basic computational skills. The clear text includes numerous practice exercises that build basic competency skills. 416 pp., $9.95, Can. **$13.95 (2012-X)**

Books may be purchased at your bookstore, or by mail from Barron's. Enclose check or money order for total amount plus sales tax where applicable and 10% for postage (minimum charge $1.50). All books are paperback editions. Prices subject to change without notice.

Barron's Educational Series, Inc.
250 Wireless Boulevard
Hauppauge, New York 11788
In Canada:
Georgetown Book Warehouse
34 Armstrong Ave.
Georgetown, Ontario L7G 4R9